清华版双语教学用书

工程电磁场
（第9版）

Engineering
Electromagnetics
（Ninth Edition）

［美］ 威廉姆·H.哈特（William H. Hayt, Jr.）
约翰·A.巴克（John A. Buck） 著

袁建生 选译

清华大学出版社
北 京

内容简介

    本书系统论述了工程电磁场的理论及其分析方法,首先介绍了电磁场中所用到的数学基础——矢量分析,然后分五章介绍静电场,包括电场的定义(库仑定律)、场与源的关系即电场强度与电荷的关系(高斯定理)、电场能量与电场做功(电位的定义)、静电场的两种材料(导体与介质)、电场范畴的元件(电容)。其后关于恒定磁场的论述也包含这五个方面的内容,只不过仅划分成两章,包括磁场的定义(毕奥-萨伐尔定律)、场与源的关系(安培环路定律)、磁场能与磁场力、磁材料、磁场范畴的元件(电感与互感)。磁场中还包含一部分内容:磁路。在介绍了非时变、各自独立的电场与磁场之后,讨论时变电磁场,即电场与磁场相互耦合,相互作用,先介绍变化的磁场产生电场的现象与定量关系(法拉第电磁感应定律),然后介绍变化的电场产生磁场的现象(位移电流与全电流定律),在此基础上结合两种现象与矢量函数的特性等,给出描述时变电磁场特性的麦克斯韦方程组;并进一步介绍可表征麦克斯韦方程组的位函数及其满足的方程。在介绍完电磁场的基本特性后,插入一章介于电磁场与电路原理知识之间的内容——传输线。后面四章内容都属于高频时变电磁场的场特性分析与应用特性。

    本书适合作为高校电气与电子信息类专业本科生与研究生的基本教材。

图书在版编目(CIP)数据

    工程电磁场:第9版/(美)威廉姆·H.哈特(William H. Hayt, Jr.),(美)约翰·A.巴克(John A. Buck)著;袁建生选译.—北京:清华大学出版社,2019(2024.1 重印)
    书名原文:Engineering Electromagnetics, Ninth Edition
    (清华版双语教学用书)
    ISBN 978-7-302-53027-5

    Ⅰ.①工… Ⅱ.①威…②约…③袁… Ⅲ.①电磁场–双语教学–教材–英、汉 Ⅳ.①O441.4

    中国版本图书馆 CIP 数据核字(2019)第 093923 号

策划编辑:盛东亮
封面设计:常雪影
责任校对:白 蕾
责任印制:沈 露

出版发行:清华大学出版社
    网   址:https://www.tup.com.cn, https://www.wqxuetang.com
    地   址:北京清华大学学研大厦 A 座   邮   编:100084
    社 总 机:010-83470000   邮   购:010-62786544
    投稿与读者服务:010-62776969, c-service@tup.tsinghua.edu.cn
    质量反馈:010-62772015, zhiliang@tup.tsinghua.edu.cn
印 装 者:三河市龙大印装有限公司
经   销:全国新华书店
开   本:186mm×240mm   印   张:37.75   字   数:675 千字
版   次:2019 年 7 月第 1 版   印   次:2024 年 1 月第 5 次印刷
定   价:109.00 元

产品编号:080484-01

ABOUT THE AUTHORS

William H. Hayt, Jr. (deceased) received his B.S. and M.S. degrees at Purdue University and his Ph.D. from the University of Illinois. After spending four years in industry, Professor Hayt joined the faculty of Purdue University, where he served as professor and head of the School of Electrical Engineering, and as professor emeritus after retiring in 1986. Professor Hayt's professional society memberships included Eta Kappa Nu, Tau Beta Pi, Sigma Xi, Sigma Delta Chi, Fellow of IEEE, ASEE, and NAEB. While at Purdue, he received numerous teaching awards, including the university's Best Teacher Award. He is also listed in Purdue's Book of Great Teachers, a permanent wall display in the Purdue Memorial Union, dedicated on April 23, 1999. The book bears the names of the inaugural group of 225 faculty members, past and present, who have devoted their lives to excellence in teaching and scholarship. They were chosen by their students and their peers as Purdue's finest educators.

A native of Los Angeles, California, **John A. Buck** received his B.S. in Engineering from UCLA in 1975, and the M.S. and Ph.D. degrees in Electrical Engineering from the University of California at Berkeley in 1977 and 1982. In 1982, he joined the faculty of the School of Electrical and Computer Engineering at Georgia Tech, and is now Professor Emeritus. His research areas and publications have centered within the fields of ultrafast switching, nonlinear optics, and optical fiber communications. He is the author of the graduate text *Fundamentals of Optical Fibers* (Wiley Interscience), which is in its second edition. Dr. Buck is the recipient of four institute teaching awards and the IEEE Third Millenium Medal.

序

William H. Hayt, Jr. 教授生前曾任教于美国普渡大学,其一生获得诸多荣誉,堪称教学名师之典范。在学习本教材之前,请读者一定要阅读作者简介,相信这一定会激发您对 William H. Hayt. Jr. 教授的敬仰和对电磁场学习的兴趣。第二作者 John A. Buck 为乔治亚理工学院电气与计算机工程系荣誉退休教授。William H. Hayt. Jr. 编写的第一版距今已 60 多年。自该书的第六版起主要由 John A. Buck 教授修订。该版本称为国际版,在美国之外的国家发行使用。第九版与第八版相比改动不大。

本教材与我国电子与电气工程专业的本科教材内容基本一致,满足我国高校相关教学大纲要求。其基本概念讲述清晰、详尽,注重物理概念,淡化公式推导,图文并茂,章节划分细致,每个基本概念和问题分为一小节;每章后面都配有约 30 道习题。

本书文笔流畅、可读性强、易理解,学生可以使用该教材进行自主学习,是电磁场或电磁场与电磁波课程的理想教材或参考书。

本书在内容编排上有独到之处。例如,在静电场部分,将能量与电位的引入结合在一起(第 4 章的题目为"能量与电位"),突出了电位是电场对单位电荷做功的物理属性,而不是像国内教材强调电位梯度等于电场强度的数学关系;在磁场部分,相对于静电场部分的总共 5 章编排,该部分仅以两章出现,一章的题目为"恒定磁场",另一章的题目为"磁场力、材料和电感",这种编排的综合性与静电场内容编排的分散性形成了明显反差,且在叙述中多处是通过与静电场类比的形式直接给出结果或结论。

相对于我国相应教材,本书静电场中电容的计算内容较丰富;磁场中介绍了磁路概念,特别介绍了铁磁材料的非线性与磁滞回线概念,铁磁材料的非线性现象确实是需要学生牢记的一种现象和一定要掌握的基本概念;另外就是利用一章介绍了传输线问题(第 10 章)。传输线分析内容属于电磁场课程还是电路课程内容并无明确定论,但从场的根本属性和空间分布看,传输线内容应该归在电路课程中。无论归为哪个课程内容,传输线概念是学生应具备的电工基础知识。

在使用本教材时应注意三点:

① 其时变电磁场部分仅一章(第 9 章:时变场与麦克斯韦方程组),内容偏少、偏简单、过于数学化,没有准静态场的概念,没有涡流、趋肤效应与邻近效应等较重要的实用性内容,这不利于以低频电磁场为基础的电气工程类学生的知识构成。

② 5.1～5.3 节的电流与电阻内容插在静电场内容中容易使学生混淆不同类型的场,混淆静电场是分析介质中的电场,而电流场是分析导体内的电场。静电场中导体内场强为零,电容属于

静电场的概念,而电流场中导体内场强不为零,电阻属于电流场的概念,两种场研究的问题截然不同。

③ 边值问题的内容比较薄弱,在电位与磁矢量位的泊松方程相关内容中,没有给出位函数应满足的交界面条件和边界条件,也没有给出多媒质中通过解边值问题得到场分布的内容和例题。这不利于学生建立电磁场数值计算所需的电磁场基本概念。

本书配有学习网站(网址为 www. mhhe. com/hayt/buck9e),网站提供了丰富的学习资源和内容,包括小测验问答题,一些场图、动画、互动等内容,书中页面空白处的 WWW 图标指明了网站对应的内容,每当学到图标处建议查阅网站上的相应内容。另外,每章后面的习题采用三级难度分类标识给出了每道题的难易程度(见题号后的竖条),便于不同水平的学生选择习题。

本人对该书的部分词语进行了翻译和注释,并写了一些导读,不当之处请批评指正!

袁建生

清华大学电机与应用电子技术系

2019 年 3 月

PREFACE

The printing of this book occurs one year short of 60 years since its first edition, which was at that time under the sole authorship of William H. Hayt, Jr. In a sense, I grew up with the book, having used the second edition in a basic electromagnetics course as a college junior. The reputation of the subject matter precedes itself. The prospect of taking the first course in electromagnetics was then, as now, a matter of dread to many if not most. One professor of mine at Berkeley put it succinctly through the rather negative observation that electromagetics is "a test of your ability to bend your mind". But on entering the course and first opening the book, I was surprised and relieved to find the friendly writing style and the measured approach to the subject. This for me made it a very readable book, out of which I was able to learn with little help from my instructor. I referred to the book often while in graduate school, taught from the fourth and fifth editions as a faculty member, and then became coauthor for the sixth edition on the retirement (and subsequent untimely death) of Bill Hayt. To this day, the memories of my time as a beginner are vivid, and in preparing the sixth and subsequent editions, I have tried to maintain the accessible style that I found so encouraging and useful then.

Over the 60-year span, the subject matter has not changed, but emphases have. In universities, the trend continues toward reducing electrical engineering core course allocations to electromagnetics. This is a matter of economy, rather than any belief in diminished relevance. Quite the contrary: A knowledge of electromagnetic field theory is in the present day more important than ever for the electrical engineer. Examples that demonstrate this include the continuing expansion of high-speed wireless and optical fiber communication. Additionally, the need continues for ever-smaller and denser microcircuitry, in which a command of field theory is essential for successful designs. The more traditional applications of electrical power generation and distribution remain as important as ever.

I have made efforts to further improve the presentation in this new edition. Most changes occur in the earlier chapters, in which much of the wording has been shortened, and several explanations were improved. Additional introductory material has been added in several places to provide perspective. In addition, all chapters are now subsectioned, to improve the organization and to make topics easier to locate.

Some 100 new end-of-chapter problems have been added throughout, all of which replaced older problems that I considered well-worn. I have retained the previous system in which the approximate level of difficulty is indicated beside each problem on a three-level scale. The lowest level is considered a fairly straightforward problem, requiring little work assuming the material is understood; a level 2 problem is conceptually more difficult, and/or may require more work to solve; a level 3 problem is considered either difficult conceptually, or may require extra effort (including possibly the help of a computer) to solve.

As in the previous edition, the transmission lines chapter (10) is stand-alone, and can be read or covered in any part of a course, including the beginning. In it, transmission lines are treated entirely within the context of circuit theory; wave phenomena are introduced and used exclusively in the form of voltages and currents. Inductance and capacitance concepts are treated as known parameters, and so there is no reliance on any other chapter. Field concepts and parameter computation in transmission lines appear in the early part of the waveguides chapter (13), where they play additional roles of helping to introduce waveguiding concepts. The chapters on electromagnetic waves, 11 and 12, retain their independence of transmission line theory in that one can progress from Chapter 9 directly to Chapter 11. By doing this, wave phenomena are introduced from first principles but within the context of the uniform plane wave. Chapter 11 refers to Chapter 10 in places where the latter may give additional perspective, along with a little more detail. Nevertheless, all necessary material to learn plane waves without previously studying transmission line waves is found in Chapter 11, should the student or instructor wish to proceed in that order.

The antennas chapter covers radiation concepts, building on the retarded potential discussion in Chapter 9. The discussion focuses on the dipole antenna, individually and in simple arrays. The last section covers elementary transmit-receive systems, again using the dipole as a vehicle.

The book is designed optimally for a two-semester course. As is evident, statics concepts are emphasized and occur first in the presentation, but again Chapter 10 (transmission lines) can be read first. In a single course that emphasizes dynamics, the transmission lines chapter can be covered initially as mentioned or at any point in the course. One way to cover the statics material more rapidly is by deemphasizing materials properties (assuming these are covered in other courses) and some of the advanced topics. This involves omitting Chapter 1 (assigned to be read as a review), and omitting Sections 2.5, 2.6, 4.7, 4.8, 5.5–5.7, 6.3, 6.4, 6.7, 7.6, 7.7, 8.5, 8.6, 8.8, 8.9, and 9.5.

A supplement to this edition is web-based material consisting of articles on special topics in addition to animated demonstrations and interactive programs developed by Natalya Nikolova of McMaster University and Vikram Jandhyala of the University of Washington. Their excellent contributions are geared to the text, and icons appear in the margins whenever an exercise that pertains to the narrative exists. In addition, quizzes are provided to aid in further study.

The theme of the text is the same as it has been since the first edition of 1958. An inductive approach is used that is consistent with the historical development. In it, the experimental laws are presented as individual concepts that are later unified in Maxwell's equations. After the first chapter on vector analysis, additional mathematical tools are introduced in the text on an as-needed basis. Throughout every edition, as well as this one, the primary goal has been to enable students to learn independently. Numerous examples, drill problems (usually having multiple parts), end-of-chapter problems, and material on the web site, are provided to facilitate this.

Answers to the drill problems are given below each problem. Answers to odd-numbered end-of-chapter problems are found in Appendix F. A solutions manual and a set of PowerPoint slides, containing pertinent figures and equations, are available to instructors. These, along with all other material mentioned previously, can be accessed on the website:

www.mhhe.com/haytbuck

I would like to acknowledge the valuable input of several people who helped to make this a better edition. They include:

Philip D. Loewen – University of British Columbia
Gerald Whitman – New Jersey Institute of Technology
Andrew F. Peterson – Georgia Institute of Technology
M. Chris Wernicki, Ph.D. – NYIT
David Baumann – Lake Superior State University
Jesmin Khan – Tuskegee University
Dr. S. Hossein Mousavinezhad – Idaho State University
Kiyun Han – Wichita State University
Anand Gopinath – University of Minnesota
Donald M. Keller – Point Park University
Argyrios VAronides – University of Scranton
Otsebele Nare – Hampton University
Robert Wayne Scharstein – University of Alabama
Virgil Thomason – University of Tennessee at Chattanooga
Gregory M. Wilkins, Ph.D. – Morgan State University
Mark A. Jerabek – West Virginia University
James Richie – Marquette University
Dean Johnson – Western Michigan
David A. Rogers – North Dakota State University
Tomasz Petelenz – University of Utah
Surendra Singh – The University of Tulsa
Tom Vandervelde – Tufts
John Zwart – Dordt College
Taan ElAli – Embry-Riddle Aeronautical University
R. Clive Woods – Louisiana State University
Jack Adams – Merrimack College

I also acknowledge the feedback and many comments from students, too numerous to name, including several who have contacted me from afar. I continue to be open and grateful for this feedback and can be reached at john.buck@ece.gatech.edu. Many suggestions were made that I considered constructive and actionable. I regret that not all could be incorporated because of time restrictions. Creating this book was a team effort, involving several outstanding people at McGraw-Hill. These include my editors, Raghu Srinivasan and Tomm Scaife, whose vision and encouragement were invaluable. Jenilynn McAtee and Lora Neyens deftly coordinated the production phase with excellent ideas and enthusiasm, and Tina Bower, who was my guide and conscience from the beginning, providing valuable insights, and jarring me into action when necessary. I am, as usual in these projects, grateful to a patient and supportive family.

John A. Buck
Marietta, Georgia
May, 2017

On the cover: Three-dimensional field map for an electric dipole.

McGraw-Hill Digital Offerings Include:

This text is available as an eBook at www.CourseSmart.com. At CourseSmart, students can take advantage of significant savings of the printed textbook, reduce their impact on the environment, and gain access to powerful web tools for learning. CourseSmart eBooks can be viewed online or downloaded to a computer. The eBooks allow students to do full text searches, add highlighting and notes, and share notes with classmates.Contact your McGraw-Hill sales representative or visit www .CourseSmart.com to learn more.

Professors can benefit from McGraw-Hill's Complete Online Solutions Manual Organization System (COSMOS). COSMOS enables instructors to generate a limitless supply of problem material for assignment, as well as transfer and integrate their own problems into the software. For additional information, contact your McGraw-Hill sales representative.

McGraw-Hill Create™
Craft your teaching resources to match the way you teach! With McGraw-Hill Create, www.mcgrawhillcreate.com, you can easily rearrange chapters, combine material from other content sources, and quickly upload content you have written like your course syllabus or teaching notes. Find the content you need in Create by searching through thousands of leading McGraw-Hill textbooks. Arrange your book to fit your teaching style. Create even allows you to personalize your book's appearance by selecting the cover and adding your name, school, and course information. Order a Create book and you'll receive a complimentary print review copy in 3–5 business days or a complimentary electronic review copy (eComp) via email in minutes. Go to www.mcgrawhillcreate.com today and register to experience how McGraw-Hill Create empowers you to teach *your* students *your* way.

This partnership allows you and your students access to McGraw-Hill's Create right from within your Blackboard course—all with one single sign-on. McGraw-Hill and Blackboard can now offer you easy access to industry leading technology and content, whether your campus hosts it, or we do. Be sure to ask your local McGraw-Hill representative for details.

CONTENTS

CHAPTER 1 Vector Analysis

术语

vector 矢量，既有大小又有方向的量。矢量字母用黑体，若手写，则字母上面要加箭头。

scalar 标量，仅有大小的量。

algebra 代数。

coplanar vector 共面矢量，位于一个平面上的矢量。

right-handed rectangular coordinate system 右手直角坐标系，四个手指从x轴绕向y轴，大拇指指向z轴。

component 矢量的分量，一个矢量是由三个分量表示的，三个分量分别是该矢量在三个坐标轴上的投影的长度。

magnitude of vector 矢量的模，矢量的长度或大小。

vector field 矢量场（参见导读）。

unit vector 单位矢量，模为1的矢量。

dot product 点乘，也叫投影积。两个矢量点积的结果是标量。

cross product 叉乘，两个矢量的叉乘结果为一个矢量，该矢量与原两矢量构成右手螺旋关系。

circular cylindrical coordinate 圆柱坐标系。

spherical coordinate system 球坐标系，对特定的场分布选择合适的坐标系表示或在合适的坐标系下分析会使问题大大简化。

Vector analysis is a subject that is better taught by mathematicians than by engineers. Most junior and senior engineering students have not had the time (or the inclination) to take a course in vector analysis, although it is likely that vector concepts and operations were introduced in the calculus courses. These are covered in this chapter, and the time devoted to them now should depend on past exposure.

The viewpoint here is that of the engineer or physicist and not that of the mathematician. Proofs are indicated rather than rigorously expounded, and physical interpretation is stressed. It is easier for engineers to take a more rigorous course in the mathematics department after they have been presented with a few physical pictures and applications.

Vector analysis is a mathematical shorthand. It has some new symbols and some new rules, and it demands concentration and practice. The drill problems, first found at the end of Section 1.4, should be considered part of the text and should all be worked. They should not prove to be difficult if the material in the accompanying section of the text has been thoroughly understood. ∎

1.1 SCALARS AND VECTORS

The term *scalar* refers to a quantity whose value may be represented by a single (positive or negative) real number. The $x, y,$ and z we use in basic algebra are scalars, as are the quantities they represent. If we speak of a body falling a distance L in a time t, or the temperature T at any point whose coordinates are $x, y,$ and z, then $L, t, T, x, y,$ and z are all scalars. Other scalar quantities are mass, density, pressure (but not force), volume, volume resistivity, and voltage.

A *vector* quantity has both a magnitude[1] and a direction in space. We are concerned with two- and three-dimensional spaces only, but vectors may be defined in *n*-dimensional space in more advanced applications. Force, velocity, acceleration, and a straight line from the positive to the negative terminal of a storage battery are examples of vectors. Each quantity is characterized by both a magnitude and a direction.

Our work will mainly concern scalar and vector *fields*. A field (scalar or vector) may be defined mathematically as some function that connects an arbitrary origin to a general point in space. We usually associate some physical effect with a field, such as the force on a compass needle in the earth's magnetic field, or the movement of smoke particles in the field defined by the vector velocity of air in some region of space. Note that the field concept invariably is related to a region. Some quantity is defined at every point in a region. Both *scalar fields* and *vector fields* exist. The temperature and the density at any point in the earth are examples of scalar fields. The gravitational and magnetic fields of the earth, the voltage gradient in a cable, and the temperature gradient in a soldering-iron tip are examples of vector fields. The value of a field varies in general with both position and time.

In this book, as in most others using vector notation, vectors will be indicated by boldface type, for example, **A**. Scalars are printed in italic type, for example, *A*. When writing longhand, it is customary to draw a line or an arrow over a vector quantity to show its vector character. (CAUTION: This is the first pitfall. Sloppy notation, such as the omission of the line or arrow symbol for a vector, is the major cause of errors in vector analysis.)

1.2 VECTOR ALGEBRA

In this section, the rules of vector arithmetic, vector algebra, and (later) vector calculus are defined. Some of the rules will be similar to those of scalar algebra, some will differ slightly, and some will be entirely new.

1.2.1 Addition and Subtraction

The addition of vectors follows the parallelogram law. Figure 1.1 shows the sum of two vectors, **A** and **B**. It is easily seen that $\mathbf{A} + \mathbf{B} = \mathbf{B} + \mathbf{A}$, or that vector addition obeys the commutative law. Vector addition also obeys the associative law,

$$\mathbf{A} + (\mathbf{B} + \mathbf{C}) = (\mathbf{A} + \mathbf{B}) + \mathbf{C}$$

Note that when a vector is drawn as an arrow of finite length, its location is defined to be at the tail end of the arrow.

Coplanar vectors are vectors lying in a common plane, such as those shown in Figure 1.1. Both lie in the plane of the paper and may be added by expressing each vector in terms of "horizontal" and "vertical" components and then adding the corresponding components.

Vectors in three dimensions may likewise be added by expressing the vectors

---

[1] We adopt the convention that magnitude infers absolute value; the magnitude of any quantity is therefore always positive.

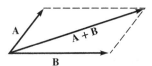 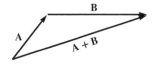

Figure 1.1 Two vectors may be added graphically either by drawing both vectors from a common origin and completing the parallelogram or by beginning the second vector from the head of the first and completing the triangle; either method is easily extended to three or more vectors.

in terms of three components and adding the corresponding components. Examples of this process of addition will be given after vector components are discussed in Section 1.4.

The rule for the subtraction of vectors follows easily from that for addition, for we may always express $\mathbf{A} - \mathbf{B}$ as $\mathbf{A} + (-\mathbf{B})$; the sign, or direction, of the second vector is reversed, and this vector is then added to the first by the rule for vector addition.

1.2.2 Multiplication and Division

Vectors may be multiplied by scalars. The magnitude of the vector changes, but its direction does not when the scalar is positive, although it reverses direction when multiplied by a negative scalar. Multiplication of a vector by a scalar also obeys the associative and distributive laws of algebra, leading to

$$(r + s)(\mathbf{A} + \mathbf{B}) = r(\mathbf{A} + \mathbf{B}) + s(\mathbf{A} + \mathbf{B}) = r\mathbf{A} + r\mathbf{B} + s\mathbf{A} + s\mathbf{B}$$

Division of a vector by a scalar is merely multiplication by the reciprocal of that scalar. The multiplication of a vector by a vector is discussed in Sections 1.6 and 1.7. Two vectors are said to be equal if their difference is zero, or $\mathbf{A} = \mathbf{B}$ if $\mathbf{A} - \mathbf{B} = 0$.

In our use of vector fields we always add and subtract vectors that are defined at the same point. For example, the *total* magnetic field about a small horseshoe magnet will be shown to be the sum of the fields produced by the earth and the permanent magnet; the total field at any point is the sum of the individual fields *at that point*.

1.3 THE RECTANGULAR COORDINATE SYSTEM

To describe a vector accurately, some specific lengths, directions, angles, projections, or components must be given. There are three simple coordinate systems by which this is done, and about eight or ten other systems that are useful in very special cases. We are going to use only the three simple systems, the simplest of which is the *rectangular*, or *rectangular cartesian*, coordinate system.

1.3.1 Right-Handed Coordinate Systems

In the rectangular coordinate system we set up three coordinate axes mutually at right angles to each other and call them the x, y, and z axes. It is customary to choose a *right-handed* coordinate system, in which a rotation (through the smaller angle) of the x axis into the y axis would cause a right-handed screw to progress in the direction of the z axis. If the right hand is used, then the thumb, forefinger, and

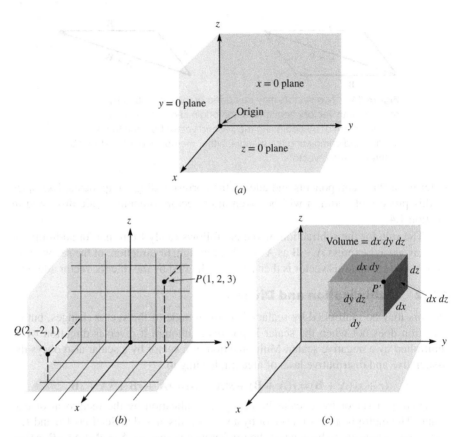

Figure 1.2   (a) A right-handed rectangular coordinate system. If the curved fingers of the right hand indicate the direction through which the x axis is turned into coincidence with the y axis, the thumb shows the direction of the z axis. (b) The location of points P(1, 2, 3) and Q(2, −2, 1). (c) The differential volume element in rectangular coordinates; dx, dy, and dz are, in general, independent differentials.

middle finger may be identified, respectively, as the x, y, and z axes. Figure 1.2a shows a right-handed rectangular coordinate system. A point is located by giving its x, y, and z coordinates. These are, respectively, the distances from the origin to the intersection of perpendicular lines dropped from the point to the x, y, and z axes.

1.3.2   Point Locations as Intersections of Planes

An alternative method of interpreting coordinate values, which *must* be used in all other coordinate systems, is to consider a point as being at the common intersection of three surfaces. In rectangular coordinates, these are the planes x = constant, y = constant, and z = constant, where the constants are the coordinate values of the point.

Figure 1.2b shows points P and Q whose coordinates are (1, 2, 3) and (2, −2, 1), respectively. Point P is therefore located at the common point of intersection of the planes $x = 1$, $y = 2$, and $z = 3$, whereas point Q is located at the intersection of the planes $x = 2$, $y = -2$, and $z = 1$.

In other coordinate systems, as discussed in Sections 1.8 and 1.9, we expect points to be located at the common intersection of three surfaces, not necessarily planes, but still mutually perpendicular at the point of intersection.

If we visualize three planes intersecting at the general point P, whose coordinates are x, y, and z, we may increase each coordinate value by a differential amount and obtain three slightly displaced planes intersecting at point P', whose coordinates are $x + dx$, $y + dy$, and $z + dz$. The six planes define a rectangular parallelepiped whose volume is $dv = dxdydz$; the surfaces have differential areas dS of $dxdy$, $dydz$, and $dzdx$. Finally, the distance dL from P to P' is the diagonal of the parallelepiped and has a length of $\sqrt{(dx)^2 + (dy)^2 + (dz)^2}$. The volume element is shown in Figure 1.2c; point P' is indicated, but point P is located at the only invisible corner.

All this is familiar from trigonometry or solid geometry and as yet involves only scalar quantities. We will describe vectors in terms of a coordinate system in the next section.

1.4 VECTOR COMPONENTS AND UNIT VECTORS

To describe a vector in the rectangular coordinate system, first consider a vector **r** extending outward from the origin. A logical way to identify this vector is by giving the three *component vectors*, lying along the three coordinate axes, whose vector sum must be the given vector. If the component vectors of the vector **r** are **x**, **y**, and **z**, then **r** = **x** + **y** + **z**. The component vectors are shown in Figure 1.3a. Instead of one vector, we now have three, but this is a step forward because the three vectors are of a very simple nature; each is always directed along one of the coordinate axes.

The component vectors in Figure 1.3 have magnitudes that depend on the given vector (such as **r**), but they each have a known and constant direction. This suggests the use of *unit vectors* having unit magnitude by definition; these are parallel to the coordinate axes and they point in the direction of increasing coordinate values. We reserve the symbol **a** for a unit vector and identify its direction by an appropriate subscript. Thus $\mathbf{a}_x$, $\mathbf{a}_y$, and $\mathbf{a}_z$ are the unit vectors in the rectangular coordinate system.[2] They are directed along the x, y, and z axes, respectively, as shown in Figure 1.3b.

If the component vector **y** happens to be two units in magnitude and directed toward increasing values of y, we then write $\mathbf{y} = 2\mathbf{a}_y$. A vector $\mathbf{r}_P$ pointing from the origin to point $P(1, 2, 3)$ is written $\mathbf{r}_P = \mathbf{a}_x + 2\mathbf{a}_y + 3\mathbf{a}_z$. The vector from P to Q is obtained by applying the rule of vector addition. This rule shows that the vector from the origin to P plus the vector from P to Q is equal to the vector from the origin to Q. The desired vector from $P(1, 2, 3)$ to $Q(2, -2, 1)$ is therefore

$$\mathbf{R}_{PQ} = \mathbf{r}_Q - \mathbf{r}_P = (2 - 1)\mathbf{a}_x + (-2 - 2)\mathbf{a}_y + (1 - 3)\mathbf{a}_z$$
$$= \mathbf{a}_x - 4\mathbf{a}_y - 2\mathbf{a}_z$$

The vectors $\mathbf{r}_P$, $\mathbf{r}_Q$, and $\mathbf{R}_{PQ}$ are shown in Figure 1.3c.

---

[2] The symbols **i**, **j**, and **k** are also commonly used for the unit vectors in rectangular coordinates.

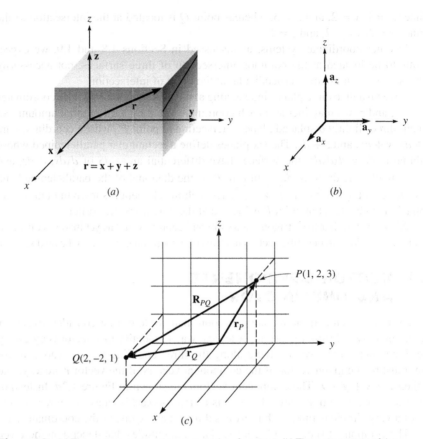

Figure 1.3   (*a*) The component vectors **x**, **y**, and **z** of vector **r**. (*b*) The unit vectors of the rectangular coordinate system have unit magnitude and are directed toward increasing values of their respective variables. (*c*) The vector $\mathbf{R}_{PQ}$ is equal to the vector difference $\mathbf{r}_Q - \mathbf{r}_P$.

The last vector does not extend outward from the origin, as did the vector **r** we initially considered. However, we have already learned that vectors having the same magnitude and pointing in the same direction are equal, so we see that to help our visualization processes we are at liberty to slide any vector over to the origin before determining its component vectors. Parallelism must, of course, be maintained during the sliding process.

In discussing a force vector **F**, or any vector other than a displacement-type vector such as **r**, the problem arises of providing suitable letters for the three component vectors. It would not do to call them **x**, **y**, and **z**, for these are displacements, or directed distances, and are measured in meters (abbreviated m) or some other unit of length. The problem is most often avoided by using *component scalars*, simply called *components*, F_x, F_y, and F_z. The components are the signed magnitudes of the component vectors. We may then write $\mathbf{F} = F_x\mathbf{a}_x + F_y\mathbf{a}_y + F_z\mathbf{a}_z$. The component vectors are $F_x\mathbf{a}_x, F_y\mathbf{a}_y$, and $F_z\mathbf{a}_z$.

Any vector **B** then may be described by $\mathbf{B} = B_x\mathbf{a}_x + B_y\mathbf{a}_y + B_z\mathbf{a}_z$. The magnitude of **B** written $|\mathbf{B}|$ or simply B, is given by

$$|\mathbf{B}| = \sqrt{B_x^2 + B_y^2 + B_z^2} \qquad (1)$$

Each of the three coordinate systems we discuss will have its three fundamental and mutually perpendicular unit vectors that are used to resolve any vector into its component vectors. Unit vectors are not limited to this application. It is helpful to write a unit vector having a specified direction. This is easily done, for a unit vector in a given direction is merely a vector in that direction divided by its magnitude. A unit vector in the **r** direction is $\mathbf{r}/\sqrt{x^2 + y^2 + z^2}$, and a unit vector in the direction of the vector **B** is

$$\mathbf{a}_B = \frac{\mathbf{B}}{\sqrt{B_x^2 + B_y^2 + B_z^2}} = \frac{\mathbf{B}}{|\mathbf{B}|} \qquad (2)$$

EXAMPLE 1.1

Specify the unit vector extending from the origin toward the point $G(2, -2, -1)$.

Solution. We first construct the vector extending from the origin to point G,

$$\mathbf{G} = 2\mathbf{a}_x - 2\mathbf{a}_y - \mathbf{a}_z$$

We continue by finding the magnitude of **G**,

$$|\mathbf{G}| = \sqrt{(2)^2 + (-2)^2 + (-1)^2} = 3$$

and finally expressing the desired unit vector as the quotient,

$$\mathbf{a}_G = \frac{\mathbf{G}}{|\mathbf{G}|} = \frac{2}{3}\mathbf{a}_x - \frac{2}{3}\mathbf{a}_y - \frac{1}{3}\mathbf{a}_z = 0.667\,\mathbf{a}_x - 0.667\,\mathbf{a}_y - 0.333\,\mathbf{a}_z$$

A special symbol is desirable for a unit vector so that its character is immediately apparent. Symbols that have been used are $\mathbf{u}_B$, $\mathbf{a}_B$, $\mathbf{1}_B$, or even **b**. We will consistently use the lowercase **a** with an appropriate subscript.

[NOTE: Throughout the text, drill problems appear following sections in which a new principle is introduced in order to allow students to test their understanding of the basic fact itself. The problems are useful in gaining familiarity with new terms and ideas and should all be worked. More general problems appear at the ends of the chapters. The answers to the drill problems are given in the same order as the parts of the problem.]

D1.1. Given points $M(-1, 2, 1)$, $N(3, -3, 0)$, and $P(-2, -3, -4)$, find: (a) $\mathbf{R}_{MN}$; (b) $\mathbf{R}_{MN} + \mathbf{R}_{MP}$; (c) $|\mathbf{r}_M|$; (d) $\mathbf{a}_{MP}$; (e) $|2\mathbf{r}_P - 3\mathbf{r}_N|$.

Ans. (a) $4\mathbf{a}_x - 5\mathbf{a}_y - \mathbf{a}_z$; (b) $3\mathbf{a}_x - 10\mathbf{a}_y - 6\mathbf{a}_z$; (c) 2.45; (d) $-0.14\mathbf{a}_x - 0.7\mathbf{a}_y - 0.7\mathbf{a}_z$; (e) 15.56

1.5 THE VECTOR FIELD

We have defined a vector field as a vector function of a position vector. In general, the magnitude and direction of the function will change as we move throughout the region, and the value of the vector function must be determined using the coordinate values of the point in question. In the rectangular coordinate system, the vector will be a function of the variables x, y, and z.

Again, representing the position vector as $\mathbf{r}$, a vector field $\mathbf{G}$ can be expressed in functional notation as $\mathbf{G}(\mathbf{r})$; a scalar field T is written as $T(\mathbf{r})$.

If we inspect the velocity of the water in the ocean in some region near the surface where tides and currents are important, we might decide to represent it by a velocity vector that is in any direction, even up or down. If the z axis is taken as upward, the x axis in a northerly direction, the y axis to the west, and the origin at the surface, we have a right-handed coordinate system and may write the velocity vector as $\mathbf{v} = v_x\mathbf{a}_x + v_y\mathbf{a}_y + v_z\mathbf{a}_z$, or $\mathbf{v}(\mathbf{r}) = v_x(\mathbf{r})\mathbf{a}_x + v_y(\mathbf{r})\mathbf{a}_y + v_z(\mathbf{r})\mathbf{a}_z$; each of the components v_x, v_y, and v_z may be a function of the three variables x, y, and z. If we are in some portion of the Gulf Stream where the water is moving only to the north, then v_y and v_z are zero. Further simplifying assumptions might be made if the velocity falls off with depth and changes very slowly as we move north, south, east, or west. A suitable expression could be $\mathbf{v} = 2e^{z/100}\mathbf{a}_x$. We have a velocity of 2 m/s (meters per second) at the surface and a velocity of 0.368×2, or 0.736 m/s, at a depth of 100 m ($z = -100$). The velocity continues to decrease with depth while maintaining a constant direction.

D1.2. A vector field $\mathbf{S}$ is expressed in rectangular coordinates as $\mathbf{S} = \{125/[(x-1)^2 + (y-2)^2 + (z+1)^2]\}\{(x-1)\mathbf{a}_x + (y-2)\mathbf{a}_y + (z+1)\mathbf{a}_z\}$. (a) Evaluate $\mathbf{S}$ at $P(2, 4, 3)$. (b) Determine a unit vector that gives the direction of $\mathbf{S}$ at P. (c) Specify the surface $f(x, y, z)$ on which $|\mathbf{S}| = 1$.

Ans. (a) $5.95\mathbf{a}_x + 11.90\mathbf{a}_y + 23.8\mathbf{a}_z$; (b) $0.218\mathbf{a}_x + 0.436\mathbf{a}_y + 0.873\mathbf{a}_z$; (c) $\sqrt{(x-1)^2 + (y-2)^2 + (z+1)^2} = 125$

1.6 THE DOT PRODUCT

The *dot product* (or scalar product) is used to multiply a given vector field by the component of another field that is *parallel* to the first. This gives the same result when the roles of the fields are reversed. In that sense, the dot product is a projection operation, which can be used to obtain the magnitude of a given field in a specific direction in space.

1.6.1 Geometric Definition

Given two vectors $\mathbf{A}$ and $\mathbf{B}$, the dot product is geometrically defined as the product of the magnitude of $\mathbf{A}$, the magnitude of $\mathbf{B}$, and the cosine of the smaller angle between them, thus projecting one field onto the other:

$$\boxed{\mathbf{A} \cdot \mathbf{B} = |\mathbf{A}||\mathbf{B}| \cos\theta_{AB}} \qquad (3)$$

The dot appears between the two vectors and should be made heavy for emphasis. The dot, or scalar, product is a scalar, as one of the names implies, and it obeys the commutative law,

$$\mathbf{A} \cdot \mathbf{B} = \mathbf{B} \cdot \mathbf{A} \qquad (4)$$

for the sign of the angle does not affect the cosine term. The expression $\mathbf{A} \cdot \mathbf{B}$ is read "**A** dot **B**."

A common application of the dot product is in mechanics, where a constant force **F** applied over a straight displacement **L** does an amount of work $FL \cos \theta$, which is more easily written $\mathbf{F} \cdot \mathbf{L}$. If the force varies along the path, integration is necessary to find the total work (as is taken up in Chapter 4), and the result becomes

$$\text{Work} = \int \mathbf{F} \cdot d\mathbf{L}$$

Another example occurs in magnetic fields. The total flux Φ crossing a surface of area S is given by BS if the magnetic flux density B is perpendicular to the surface and uniform over it. We define a *vector surface* **S** as having area for its magnitude and having a direction *normal* to the surface (avoiding for the moment the problem of which of the two possible normals to take). The flux crossing the surface is then $\mathbf{B} \cdot \mathbf{S}$. This expression is valid for any direction of the uniform magnetic flux density. If the flux density is not constant over the surface, the total flux is $\Phi = \int \mathbf{B} \cdot d\mathbf{S}$. Integrals of this general form appear in Chapter 3 in the context of electric flux density.

1.6.2 Operational Definition

Finding the angle between two vectors in three-dimensional space is often a job we would prefer to avoid, and for that reason the definition of the dot product is usually not used in its basic form. A more helpful result is obtained by considering two vectors whose rectangular components are given, such as $\mathbf{A} = A_x \mathbf{a}_x + A_y \mathbf{a}_y + A_z \mathbf{a}_z$ and $\mathbf{B} = B_x \mathbf{a}_x + B_y \mathbf{a}_y + B_z \mathbf{a}_z$. The dot product also obeys the distributive law, and, therefore, $\mathbf{A} \cdot \mathbf{B}$ yields the sum of nine scalar terms, each involving the dot product of two unit vectors. Because the angle between two different unit vectors of the rectangular coordinate system is $90°$, we then have

$$\mathbf{a}_x \cdot \mathbf{a}_y = \mathbf{a}_y \cdot \mathbf{a}_x = \mathbf{a}_x \cdot \mathbf{a}_z = \mathbf{a}_z \cdot \mathbf{a}_x = \mathbf{a}_y \cdot \mathbf{a}_z = \mathbf{a}_z \cdot \mathbf{a}_y = 0$$

The remaining three terms involve the dot product of a unit vector with itself, which is unity, giving finally the operational definition:

$$\mathbf{A} \cdot \mathbf{B} = A_x B_x + A_y B_y + A_z B_z \qquad (5)$$

which is an expression involving no angles.

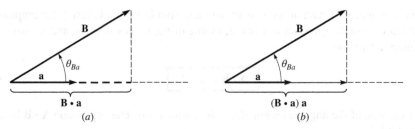

Figure 1.4 (*a*) The scalar component of **B** in the direction of the unit vector **a** is **B** · **a**. (*b*) The vector component of **B** in the direction of the unit vector **a** is (**B** · **a**)**a**.

A vector dotted with itself yields the magnitude squared, or

$$\mathbf{A} \cdot \mathbf{A} = A^2 = |\mathbf{A}|^2 \tag{6}$$

and any unit vector dotted with itself is unity,

$$\mathbf{a}_A \cdot \mathbf{a}_A = 1$$

One of the most important applications of the dot product is that of finding the component of a vector in a given direction. Referring to Figure 1.4*a*, we can obtain the component (scalar) of **B** in the direction specified by the unit vector **a** as

$$\mathbf{B} \cdot \mathbf{a} = |\mathbf{B}|\,|\mathbf{a}|\cos\theta_{Ba} = |\mathbf{B}|\cos\theta_{Ba}$$

The sign of the component is positive if $0 \le \theta_{Ba} \le 90°$ and negative whenever $90° \le \theta_{Ba} \le 180°$.

To obtain the component *vector* of **B** in the direction of **a**, we multiply the component (scalar) by **a**, as illustrated by Figure 1.4*b*. For example, the component of **B** in the direction of $\mathbf{a}_x$ is $\mathbf{B} \cdot \mathbf{a}_x = B_x$, and the component vector is $B_x\mathbf{a}_x$, or $(\mathbf{B} \cdot \mathbf{a}_x)\mathbf{a}_x$. Hence, the problem of finding the component of a vector in any direction becomes the problem of finding a unit vector in that direction, and that we can do.

The geometrical term *projection* is also used with the dot product. Thus, **B** · **a** is the projection of **B** in the **a** direction.

EXAMPLE 1.2

In order to illustrate these definitions and operations, consider the vector field $\mathbf{G} = y\mathbf{a}_x - 2.5x\mathbf{a}_y + 3\mathbf{a}_z$ and the point $Q(4, 5, 2)$. We wish to find: **G** at Q; the scalar component of **G** at Q in the direction of $\mathbf{a}_N = \frac{1}{3}(2\,\mathbf{a}_x + \mathbf{a}_y - 2\,\mathbf{a}_z)$; the vector component of **G** at Q in the direction of $\mathbf{a}_N$; and finally, the angle θ_{Ga} between $\mathbf{G}(\mathbf{r}_Q)$ and $\mathbf{a}_N$.

Solution. Substituting the coordinates of point Q into the expression for **G**, we have

$$\mathbf{G}(\mathbf{r}_Q) = 5\mathbf{a}_x - 10\mathbf{a}_y + 3\mathbf{a}_z$$

Next we find the scalar component. Using the dot product, we have

$$\mathbf{G} \cdot \mathbf{a}_N = (5\mathbf{a}_x - 10\mathbf{a}_y + 3\mathbf{a}_z) \cdot \frac{1}{3}(2\mathbf{a}_x + \mathbf{a}_y - 2\mathbf{a}_z) = \frac{1}{3}(10 - 10 - 6) = -2$$

The vector component is obtained by multiplying the scalar component by the unit vector in the direction of $\mathbf{a}_N$,

$$(\mathbf{G} \cdot \mathbf{a}_N)\mathbf{a}_N = -(2)\frac{1}{3}(2\mathbf{a}_x + \mathbf{a}_y - 2\mathbf{a}_z) = -1.333\,\mathbf{a}_x - 0.667\,\mathbf{a}_y + 1.333\,\mathbf{a}_z$$

The angle between $\mathbf{G}(\mathbf{r}_Q)$ and $\mathbf{a}_N$ is found from

$$\mathbf{G} \cdot \mathbf{a}_N = |\mathbf{G}|\cos\theta_{Ga}$$
$$-2 = \sqrt{25 + 100 + 9}\cos\theta_{Ga}$$

and

$$\theta_{Ga} = \cos^{-1}\frac{-2}{\sqrt{134}} = 99.9°$$

D1.3. The three vertices of a triangle are located at $A(6, -1, 2)$, $B(-2, 3, -4)$, and $C(-3, 1, 5)$. Find: (*a*) $\mathbf{R}_{AB}$; (*b*) $\mathbf{R}_{AC}$; (*c*) the angle θ_{BAC} at vertex A; (*d*) the (vector) projection of $\mathbf{R}_{AB}$ on $\mathbf{R}_{AC}$.

Ans. (*a*) $-8\mathbf{a}_x + 4\mathbf{a}_y - 6\mathbf{a}_z$; (*b*) $-9\mathbf{a}_x + 2\mathbf{a}_y + 3\mathbf{a}_z$; (*c*) $53.6°$; (*d*) $-5.94\mathbf{a}_x + 1.319\mathbf{a}_y + 1.979\mathbf{a}_z$

1.7 THE CROSS PRODUCT

Here, we introduce the second special vector operation, the *cross product* (or vector product). This operation effectively multiplies one vector field by the component of another field that is *perpendicular* to the first. The result is specified as a third field that is normal to both fields at every point. In that sense, the operation is slightly more complicated than the dot product in that the result is a vector, rather than a scalar. Many physical phenomena are well described by a cross product of some kind; this is particularly true in electromagnetics as we will see.

1.7.1 Geometric Definition

Given two vectors **A** and **B**, we now define the cross product, or vector product, of **A** and **B**, written with a cross between the two vectors as **A** × **B** and read "A cross B." The cross product **A** × **B** is a vector; the magnitude of **A** × **B** is equal to the product of the magnitudes of **A**, **B**, and the sine of the smaller angle between **A** and **B**; the direction of **A** × **B** is perpendicular to the plane containing **A** and **B** and is along one of the two possible perpendiculars which is in the direction of advance of a right-handed screw as **A** is turned into **B**. This direction is illustrated in Figure 1.5. Remember that either vector may be moved about at will, maintaining its direction constant, until the two vectors have a "common origin." This determines the plane containing both. However, in most of our applications we will be concerned with vectors defined at the same point.

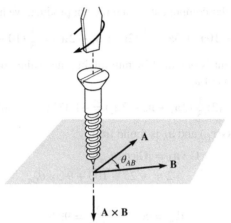

Figure 1.5   The direction of **A** × **B** is in the direction of advance of a right-handed screw as **A** is turned into **B**.

As an equation we can write

$$\mathbf{A} \times \mathbf{B} = \mathbf{a}_N \,|\mathbf{A}|\,|\mathbf{B}|\,\sin\theta_{AB}$$

(7)

where an additional statement, such as that given above, is required to explain the direction of the unit vector $\mathbf{a}_N$. The subscript N stands for "normal."

Reversing the order of the vectors **A** and **B** results in a unit vector in the opposite direction, and we see that the cross product is not commutative, for $\mathbf{B} \times \mathbf{A} = -(\mathbf{A} \times \mathbf{B})$. If the definition of the cross product is applied to the unit vectors $\mathbf{a}_x$ and $\mathbf{a}_y$, we find $\mathbf{a}_x \times \mathbf{a}_y = \mathbf{a}_z$, for each vector has unit magnitude, the two vectors are perpendicular, and the rotation of $\mathbf{a}_x$ into $\mathbf{a}_y$ indicates the positive z direction by the definition of a right-handed coordinate system. In a similar way, $\mathbf{a}_y \times \mathbf{a}_z = \mathbf{a}_x$ and $\mathbf{a}_z \times \mathbf{a}_x = \mathbf{a}_y$. Note the alphabetic symmetry. As long as the three vectors $\mathbf{a}_x$, $\mathbf{a}_y$, and $\mathbf{a}_z$ are written in order (and assuming that $\mathbf{a}_x$ follows $\mathbf{a}_z$, like three elephants in a circle holding tails, so that we could also write $\mathbf{a}_y$, $\mathbf{a}_z$, $\mathbf{a}_x$ or $\mathbf{a}_z$, $\mathbf{a}_x$, $\mathbf{a}_y$), then the cross and equal sign may be placed in either of the two vacant spaces. As a matter of fact, it is now simpler to define a right-handed rectangular coordinate system by saying that $\mathbf{a}_x \times \mathbf{a}_y = \mathbf{a}_z$.

A simple example of the use of the cross product may be taken from geometry or trigonometry. To find the area of a parallelogram, the product of the lengths of two adjacent sides is multiplied by the sine of the angle between them. Using vector notation for the two sides, we then may express the (scalar) area as the *magnitude* of **A** × **B**, or |**A** × **B**|.

The cross product may be used to replace the right-hand rule familiar to all electrical engineers. Consider the force on a straight conductor of length **L**, where the direction assigned to **L** corresponds to the direction of the steady current I, and a uniform magnetic field of flux density **B** is present. Using vector notation, we may write the result neatly as $\mathbf{F} = I\,\mathbf{L} \times \mathbf{B}$. This relationship will be obtained later in Chapter 8.

1.7.2 Operational Definition

The evaluation of a cross product by means of its definition turns out to be more work than the evaluation of the dot product from its definition, for not only must we find the angle between the vectors, but we must also find an expression for the unit vector $\mathbf{a}_N$. This work may be avoided by using rectangular components for the two vectors **A** and **B** and expanding the cross product as a sum of nine simpler cross products, each involving two unit vectors,

$$\begin{aligned}
\mathbf{A} \times \mathbf{B} = {} & A_x B_x \mathbf{a}_x \times \mathbf{a}_x + A_x B_y \mathbf{a}_x \times \mathbf{a}_y + A_x B_z \mathbf{a}_x \times \mathbf{a}_z \\
& + A_y B_x \mathbf{a}_y \times \mathbf{a}_x + A_y B_y \mathbf{a}_y \times \mathbf{a}_y + A_y B_z \mathbf{a}_y \times \mathbf{a}_z \\
& + A_z B_x \mathbf{a}_z \times \mathbf{a}_x + A_z B_y \mathbf{a}_z \times \mathbf{a}_y + A_z B_z \mathbf{a}_z \times \mathbf{a}_z
\end{aligned}$$

We have already found that $\mathbf{a}_x \times \mathbf{a}_y = \mathbf{a}_z$, $\mathbf{a}_y \times \mathbf{a}_z = \mathbf{a}_x$, and $\mathbf{a}_z \times \mathbf{a}_x = \mathbf{a}_y$. The three remaining terms are zero, for the cross product of any vector with itself is zero, since the included angle is zero. These results may be combined to give the operational definition in rectangular coordinates:

$$\mathbf{A} \times \mathbf{B} = (A_y B_z - A_z B_y)\mathbf{a}_x + (A_z B_x - A_x B_z)\mathbf{a}_y + (A_x B_y - A_y B_x)\mathbf{a}_z \qquad (8)$$

This can be written as a determinant in a more easily remembered form:

$$\mathbf{A} \times \mathbf{B} = \begin{vmatrix} \mathbf{a}_x & \mathbf{a}_y & \mathbf{a}_z \\ A_x & A_y & A_z \\ B_x & B_y & B_z \end{vmatrix} \qquad (9)$$

Thus, if $\mathbf{A} = 2\mathbf{a}_x - 3\mathbf{a}_y + \mathbf{a}_z$ and $\mathbf{B} = -4\mathbf{a}_x - 2\mathbf{a}_y + 5\mathbf{a}_z$, we have

$$\begin{aligned}
\mathbf{A} \times \mathbf{B} &= \begin{vmatrix} \mathbf{a}_x & \mathbf{a}_y & \mathbf{a}_z \\ 2 & -3 & 1 \\ -4 & -2 & 5 \end{vmatrix} \\
&= [(-3)(5) - (1)(-2)]\mathbf{a}_x - [(2)(5) - (1)(-4)]\mathbf{a}_y + [(2)(-2) - (-3)(-4)]\mathbf{a}_z \\
&= -13\mathbf{a}_x - 14\mathbf{a}_y - 16\mathbf{a}_z
\end{aligned}$$

D1.4. The three vertices of a triangle are located at $A(6, -1, 2)$, $B(-2, 3, -4)$, and $C(-3, 1, 5)$. Find: (*a*) $\mathbf{R}_{AB} \times \mathbf{R}_{AC}$; (*b*) the area of the triangle; (*c*) a unit vector perpendicular to the plane in which the triangle is located.

Ans. (*a*) $24\mathbf{a}_x + 78\mathbf{a}_y + 20\mathbf{a}_z$; (*b*) 42.0; (*c*) $0.286\mathbf{a}_x + 0.928\mathbf{a}_y + 0.238\mathbf{a}_z$

1.8   OTHER COORDINATE SYSTEMS: CIRCULAR CYLINDRICAL COORDINATES

The rectangular coordinate system is generally the one in which students prefer to work every problem. This often means a lot more work, because many problems possess a type of symmetry that pleads for a more logical treatment. It is easier to do now, once and for all, the work required to become familiar with cylindrical and spherical coordinates, instead of applying an equal or greater effort to every problem involving cylindrical or spherical symmetry later. With this in mind, we will take a careful and unhurried look at cylindrical and spherical coordinates.

1.8.1   Point Coordinates

The circular cylindrical coordinate system is the three-dimensional version of the polar coordinates of analytic geometry. In polar coordinates, a point is located in a plane by giving both its distance ρ from the origin and the angle ϕ between the line from the point to the origin and an arbitrary radial line, taken as $\phi = 0$.[3] In circular cylindrical coordinates, we also specify the distance z of the point from an arbitrary $z = 0$ reference plane. For simplicity, we usually refer to circular cylindrical coordinates simply as cylindrical coordinates. This will not cause any confusion in reading this book, but it is only fair to point out that there are such systems as elliptic cylindrical coordinates, hyperbolic cylindrical coordinates, parabolic cylindrical coordinates, and others.

We no longer set up three axes as with rectangular coordinates, but we must instead consider any point as the intersection of three mutually perpendicular surfaces. These surfaces are a circular cylinder (ρ = constant), a plane (ϕ = constant), and another plane (z = constant). This corresponds to the location of a point in a rectangular coordinate system by the intersection of three planes (x = constant, y = constant, and z = constant). The three surfaces of circular cylindrical coordinates are shown in Figure 1.6a. Note that three such surfaces may be passed through any point, unless it lies on the z axis, in which case one plane suffices.

1.8.2   Unit Vectors

Three unit vectors are defined for the cylindrical system, but we may no longer direct them along the "coordinate axes," for such axes exist only in rectangular coordinates. Instead, we take a broader view of the unit vectors in rectangular coordinates and realize that they are directed toward increasing coordinate values and are perpendicular to the surface on which that coordinate value is constant (i.e., the unit vector $\mathbf{a}_x$ is normal to the plane x = constant and points toward larger values of x). In a corresponding way we may now define three unit vectors in cylindrical coordinates, $\mathbf{a}_\rho$, $\mathbf{a}_\phi$, and $\mathbf{a}_z$.

---

[3] The two variables of polar coordinates are commonly called r and θ. With three coordinates, however, it is more common to use ρ for the radius variable of cylindrical coordinates and r for the (different) radius variable of spherical coordinates. Also, the angle variable of cylindrical coordinates is customarily called ϕ because everyone uses θ for a different angle in spherical coordinates. The angle ϕ is common to both cylindrical and spherical coordinates. See?

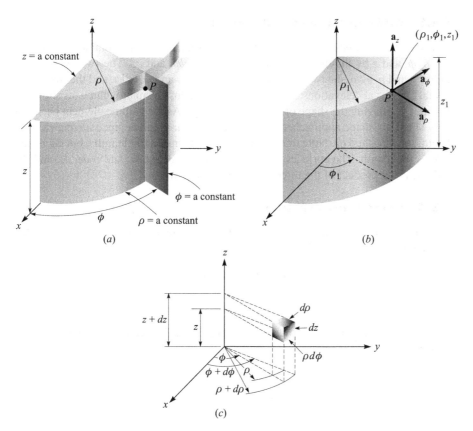

Figure 1.6 (*a*) The three mutually perpendicular surfaces of the circular cylindrical co-ordinate system. (*b*) The three unit vectors of the circular cylindrical coordinate system. (*c*) The differential volume unit in the circular cylindrical coordinate system; $d\rho$, $\rho d\phi$, and dz are all elements of length.

The unit vector $\mathbf{a}_\rho$ at a point $P(\rho_1, \phi_1, z_1)$ is directed radially outward, normal to the cylindrical surface $\rho = \rho_1$. It lies in the planes $\phi = \phi_1$ and $z = z_1$. The unit vector $\mathbf{a}_\phi$ is normal to the plane $\phi = \phi_1$, points in the direction of increasing ϕ, lies in the plane $z = z_1$, and is tangent to the cylindrical surface $\rho = \rho_1$. The unit vector $\mathbf{a}_z$ is the same as the unit vector $\mathbf{a}_z$ of the rectangular coordinate system. Figure 1.6*b* shows the three vectors in cylindrical coordinates.

In rectangular coordinates, the unit vectors are not functions of the coordinates. Two of the unit vectors in cylindrical coordinates, $\mathbf{a}_\rho$ and $\mathbf{a}_\phi$, however, *do* vary with the coordinate ϕ, as their directions change. In integration or differentiation with respect to ϕ, then, $\mathbf{a}_\rho$ and $\mathbf{a}_\phi$ must not be treated as constants.

The unit vectors are again mutually perpendicular, for each is normal to one of the three mutually perpendicular surfaces, and we may define a right-handed cylindrical coordinate system as one in which $\mathbf{a}_\rho \times \mathbf{a}_\phi = \mathbf{a}_z$, or (for those who have flexible fingers) as one in which the thumb, forefinger, and middle finger point in the direction of increasing ρ, ϕ, and z, respectively.

1.8.3   Differential Area and Volume

A differential volume element in cylindrical coordinates may be obtained by increasing ρ, ϕ, and z by the differential increments $d\rho$, $d\phi$, and dz. The two cylinders of radius ρ and $\rho + d\rho$, the two radial planes at angles ϕ and $\phi + d\phi$, and the two "horizontal" planes at "elevations" z and $z + dz$ now enclose a small volume, as shown in Figure 1.6c, having the shape of a truncated wedge. As the volume element becomes very small, its shape approaches that of a rectangular parallelepiped having sides of length $d\rho$, $\rho d\phi$, and dz. Note that $d\rho$ and dz are dimensionally lengths, but $d\phi$ is not; $\rho d\phi$ is the length. The surfaces have areas of $\rho \, d\rho \, d\phi$, $d\rho \, dz$, and $\rho \, d\phi \, dz$, and the volume is the product of the three side lengths, or $\rho \, d\rho \, d\phi \, dz$.

1.8.4   Point Transformations

The variables of the rectangular and cylindrical coordinate systems are easily related to each other. Referring to Figure 1.7, we see that

$$\begin{aligned} x &= \rho \, \cos\phi \\ y &= \rho \, \sin\phi \\ z &= z \end{aligned} \qquad (10)$$

From the other viewpoint, we may express the cylindrical variables in terms of x, y, and z:

$$\begin{aligned} \rho &= \sqrt{x^2 + y^2} \quad (\rho \geq 0) \\ \phi &= \tan^{-1}\frac{y}{x} \\ z &= z \end{aligned} \qquad (11)$$

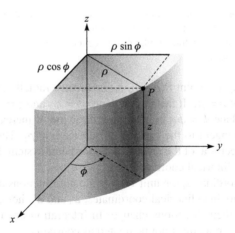

Figure 1.7 The relationship between the rectangular variables x, y, z and the cylindrical coordinate variables ρ, ϕ, z. There is no change in the variable z between the two systems.

We consider the variable ρ to be positive or zero, thus using only the positive sign for the radical in (11). The proper value of the angle ϕ is determined by inspecting the signs of x and y. Thus, if $x = -3$ and $y = 4$, we find that the point lies in the second quadrant so that $\rho = 5$ and $\phi = 126.9°$. For $x = 3$ and $y = -4$, we have $\phi = -53.1°$ or $306.9°$, whichever is more convenient.

Using (10) or (11), scalar functions given in one coordinate system are easily transformed into the other system.

1.8.5 Vector Component Transformations

A vector function in one coordinate system requires two steps in order to transform it to another coordinate system, because a different set of component vectors is generally required. That is, we may be given a rectangular vector

$$\mathbf{A} = A_x \mathbf{a}_x + A_y \mathbf{a}_y + A_z \mathbf{a}_z$$

where each component is given as a function of x, y, and z, and we need a vector in cylindrical coordinates

$$\mathbf{A} = A_\rho \mathbf{a}_\rho + A_\phi \mathbf{a}_\phi + A_z \mathbf{a}_z$$

where each component is given as a function of ρ, ϕ, and z.

To find any desired component of a vector, we recall from the discussion of the dot product that a component in a desired direction may be obtained by taking the dot product of the vector and a unit vector in the desired direction. Hence,

$$A_\rho = \mathbf{A} \cdot \mathbf{a}_\rho \qquad \text{and} \qquad A_\phi = \mathbf{A} \cdot \mathbf{a}_\phi$$

In other words, A_ρ and A_ϕ are the *projections* of $\mathbf{A}$ into the $\mathbf{a}_\rho$ and $\mathbf{a}_\phi$ directions. Expanding these dot products, we have

$$A_\rho = (A_x \mathbf{a}_x + A_y \mathbf{a}_y + A_z \mathbf{a}_z) \cdot \mathbf{a}_\rho = A_x \mathbf{a}_x \cdot \mathbf{a}_\rho + A_y \mathbf{a}_y \cdot \mathbf{a}_\rho \tag{12}$$

$$A_\phi = (A_x \mathbf{a}_x + A_y \mathbf{a}_y + A_z \mathbf{a}_z) \cdot \mathbf{a}_\phi = A_x \mathbf{a}_x \cdot \mathbf{a}_\phi + A_y \mathbf{a}_y \cdot \mathbf{a}_\phi \tag{13}$$

and

$$A_z = (A_x \mathbf{a}_x + A_y \mathbf{a}_y + A_z \mathbf{a}_z) \cdot \mathbf{a}_z = A_z \mathbf{a}_z \cdot \mathbf{a}_z = A_z \tag{14}$$

since $\mathbf{a}_z \cdot \mathbf{a}_\rho$ and $\mathbf{a}_z \cdot \mathbf{a}_\phi$ are zero.

In order to complete the transformation of the components, it is necessary to know the dot products $\mathbf{a}_x \cdot \mathbf{a}_\rho$, $\mathbf{a}_y \cdot \mathbf{a}_\rho$, $\mathbf{a}_x \cdot \mathbf{a}_\phi$, and $\mathbf{a}_y \cdot \mathbf{a}_\phi$. Applying the definition of the dot product, we see that since we are concerned with unit vectors, the result is merely the cosine of the angle between the two unit vectors in question. Referring to Figure 1.7 and thinking mightily, we identify the angle between $\mathbf{a}_x$ and $\mathbf{a}_\rho$ as ϕ, and thus $\mathbf{a}_x \cdot \mathbf{a}_\rho = \cos \phi$, but the angle between $\mathbf{a}_y$ and $\mathbf{a}_\rho$ is $90° - \phi$, and $\mathbf{a}_y \cdot \mathbf{a}_\rho = \cos(90° - \phi) = \sin \phi$. The remaining dot products of the unit vectors are found in a similar manner, and the results are tabulated as functions of ϕ in Table 1.1.

Table 1.1   Dot products of unit vectors in cylindrical and rectangular coordinate systems

	$\mathbf{a}_\rho$	$\mathbf{a}_\phi$	$\mathbf{a}_z$
$\mathbf{a}_x \cdot$	$\cos\phi$	$-\sin\phi$	0
$\mathbf{a}_y \cdot$	$\sin\phi$	$\cos\phi$	0
$\mathbf{a}_z \cdot$	0	0	1

Transforming vectors from rectangular to cylindrical coordinates or vice versa is therefore accomplished by using (10) or (11) to change variables, and by using the dot products of the unit vectors given in Table 1.1 to change components. The two steps may be taken in either order.

EXAMPLE 1.3

Transform the vector $\mathbf{B} = y\mathbf{a}_x - x\mathbf{a}_y + z\mathbf{a}_z$ into cylindrical coordinates.

Solution. The new components are

$$B\rho = \mathbf{B} \cdot \mathbf{a}_\rho = y(\mathbf{a}_x \cdot \mathbf{a}_\rho) - x(\mathbf{a}_y \cdot \mathbf{a}_\rho)$$
$$= y \cos\phi - x \sin\phi = \rho \sin\phi \cos\phi - \rho \cos\phi \sin\phi = 0$$
$$B_\phi = \mathbf{B} \cdot \mathbf{a}_\phi = y(\mathbf{a}_x \cdot \mathbf{a}_\phi) - x(\mathbf{a}_y \cdot \mathbf{a}_\phi)$$
$$= -y \sin\phi - x \cos\phi = -\rho \sin^2\phi - \rho\cos^2\phi = -\rho$$

Thus,

$$\mathbf{B} = -\rho\,\mathbf{a}_\phi + z\,\mathbf{a}_z$$

D1.5. (a) Give the rectangular coordinates of the point C ($\rho = 4.4$, $\phi = -115°$, $z = 2$). (b) Give the cylindrical coordinates of the point D ($x = -3.1$, $y = 2.6$, $z = -3$). (c) Specify the distance from C to D.

Ans. (a) $C(x = -1.860, y = -3.99, z = 2)$; (b) $D(\rho = 4.05, \phi = 140.0°, z = -3)$; (c) 8.36

D1.6. Transform to cylindrical coordinates: (a) $\mathbf{F} = 10\mathbf{a}_x - 8\mathbf{a}_y + 6\mathbf{a}_z$ at point $P(10, -8, 6)$; (b) $\mathbf{G} = (2x + y)\mathbf{a}_x - (y - 4x)\mathbf{a}_y$ at point Q (ρ, ϕ, z). (c) Give the rectangular components of the vector $\mathbf{H} = 20\mathbf{a}_\rho - 10\mathbf{a}_\phi + 3\mathbf{a}_z$ at $P(x = 5$, $y = 2, z = -1)$.

Ans. (a) $12.81\mathbf{a}_\rho + 6\mathbf{a}_z$; (b) $(2\rho \cos^2\phi - \rho \sin^2\phi + 5\rho \sin\phi \cos\phi)\mathbf{a}_\rho + (4\rho \cos^2\phi - \rho \sin^2\phi - 3\rho \sin\phi \cos\phi)\mathbf{a}_\phi$; (c) $H_x = 22.3, H_y = -1.857, H_z = 3$

1.9   THE SPHERICAL COORDINATE SYSTEM

We have no two-dimensional coordinate system to help us understand the three-dimensional spherical coordinate system, as we have for the circular cylindrical coordinate system. In certain respects we can draw on our knowledge of the

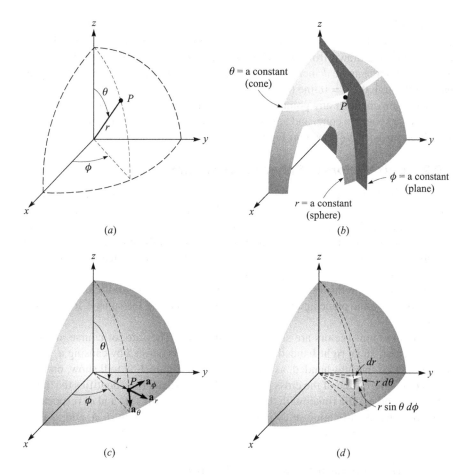

Figure 1.8 (*a*) The three spherical coordinates. (*b*) The three mutually perpendicular surfaces of the spherical coordinate system. (*c*) The three unit vectors of spherical coordinates: $\mathbf{a}_r \times \mathbf{a}_\theta = \mathbf{a}_\phi$. (*d*) The differential volume element in the spherical coordinate system.

latitude-and-longitude system of locating a place on the surface of the earth, but usually we consider only points on the surface and not those below or above ground.

1.9.1 Coordinates of a Point

We begin by building a spherical coordinate system on the three rectangular axes (Figure 1.8*a*). The distance from the origin to any point is defined as r. The surface $r =$ constant is a sphere.

The second coordinate is an angle θ between the z axis and the line drawn from the origin to the point in question. The surface $\theta =$ constant is a cone, and the two surfaces, cone and sphere, are everywhere perpendicular along their intersection, which is a circle of radius $r \sin \theta$. The coordinate θ corresponds to latitude, except that latitude is measured from the equator and θ is measured from the "North Pole."

The third coordinate ϕ is also an angle and is exactly the same as the angle ϕ of cylindrical coordinates. It is the angle between the x axis and the projection in the $z = 0$ plane of the line drawn from the origin to the point. It corresponds to the angle of longitude, but the angle ϕ increases to the "east." The surface $\phi =$ constant is a plane passing through the $\theta = 0$ line (or the z axis).

We again consider any point as the intersection of three mutually perpendicular surfaces—a sphere, a cone, and a plane—each oriented in the manner just described. The three surfaces are shown in Figure 1.8b.

1.9.2 Unit Vectors in Spherical Coordinates

Three unit vectors may again be defined at any point. Each unit vector is perpendicular to one of the three mutually perpendicular surfaces and is oriented in that direction in which the coordinate increases. The unit vector $\mathbf{a}_r$ is directed radially outward, normal to the sphere $r =$ constant, and lies in the cone $\theta =$ constant and the plane $\phi =$ constant. The unit vector $\mathbf{a}_\theta$ is normal to the conical surface, lies in the plane, and is tangent to the sphere. It is directed along a line of "longitude" and points "south." The third unit vector $\mathbf{a}_\phi$ is the same as in cylindrical coordinates, being normal to the plane and tangent to both the cone and the sphere. It is directed to the "east."

The three unit vectors are shown in Figure 1.8c. They are, of course, mutually perpendicular, and a right-handed coordinate system is defined by causing $\mathbf{a}_r \times \mathbf{a}_\theta = \mathbf{a}_\phi$. Our system is right-handed, as an inspection of Figure 1.8c will show, on application of the definition of the cross product. The right-hand rule identifies the thumb, forefinger, and middle finger with the direction of increasing r, θ, and ϕ, respectively. (Note that the identification in cylindrical coordinates was with ρ, ϕ, and z, and in rectangular coordinates with x, y, and z.)

1.9.3 Differential Surfaces and Volume

A differential volume element may be constructed in spherical coordinates by increasing r, θ, and ϕ by dr, $d\theta$, and $d\phi$, as shown in Figure 1.8d. The distance between the two spherical surfaces of radius r and $r + dr$ is dr; the distance between the two cones having generating angles of θ and $\theta + d\theta$ is $r d\theta$; and the distance between the two radial planes at angles ϕ and $\phi + d\phi$ is found to be $r \sin \theta \, d\phi$, after a few moments of trigonometric thought. The surfaces have areas of $r \, dr \, d\theta$, $r \sin \theta \, dr \, d\phi$, and $r^2 \sin \theta \, d\theta \, d\phi$, and the volume is $r^2 \sin \theta \, dr \, d\theta \, d\phi$.

1.9.4 Point Transformations

The transformation of scalars from the rectangular to the spherical coordinate system is easily made by using Figure 1.8a to relate the two sets of variables:

$$
\begin{aligned}
x &= r \sin\theta \, \cos\phi \\
y &= r \sin\theta \, \sin\phi \\
z &= r \cos\theta
\end{aligned}
\tag{15}
$$

Table 1.2 Dot products of unit vectors in spherical and rectangular coordinate systems

	$\mathbf{a}_r$	$\mathbf{a}_\theta$	$\mathbf{a}_\phi$
$\mathbf{a}_x \cdot$	$\sin\theta\cos\phi$	$\cos\theta\cos\phi$	$-\sin\phi$
$\mathbf{a}_y \cdot$	$\sin\theta\sin\phi$	$\cos\theta\sin\phi$	$\cos\phi$
$\mathbf{a}_z \cdot$	$\cos\theta$	$-\sin\theta$	0

The transformation in the reverse direction is achieved with the help of

$$
\begin{aligned}
r &= \sqrt{x^2 + y^2 + z^2} && (r \geq 0) \\
\theta &= \cos^{-1}\frac{z}{\sqrt{x^2 + y^2 + z^2}} && (0° \leq \theta \leq 180°) \\
\phi &= \tan^{-1}\frac{y}{x}
\end{aligned}
\tag{16}
$$

The radius variable r is nonnegative, and θ is restricted to the range from 0° to 180°, inclusive. The angles are placed in the proper quadrants by inspecting the signs of x, y, and z.

1.9.5　Vector Component Transformations

The transformation of vectors requires us to determine the products of the unit vectors in rectangular and spherical coordinates. We work out these products from Figure 1.8c and a pinch of trigonometry. Because the dot product of any spherical unit vector with any rectangular unit vector is the component of the spherical vector in the direction of the rectangular vector, the dot products with $\mathbf{a}_z$ are found to be

$$
\begin{aligned}
\mathbf{a}_z \cdot \mathbf{a}_r &= \cos\theta \\
\mathbf{a}_z \cdot \mathbf{a}_\theta &= -\sin\theta \\
\mathbf{a}_z \cdot \mathbf{a}_\phi &= 0
\end{aligned}
$$

The dot products involving $\mathbf{a}_x$ and $\mathbf{a}_y$ require first the projection of the spherical unit vector on the xy plane and then the projection onto the desired axis. For example, $\mathbf{a}_r \cdot \mathbf{a}_x$ is obtained by projecting $\mathbf{a}_r$ onto the xy plane, giving $\sin\theta$, and then projecting $\sin\theta$ on the x axis, which yields $\sin\theta\cos\phi$. The other dot products are found in a like manner, and all are shown in Table 1.2.

EXAMPLE 1.4

We illustrate this procedure by transforming the vector field $\mathbf{G} = (xz/y)\mathbf{a}_x$ into spherical components and variables.

Solution. We find the three spherical components by dotting $\mathbf{G}$ with the appropriate unit vectors, and we change variables during the procedure:

$$
\begin{aligned}
G_r = \mathbf{G} \cdot \mathbf{a}_r &= \frac{xz}{y}\,\mathbf{a}_x \cdot \mathbf{a}_r = \frac{xz}{y}\,\sin\theta\,\cos\phi \\
&= r\sin\theta\,\cos\theta\,\frac{\cos^2\phi}{\sin\phi}
\end{aligned}
$$

$$G_\theta = \mathbf{G} \cdot \mathbf{a}_\theta = \frac{xz}{y}\, \mathbf{a}_x \cdot \mathbf{a}_\theta = \frac{xz}{y}\cos\theta\,\cos\phi$$

$$= r\cos^2\theta\,\frac{\cos^2\phi}{\sin\phi}$$

$$G_\phi = \mathbf{G} \cdot \mathbf{a}_\phi = \frac{xz}{y}\mathbf{a}_x \cdot \mathbf{a}_\phi = \frac{xz}{y}(-\sin\phi)$$

$$= -r\cos\theta\,\cos\phi$$

Collecting these results, we have

$$\mathbf{G} = r\cos\theta\,\cos\phi\,(\sin\theta\,\cot\phi\,\mathbf{a}_r + \cos\theta\,\cot\phi\,\mathbf{a}_\theta - \mathbf{a}_\phi)$$

Appendix A describes the general curvilinear coordinate system of which the rectangular, circular cylindrical, and spherical coordinate systems are special cases. The first section of this appendix could well be scanned now.

D1.7. Given the two points, $C(-3, 2, 1)$ and $D(r = 5, \theta = 20°, \phi = -70°)$, find: (*a*) the spherical coordinates of C; (*b*) the rectangular coordinates of D; (*c*) the distance from C to D.

Ans. (*a*) $C(r = 3.74, \theta = 74.5°, \phi = 146.3°)$; (*b*) $D(x = 0.585, y = -1.607, z = 4.70)$; (*c*) 6.29

D1.8. Transform the following vectors to spherical coordinates at the points given: (*a*) $10\mathbf{a}_x$ at $P(x = -3, y = 2, z = 4)$; (*b*) $10\mathbf{a}_y$ at $Q(\rho = 5, \phi = 30°, z = 4)$; (*c*) $10\mathbf{a}_z$ at $M(r = 4, \theta = 110°, \phi = 120°)$.

Ans. (*a*) $-5.57\mathbf{a}_r - 6.18\mathbf{a}_\theta - 5.55\mathbf{a}_\phi$; (*b*) $3.90\mathbf{a}_r + 3.12\mathbf{a}_\theta + 8.66\mathbf{a}_\phi$; (*c*) $-3.42\mathbf{a}_r - 9.40\mathbf{a}_\theta$

REFERENCES

1. Grossman, S. I. *Calculus*. 3d ed. Orlando, Fla.: Academic Press and Harcourt Brace Jovanovich, 1984. Vector algebra and cylindrical and spherical coordinates appear in Chapter 17, and vector calculus is introduced in Chapter 20.

2. Spiegel, M. R. *Vector Analysis*. Schaum Outline Series. New York: McGraw-Hill, 1959. A large number of examples and problems with answers are provided in this concise, inexpensive member of an outline series.

3. Swokowski, E. W. *Calculus with Analytic Geometry*. 3d ed. Boston: Prindle, Weber, & Schmidt, 1984. Vector algebra and the cylindrical and spherical coordinate systems are discussed in Chapter 14, and vector calculus appears in Chapter 18.

4. Thomas, G. B., Jr., and R. L. Finney. *Calculus and Analytic Geometry*. 6th ed. Reading, Mass.: Addison-Wesley Publishing Company, 1984. Vector algebra and the three coordinate systems we use are discussed in Chapter 13. Other vector operations are discussed in Chapters 15 and 17.

CHAPTER 1 PROBLEMS

Quizzes

1.1 If **A** represents a vector two units in length directed due west, **B** represents a vector three units in length directed due north, and $\mathbf{A} + \mathbf{B} = \mathbf{C} - \mathbf{D}$ and

$2\mathbf{B} - \mathbf{A} = \mathbf{C} + \mathbf{D}$, find the magnitudes and directions of $\mathbf{C}$ and $\mathbf{D}$. Take north as the positive y direction.

1.2 Vector $\mathbf{A}$ extends from the origin to $(1, 2, 3)$, and vector $\mathbf{B}$ extends from the origin to $(2, 3, -2)$. Find (a) the unit vector in the direction of $(\mathbf{A} - \mathbf{B})$; (b) the unit vector in the direction of the line extending from the origin to the midpoint of the line joining the ends of $\mathbf{A}$ and $\mathbf{B}$.

1.3 The vector from the origin to point A is given as $(6, -2, -4)$, and the unit vector directed from the origin toward point B is $(2, -2, 1)/3$. If points A and B are ten units apart, find the coordinates of point B.

1.4 A circle, centered at the origin with a radius of 2 units, lies in the xy plane. Determine the unit vector in rectangular components that lies in the xy plane, is tangent to the circle at $(\sqrt{3}, -1, 0)$, and is in the general direction of increasing values of x.

1.5 An equilateral triangle lies in the xy plane with its centroid at the origin. One vertex lies on the positive y axis. (a) Find unit vectors that are directed from the origin to the three vertices. (b) Find unit vectors that are directed from the origin to the three sides, intersecting these at right angles.

1.6 Find the acute angle between the two vectors $\mathbf{A} = 2\mathbf{a}_x + \mathbf{a}_y + 3\mathbf{a}_z$ and $\mathbf{B} = \mathbf{a}_x - 3\mathbf{a}_y + 2\mathbf{a}_z$ by using the definition of (a) the dot product; (b) the cross product.

1.7 Given the field $\mathbf{F} = x\mathbf{a}_x + y\mathbf{a}_y$. If $\mathbf{F} \cdot \mathbf{G} = 2xy$ and $\mathbf{F} \times \mathbf{G} = \left(x^2 - y^2\right)\mathbf{a}_z$, find $\mathbf{G}$.

1.8 Demonstrate the ambiguity that results when the cross product is used to find the angle between two vectors by finding the angle between $\mathbf{A} = 3\mathbf{a}_x - 2\mathbf{a}_y + 4\mathbf{a}_z$ and $\mathbf{B} = 2\mathbf{a}_x + \mathbf{a}_y - 2\mathbf{a}_z$. Does this ambiguity exist when the dot product is used?

1.9 A field is given as $\mathbf{G} = [25/(x^2 + y^2)](x\mathbf{a}_x + y\mathbf{a}_y)$. Find (a) a unit vector in the direction of $\mathbf{G}$ at $P(3, 4, -2)$; (b) the angle between $\mathbf{G}$ and $\mathbf{a}_x$ at P; (c) the value of the following double integral on the plane $y = 7$.

$$\int_0^4 \int_0^2 \mathbf{G} \cdot \mathbf{a}_y \, dz \, dx$$

1.10 By expressing diagonals as vectors and using the definition of the dot product, find the smaller angle between any two diagonals of a cube, where each diagonal connects diametrically opposite corners and passes through the center of the cube.

1.11 Given the points $M(0.1, -0.2, -0.1)$, $N(-0.2, 0.1, 0.3)$, and $P(0.4, 0, 0.1)$, find (a) the vector $\mathbf{R}_{MN}$; (b) the dot product $\mathbf{R}_{MN} \cdot \mathbf{R}_{MP}$; (c) the scalar projection of $\mathbf{R}_{MN}$ on $\mathbf{R}_{MP}$; (d) the angle between $\mathbf{R}_{MN}$ and $\mathbf{R}_{MP}$.

1.12 Write an expression in rectangular components for the vector that extends from (x_1, y_1, z_1) to (x_2, y_2, z_2) and determine the magnitude of this vector.

1.13 Find (a) the vector component of $\mathbf{F} = 10\mathbf{a}_x - 6\mathbf{a}_y + 5\mathbf{a}_z$ that is parallel to $\mathbf{G} = 0.1\mathbf{a}_x + 0.2\mathbf{a}_y + 0.3\mathbf{a}_z$; (b) the vector component of $\mathbf{F}$ that is perpendicular to $\mathbf{G}$; (c) the vector component of $\mathbf{G}$ that is perpendicular to $\mathbf{F}$.

1.14 Given that $\mathbf{A} + \mathbf{B} + \mathbf{C} = 0$, where the three vectors represent line segments and extend from a common origin, must the three vectors be coplanar? If $\mathbf{A} + \mathbf{B} + \mathbf{C} + \mathbf{D} = 0$, are the four vectors coplanar?

1.15 Three vectors extending from the origin are given as $\mathbf{r}_1 = (7, 3, -2)$, $\mathbf{r}_2 = (-2, 7, -3)$, and $\mathbf{r}_3 = (0, 2, 3)$. Find (a) a unit vector perpendicular to both $\mathbf{r}_1$ and $\mathbf{r}_2$; (b) a unit vector perpendicular to the vectors $\mathbf{r}_1 - \mathbf{r}_2$ and $\mathbf{r}_2 - \mathbf{r}_3$; (c) the area of the triangle defined by $\mathbf{r}_1$ and $\mathbf{r}_2$; (d) the area of the triangle defined by the heads of $\mathbf{r}_1$, $\mathbf{r}_2$, and $\mathbf{r}_3$.

1.16 In geometrical optics, the path of a light ray can be treated as a vector having the usual three components in a rectangular coordinate system. When light reflects from a plane surface, the effect is to reverse the vector component of the ray that is normal to that surface. This yields a reflection angle that is equal to the incidence angle. Using a vector representation in rectangular coordinates, demonstrate what happens when a light ray reflects from a *corner cube reflector*, consisting of three mutually orthogonal surfaces that occupy, for example, the *xy*, *xz*, and *yz* planes. The ray is incident in an arbitrary direction within the first octant of the coordinate system and has initially negative *x*, *y*, and *z* components of travel.

1.17 Point $A(-4, 2, 5)$ and the two vectors, $\mathbf{R}_{AM} = (20, 18, -10)$? and $\mathbf{R}_{AN} = (-10, 8, 15)$, define a triangle. Find (a) a unit vector perpendicular to the triangle; (b) a unit vector in the plane of the triangle and perpendicular to $\mathbf{R}_{AN}$; (c) a unit vector in the plane of the triangle that bisects the interior angle at A.

1.18 Given two fields in spherical coordinates, $\mathbf{E} = (A/r) \sin \theta \, \mathbf{a}_\theta$ and $\mathbf{H} = (B/r) \sin \theta \, \mathbf{a}_\phi$ (a) evaluate $\mathbf{S} = \mathbf{E} \times \mathbf{H}$ and express the result in rectangular coordinates. (b) Determine $\mathbf{S}$ along the *x*, *y*, and *z* axes.

1.19 Consider the important inverse-square radial field, expressed in spherical coordinates: $\mathbf{F} = Ar^{-2} \, \mathbf{a}_r$ where A is a constant. (a) Transform the field into cylindrical coordinates. (b) Transform the field into rectangular coordinates.

1.20 If the three sides of a triangle are represented by vectors $\mathbf{A}$, $\mathbf{B}$, and $\mathbf{C}$, all directed counterclockwise, show that $|\mathbf{C}|^2 = (\mathbf{A} + \mathbf{B}) \cdot (\mathbf{A} + \mathbf{B})$ and expand the product to obtain the law of cosines.

1.21 Express in cylindrical components: (a) the vector from $C(3, 2, -7)$ to $D(-1, -4, 2)$; (b) a unit vector at D directed toward C; (c) a unit vector at D directed toward the origin.

1.22 A sphere of radius a, centered at the origin, rotates about the z axis at angular velocity Ω rad/s. The rotation direction is clockwise when one is looking in the positive z direction. (a) Using spherical components, write an expression for the velocity field, $\mathbf{v}$, that gives the tangential velocity at any point on the sphere surface. (b) Derive an expression for the difference in velocities between two points on the surface that have different latitudes, where the latitude difference is $\Delta\theta$ in radians. Assume $\Delta\theta$ is small. (c) Find the difference in velocities one degree on either side of $45°$ north latitude on earth. Take the earth's radius as 6370 km at $45°$.

1.23 The surfaces $\rho = 3$, $\rho = 5$, $\phi = 100°$, $\phi = 130°$, $z = 3$, and $z = 4.5$ define a closed surface. Find (a) the enclosed volume; (b) the total area of the enclosing surface; (c) the total length of the 12 edges of the surfaces; (d) the length of the longest straight line that lies entirely within the volume.

1.24 Two unit vectors, $\mathbf{a}_1$ and $\mathbf{a}_2$, lie in the xy plane and pass through the origin. They make angles ϕ_1 and ϕ_2, respectively, with the x axis (a) Express each vector in rectangular components; (b) take the dot product and verify the trigonometric identity, $\cos(\phi_1 - \phi_2) = \cos\phi_1 \cos\phi_2 + \sin\phi_1 \sin\phi_2$; (c) take the cross product and verify the trigonometric identity $\sin(\phi_2 - \phi_1) = \sin\phi_2 \cos\phi_1 - \cos\phi_2 \sin\phi_1$.

1.25 Convert the vector field $\mathbf{H} = A(x^2 + y^2)^{-1} [x\mathbf{a}_y - y\mathbf{a}_x]$ into cylindrical coordinates. A is a constant.

1.26 Express the uniform vector field $\mathbf{F} = 10\mathbf{a}_y$ in (a) cylindrical components; (b) spherical components.

1.27 The important dipole field (to be addressed in Chapter 4) is expressed in spherical coordinates as

$$E = \frac{A}{r^3}(2\cos\theta\,\mathbf{a}_r + \sin\theta\,\mathbf{a}_\theta)$$

where A is a constant, and where $r > 0$. See Figure 4.9 for a sketch. (a) Identify the surface on which the field is entirely perpendicular to the xy plane and express the field on that surface in cylindrical coordinates. (b) Identify the coordinate axis on which the field is entirely perpendicular to the xy plane and express the field there in cylindrical coordinates. (c) Specify the surface on which the field is entirely parallel to the xy plane.

1.28 State whether or not $\mathbf{A} = \mathbf{B}$ and, if not, what conditions are imposed on $\mathbf{A}$ and $\mathbf{B}$ when (a) $\mathbf{A} \cdot \mathbf{a}_x = \mathbf{B} \cdot \mathbf{a}_x$; (b) $\mathbf{A} \times \mathbf{a}_x = \mathbf{B} \times \mathbf{a}_x$; (c) $\mathbf{A} \cdot \mathbf{a}_x = \mathbf{B} \cdot \mathbf{a}_x$ and $\mathbf{A} \times \mathbf{a}_x = \mathbf{B} \times \mathbf{a}_x$; (d) $\mathbf{A} \cdot \mathbf{C} = \mathbf{B} \cdot \mathbf{C}$ and $\mathbf{A} \times \mathbf{C} = \mathbf{B} \times \mathbf{C}$ where $\mathbf{C}$ is any vector except $\mathbf{C} = 0$.

1.29 A vector field is expressed as $\mathbf{F} = Az\,\mathbf{a}_z$ where A is a constant. Evaluate the components of this field that are (a) normal and (b) tangent to a spherical surface of radius a. Express all results in spherical coordinates and components.

1.30 Consider a problem analogous to the varying wind velocities encountered by transcontinental aircraft. We assume a constant altitude, a plane earth, a flight along the x axis from 0 to 10 units, no vertical velocity component, and no change in wind velocity with time. Assume $\mathbf{a}_x$ to be directed to the east and $\mathbf{a}_y$ to the north. The wind velocity at the operating altitude is assumed to be:

$$\mathbf{v}(x,y) = \frac{(0.01\,x^2 - 0.08x + 0.66)\,\mathbf{a}_x - (0.05x - 0.4)\,\mathbf{a}_y}{1 + 0.5\,y^2}$$

Determine the location and magnitude of (a) the maximum tailwind encountered; (b) repeat for headwind; (c) repeat for crosswind; (d) Would more favorable tailwinds be available at some other latitude? If so, where?

CHAPTER 2   Coulomb's Law and Electric Field Intensity

术语

Coulomb's law 库仑定律，描述了两个点电荷之间的作用力；力的单位是牛［顿］（N）。

electric field intensity 电场强度，电场强度定义为单位正点电荷所受的电场力；单位是伏［特］/ 米（V / m），这个单位也表明电场强度等于电位对空间的变化率。

permittivity 介电常数、电容率，单位是法［拉］/ 米（F / m）；电容值与电容器中电介质的介电常数（电容率）成正比，所以介电常数也可认为是电容率。

streamlines of field 电力线，是用一簇曲线描述场的分布，曲线上每一点的切向即为该点场的方向。

sketches of field 场图，可用箭头描述场的分布，箭头的方向为场的方向，箭头的长度正比于场的大小。

In this chapter, we introduce Coulomb's electrostatic force law and then formulate this in a general way using field theory. The tools that will be developed can be used to solve any problem in which forces between static charges are to be evaluated or to determine the electric field that is associated with any charge distribution. Initially, we will restrict the study to fields in *vacuum* or *free space*; this would apply to media such as air and other gases. Other materials are introduced in Chapter 5 and time-varying fields are introduced in Chapter 9. ■

2.1   THE EXPERIMENTAL LAW OF COULOMB

Records from at least 600 B.C. show evidence of the knowledge of static electricity. The Greeks were responsible for the term *electricity*, derived from their word for amber, and they spent many leisure hours rubbing a small piece of amber on their sleeves and observing how it would then attract pieces of fluff and stuff. However, their main interest lay in philosophy and logic, not in experimental science, and it was many centuries before the attracting effect was considered to be anything other than magic or a "life force."

Dr. Gilbert, physician to Her Majesty the Queen of England, was the first to do any true experimental work with this effect, and in 1600 he stated that glass, sulfur, amber, and other materials, which he named, would "not only draw to themselves straws and chaff, but all metals, wood, leaves, stone, earths, even water and oil."

Shortly thereafter, an officer in the French Army Engineers, Colonel Charles Coulomb, performed an elaborate series of experiments using a delicate torsion balance, invented by himself, to determine quantitatively the force exerted between two objects, each having a static charge of electricity. His published result is very similar to Newton's gravitational law (discovered about a hundred years earlier). Coulomb stated that the force between two very small objects separated in a vacuum or free space by a distance, which is large compared to their size, is proportional to the charge on each and inversely proportional to the square of the distance between them, or

$$F = k\frac{Q_1 Q_2}{R^2}$$

where Q_1 and Q_2 are the positive or negative quantities of charge, R is the separation, and k is a proportionality constant. If the International System of Units[1] (SI) is used, Q is measured in coulombs (C), R is in meters (m), and the force should be newtons (N). This will be achieved if the constant of proportionality k is written as

$$k = \frac{1}{4\pi\epsilon_0}$$

The new constant ϵ_0 is called the *permittivity of free space* and has magnitude, measured in farads per meter (F/m),

$$\epsilon_0 = 8.854 \times 10^{-12} = \frac{1}{36\pi}10^{-9} \text{ F/m} \tag{1}$$

The quantity ϵ_0 is not dimensionless, for Coulomb's law shows that it has the label $C^2/N \cdot m^2$. We will later define the farad and show that it has the dimensions $C^2/N \cdot m$; we have anticipated this definition by using the unit F/m in Eq. (1).

Coulomb's law is now

$$F = \frac{Q_1 Q_2}{4\pi\epsilon_0 R^2} \tag{2}$$

The coulomb is an extremely large unit of charge, for the smallest known quantity of charge is that of the electron (negative) or proton (positive), given in SI units as 1.602×10^{-19} C; hence a negative charge of one coulomb represents about 6×10^{18} electrons.[2] Coulomb's law shows that the force between two charges of one coulomb each, separated by one meter, is 9×10^9 N, or about one million tons. The electron has a rest mass of 9.109×10^{-31} kg and has a radius on the order of magnitude of 3.8×10^{-15} m. This does not mean that the electron is spherical, but it merely describes the size of the region in which a slowly moving electron has the greatest probability of being found. All other known charged particles, including the proton, have larger masses and larger radii, and they occupy a probabilistic volume larger than does the electron.

In order to write the vector form of (2), we need the additional fact (furnished also by Colonel Coulomb) that the force acts along the line joining the two charges and is repulsive if the charges are alike in sign or attractive if they are of opposite sign. Let the vector $\mathbf{r}_1$ locate Q_1, whereas $\mathbf{r}_2$ locates Q_2. Then the vector $\mathbf{R}_{12} = \mathbf{r}_2 - \mathbf{r}_1$ represents the directed line segment from Q_1 to Q_2, as shown in Figure 2.1. The

1. 一个电荷所受的电场力等于场强与该电荷值的乘积。但要注意,该场强是去除该电荷后其他电荷所产生的电场。

2. 已知无限大均匀介质空间中的电荷分布求场强的问题,是一个基于点电荷场强表达式进行叠加或积分的问题,要注意矢量叠加。

3. 无限大均匀介质空间中已知电荷分布求电场的问题是一大类计算问题,下面几章将要介绍的有限空间(也可以是无限大空间)中具有不同电介质或导体的问题是另一大类计算问题。

[1] The International System of Units (an mks system) is described in Appendix B. Abbreviations for the units are given in Table B.1. Conversions to other systems of units are given in Table B.2, while the prefixes designating powers of ten in SI appear in Table B.3.

[2] The charge and mass of an electron and other physical constants are tabulated in Table C.4 of Appendix C.

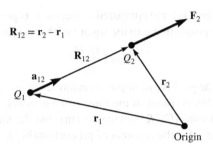

Figure 2.1 If Q_1 and Q_2 have like signs, the vector force $\mathbf{F}_2$ on Q_2 is in the same direction as the vector $\mathbf{R}_{12}$.

vector $\mathbf{F}_2$ is the force on Q_2 and is shown for the case where Q_1 and Q_2 have the same sign. The vector form of Coulomb's law is

$$\mathbf{F}_2 = \frac{Q_1 Q_2}{4\pi \epsilon_0 R_{12}^2} \mathbf{a}_{12} \tag{3}$$

where $\mathbf{a}_{12} =$ a unit vector in the direction of $\mathbf{R}_{12}$, or

$$\mathbf{a}_{12} = \frac{\mathbf{R}_{12}}{|\mathbf{R}_{12}|} = \frac{\mathbf{R}_{12}}{R_{12}} = \frac{\mathbf{r}_2 - \mathbf{r}_1}{|\mathbf{r}_2 - \mathbf{r}_1|} \tag{4}$$

EXAMPLE 2.1

We illustrate the use of the vector form of Coulomb's law by locating a charge of $Q_1 = 3 \times 10^{-4}$ C at $M(1, 2, 3)$ and a charge of $Q_2 = -10^{-4}$ C at $N(2, 0, 5)$ in a vacuum. We want to find the force exerted on Q_2 by Q_1.

Solution. We use (3) and (4) to obtain the vector force. The vector $\mathbf{R}_{12}$ is

$$\mathbf{R}_{12} = \mathbf{r}_2 - \mathbf{r}_1 = (2 - 1)\mathbf{a}_x + (0 - 2)\mathbf{a}_y + (5 - 3)\mathbf{a}_z = \mathbf{a}_x - 2\mathbf{a}_y + 2\mathbf{a}_z$$

leading to $|\mathbf{R}_{12}| = 3$, and the unit vector, $\mathbf{a}_{12} = \frac{1}{3}(\mathbf{a}_x - 2\mathbf{a}_y + 2\mathbf{a}_z)$. Thus,

$$\mathbf{F}_2 = \frac{3 \times 10^{-4}(-10^{-4})}{4\pi(1/36\pi)\,10^{-9} \times 3^2}\left(\frac{\mathbf{a}_x - 2\mathbf{a}_y + 2\mathbf{a}_z}{3}\right)$$
$$= -30\left(\frac{\mathbf{a}_x - 2\mathbf{a}_y + 2\mathbf{a}_z}{3}\right) \text{N}$$

The magnitude of the force is 30 N, and the direction is specified by the unit vector, which has been left in parentheses to display the magnitude of the force. The force on Q_2 may also be considered as three component forces,

$$\mathbf{F}_2 = -10\mathbf{a}_x + 20\mathbf{a}_y - 20\mathbf{a}_z$$

The force expressed by Coulomb's law is a mutual force, for each of the two charges experiences a force of the same magnitude, although of opposite direction. We might equally well have written

$$\mathbf{F}_1 = -\mathbf{F}_2 = \frac{Q_1 Q_2}{4\pi \epsilon_0 R_{12}^2} \mathbf{a}_{21} = -\frac{Q_1 Q_2}{4\pi \epsilon_0 R_{12}^2} \mathbf{a}_{12} \tag{5}$$

Coulomb's law is linear, for if we multiply Q_1 by a factor n, the force on Q_2 is also multiplied by the same factor n. It is also true that the force on a charge in the presence of several other charges is the sum of the forces on that charge arising from each of the other charges acting alone.

D2.1. A charge $Q_A = -20\ \mu C$ is located at $A(-6, 4, 7)$, and a charge $Q_B = 50\ \mu C$ is at $B(5, 8, -2)$ in free space. If distances are given in meters, find: (a) $\mathbf{R}_{AB}$; (b) R_{AB}. Determine the vector force exerted on Q_A by Q_B if $\epsilon_0 =$ (c) $10^{-9}/(36\pi)$ F/m; (d) 8.854×10^{-12} F/m.

Ans. (a) $11\mathbf{a}_x + 4\mathbf{a}_y - 9\mathbf{a}_z$ m; (b) 14.76 m; (c) $30.76\mathbf{a}_x + 11.184\mathbf{a}_y - 25.16\mathbf{a}_z$ mN; (d) $30.72\mathbf{a}_x + 11.169\mathbf{a}_y - 25.13\mathbf{a}_z$ mN

2.2 ELECTRIC FIELD INTENSITY

Here, we introduce the first of several field quantities that we will use throughout our study. The *electric field intensity* gives the magnitude and direction of electrostatic force that would be applied to a point charge of unit magnitude that resides in the field, and as a function of its location. Emphasized here is the notion of the force acting *at a point,* and as such, the electric field intensity, like all other field quantities we will encounter, is a *point function.* Forces on larger objects, or charge distributions, must be found by summing contributions at all points that make up the object by way of a *superposition integral.* Such procedures are used in every aspect of applied electromagnetics and are introduced in later sections.

2.2.1 Electric Field Definition for a Point Charge

Consider a single point charge fixed in position, say Q_1, and move a second charge slowly around. It will be found that there exists everywhere a force on this second charge; in other words, this second charge is displaying the existence of a force *field* that is associated with charge Q_1. Call this second charge a test charge Q_t. The force on it is given by Coulomb's law, expressed by adapting Eq. (3):

$$\mathbf{F}_t = \frac{Q_1 Q_t}{4\pi\epsilon_0 R_{1t}^2}\mathbf{a}_{1t}$$

Writing this force as a force per unit charge gives the electric field intensity $\mathbf{E}_1$ arising from Q_1:

$$\mathbf{E}_1 = \frac{\mathbf{F}_t}{Q_1} = \frac{Q_1}{4\pi\epsilon_0 R_{1t}^2}\mathbf{a}_{1t} \tag{6}$$

$\mathbf{E}_1$ is interpreted as the vector force, arising from charge Q_1, that acts on a unit positive test charge. More generally, we write the defining expression:

$$\boxed{\mathbf{E} = \frac{\mathbf{F}_t}{Q_t}} \tag{7}$$

in which $\mathbf{E}$, a vector function, is the electric field intensity *evaluated at the test charge location* that arises from all *other* charges in the vicinity—meaning the electric field arising from the test charge itself is not included in $\mathbf{E}$.

The units of **E** would be in force per unit charge (newtons per coulomb). Again anticipating a new dimensional quantity, the *volt* (V), having the label of joules per coulomb (J/C), or newton-meters per coulomb (N · m/C), we measure electric field intensity in the practical units of volts per meter (V/m).

Most of the subscripts in (6) are now removed, reserving the right to use them again any time there is a possibility of misunderstanding. The electric field of a single point charge becomes:

$$\mathbf{E} = \frac{Q}{4\pi\epsilon_0 R^2}\mathbf{a}_R \tag{8}$$

We remember that R is the magnitude of the vector **R**, the directed line segment from the point at which the point charge Q is located to the point at which **E** is desired, and $\mathbf{a}_R$ is a unit vector in the **R** direction.[3]

We arbitrarily locate Q_1 at the center of a spherical coordinate system. The unit vector $\mathbf{a}_R$ then becomes the radial unit vector $\mathbf{a}_r$, and R is r. Hence

$$\boxed{\mathbf{E} = \frac{Q_1}{4\pi\epsilon_0 r^2}\mathbf{a}_r} \tag{9}$$

The field has a single radial component, and its inverse-square-law relationship is quite obvious.

2.2.2 Fields Associated with Charges at General Locations

For a charge that is *not* at the origin of our coordinate system, the field no longer possesses spherical symmetry, and we might as well use rectangular coordinates. For a charge Q located at the source point $\mathbf{r}' = x'\mathbf{a}_x + y'\mathbf{a}_y + z'\mathbf{a}_z$, as illustrated in Figure 2.2, the field at a general point $\mathbf{r} = x\mathbf{a}_x + y\mathbf{a}_y + z\mathbf{a}_z$ can be found by expressing **R** as $\mathbf{r} - \mathbf{r}'$:

$$\mathbf{E}(\mathbf{r}) = \frac{Q}{4\pi\epsilon_0|\mathbf{r} - \mathbf{r}'|^2}\frac{\mathbf{r} - \mathbf{r}'}{|\mathbf{r} - \mathbf{r}'|} = \frac{Q(\mathbf{r} - \mathbf{r}')}{4\pi\epsilon_0|\mathbf{r} - \mathbf{r}'|^3}$$

$$= \frac{Q[(x - x')\mathbf{a}_x + (y - y')\mathbf{a}_y + (z - z')\mathbf{a}_z]}{4\pi\epsilon_0[(x - x')^2 + (y - y')^2 + (z - z')^2]^{3/2}} \tag{10}$$

Earlier, we defined a vector field as a vector function of a position vector, and this is emphasized by letting **E** be symbolized in functional notation by **E**(**r**).

Because the coulomb forces are linear, the electric field intensity arising from two point charges, Q_1 at $\mathbf{r}_1$ and Q_2 at $\mathbf{r}_2$, is the sum of the forces on Q_t caused by Q_1 and Q_2 acting alone, or

$$\mathbf{E}(\mathbf{r}) = \frac{Q_1}{4\pi\epsilon_0|\mathbf{r} - \mathbf{r}_1|^2}\mathbf{a}_1 + \frac{Q_2}{4\pi\epsilon_0|\mathbf{r} - \mathbf{r}_2|^2}\mathbf{a}_2$$

where $\mathbf{a}_1$ and $\mathbf{a}_2$ are unit vectors in the direction of $(\mathbf{r} - \mathbf{r}_1)$ and $(\mathbf{r} - \mathbf{r}_2)$, respectively. The vectors $\mathbf{r}$, $\mathbf{r}_1$, $\mathbf{r}_2$, $\mathbf{r} - \mathbf{r}_1$, $\mathbf{r} - \mathbf{r}_2$, $\mathbf{a}_1$, and $\mathbf{a}_2$ are shown in Figure 2.3.

---

[3] We firmly intend to avoid confusing r and $\mathbf{a}_r$ with R and $\mathbf{a}_R$. The first two refer specifically to the spherical coordinate system, whereas R and $\mathbf{a}_R$ do not refer to any coordinate system—the choice is still available to us.

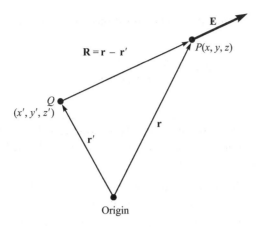

Figure 2.2 The vector **r'** locates the point charge Q, the vector **r** identifies the general point in space $P(x, y, z)$, and the vector **R** from Q to $P(x, y, z)$ is then $\mathbf{R} = \mathbf{r} - \mathbf{r}'$.

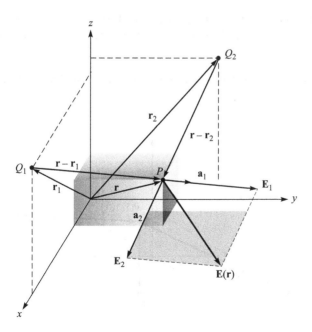

Figure 2.3 The vector addition of the total electric field intensity at P due to Q_1 and Q_2 is made possible by the linearity of Coulomb's law.

If more charges are added at other positions, the field arising from n point charges is

$$\mathbf{E}(\mathbf{r}) = \sum_{m=1}^{n} \frac{Q_m}{4\pi\epsilon_0 |\mathbf{r} - \mathbf{r}_m|^2} \mathbf{a}_m \qquad (11)$$

EXAMPLE 2.2

In order to illustrate the application of (11), we find $\mathbf{E}$ at $P(1, 1, 1)$ caused by four identical 3-nC (nanocoulomb) charges located at $P_1(1, 1, 0), P_2(-1, 1, 0), P_3(-1, -1, 0)$, and $P_4(1, -1, 0)$, as shown in Figure 2.4.

Solution. We find that $\mathbf{r} = \mathbf{a}_x + \mathbf{a}_y + \mathbf{a}_z$, $\mathbf{r}_1 = \mathbf{a}_x + \mathbf{a}_y$, and thus $\mathbf{r} - \mathbf{r}_1 = \mathbf{a}_z$. The magnitudes are: $|\mathbf{r} - \mathbf{r}_1| = 1$, $|\mathbf{r} - \mathbf{r}_2| = \sqrt{5}$, $|\mathbf{r} - \mathbf{r}_3| = 3$, and $|\mathbf{r} - \mathbf{r}_4| = \sqrt{5}$. Because $Q/4\pi\epsilon_0 = 3 \times 10^{-9}/(4\pi \times 8.854 \times 10^{-12}) = 26.96$ V · m, we may now use (11) to obtain

$$\mathbf{E} = 26.96 \left[\frac{\mathbf{a}_z}{1} \frac{1}{1^2} + \frac{2\mathbf{a}_x + \mathbf{a}_z}{\sqrt{5}} \frac{1}{(\sqrt{5})^2} + \frac{2\mathbf{a}_x + 2\mathbf{a}_y + \mathbf{a}_z}{3} \frac{1}{3^2} + \frac{2\mathbf{a}_y + \mathbf{a}_z}{\sqrt{5}} \frac{1}{(\sqrt{5})^2} \right]$$

or

$$\mathbf{E} = 6.82\,\mathbf{a}_x + 6.82\,\mathbf{a}_y + 32.8\,\mathbf{a}_z \text{ V/m}$$

D2.2. A charge of $-0.3\ \mu$C is located at $A(25, -30, 15)$ (in cm), and a second charge of $0.5\ \mu$C is at $B(-10, 8, 12)$ cm. Find $\mathbf{E}$ at: (*a*) the origin; (*b*) $P(15, 20, 50)$ cm.

Ans. (*a*) $92.3\mathbf{a}_x - 77.6\mathbf{a}_y - 94.2\mathbf{a}_z$ kV/m; (*b*) $11.9\mathbf{a}_x - 0.519\mathbf{a}_y + 12.4\mathbf{a}_z$ kV/m

---

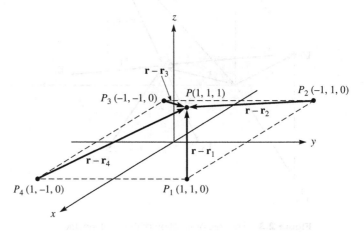

Figure 2.4 A symmetrical distribution of four identical 3-nC point charges produces a field at P, $\mathbf{E} = 6.82a_x + 6.82a_y + 32.8a_z$ V/m.

D2.3. Evaluate the sums: (a) $\sum_{m=0}^{5} \dfrac{1 + (-1)^m}{m^2 + 1}$; (b) $\sum_{m=1}^{4} \dfrac{(0.1)^m + 1}{(4 + m^2)^{1.5}}$

Ans. (a) 2.52; (b) 0.176

2.3 FIELD ARISING FROM A CONTINUOUS VOLUME CHARGE DISTRIBUTION

If we now visualize a region of space filled with a tremendous number of charges separated by minute distances, we see that we can replace this distribution of very small particles with a smooth continuous distribution described by a *volume charge density*, just as we describe water as having a density of 1 g/cm³ (gram per cubic centimeter) even though it consists of atomic- and molecular-sized particles. This can be done only if we are uninterested in the small irregularities (or ripples) in the field as we move from electron to electron or if we care little that the mass of the water actually increases in small but finite steps as each new molecule is added.

This is really no limitation at all, because the end results for electrical engineers are almost always in terms of a current in a receiving antenna, a voltage in an electronic circuit, or a charge on a capacitor, or in general in terms of some large-scale *macroscopic* phenomenon. It is very seldom that we must know a current electron by electron.[4]

2.3.1 Volume Charge Density Definition

Volume charge density is denoted by ρ_v, having the units of coulombs per cubic meter (C/m³).

The small amount of charge ΔQ in a small volume Δv is

$$\Delta Q = \rho_v \Delta v \tag{12}$$

and ρ_v may be defined mathematically by using a limiting process on (12),

$$\boxed{\rho_v = \lim_{\Delta v \to 0} \frac{\Delta Q}{\Delta v}} \tag{13}$$

The total charge within some finite volume is obtained by integrating throughout that volume,

$$\boxed{Q = \int_{\text{vol}} \rho_v dv} \tag{14}$$

Only one integral sign is customarily indicated, but the differential dv signifies integration throughout a volume, and hence a triple integration.

---

[4] A study of the noise generated by electrons in semiconductors and resistors, however, requires just such an examination of the charge through statistical analysis.

EXAMPLE 2.3

As an example of the evaluation of a volume integral, we find the total charge contained in a 2-cm length of the electron beam shown in Figure 2.5.

Solution. From the illustration, we see that the charge density is

$$\rho_v = -5 \times 10^{-6} e^{-10^5 \rho z} \text{ C/m}^2$$

The volume differential in cylindrical coordinates is given in Section 1.8; therefore,

$$Q = \int_{0.02}^{0.04} \int_0^{2\pi} \int_0^{0.01} -5 \times 10^{-6} e^{-10^5 \rho z} \rho \, d\rho \, d\phi \, dz$$

We integrate first with respect to ϕ because it is so easy,

$$Q = \int_{0.02}^{0.04} \int_0^{0.01} -10^{-5} \pi e^{-10^5 \rho z} \rho \, d\rho \, dz$$

and then with respect to z, because this will simplify the last integration with respect to ρ,

$$Q = \int_0^{0.01} \left(\frac{-10^{-5} \pi}{-10^5 \rho} e^{-10^5 \rho z} \rho \, d\rho \right) \Bigg|_{z=0.02}^{z=0.04}$$

$$= \int_0^{0.01} -10^{-5} \pi (e^{-2000\rho} - e^{-4000\rho}) \, d\rho$$

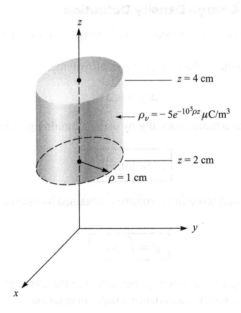

Figure 2.5 The total charge contained within the right circular cylinder may be obtained by evaluating $Q = \int_{\text{vol}} \rho_v dv$.

Finally,

$$Q = -10^{-10}\pi \left(\frac{e^{-2000\rho}}{-2000} - \frac{e^{-4000\rho}}{-4000} \right)_0^{0.01}$$

$$Q = -10^{-10}\pi \left(\frac{1}{2000} - \frac{1}{4000} \right) = \frac{-\pi}{40} = 0.0785\,\text{pC}$$

where pC indicates picocoulombs.

2.3.2 Electric Field Associated with a Volume Charge Distribution

Consider an incremental charge, ΔQ at $\mathbf{r}'$ that represents a small portion of a larger charge volume of density ρ_v, which in general may vary with position. ΔQ lies within a small volume Δv, and is thus treated as a point charge, where $\Delta Q = \rho_v \Delta v$ as before. The incremental contribution to the electric field intensity at $\mathbf{r}$ associated with this charge is written, using (10):

$$\Delta\mathbf{E}(\mathbf{r}) = \frac{\Delta Q}{4\pi\epsilon_0|\mathbf{r}-\mathbf{r}'|^2}\frac{\mathbf{r}-\mathbf{r}'}{|\mathbf{r}-\mathbf{r}'|} = \frac{\rho_v \Delta v}{4\pi\epsilon_0|\mathbf{r}-\mathbf{r}'|^2}\frac{\mathbf{r}-\mathbf{r}'}{|\mathbf{r}-\mathbf{r}'|}$$

The above gives the field contribution at $\mathbf{r}$ for the small volume of charge within the larger distribution. To find the total field at $\mathbf{r}$, we sum the contributions at that point of all the charges in the distribution. This is done by first letting the volume element Δv approach zero. The effect of this is twofold: First, it provides essentially infinite spatial resolution (as the volume charge density may vary from point to point); second, the summation becomes an integral over the charge volume:

$$\mathbf{E}(\mathbf{r}) = \int_{\text{vol}} \frac{\rho_v(\mathbf{r}')\,dv'}{4\pi\epsilon_0|\mathbf{r}-\mathbf{r}'|^2}\frac{\mathbf{r}-\mathbf{r}'}{|\mathbf{r}-\mathbf{r}'|} \tag{15}$$

This is again a triple integral, and (except in Drill Problem 2.4) we will do our best to avoid actually performing the integration.

The significance of the various quantities under the integral sign of (15) might stand a little review. The vector $\mathbf{r}$ from the origin locates the field point where $\mathbf{E}$ is being determined, whereas the vector $\mathbf{r}'$ extends from the origin to the source point where $\rho_v(\mathbf{r}')dv'$ is located. The scalar distance between the source point and the field point is $|\mathbf{r}-\mathbf{r}'|$, and the fraction $(\mathbf{r}-\mathbf{r}')/|\mathbf{r}-\mathbf{r}'|$ is a unit vector directed from source point to field point. The variables of integration are x', y', and z' in rectangular coordinates.

D2.4. Calculate the total charge within each of the indicated volumes: (a) $0.1 \le |x|, |y|, |z| \le 0.2$: $\rho_v = \frac{1}{x^3 y^3 z^3}$; (b) $0 \le \rho \le 0.1$, $0 \le \phi \le \pi$, $2 \le z \le 4$; $\rho_v = \rho^2 z^2 \sin 0.6\phi$; (c) universe: $\rho_v = e^{-2r}/r^2$.

Ans. (a) 0; (b) 1.018 mC; (c) 6.28 C

2.4 FIELD OF A LINE CHARGE

Up to this point we have considered two types of charge distribution, the point charge and continuous charge distributed throughout a volume with a density ρ_v C/m^3. We now consider a filamentlike distribution of volume charge density, such as a charged conductor of very small radius. It is convenient to treat the charge as a line charge of density ρ_L C/m.

Consider a straight-line charge extending along the z axis in a cylindrical coordinate system from $-\infty$ to ∞, as shown in Figure 2.6. We will find the electric field intensity **E** at any and every point resulting from a *uniform* line charge density ρ_L.

2.4.1 Setting Up the Problem: The Importance of Symmetry

Symmetry should always be considered first in order to determine two specific factors: (1) with which coordinates the field does *not* vary, and (2) which components of the field are *not* present. The answers to these questions then tell us which components are present and with which coordinates they *do* vary.

Referring to Figure 2.6, we realize that as we move around the line charge, varying ϕ while keeping ρ and z constant, the line charge appears the same from every angle. In other words, *azimuthal* symmetry is present, and no field component may vary with ϕ.

Again, if we maintain ρ and ϕ constant while moving up and down the line charge by changing z, the line charge still recedes into infinite distance in both directions and the problem is unchanged. This is *axial* symmetry and leads to fields that are not functions of z.

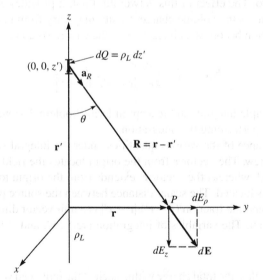

Figure 2.6 The contribution $d\mathbf{E} = dE_\rho\, \mathbf{a}_\rho + dE_z$ $\mathbf{a}_z$ to the electric field intensity produced by an element of charge $dQ = \rho_L\, dz'$ located a distance z' from the origin. The linear charge density is uniform and extends along the entire z axis.

If we maintain ϕ and z constant and vary ρ, the problem changes, and Coulomb's law leads us to expect the field to become weaker as ρ increases. Hence, by a process of elimination we are led to the fact that the field varies only with ρ.

Now, which components are present? Each incremental length of line charge acts as a point charge and produces an incremental contribution to the electric field intensity that is directed away from the bit of charge (assuming a positive line charge). No element of charge produces a ϕ component of electric intensity; E_ϕ is zero. However, each element does produce an E_ρ and an E_z component, but the contribution to E_z by elements of charge that are equal distances above and below the point at which we are determining the field will cancel. Therefore only an $E\rho$ component is expected, and this will vary only with ρ. Now to find this component.

We choose a point $P(0, y, 0)$ on the y axis at which to determine the field. This is a perfectly general point in view of the lack of variation of the field with ϕ and z. Applying (10) to find the incremental field at P due to the incremental charge $dQ = \rho_L dz'$, we have

$$d\mathbf{E} = \frac{\rho_L dz(\mathbf{r} - \mathbf{r}')}{4\pi\epsilon_0 |\mathbf{r} - \mathbf{r}'|^3}$$

where $\mathbf{r}' = z'\mathbf{a}_z$ and $\mathbf{r} = y\mathbf{a}_y = \rho\mathbf{a}_\rho$. The replacement of y with ρ in the last equality arises from the symmetry, in that the field in the xy plane will vary only with distance from the origin, expressed as the more general ρ direction in cylindrical coordinates. We now have

$$\mathbf{r} - \mathbf{r}' = \rho\mathbf{a}_\rho - z'\mathbf{a}_z$$

and therefore,

$$d\mathbf{E} = \frac{\rho_L dz'(\rho\mathbf{a}_\rho - z'\mathbf{a}_z)}{4\pi\epsilon_0(\rho^2 + z'^2)^{3/2}}$$

The differential field contributions to a point in the xy plane are now summed by integrating the preceding differential field over the infinite line charge:

$$\mathbf{E}_\rho = \int_{-\infty}^{\infty} \frac{\rho_L dz'(\rho\mathbf{a}_\rho - z'\mathbf{a}_z)}{4\pi\epsilon_0(\rho^2 + z'^2)^{3/2}}$$

At this point we note that the second term in the integral, involving $z'\mathbf{a}_{z'}$, integrates to zero because it gives equal and opposite contributions that cancel each other as z' changes sign at the origin. This is an example of a function that exhibits *odd parity*. This result demonstrates mathematically what was already discussed, that the z contributions to the field from the symmetric charge will cancel out. The remaining part of the integral, involving $\mathbf{a}_\rho$, is evaluated by integral tables or by a change of variable, $z' = \rho \cot \theta$, leading to:

$$E_\rho = \frac{\rho_L}{4\pi\epsilon_0}\rho\left(\frac{1}{\rho^2}\frac{z'}{\sqrt{\rho^2 + z'^2}}\right)_{-\infty}^{\infty}$$

so that

$$E_\rho = \frac{\rho_L}{2\pi\epsilon_0\rho}$$

or finally,

$$\mathbf{E} = \frac{\rho_L}{2\pi\epsilon_0\rho}\mathbf{a}_\rho \qquad (16)$$

We note that the field falls off inversely with the distance to the charged line, as compared with the point charge, where the field decreased with the *square* of the distance. Moving 10 times as far from a point charge leads to a field only 1 percent the previous strength, but moving 10 times as far from a line charge only reduces the field to 10 percent of its former value. An analogy can be drawn with a source of illumination, for the light intensity from a point source of light also falls off inversely as the square of the distance to the source. The field of an infinitely long fluorescent tube thus decays inversely as the first power of the radial distance to the tube, and we should expect the light intensity about a finite-length tube to obey this law near the tube. As our point recedes farther and farther from a finite-length tube, however, it eventually looks like a point source, and the field obeys the inverse-square relationship.

2.4.2 Field of an Off-Axis Line Charge

Before leaving this introductory look at the field of the infinite line charge, it should be recognized that not all line charges are located along the z axis. As an example, consider an infinite line charge parallel to the z axis at $x = 6$, $y = 8$, shown in Figure 2.7. $\mathbf{E}$ is to be found at the general field point $P(x, y, z)$.

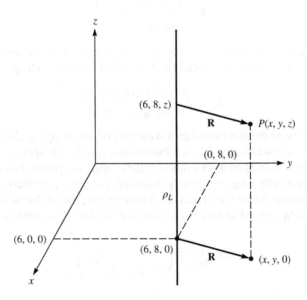

Figure 2.7 A point $P(x, y, z)$ is identified near an infinite uniform line charge located at $x = 6$, $y = 8$.

ρ is replaced in (16) by the radial distance between the line charge and point, $P, R = \sqrt{(x-6)^2 + (y-8)^2}$, and let $\mathbf{a}_\rho$ be $\mathbf{a}_R$. Thus,

$$\mathbf{E} = \frac{\rho_L}{2\pi\epsilon_0\sqrt{(x-6)^2 + (y-8)^2}}\mathbf{a}_R$$

where

$$\mathbf{a}_R = \frac{\mathbf{R}}{|\mathbf{R}|} = \frac{(x-6)\mathbf{a}_x + (y-8)\mathbf{a}_y}{\sqrt{(x-6)^2 + (y-8)^2}}$$

Therefore,

$$\mathbf{E} = \frac{\rho_L}{2\pi\epsilon_0}\frac{(x-6)\mathbf{a}_x + (y-8)\mathbf{a}_y}{(x-6)^2 + (y-8)^2}$$

We again note that the field is not a function of z.

In Section 2.6, we describe how fields may be sketched, and the field of the line charge is one example.

D2.5. Infinite uniform line charges of 5 nC/m lie along the (positive and negative) x and y axes in free space. Find $\mathbf{E}$ at: (a) $P_A(0, 0, 4)$; (b) $P_B(0, 3, 4)$.

Ans. (a) $45\mathbf{a}_z$ V/m; (b) $10.8\mathbf{a}_y + 36.9\mathbf{a}_z$ V/m

---

2.5 FIELD OF A SHEET OF CHARGE

Another basic charge configuration is the infinite sheet of charge having a uniform density of ρ_S C/m². Such a charge distribution may often be used to approximate that found on the conductors of a strip transmission line or a parallel-plate capacitor. As will be seen in Chapter 5, static charge resides on conductor surfaces and not in their interiors; for this reason, ρ_S is commonly known as *surface charge density*. The charge-distribution family now is complete—point, line, surface, and volume, or Q, ρ_L, ρ_S, and ρ_v.

2.5.1 Symmetry

Consider an infinite sheet of charge in the yz plane and again be aware of symmetry (Figure 2.8). We observe first that the field cannot vary with y or with z, and that the y and z components arising from differential elements of charge symmetrically located with respect to the point at which we evaluate the field will cancel. Therefore only E_x is present, and, as will be demonstrated, will not vary in any direction. We are again faced with a choice of many methods by which to evaluate this component, and this time we use only one method and leave the others as exercises for a quiet Sunday afternoon.

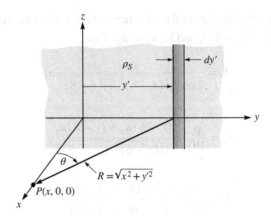

Figure 2.8   An infinite sheet of charge in the *yz* plane, a general point *P* on the *x* axis, and the differential-width line charge used as the element in determining the field at *P* by $d\mathbf{E} = \rho_S dy' \mathbf{a}_R/(2\pi\epsilon_0 R)$.

2.5.2   The Sheet Charge as an Ensemble of Line Charges

The field of the infinite line charge (16) is implemented here by dividing the infinite sheet into differential-width strips. One such strip is shown in Figure 2.8. The line charge density, or charge per unit length, is $\rho_L = \rho_S dy'$, and the distance from this line charge to our general point P on the x axis is $R = \sqrt{x^2 + y^2}$. The contribution to E_x at P from this differential-width strip is then

$$dE_x = \frac{\rho_S dy'}{2\pi\epsilon_0 \sqrt{x^2 + y^2}}\cos\theta = \frac{\rho_S}{2\pi\epsilon_0}\frac{x\,dy'}{x^2 + y^2}$$

Adding the effects of all the strips,

$$E_x = \frac{\rho_S}{2\epsilon_0}\int_{-\infty}^{\infty}\frac{x\,dy'}{x^2 + y'^2} = \frac{\rho_S}{2\pi\epsilon_0}\tan^{-1}\frac{y'}{x}\bigg]_{-\infty}^{\infty} = \frac{\rho_S}{2\pi\epsilon_0}$$

If the point P were chosen on the negative x axis, then

$$E_x = -\frac{\rho_S}{2\epsilon_0}$$

for the field is always directed away from the positive charge. This difficulty in sign is usually overcome by specifying a unit vector $\mathbf{a}_N$, which is normal to the sheet and directed outward, or away from it. Then

$$\boxed{\mathbf{E} = \frac{\rho_S}{2\epsilon_0}\mathbf{a}_N} \tag{17}$$

This is a startling answer, for the field is constant in magnitude and direction. It is just as strong a million miles away from the sheet as it is right off the surface. Returning to our light analogy, we see that a uniform source of light on the ceiling of a very large room leads to just as much illumination on a square foot on the floor as it does on a square foot a few inches below the ceiling. If you desire greater illumination on this subject, it will do you no good to hold the book closer to such a light source.

2.5.3 Capacitor Model

If a second infinite sheet of charge, having a *negative* charge density $-\rho_S$, is located in the plane $x = a$, the total field may be found by adding the contribution of each sheet. In the region $x > a$,

$$\mathbf{E}_+ = \frac{\rho_S}{2\epsilon_0}\mathbf{a}_x \qquad \mathbf{E}_- = -\frac{\rho_S}{2\epsilon_0}\mathbf{a}_x \qquad \mathbf{E} = \mathbf{E}_+ + \mathbf{E}_- = 0$$

and for $x < 0$,

$$\mathbf{E}_+ = -\frac{\rho_S}{2\epsilon_0}\mathbf{a}_x \qquad \mathbf{E}_- = \frac{\rho_S}{2\epsilon_0}\mathbf{a}_x \qquad \mathbf{E} = \mathbf{E}_+ + \mathbf{E}_- = 0$$

and when $0 < x < a$,

$$\mathbf{E}_+ = \frac{\rho_S}{2\epsilon_0}\mathbf{a}_x \qquad \mathbf{E}_- = \frac{\rho_S}{2\epsilon_0}\mathbf{a}_x$$

and

$$\boxed{\mathbf{E} = \mathbf{E}_+ + \mathbf{E}_- = \frac{\rho_S}{\epsilon_0}\mathbf{a}_x} \tag{18}$$

This is an important practical answer, for it is the field between the parallel plates of an air capacitor, provided the linear dimensions of the plates are very much greater than their separation and provided also that we are considering a point well removed from the edges. The field outside the capacitor, while not zero, as we found for the preceding ideal case, is usually negligible.

D2.6. Three infinite uniform sheets of charge are located in free space as follows: 3 nC/m² at $z = -4$, 6 nC/m² at $z = 1$, and -8 nC/m² at $z = 4$. Find **E** at the point: (*a*) $P_A(2, 5, -5)$; (*b*) $P_B(4, 2, -3)$; (*c*) $P_C(-1, -5, 2)$; (*d*) $P_D(-2, 4, 5)$.

Ans. (*a*) $-56.5\mathbf{a}_z$; (*b*) $283\mathbf{a}_z$; (*c*) $961\mathbf{a}_z$; (*d*) $56.5\mathbf{a}_z$ all V/m

---

2.6 STREAMLINES AND SKETCHES OF FIELDS

We now have vector equations for the electric field intensity resulting from several different charge configurations, and we have had little difficulty in interpreting the magnitude and direction of the field from the equations. Unfortunately, this simplicity cannot last much longer, for we have solved most of the simple cases and our new charge distributions must lead to more complicated expressions for the fields and more difficulty in visualizing the fields through the equations. However, it is true that one picture would be worth about a thousand words, if we just knew what picture to draw.

Consider the field about the line charge,

$$\mathbf{E} = \frac{\rho_L}{2\pi\epsilon_0\rho}\mathbf{a}_\rho$$

Figure 2.9*a* shows a cross-sectional view of the line charge and presents what might be our first effort at picturing the field—short line segments drawn here and there having lengths proportional to the magnitude of **E** and pointing in the direction of **E**. The figure fails to show the symmetry with respect to ϕ, so we try again in Figure 2.9*b* with a symmetrical location of the line segments. The real trouble now appears—the longest lines must be drawn in the most crowded region, and this also plagues us if we use line segments of equal length but of a thickness that is proportional to **E** (Figure 2.9*c*). Other schemes include drawing shorter lines to represent stronger fields (inherently misleading) and using intensity of color or different colors to represent stronger fields.

For the present, we will show only the *direction* of **E** by drawing continuous lines, which are everywhere tangent to **E**, from the charge. Figure 2.9*d* shows this compromise. A symmetrical distribution of lines (one every 45°) indicates azimuthal symmetry, and arrowheads are used to show direction.

These lines are usually called *streamlines*, although other terms such as flux lines and direction lines are also used. A small positive test charge placed at any point in this field and free to move would accelerate in the direction of the streamline passing through that point. If the field represented the velocity of a liquid or a gas (which, incidentally, would have to have a source at $\rho = 0$), small suspended particles in the liquid or gas would trace out the streamlines.

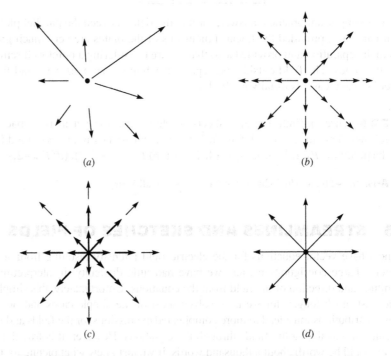

Figure 2.9   (*a*) One very poor sketch, (*b*) and (*c*) two fair sketches, and (*d*) the usual form of a streamline sketch. In the last form, the arrows show the direction of the field at every point along the line, and the spacing of the lines is inversely proportional to the strength of the field.

We will find out later that a bonus accompanies this streamline sketch, for the magnitude of the field can be shown to be inversely proportional to the spacing of the streamlines for some important special cases. The closer they are together, the stronger is the field. At that time we will also find an easier, more accurate method of making that type of streamline sketch.

If we tried to sketch the field of the point charge, the variation of the field into and away from the page would cause essentially insurmountable difficulties; for this reason sketching is usually limited to two-dimensional fields.

In the case of the two-dimensional field, we may arbitrarily set $E_z = 0$. The streamlines are thus confined to planes for which z is constant, and the sketch is the same for any such plane. Several streamlines are shown in Figure 2.10, and the E_x and E_y components are indicated at a general point. It is apparent from the geometry that

$$\boxed{\frac{E_y}{E_x} = \frac{dy}{dx}} \tag{19}$$

A knowledge of the functional form of E_x and E_y (and the ability to solve the resultant differential equation) will enable us to obtain the equations of the streamlines.

As an illustration of this method, consider the field of the uniform line charge with $\rho_L = 2\pi\epsilon_0$,

$$\mathbf{E} = \frac{1}{\rho}\mathbf{a}_\rho$$

In rectangular coordinates,

$$\mathbf{E} = \frac{x}{x^2 + y^2}\mathbf{a}_x + \frac{y}{x^2 + y^2}\mathbf{a}_y$$

Thus we form the differential equation

$$\frac{dy}{dx} = \frac{E_y}{E_x} = \frac{y}{x} \qquad \text{or} \qquad \frac{dy}{y} = \frac{dx}{x}$$

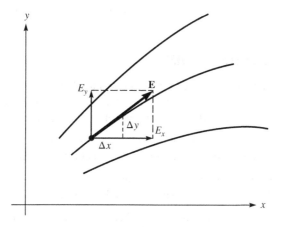

Figure 2.10 The equation of a streamline is obtained by solving the differential equation $E_y/E_x = dy/dx$.

Therefore,

$$\ln y = \ln x + C_1 \quad \text{or} \quad \ln y = \ln x + \ln C$$

from which the equations of the streamlines are obtained,

$$y = Cx$$

If we want to find the equation of one particular streamline, say one passing through $P(-2, 7, 10)$, we merely substitute the coordinates of that point into our equation and evaluate C. Here, $7 = C(-2)$, and $C = -3.5$, so $y = -3.5x$.

Each streamline is associated with a specific value of C, and the radial lines shown in Figure 2.9d are obtained when $C = 0, 1, -1$, and $1/C = 0$.

The equations of streamlines may also be obtained directly in cylindrical or spherical coordinates. A spherical coordinate example will be examined in Section 4.7.

D2.7. Find the equation of the streamline that passes through the point $P(1, 4, -2)$ in the field

$$\mathbf{E} = (a)\ \frac{-8x}{y}\mathbf{a}_x + \frac{4x^2}{y^2}\mathbf{a}_y;\ (b)\ 2\,e^{5x}[y(5x+1)\mathbf{a}_x + x\mathbf{a}_y].$$

Ans. $(a)\ x^2 + 2y^2 = 33;\ (b)\ y^2 = 15.7 + 0.4x - 0.08\ln(5x+1)$

REFERENCES

1. Boast, W. B. *Vector Fields*. New York: Harper and Row, 1964. This book contains many examples and sketches of fields.

2. Della Torre, E., and Longo, C. L. *The Electromagnetic Field*. Boston: Allyn and Bacon, 1969. The authors introduce all of electromagnetic theory with a careful and rigorous development based on a single experimental law—that of Coulomb. It begins in Chapter 1.

3. Schelkunoff, S. A. *Electromagnetic Fields*. New York: Blaisdell Publishing Company, 1963. Many of the physical aspects of fields are discussed early in this text without advanced mathematics.

CHAPTER 2 PROBLEMS

Quizzes

2.1 Three point charges of equal magnitude q are located at $x = -2$, $y = +2$, and $y = -\sqrt{2}$. Find the coordinates of a fourth positive charge, also of magnitude q, that will yield a zero net electric field at the origin.

2.2 Point charges of 1 nC and -2 nC are located at $(0, 0, 0)$ and $(1, 1, 1)$, respectively, in free space. Determine the vector force acting on each charge.

2.3 Point charges of 50 nC each are located at $A(1, 0, 0)$, $B(-1, 0, 0)$, $C(0, 1, 0)$, and $D(0, -1, 0)$ in free space. Find the total force on the charge at A.

2.4 Eight identical point charges of Q C each are located at the corners of a cube of side length a, with one charge at the origin, and with the three nearest charges at $(a, 0, 0)$, $(0, a, 0)$, and $(0, 0, a)$. Find an expression for the total vector force on the charge at $P(a, a, a)$, assuming free space.

2.5 A point charge of 3 nC is located at the point $(1, 1, 1)$ in free space. What charge must be located at $(1, 3, 2)$ to cause the y component of $\mathbf{E}$ to be zero at the origin?

2.6 Two point charges of equal magnitude q are positioned at $z = \pm d/2$. (a) Find the electric field everywhere on the z axis; (b) find the electric field everywhere on the xy plane.

2.7 Two point charges of equal magnitude but of opposite sign are positioned with charge $+q$ at $z = +d/2$ and charge $-q$ at $z = -d/2$. The charges in this configuration form an *electric dipole*. (a) Find the electric field intensity $\mathbf{E}$ everywhere on the z axis. (b) Evaluate your part a result at the origin. (c) Find the electric field intensity everywhere on the xy plane, expressing your result as a function of radius ρ in cylindrical coordinates. (d) Evaluate your part c result at the origin. (e) Simplify your part c result for the case in which $\rho >> d$.

2.8 A crude device for measuring charge consists of two small insulating spheres of radius a, one of which is fixed in position. The other is movable along the x axis and is subject to a restraining force kx, where k is a spring constant. The uncharged spheres are centered at $x = 0$ and $x = d$, the latter fixed. If the spheres are given equal and opposite charges of Q/C, obtain the expression by which Q may be found as a function of x. Determine the maximum charge that can be measured in terms of ϵ_0, k, and d, and state the separation of the spheres then. What happens if a larger charge is applied?

2.9 A 100-nC point charge is located at $A(-1, 1, 3)$ in free space. (a) Find the locus of all points $P(x, y, z)$ at which $E_x = 500$ V/m. (b) Find y_1 if $P(0, y_1, 3)$ lies on that locus.

2.10 A configuration of point charges consists of a single charge of value $-2q$ at the origin, and two charges of value $+q$ at locations $z = -d$ and $+d$. The charges as positioned form an *electric quadrupole,* equivalent to two dipoles of opposite orientation that are separated by distance d along the z axis. (a) Find the electric field intensity $\mathbf{E}$ everywhere in the xy plane, expressing your result as a function of cylindrical radius ρ. (b) Specialize your part a result for large distances, $\rho >> d$.

2.11 A charge Q_0 located at the origin in free space produces a field for which $E_z = 1$ kV/m at point $P(-2, 1, -1)$. (a) Find Q_0. Find $\mathbf{E}$ at $M(1, 6, 5)$ in (b) rectangular coordinates; (c) cylindrical coordinates; (d) spherical coordinates.

2.12 Electrons are in random motion in a fixed region in space. During any 1 μs interval, the probability of finding an electron in a subregion of volume 10^{-15} m^2 is 0.27. What volume charge density, appropriate for such time durations, should be assigned to that subregion?

2.13 A uniform volume charge density of 0.2 μC/m^3 is present throughout the spherical shell extending from $r = 3$ cm to $r = 5$ cm. If $\rho_v = 0$ elsewhere, find (a) the total charge present throughout the shell, and (b) r_1 if half the total charge is located in the region 3 cm $< r < r_1$.

2.14 The electron beam in a certain cathode ray tube possesses cylindrical symmetry, and the charge density is represented by $\rho_v = -0.1/(\rho^2 + 10^{-8})$ pC/m^3 for $0 < \rho < 3 \times 10^{-4}$ m, and $\rho_v = 0$ for $\rho > 3 \times 10^{-4}$ m. (a) Find the total charge per meter along the length of the beam. (b) If the electron velocity is 5×10^7 m/s, and with one ampere defined as 1 C/s, find the beam current.

2.15 A spherical volume having a 2-μm radius contains a uniform volume charge density of 10^5 C/m^3. (a) What total charge is enclosed in the spherical volume? (b) Now assume that a large region contains one of these little spheres at every corner of a cubical grid 3 mm on a side and that there is no charge between the spheres. What is the average volume charge density throughout this large region?

2.16 Within a region of free space, charge density is given as $\rho_v = \dfrac{\rho_0 r \cos\theta}{a}$ C/m^3, where ρ_0 and a are constants. Find the total charge lying within (a) the sphere, $r \leq a$; (b) the cone, $r \leq a$, $0 \leq \theta \leq 0.1\pi$; (c) the region, $r \leq a$, $0 \leq \theta \leq 0.1\pi$, $0 \leq \phi \leq 0.2\pi$.

2.17 A length d of line charge lies on the z axis in free space. The charge density on the line is $\rho_L = +\rho_0$ C/m ($0 < z < d/2$) and $\rho_L = -\rho_0$ C/m ($-d/2 < z < 0$) where ρ_0 is a positive constant. (a) Find the electric field intensity **E** everywhere in the xy plane, expressing your result as a function of cylindrical radius ρ. (b) Simplify your part a result for the case in which radius $\rho >> d$, and express this result in terms of charge $q = \rho_0\, d/2$.

2.18 (a) Find **E** in the plane $z = 0$ that is produced by a uniform line charge, ρ_L, extending along the z axis over the range $-L < z < L$ in a cylindrical coordinate system. (b) If the finite line charge is approximated by an infinite line charge ($L \to \infty$), by what percentage is E_ρ in error if $\rho = 0.5L$? (c) Repeat (b) with $\rho = 0.1L$.

2.19 A line having charge density $\rho_0 |z|$ C/m and of length ℓ is oriented along the z axis at $-\ell/2 < z < \ell/2$. (a) Find the electric field intensity **E** everywhere in the xy plane, expressing your result in cylindrical coordinates. (b) Evaluate your part a result in the limit as z approaches infinity.

2.20 A line charge of uniform charge density ρ_0 C/m and of length ℓ is oriented along the z axis at $-\ell/2 < z < \ell/2$. (a) Find the electric field strength, **E**, in magnitude and direction at any position along the x axis. (b) With the given line charge in position, find the force acting on an identical line charge that is oriented along the x axis at $\ell/2 < x < 3\ell/2$.

2.21 A charged filament forms a circle of radius a in the xy plane with its center at the origin. The filament carries uniform line charge density $+\rho_0$ C/m for $-\pi/2 < \phi < \pi/2$ and $-\rho_0$ C/m for $\pi/2 < \phi < 3\pi/2$. Find the electric field intensity **E** at the origin.

2.22 Two identical uniform sheet charges with $\rho_s = 100$ nC/m^2 are located in free space at $z = \pm 2.0$ cm. What force per unit area does each sheet exert on the other?

2.23 A disk of radius a in the xy plane carries surface charge of density $\rho_s = \rho_{s0}/\rho$ C/m^2 where ρ_{s0} is a constant. Find the electric field intensity $\mathbf{E}$ everywhere on the z axis.

2.24 (a) Find the electric field on the z axis produced by an annular ring of uniform surface charge density ρ_s in free space. The ring occupies the region $z = 0$, $a \le \rho \le b$, $0 \le \phi \le 2\pi$ in cylindrical coordinates. (b) From your part a result, obtain the field of an infinite uniform sheet charge by taking appropriate limits.

2.25 A disk of radius a in the xy plane carries surface charge of density $\rho_{s1} = +\rho_{s0}/\rho$ C/m^2 for $0 < \phi < \pi$, and $\rho_{s2} = -\rho_{s0}/\rho$ C/m^2 for $\pi < \phi < 2\pi$, where ρ_{s0} is a constant. (a) Find the electric field intensity $\mathbf{E}$ everywhere on the z axis. (b) Specialize your part a result for distances $z >> a$.

2.26 (a) Find the electric field intensity on the z axis produced by a cone surface that carries charge density $\rho_s(r) = \rho_0/r$ C/m^2 in free space. The cone has its vertex at the origin and occupies the region $\theta = \alpha$, $0 < r < a$, and $0 < \phi < 2\pi$ in spherical coordinates. Differential area for a cone is given in spherical coordinates as $da = r \sin \alpha \, dr \, d\phi$. (b) Find the total charge on the cone. (c) Specialize your result of part a to the case in which $\alpha = 90°$, at which the cone flattens to a disk in the xy plane. Compare this result to the answer to problem 2.23. (d) Show that your part a result becomes a point charge field when $z >> a$. (e) Show that your part a result becomes an inverse-z-dependent $\mathbf{E}$ field when $z << a$.

2.27 Given the electric field $\mathbf{E} = (4x - 2y)\mathbf{a}_x - (2x + 4y)\mathbf{a}_y$, find (a) the equation of the streamline that passes through the point $P(2, 3, -4)$; (b) a unit vector specifying the direction of $\mathbf{E}$ at $Q(3, -2, 5)$.

2.28 An electric dipole (introduced in Problem 2.7, and discussed in detail in Section 4.7) consists of two point charges of equal and opposite magnitude $\pm q$ spaced by distance d. With the charges along the z axis at positions $z = \pm d/2$ (with the positive charge at the positive z location), the electric field in spherical coordinates is given by $\mathbf{E}(r, \theta) = [qd/(4\pi\epsilon_0 r^3)][2 \cos\theta \, \mathbf{a}_r + \sin \theta \, \mathbf{a}_\theta]$, where $r >> d$. Using rectangular coordinates, determine expressions for the vector force on a point charge of magnitude q (a) at $(0, 0, z)$; (b) at $(0, y, 0)$.

2.29 If $\mathbf{E} = 20e^{-5y} (\cos 5x\mathbf{a}_x - \sin 5x\mathbf{a}_y)$, find (a) $|\mathbf{E}|$ at $P(\pi/6, 0.1, 2)$; (b) a unit vector in the direction of $\mathbf{E}$ at P; (c) the equation of the direction line passing through P.

2.30 For fields that do not vary with z in cylindrical coordinates, the equations of the streamlines are obtained by solving the differential equation $E_\rho/E_\phi = d\rho/(\rho d\phi)$. Find the equation of the line passing through the point $(2, 30°, 0)$ for the field $\mathbf{E} = \rho \cos 2\phi \, \mathbf{a}_\rho - \rho \sin 2\phi\mathbf{a}_\phi$.

CHAPTER 3 Electric Flux Density, Gauss's Law, and Divergence

术语

electric flux density 电通量密度。

displacement density 电位移密度，简称电位移，即为电通量密度，用首字母 *D* 表示，它是一个矢量，单位是库［仑］/ 平方米（C / m²）。

gauss' law 高斯通量定理，它从另一个角度描述了场与源的关系。

divergence 散度，对一个矢量函数求散度，结果是一个标量函数。

After drawing the fields described in the previous chapter and becoming familiar with the concept of the streamlines that show the direction of the force on a test charge at every point, it is appropriate to give these lines a physical significance and to think of them as *flux* lines. No physical particle is projected radially outward from the point charge, and there are no steel tentacles reaching out to attract or repel an unwary test charge, but as soon as the streamlines are drawn on paper there seems to be a picture showing "something" is present.

It is very helpful to invent an *electric flux* that streams away symmetrically from a point charge and is coincident with the streamlines and to visualize this flux wherever an electric field is present.

This chapter introduces and uses the concept of electric flux and electric flux density to again solve several of the problems presented in Chapter 2. The work here turns out to be much easier for problems that possess a high degree of spatial symmetry. ■

3.1 ELECTRIC FLUX DENSITY

At this stage, the second basic field quantity in our study is introduced: the *electric flux density*, or *electric displacement*, given the symbol, **D**. This field can be considered a companion field to **E**, the electric field intensity, as it is usually (but not always) parallel to **E**, and in the basic sense, they are both associated with electric charge. The two fields are related, but they have totally different meanings: We have already defined **E** in the context of finding forces on charges, but **D** is defined in a more direct way to the charge that is generating **E**. The two fields are related to each other through the properties of the medium in which they exist. This relation is a study on its own and will be taken up in detail in Chapter 5. In this section, we explain **D** in a historical context.

3.1.1 Faraday's Experiments on Electric Displacement

About 1837, the director of the Royal Society in London, Michael Faraday, became very interested in static electric fields and the effect of various insulating materials on these fields. This problem had been bothering him during the previous ten years when he was experimenting in his now-famous work on induced electromotive force, which we will discuss in Chapter 9. With that subject completed, he had a pair of concentric metallic spheres constructed, the outer one consisting of two hemispheres that could be firmly clamped together. He also prepared shells of insulating material (or dielectric material, or simply *dielectric*) that would occupy the entire volume between the concentric spheres.

His experiment, then, consisted essentially of the following steps:

1. With the equipment dismantled, the inner sphere was given a known positive charge.
2. The hemispheres were then clamped together around the charged sphere with about 2 cm of dielectric material between them.
3. The outer sphere was discharged by connecting it momentarily to ground.
4. The outer sphere was separated carefully, using tools made of insulating material in order not to disturb the induced charge on it, and the negative induced charge on each hemisphere was measured.

Faraday found that the total charge on the outer sphere was equal in *magnitude* to the original charge placed on the inner sphere and that this was true regardless of the dielectric material separating the two spheres. He concluded that there was some sort of "displacement" from the inner sphere to the outer which was independent of the medium; we now refer to this as *displacement, displacement flux,* or simply *electric flux.*

Faraday's experiments also showed, of course, that a larger positive charge on the inner sphere induced a correspondingly larger negative charge on the outer sphere, leading to a direct proportionality between the electric flux and the charge on the inner sphere. The constant of proportionality is dependent on the system of units involved, and we are fortunate in our use of SI units, because the constant is unity. If electric flux is denoted by Ψ (psi) and the total charge on the inner sphere by Q, then for Faraday's experiment

$$\boxed{\Psi = Q}$$

and the electric flux Ψ is measured in coulombs.

3.1.2 Electric Flux Density

More quantitative information can be obtained by considering an inner sphere of radius a and an outer sphere of radius b, with charges of Q and $-Q$, respectively (Figure 3.1). The paths of electric flux Ψ extending from the inner sphere to the outer sphere are indicated by the symmetrically distributed streamlines drawn radially from one sphere to the other.

At the surface of the inner sphere, Ψ coulombs of electric flux are produced by the charge $Q(=\Psi)$ coulombs distributed uniformly over a surface having an area of $4\pi a^2$ m^2. The density of the flux at this surface is $\Psi/4\pi a^2$ or $Q/4\pi a^2$ C/m^2, and this is an important new quantity.

导读

　　1. 综合第1与第2章的内容，已知电荷分布求场的方法主要有两种：（1）基于点电荷表达式采用积分的方法，该方法适用于无限大均匀介质空间的所有问题，但有时积分不易求解，需要采用数值积分方法；（2）利用高斯通量定理的积分形式求解，该方法仅能求解一些具有对称结构的问题，其关键是选择合适的高斯面。另外还有一种求解方法是第6章介绍的先求电位再求场强法，可称为求解电位的边值问题方法。电位的边值问题是指电位所满足的微分方程加上场域的外边界的边界条件和不同介质交界面的界面条件所构成的电位函数求解问题。

　　2. 高斯通量定理的微分形式，即式（15）的point form of Gauss' law描述了场域中各点上的电荷密度与场（电位移）在该点的散度之间的关系，是描述静电场的基本方程之一。电场强度的旋度为零是描述静电场的另一个基本方程，两者结合确定了静电场的特性。

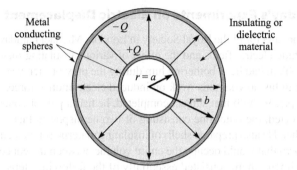

Figure 3.1 The electric flux in the region between a pair of charged concentric spheres. The direction and magnitude of **D** are not functions of the dielectric between the spheres.

Electric flux density, measured in coulombs per square meter (sometimes described as "lines per square meter," for each line is due to one coulomb), is given the letter **D**, which was originally chosen because of the alternate names of *displacement flux density* or *displacement density*. Electric flux density is more descriptive, however, and we will use the term consistently.

The electric flux density **D** is a vector field and is a member of the "flux density" class of vector fields, as opposed to the "force fields" class, which includes the electric field intensity **E**. The direction of **D** at a point is the direction of the flux lines at that point, and the magnitude is given by the number of flux lines crossing a surface normal to the lines divided by the surface area.

Referring again to Figure 3.1, the electric flux density is in the radial direction and has a value of

$$\mathbf{D}\Big|_{r=a} = \frac{Q}{4\pi a^2}\mathbf{a}_r \qquad \text{(inner sphere)}$$

$$\mathbf{D}\Big|_{r=b} = \frac{Q}{4\pi b^2}\mathbf{a}_r \qquad \text{(outer sphere)}$$

and at a radial distance r, where $a \leq r \leq b$,

$$\mathbf{D} = \frac{Q}{4\pi r^2}\mathbf{a}_r \tag{1}$$

If we now let the inner sphere become smaller and smaller, while still retaining a charge of Q, it becomes a point charge in the limit, but the electric flux density at a point r meters from the point charge is still given by (1), for Q lines of flux are symmetrically directed outward from the point and pass through an imaginary spherical surface of area $4\pi r^2$.

This result should be compared with Section 2.2, Eq. (9), the radial electric field intensity of a point charge in free space,

$$\mathbf{E} = \frac{Q}{4\pi\epsilon_0 r^2}\mathbf{a}_r$$

In free space, therefore,

$$\mathbf{D} = \epsilon_0\mathbf{E} \qquad \text{(free space only)} \tag{2}$$

Although (2) is applicable only to a vacuum, it is not restricted solely to the field of a point charge. For a general volume charge distribution in free space, the discussion in Section 2.3.2 resulted in

$$\mathbf{E} = \int_{\text{vol}} \frac{\rho_v \, dv}{4\pi\epsilon_0 R^2}\,\mathbf{a}_R \qquad \text{(free space only)} \tag{3}$$

This relationship was developed from the field of a single point charge. In a similar manner, (1) leads to

$$\mathbf{D} = \int_{\text{vol}} \frac{\rho_v \, dv}{4\pi R^2}\,\mathbf{a}_R \tag{4}$$

and (2) is therefore true for any free-space charge configuration; we will consider (2) as defining **D** in free space.

As a preparation for the study of dielectrics later, it might be well to point out now that, for a point charge embedded in an infinite ideal dielectric medium, Faraday's results show that (1) is still applicable, and thus so is (4). Equation (3) is not applicable, however, and so the relationship between **D** and **E** will be slightly more complicated than (2).

Because **D** is directly proportional to **E** in free space, it does not seem that it should really be necessary to introduce a new symbol. We do so for a few reasons. First, **D** is associated with the flux concept, which is an important new idea. Second, the **D** fields we obtain will be a little simpler than the corresponding **E** fields because ϵ_0 does not appear.

D3.1. Given a 60-μC point charge located at the origin, find the total electric flux passing through: (a) that portion of the sphere $r = 26$ cm bounded by $0 < \theta < \frac{\pi}{2}$ and $0 < \phi < \frac{\pi}{2}$; (b) the closed surface defined by $\rho = 26$ cm and $z = \pm 26$ cm; (c) the plane $z = 26$ cm.

Ans. (a) 7.5 μC; (b) 60 μC; (c) 30 μC

D3.2. Calculate **D** in rectangular coordinates at point $P(2, -3, 6)$ produced by: (*a*) a point charge $Q_A = 55$ mC at $Q(-2, 3, -6)$; (*b*) a uniform line charge $\rho_{LB} = 20$ mC/m on the x axis; (*c*) a uniform surface charge density $\rho_{SC} = 120$ μC/m² on the plane $z = -5$ m.

Ans. (*a*) $6.38\mathbf{a}_x - 9.57\mathbf{a}_y + 19.14\mathbf{a}_z$ μC/m²; (*b*) $-212\mathbf{a}_y + 424\mathbf{a}_z$ μC/m²; (*c*) $60\mathbf{a}_z$ μC/m²

3.2 GAUSS'S LAW

The results of Faraday's experiments with the concentric spheres could be summed up as an experimental law by stating that the electric flux passing through any imaginary spherical surface lying between the two conducting spheres is equal to the charge enclosed within that imaginary surface. This enclosed charge is distributed on the surface of the inner sphere, or it might be concentrated as a point charge at the center of the imaginary sphere. However, because one coulomb of electric flux is produced by one coulomb of charge, the inner conductor might just as well have been a cube or a brass door key and the total induced charge on the outer sphere would still be the same. Certainly the flux density would change from its previous symmetrical distribution to some unknown configuration, but $+Q$ coulombs on any inner conductor would produce an induced charge of $-Q$ coulombs on the surrounding sphere. Going one step further, we could now replace the two outer hemispheres with an empty (but completely closed) soup can. Q coulombs on the brass door key would produce $\Psi = Q$ lines of electric flux and would induce $-Q$ coulombs on the soup can.[1]

These generalizations of Faraday's experiment lead to the following statement, which is known as *Gauss's law*:

The electric flux passing through any closed surface is equal to the total charge enclosed by that surface.

The contribution of Gauss, one of the greatest mathematicians the world has ever produced, was actually not in stating the law as we have, but in providing a mathematical form for this statement, which we will now obtain.

Imagine a distribution of charge, shown as a cloud of point charges in Figure 3.2, surrounded by a closed surface of any shape. The closed surface may be the surface of some real material, but more generally it is any closed surface we wish to visualize. If the total charge is Q, then Q coulombs of electric flux will pass through the enclosing surface. At every point on the surface the electric-flux-density vector **D** will have some value $\mathbf{D}_S$, where the subscript S merely reminds us that **D** must be evaluated *at the surface*, and $\mathbf{D}_S$ will in general vary in magnitude and direction from one point on the surface to another.

Now, consider the nature of an incremental element of the surface. An incremental element of area ΔS is very nearly a portion of a plane surface, and the complete description of this surface element requires not only a statement of its magnitude ΔS but also of its orientation in space. In other words, the incremental surface element is a

---

[1] If it were a perfect insulator, the soup could even be left in the can without any difference in the results.

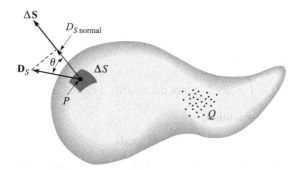

Figure 3.2 The electric flux density $\mathbf{D}_S$ at P arising from charge Q. The total flux passing through ΔS is $\mathbf{D}_S \cdot \Delta\mathbf{S}$.

vector quantity. The only unique direction that may be associated with $\Delta\mathbf{S}$ is the direction of the normal to that plane which is tangent to the surface at the point in question. There are, of course, two such normals, and the ambiguity is removed by specifying the *outward normal* whenever the surface is closed; "outward" has a specific meaning.

At any point P, consider an incremental element of surface ΔS and let $\mathbf{D}_S$ make an angle θ with $\Delta\mathbf{S}$, as shown in Figure 3.2. The flux crossing ΔS is then the product of the normal component of $\mathbf{D}_S$ and $\Delta\mathbf{S}$,

$$\Delta\Psi = \text{flux crossing } \Delta S = D_{S,\text{norm}}\,\Delta S = D_S \cos\theta\,\Delta S = \mathbf{D}_S \cdot \Delta\mathbf{S}$$

where we are able to apply the definition of the dot product developed in Chapter 1.

The *total* flux passing through the closed surface is obtained by adding the differential contributions crossing each surface element $\Delta\mathbf{S}$,

$$\Psi = \int d\Psi = \oint_{\substack{\text{closed}\\\text{surface}}} \mathbf{D}_S \cdot d\mathbf{S}$$

The resultant integral is a *closed surface integral*, and since the surface element $d\mathbf{S}$ always involves the differentials of two coordinates, such as $dx\,dy$, $\rho\,d\phi\,d\rho$, or $r^2 \sin\theta\,d\theta\,d\phi$, the integral is a double integral. Usually only one integral sign is used for brevity, and we will always place an S below the integral sign to indicate a surface integral, although this is not actually necessary, as the differential $d\mathbf{S}$ is automatically the signal for a surface integral. One last convention is to place a small circle on the integral sign itself to indicate that the integration is to be performed over a *closed* surface. Such a surface is often called a *gaussian surface*. We then have the mathematical formulation of Gauss's law,

$$\boxed{\Psi = \oint_S \mathbf{D}_S \cdot d\mathbf{S} = \text{charge enclosed} = Q} \tag{5}$$

The charge enclosed might be several point charges, in which case

$$Q = \sum Q_n$$

or a line charge,

$$Q = \int \rho_L \, dL$$

or a surface charge,

$$Q = \int_S \rho_S \, dS \quad \text{(not necessarily a closed surface)}$$

or a volume charge distribution,

$$Q = \int_{\text{vol}} \rho_v \, dv$$

The last form is usually used, and we should agree now that it represents any or all of the other forms. With this understanding, Gauss's law may be written in terms of the charge distribution as

$$\boxed{\oint_S \mathbf{D}_S \cdot d\mathbf{S} = \int_{\text{vol}} \rho_v \, dv} \tag{6}$$

a mathematical statement meaning simply that the total electric flux through any closed surface is equal to the charge enclosed.

EXAMPLE 3.1

To illustrate the application of Gauss's law, let us check the results of Faraday's experiment by placing a point charge Q at the origin of a spherical coordinate system (Figure 3.3) and by choosing our closed surface as a sphere of radius a.

Solution. We have, as before,

$$\mathbf{D} = \frac{Q}{4\pi r^2} \mathbf{a}_r$$

At the surface of the sphere,

$$\mathbf{D}_S = \frac{Q}{4\pi a^2} \mathbf{a}_r$$

The differential element of area on a spherical surface is, in spherical coordinates from Chapter 1,

$$dS = r^2 \sin\theta \, d\theta \, d\phi = a^2 \sin\theta \, d\theta \, d\phi$$

or

$$d\mathbf{S} = a^2 \sin\theta \, d\theta \, d\phi \, \mathbf{a}_r$$

The integrand is

$$\mathbf{D}_S \cdot d\mathbf{S} = \frac{Q}{4\pi a^2} a^2 \sin\theta \, d\theta \, d\phi \, \mathbf{a}_r \cdot \mathbf{a}_r = \frac{Q}{4\pi} \sin\theta \, d\theta \, d\phi$$

leading to the closed surface integral

$$\int_{\phi=0}^{\phi=2\pi} \int_{\theta=0}^{\theta=\pi} \frac{Q}{4\pi} \sin\theta \, d\theta \, d\phi$$

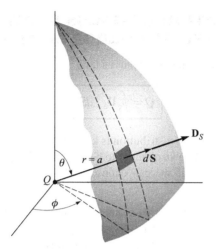

Figure 3.3 Applying Gauss's law to the field of a point charge Q on a spherical closed surface of radius a. The electric flux density **D** is everywhere normal to the spherical surface and has a constant magnitude at every point on it.

where the limits on the integrals have been chosen so that the integration is carried over the entire surface of the sphere once.[2] Integrating gives

$$\int_0^{2\pi} \frac{Q}{4\pi}(-\cos\theta)\Big|_0^{\pi}\, d\phi = \int_0^{2\pi} \frac{Q}{2\pi}\, d\phi = Q$$

and we obtain a result showing that Q coulombs of electric flux are crossing the surface, as we should since the enclosed charge is Q coulombs.

D3.3. Given the electric flux density, $\mathbf{D} = 0.3r^2\mathbf{a}_r$ nC/m^2 in free space: (a) find **E** at point $P(r = 2, \theta = 25°, \phi = 90°)$; (b) find the total charge within the sphere $r = 3$; (c) find the total electric flux leaving the sphere $r = 4$.

Ans. (a) $135.5\mathbf{a}_r$ V/m; (b) 305 nC; (c) 965 nC

D3.4. Calculate the total electric flux leaving the cubical surface formed by the six planes $x, y, z = \pm5$ if the charge distribution is: (a) two point charges, 0.1 μC at $(1, -2, 3)$ and $\frac{1}{7}$ μC at $(-1, 2, -2)$; (b) a uniform line charge of π μC/m at $x = -2, y = 3$; (c) a uniform surface charge of 0.1 μC/m^2 on the plane $y = 3x$.

Ans. (a) 0.243 μC; (b) 31.4 μC; (c) 10.54 μC

---

[2] Note that if θ and ϕ both cover the range from 0 to 2π, the spherical surface is covered twice.

3.3   APPLICATION OF GAUSS'S LAW: SOME SYMMETRICAL CHARGE DISTRIBUTIONS

We now consider how we may use Gauss's law,

$$\boxed{Q = \oint_S \mathbf{D}_S \cdot d\mathbf{S}}$$

to determine $\mathbf{D}_S$ if the charge distribution is known. This is an example of an integral equation in which the unknown quantity to be determined appears inside the integral.

The solution is easy if we can choose a closed surface which satisfies two conditions:

1. $\mathbf{D}_S$ is everywhere either normal or tangential to the closed surface, so that $\mathbf{D}_S \cdot d\mathbf{S}$ becomes either $D_S dS$ or zero, respectively.
2. On that portion of the closed surface for which $\mathbf{D}_S \cdot d\mathbf{S}$ is not zero, $D_S =$ constant.

This allows the dot product to be replaced with the product of the scalars D_S and dS, and then D_S can be brought outside the integral sign. The remaining integral is then $\int_S dS$ over that portion of the closed surface that $\mathbf{D}_S$ crosses normally, and this is simply the area of this section of that surface. Only a knowledge of the symmetry of the problem enables us to choose such a closed surface.

3.3.1   Point Charge Field

Consider a point charge Q at the origin of a spherical coordinate system and decide on a suitable closed surface that will meet the two requirements previously listed. The surface in question is obviously a spherical surface, centered at the origin *and of any radius r*. $\mathbf{D}_S$ is everywhere normal to the surface, and D_S has the same value at all points on the surface.

Then we have, in order,

$$Q = \oint_S \mathbf{D}_S \cdot d\mathbf{S} = \oint_{sph} D_S dS$$
$$= D_S \oint_{sph} dS = D_S \int_{\phi=0}^{\phi=2\pi} \int_{\theta=0}^{\theta=\pi} r^2 \sin\theta \, d\theta \, d\phi$$
$$= 4\pi r^2 D_S$$

and therefore

$$D_S = \frac{Q}{4\pi r^2}$$

Because r may have any value and because $\mathbf{D}_S$ is directed radially outward,

$$\boxed{\mathbf{D} = \frac{Q}{4\pi r^2}\mathbf{a}_r \qquad \mathbf{E} = \frac{Q}{4\pi \epsilon_0 r^2}\mathbf{a}_r}$$

which agrees with the results of Chapter 2. The example is a trivial one, and the objection could be raised that we had to know that the field was symmetrical and directed radially outward before we could obtain an answer. This is true, and that leaves the inverse-square-law relationship as the only check obtained from Gauss's law. The example does, however, serve to illustrate a method which can be applied to other problems, including several to which Coulomb's law is almost incapable of supplying an answer.

3.3.2 Line Charge Field

As a second example, consider again the uniform line charge distribution ρ_L lying along the z axis and extending from $-\infty$ to $+\infty$. We must first know the symmetry of the field, and this knowledge is complete when the answers to these two questions are known:

1. With which coordinates does the field vary (or of what variables is D a function)?
2. Which components of **D** are present?

In using Gauss's law, it is not a question of using symmetry to simplify the solution, for the application of Gauss's law depends on symmetry, and *if we cannot show that symmetry exists then we cannot use Gauss's law* to obtain a solution. The preceding two questions now become "musts."

From our previous discussion of the uniform line charge, it is evident that only the radial component of **D** is present, or

$$\mathbf{D} = D_\rho \mathbf{a}_\rho$$

and this component is a function of ρ only.

$$D_\rho = f(\rho)$$

The choice of a closed surface is now simple, for a cylindrical surface is the only surface to which D_ρ is everywhere normal, and it may be closed by plane surfaces normal to the z axis. A closed right circular cylinder of radius ρ extending from $z = 0$ to $z = L$ is shown in Figure 3.4.

We apply Gauss's law,

$$Q = \oint_{\text{cyl}} \mathbf{D}_S \cdot d\mathbf{S} = D_S \int_{\text{sides}} dS + 0\int_{\text{top}} dS + 0\int_{\text{bottom}} dS$$
$$= D_S \int_{z=0}^{L} \int_{\phi=0}^{2\pi} \rho \, d\phi \, dz = D_S 2\pi\rho L$$

and obtain

$$D_S = D_\rho = \frac{Q}{2\pi\rho L}$$

In terms of the charge density ρ_L, the total charge enclosed is

$$Q = \rho_L L$$

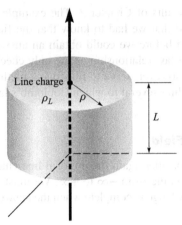

Figure 3.4 The gaussian surface for an infinite uniform line charge is a right circular cylinder of length L and radius ρ. **D** is constant in magnitude and everywhere perpendicular to the cylindrical surface; **D** is parallel to the end faces.

giving

$$D_\rho = \frac{\rho_L}{2\pi\rho}$$

or

$$E_\rho = \frac{\rho_L}{2\pi\,\epsilon_0\rho}$$

Comparing with Section 2.4, Eq. (16), shows that the correct result has been obtained and with much less work. Once the appropriate surface has been chosen, the integration usually amounts only to writing down the area of the surface at which **D** is normal.

3.3.3 Coaxial Cable Field

The problem of a coaxial cable is almost identical to that of the line charge and is an example that is extremely difficult to solve from the standpoint of Coulomb's law. Suppose that we have two coaxial cylindrical conductors, the inner of radius a and the outer of radius b, each infinite in extent (Figure 3.5). We will assume a charge distribution of ρ_S on the outer surface of the inner conductor.

Symmetry considerations show us that only the D_ρ component is present and that it can be a function only of ρ. A right circular cylinder of length L and radius ρ, where $a < \rho < b$, is necessarily chosen as the gaussian surface, and we quickly have

$$Q = D_S 2\pi\rho L$$

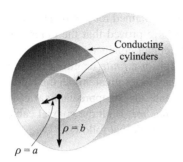

Figure 3.5 The two coaxial cylindrical conductors forming a coaxial cable provide an electric flux density within the cylinders, given by $D_\rho = a\rho_S/\rho$.

The total charge on a length L of the inner conductor is

$$Q = \int_{z=0}^{L} \int_{\phi=0}^{2\pi} \rho_S a \, d\phi \, dz = 2\pi a L \rho_S$$

from which we have

$$D_S = \frac{a\rho_S}{\rho} \qquad \mathbf{D} = \frac{a\rho_S}{\rho}\mathbf{a}_\rho \qquad (a < \rho < b)$$

This result might be expressed in terms of charge per unit length because the inner conductor has $2\pi \, a\rho_S$ coulombs on a meter length, and hence, letting $\rho_L = 2\pi \, a\rho_S$,

$$\boxed{\mathbf{D} = \frac{\rho_L}{2\pi\rho}\mathbf{a}_\rho}$$

and the solution has a form identical with that of the infinite line charge.

Because every line of electric flux starting from the charge on the inner cylinder must terminate on a negative charge on the inner surface of the outer cylinder, the total charge on that surface must be

$$Q_{\text{outer cyl}} = -2\pi a L \rho_{S,\text{inner cyl}}$$

and the surface charge on the outer cylinder is found as

$$2\pi b L \rho_{S,\text{outer cyl}} = -2\pi a L \rho_{S,\text{inner cyl}}$$

or

$$\rho_{S,\text{outer cyl}} = -\frac{a}{b}\rho_{S,\text{inner cyl}}$$

What would happen if we should use a cylinder of radius ρ, $\rho > b$, for the gaussian surface? The total charge enclosed would then be zero, for there are equal and opposite charges on each conducting cylinder. Hence

$$0 = D_S 2\pi\rho L \qquad (\rho > b)$$
$$D_S = 0 \qquad (\rho > b)$$

An identical result would be obtained for $\rho < a$. Thus the coaxial cable or capacitor has no external field (we have proved that the outer conductor is a "shield"), and there is no field within the center conductor.

Our result is also useful for a *finite* length of coaxial cable, open at both ends, provided the length L is many times greater than the radius b so that the nonsymmetrical conditions at the two ends do not appreciably affect the solution. Such a device is also termed a *coaxial capacitor*. Both the coaxial cable and the coaxial capacitor will appear frequently in the work that follows.

EXAMPLE 3.2

Let us select a 50-cm length of coaxial cable having an inner radius of 1 mm and an outer radius of 4 mm. The space between conductors is assumed to be filled with air. The total charge on the inner conductor is 30 nC. We wish to know the charge density on each conductor, and the **E** and **D** fields.

Solution. We begin by finding the surface charge density on the inner cylinder,

$$\rho_{S,\text{inner cyl}} = \frac{Q_{\text{inner cyl}}}{2\pi a L} = \frac{30 \times 10^{-9}}{2\pi(10^{-3})(0.5)} = 9.55\,\mu\text{C/m}^2$$

The negative charge density on the inner surface of the outer cylinder is

$$\rho_{S,\text{outer cyl}} = \frac{Q_{\text{outer cyl}}}{2\pi b L} = \frac{-30 \times 10^{-9}}{2\pi(4 \times 10^{-3})(0.5)} = -2.39\,\mu\text{C/m}^2$$

The internal fields may therefore be calculated easily:

$$D_\rho = \frac{a\rho_S}{\rho} = \frac{10^{-3}(9.55 \times 10^{-6})}{\rho} = \frac{9.55}{\rho}\,\text{nC/m}^2$$

and

$$E_\rho = \frac{D_\rho}{\epsilon_0} = \frac{9.55 \times 10^{-9}}{8.854 \times 10^{-12}\rho} = \frac{1079}{\rho}\,\text{V/m}$$

Both of these expressions apply to the region where $1 < \rho < 4$ mm. For $\rho < 1$ mm or $\rho > 4$ mm, **E** and **D** are zero.

D3.5. A point charge of 0.25 µC is located at $r = 0$, and uniform surface charge densities are located as follows: 2 mC/m² at $r = 1$ cm, and −0.6 mC/m² at $r = 1.8$ cm. Calculate **D** at: (a) $r = 0.5$ cm; (b) $r = 1.5$ cm; (c) $r = 2.5$ cm. (d) What uniform surface charge density should be established at $r = 3$ cm to cause **D** = 0 at $r = 3.5$ cm?

Ans. (a) $796\mathbf{a}_r$ µC/m²; (b) $977\mathbf{a}_r$ µC/m²; (c) $40.8\mathbf{a}_r$ µC/m²; (d) -28.3 µC/m²

3.4   GAUSS'S LAW IN DIFFERENTIAL FORM: DIVERGENCE

We will now apply the methods of Gauss's law to a slightly different type of problem—one that may not possess any symmetry at all. At first glance, it might seem that our case is hopeless, for without symmetry, a simple gaussian surface cannot be chosen such that the normal component of **D** is constant or zero everywhere on the surface. Without such a surface, the integral cannot be evaluated. There is only one way to circumvent these difficulties and that is to choose such a very small closed surface that **D** is *almost* constant over the surface, and the small change in **D** may be adequately represented by using the first two terms of the Taylor's-series expansion for **D**. The result will become more nearly correct as the volume enclosed by the gaussian surface decreases, and we intend eventually to allow this volume to approach zero.

This example also differs from the preceding ones in that we will not obtain the value of **D** as our answer but will instead receive some extremely valuable information about the way **D** varies in the region of our small surface. This leads directly to one of Maxwell's four equations, which are basic to all electromagnetic theory.

3.4.1   Gauss's Law Applied to a Differential Volume Element

Consider any point P, shown in Figure 3.6, located by a rectangular coordinate system. The value of **D** at the point P may be expressed in rectangular components, $\mathbf{D}_0 = D_{x0}\mathbf{a}_x + D_{y0}\mathbf{a}_y + D_{z0}\mathbf{a}_z$. We choose as our closed surface the small rectangular box, centered at P, having sides of lengths Δx, Δy, and Δz, and apply Gauss's law,

$$\oint_S \mathbf{D} \cdot d\mathbf{S} = Q$$

In order to evaluate the integral over the closed surface, the integral must be broken up into six integrals, one over each face,

$$\oint_S \mathbf{D} \cdot d\mathbf{S} = \int_{\text{front}} + \int_{\text{back}} + \int_{\text{left}} + \int_{\text{right}} + \int_{\text{top}} + \int_{\text{bottom}}$$

Consider the first of these in detail. Because the surface element is very small, **D** is essentially constant (over *this* portion of the entire closed surface) and

$$\int_{\text{front}} \doteq \mathbf{D}_{\text{front}} \cdot \Delta \mathbf{S}_{\text{front}}$$
$$\doteq \mathbf{D}_{\text{front}} \cdot \Delta y \, \Delta z \, \mathbf{a}_x$$
$$\doteq D_{x,\text{front}} \Delta y \, \Delta z$$

where we have only to approximate the value of D_x at this front face. The front face is at a distance of $\Delta x/2$ from P, and hence

$$D_{x,\text{front}} \doteq D_{x0} + \frac{\Delta x}{2} \times \text{rate of change of } D_x \text{ with } x$$
$$\doteq D_{x0} + \frac{\Delta x}{2} \frac{\partial D_x}{\partial x}$$

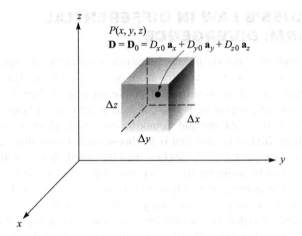

Figure 3.6 A differential-sized gaussian surface about the point P is used to investigate the space rate of change of **D** in the neighborhood of P.

where D_{x0} is the value of D_x at P, and where a partial derivative must be used to express the rate of change of D_x with x, as D_x in general also varies with y and z. This expression could have been obtained more formally by using the constant term and the term involving the first derivative in the Taylor's-series expansion for D_x in the neighborhood of P.

We now have

$$\int_{\text{front}} \doteq \left(D_{x0} + \frac{\Delta x}{2} \frac{\partial D_x}{\partial x} \right) \Delta y \, \Delta z$$

Consider now the integral over the back surface,

$$\int_{\text{back}} \doteq \mathbf{D}_{\text{back}} \cdot \Delta \mathbf{S}_{\text{back}}$$
$$\doteq \mathbf{D}_{\text{back}} \cdot (-\Delta y \, \Delta z \, \mathbf{a}_x)$$
$$\doteq - D_{x,\text{back}} \, \Delta y \, \Delta z$$

and

$$D_{x,\text{back}} \doteq D_{x0} - \frac{\Delta x}{2} \frac{\partial D_x}{\partial x}$$

giving

$$\int_{\text{back}} \doteq \left(-D_{x0} + \frac{\Delta x}{2} \frac{\partial D_x}{\partial x} \right) \Delta y \, \Delta z$$

If we combine these two integrals, we have

$$\int_{\text{front}}+\int_{\text{back}} \doteq \frac{\partial D_x}{\partial x} \Delta x \, \Delta y \, \Delta z$$

By exactly the same process we find that

$$\int_{\text{right}}+\int_{\text{left}} \doteq \frac{\partial D_y}{\partial y} \Delta x \, \Delta y \, \Delta z$$

and

$$\int_{\text{top}}+\int_{\text{bottom}} \doteq \frac{\partial D_z}{\partial z} \Delta x \, \Delta y \, \Delta z$$

and these results may be collected to yield

$$\oint_S \mathbf{D} \cdot d\mathbf{S} = Q \doteq \left(\frac{\partial D_x}{\partial x} + \frac{\partial D_y}{\partial y} + \frac{\partial D_z}{\partial z} \right) \Delta v \qquad (7)$$

where $\Delta v = \Delta x \Delta y \Delta z$. The expression is an approximation which becomes better as Δv becomes smaller, and in the following section we shall let the volume Δv approach zero. For the moment, we have applied Gauss's law to the closed surface surrounding the volume element Δv and have as a result the approximation (7) stating that

$$\boxed{\text{Charge enclosed in volume } \Delta v \doteq \left(\frac{\partial D_x}{\partial x} + \frac{\partial D_y}{\partial y} + \frac{\partial D_z}{\partial z} \right) \times \text{volume } \Delta v} \qquad (8)$$

EXAMPLE 3.3

Find an approximate value for the total charge enclosed in an incremental volume of 10^{-9} m^3 located at the origin, if $\mathbf{D} = e^{-x} \sin y \, \mathbf{a}_x - e^{-x} \cos y \, \mathbf{a}_y + 2z\mathbf{a}_z$ C/m^2.

Solution. We first evaluate the three partial derivatives in (8):

$$\frac{\partial D_x}{\partial x} = -e^{-x} \sin y$$

$$\frac{\partial D_y}{\partial y} = e^{-x} \sin y$$

$$\frac{\partial D_z}{\partial z} = 2$$

At the origin, the first two expressions are zero, and the last is 2. Thus, we find that the charge enclosed in a small volume element there must be approximately $2\Delta v$. If Δv is 10^{-9} m^3, then we have enclosed about 2 nC.

D3.6. In free space, let $\mathbf{D} = 8xyz^4\mathbf{a}_x + 4x^2z^4\mathbf{a}_y + 16x^2yz^3\,\mathbf{a}_z$ pC/m². (*a*) Find the total electric flux passing through the rectangular surface $z = 2$, $0 < x < 2$, $1 < y < 3$, in the $\mathbf{a}_z$ direction. (*b*) Find $\mathbf{E}$ at $P(2, -1, 3)$. (*c*) Find an approximate value for the total charge contained in an incremental sphere located at P $(2, -1, 3)$ and having a volume of 10^{-12} m³.

Ans. (*a*) 1365 pC; (*b*) $-146.4\mathbf{a}_x + 146.4\mathbf{a}_y - 195.2\mathbf{a}_z$ V/m; (*c*) -2.38×10^{-21} C

3.4.2 Divergence

We next obtain an exact relationship from (7), by allowing the volume element Δv to shrink to zero. We write this equation as

$$\left(\frac{\partial D_x}{\partial x} + \frac{\partial D_y}{\partial y} + \frac{\partial D_z}{\partial z}\right) = \lim_{\Delta v \to 0} \frac{\oint_S \mathbf{D} \cdot d\mathbf{S}}{\Delta v} = \lim_{\Delta v \to 0} \frac{Q}{\Delta v} = \rho_v \qquad (9)$$

in which the charge density, ρ_v, is identified in the second equality.

The methods of the previous section could have been used on any vector $\mathbf{A}$ to find $\oint_S \mathbf{A} \cdot d\mathbf{S}$ for a small closed surface, leading to

$$\left(\frac{\partial A_x}{\partial x} + \frac{\partial A_y}{\partial y} + \frac{\partial A_z}{\partial z}\right) = \lim_{\Delta v \to 0} \frac{\oint_S \mathbf{A} \cdot d\mathbf{S}}{\Delta v} \qquad (10)$$

where $\mathbf{A}$ could represent velocity, temperature gradient, force, or any other vector field.

The operation in Eq. (10) appeared so many times in physical investigations in the nineteenth century that it received a descriptive name, *divergence*. The divergence of $\mathbf{A}$ is defined as

$$\text{Divergence of } \mathbf{A} = \text{div}\,\mathbf{A} = \lim_{\Delta v \to 0} \frac{\oint_S \mathbf{A} \cdot d\mathbf{S}}{\Delta v} \qquad (11)$$

and is usually abbreviated div $\mathbf{A}$. The physical interpretation of the divergence of a vector is obtained by describing carefully the operations implied by the right-hand side of (11), where we shall consider $\mathbf{A}$ to be a member of the flux-density family of vectors in order to aid the physical interpretation.

The divergence of the vector flux density $\mathbf{A}$ is the outflow of flux from a small closed surface per unit volume as the volume shrinks to zero.

The physical interpretation of divergence afforded by this statement is often useful in obtaining qualitative information about the divergence of a vector field without resorting to a mathematical investigation. For instance, let us consider the divergence of the velocity of water in a bathtub after the drain has been opened. The net outflow of water through *any* closed surface lying entirely within the water must be zero, for water is essentially incompressible, and the water entering and leaving different regions of the closed surface must be equal. Hence the divergence of this velocity is zero.

If, however, we consider the velocity of the air in a tire that has just been punctured by a nail, we realize that the air is expanding as the pressure drops, and that consequently there is a net outflow from any closed surface lying within the tire. The divergence of this velocity is therefore greater than zero.

A positive divergence for any vector quantity indicates a *source* of that vector quantity at that point. Similarly, a negative divergence indicates a *sink*. Because the divergence of the water velocity above is zero, no source or sink exists.[3] The expanding air, however, produces a positive divergence of the velocity, and each interior point may be considered a source.

Writing (9) with our new term, we have

$$\text{div } \mathbf{D} = \left(\frac{\partial D_x}{\partial x} + \frac{\partial D_y}{\partial y} + \frac{\partial D_z}{\partial z}\right) \quad \text{(rectangular)} \tag{12}$$

This expression is again of a form that does not involve the charge density. It is the result of applying the definition of divergence (11) to a differential volume element in *rectangular coordinates*.

If a differential volume unit $\rho \, d\rho \, d\phi \, dz$ in cylindrical coordinates, or $r^2 \sin \theta \, dr \, d\theta \, d\phi$ in spherical coordinates, had been chosen, expressions for divergence involving the components of the vector in the particular coordinate system and involving partial derivatives with respect to the variables of that system would have been obtained. These expressions are obtained in Appendix A and are given here for convenience:

$$\text{div } \mathbf{D} = \frac{1}{\rho} \frac{\partial}{\partial \rho}(\rho D_\rho) + \frac{1}{\rho} \frac{\partial D_\phi}{\partial \phi} + \frac{\partial D_z}{\partial z} \quad \text{(cylindrical)} \tag{13}$$

$$\text{div } \mathbf{D} = \frac{1}{r^2} \frac{\partial}{\partial r}(r^2 D_r) + \frac{1}{r \sin\theta} \frac{\partial}{\partial \theta}(\sin\theta \, D_\theta) + \frac{1}{r \sin\theta} \frac{\partial D_\phi}{\partial \phi} \quad \text{(spherical)} \tag{14}$$

These relationships are also shown at the end of this book for easy reference.

It should be noted that the divergence is an operation which is performed on a vector, but that the result is a scalar. We should recall that, in a somewhat similar way, the dot or scalar product was a multiplication of two vectors which yielded a scalar.

For some reason, it is a common mistake on meeting divergence for the first time to impart a vector quality to the operation by scattering unit vectors around in

[3] Having chosen a differential element of volume within the water, the gradual decrease in water level with time will eventually cause the volume element to lie above the surface of the water. At the instant the surface of the water intersects the volume element, the divergence is positive and the small volume is a source. This complication is avoided above by specifying an integral point.

the partial derivatives. Divergence merely tells us *how much* flux is leaving a small volume on a per-unit-volume basis; no direction is associated with it.

We can illustrate the concept of divergence by continuing with the example at the end of Section 3.4.

EXAMPLE 3.4

Find div **D** at the origin if $\mathbf{D} = e^{-x} \sin y\, \mathbf{a}_x - e^{-x} \cos y\, \mathbf{a}_y + 2z \mathbf{a}_z$.

Solution. We use (10) to obtain

$$\text{div}\,\mathbf{D} = \frac{\partial D_x}{\partial x} + \frac{\partial D_y}{\partial y} + \frac{\partial D_z}{\partial z}$$

$$= -e^{-x} \sin y + e^{-x} \sin y + 2 = 2$$

The value is the constant 2, regardless of location.

If the units of **D** are C/m^2, then the units of div **D** are C/m^3. This is a volume charge density, a concept discussed in the next section.

D3.7. In each of the following parts, find a numerical value for div **D** at the point specified: (*a*) $\mathbf{D} = (2xyz - y^2)\mathbf{a}_x + (x^2z - 2xy)\mathbf{a}_y + x^2y\mathbf{a}_z$ C/m^2 at $P_A(2, 3, -1)$; (*b*) $\mathbf{D} = 2\rho z^2 \sin^2 \phi\, \mathbf{a}_\rho + \rho z^2 \sin 2\phi\, \mathbf{a}_\phi + 2\rho^2 z \sin^2 \phi\, \mathbf{a}_z$ C/m^2 at $P_B(\rho = 2, \phi = 110°, z = -1)$; (*c*) $\mathbf{D} = 2r \sin \theta \cos \phi\, \mathbf{a}_r + r \cos \theta \cos \phi\, \mathbf{a}_\theta - r \sin \phi\, \mathbf{a}_\phi$ C/m^2 at $P_C(r = 1.5, \theta = 30°, \phi = 50°)$.

Ans. (*a*) -10.00; (*b*) 9.06; (*c*) 1.29

3.4.3 Maxwell's First Equation: Gauss's Law in Point Form

Finally, we can combine Eqs. (9) and (12) and form the relation between electric flux density and charge density:

$$\boxed{\text{div}\,\mathbf{D} = \rho_v} \tag{15}$$

This is the first of Maxwell's four equations as they apply to electrostatics and steady magnetic fields, and it states that the electric flux per unit volume leaving a vanishingly small volume unit is exactly equal to the volume charge density there. This equation is aptly called the *point form of Gauss's law*. Gauss's law relates the flux leaving any closed surface to the charge enclosed, and Maxwell's first equation makes an identical statement on a per-unit-volume basis for a vanishingly small volume, or at a point. Because the divergence may be expressed as the sum of three partial derivatives, Maxwell's first equation is also described as the differential-equation form of Gauss's law, and conversely, Gauss's law is recognized as the integral form of Maxwell's first equation.

As a specific illustration, let us consider the divergence of **D** in the region about a point charge Q located at the origin. We have the field

$$\mathbf{D} = \frac{Q}{4\pi r^2}\mathbf{a}_r$$

and use (14), the expression for divergence in spherical coordinates:

$$\operatorname{div}\mathbf{D} = \frac{1}{r^2}\frac{\partial}{\partial r}(r^2 D_r) + \frac{1}{r\sin\theta}\frac{\partial}{\partial\theta}(D_\theta\sin\theta) + \frac{1}{r\sin\theta}\frac{\partial D_\phi}{\partial\phi}$$

Because D_θ and D_ϕ are zero, we have

$$\operatorname{div}\mathbf{D} = \frac{1}{r^2}\frac{d}{dr}\left(r^2\frac{Q}{4\pi r^2}\right) = 0 \qquad (\text{if } r \neq 0)$$

Thus, $\rho_v = 0$ everywhere except at the origin, where it is infinite.

The divergence operation is not limited to electric flux density; it can be applied to any vector field. We will apply it to several other electromagnetic fields in the coming chapters.

D3.8. Determine an expression for the volume charge density associated with each **D** field: (a) $\mathbf{D} = \frac{4xy}{z}\mathbf{a}_x + \frac{2x^2}{z}\mathbf{a}_y - \frac{2x^2 y}{z^2}\mathbf{a}_z$; (b) $\mathbf{D} = z\sin\phi\,\mathbf{a}_\rho + z\cos\phi\,\mathbf{a}_\phi + \rho\sin\phi\,\mathbf{a}_z$; (c) $\mathbf{D} = \sin\theta\sin\phi\,\mathbf{a}_r + \cos\theta\sin\phi\,\mathbf{a}_\theta + \cos\phi\,\mathbf{a}_\phi$.

Ans. $\frac{4y}{z^3}(x^2 + z^2)$; 0; 0.

3.5 DIVERGENCE THEOREM

Gauss's law for the electric field as we have used it is a specialization of what is called the *divergence theorem* in field theory. This general theorem is applied in other ways to different problems in physics, as well as to a few more in electromagnetics. In this section, we develop this broader perspective and demonstrate the use of the theorem in simplifying some otherwise complicated problems.

3.5.1 The Del Operator

As divergence is an operation on a vector yielding a scalar result, just as the dot product of two vectors gives a scalar result, it seems possible that we can find something that may be dotted formally with **D** to yield the scalar

$$\frac{\partial D_x}{\partial x} + \frac{\partial D_y}{\partial y} + \frac{\partial D_z}{\partial z}$$

Obviously, this cannot be accomplished by using a dot *product*; the process must be a dot *operation*.

With this in mind, we define the *del operator* ∇ as a *vector operator*,

$$\nabla = \frac{\partial}{\partial x}\mathbf{a}_x + \frac{\partial}{\partial y}\mathbf{a}_y + \frac{\partial}{\partial z}\mathbf{a}_z \qquad (16)$$

Similar *scalar operators* appear in several methods of solving differential equations where we often let D replace d/dx, D^2 replace d^2/dx^2, and so forth.[4] ∇ is treated in every way as an ordinary vector with the one important exception that partial derivatives result instead of products of scalars.

3.5.2 Obtaining Divergence with the Del Operator

Consider the operation $\nabla \cdot \mathbf{D}$, signifying

$$\nabla \cdot \mathbf{D} = \left(\frac{\partial}{\partial x}\mathbf{a}_x + \frac{\partial}{\partial y}\mathbf{a}_y + \frac{\partial}{\partial z}\mathbf{a}_z\right) \cdot (D_x\mathbf{a}_x + D_y\mathbf{a}_y + D_z\mathbf{a}_z)$$

We first consider the dot products of the unit vectors, discarding the six zero terms, and obtain the result that we recognize as the divergence of $\mathbf{D}$:

$$\nabla \cdot \mathbf{D} = \frac{\partial D_x}{\partial x} + \frac{\partial D_y}{\partial y} + \frac{\partial D_z}{\partial z} = \text{div}(\mathbf{D})$$

The use of $\nabla \cdot \mathbf{D}$ is much more prevalent than that of div $\mathbf{D}$, although both usages have their advantages. Writing $\nabla \cdot \mathbf{D}$ allows us to obtain simply and quickly the correct partial derivatives, but only in rectangular coordinates, as we will see. On the other hand, div $\mathbf{D}$ is an excellent reminder of the physical interpretation of divergence. We will use the operator notation $\nabla \cdot \mathbf{D}$ from now on to indicate the divergence operation.

The vector operator ∇ is used not only with divergence, but also with several other very important operations that appear later. One of these, the *gradient*, is ∇u, where u is any scalar field, and it leads to

$$\nabla u = \left(\frac{\partial}{\partial x}\mathbf{a}_x + \frac{\partial}{\partial y}\mathbf{a}_y + \frac{\partial}{\partial z}\mathbf{a}_z\right)u = \frac{\partial u}{\partial x}\mathbf{a}_x + \frac{\partial u}{\partial y}\mathbf{a}_y + \frac{\partial u}{\partial z}\mathbf{a}_z$$

The ∇ operator does not have a specific form in other coordinate systems. If we are considering $\mathbf{D}$ in cylindrical coordinates, then $\nabla \cdot \mathbf{D}$ still indicates the divergence of $\mathbf{D}$, or

$$\nabla \cdot \mathbf{D} = \frac{1}{\rho}\frac{\partial}{\partial \rho}(\rho\, D_\rho) + \frac{1}{\rho}\frac{\partial D_\phi}{\partial \phi} + \frac{\partial D_z}{\partial z}$$

where this expression has been taken from Section 3.5. We have no form for ∇ itself to help us obtain this sum of partial derivatives. This means that ∇u, as yet unnamed but easily written in rectangular coordinates, cannot be expressed by us at this time in cylindrical coordinates. Such an expression will be obtained when ∇u is defined in Chapter 4.

---

[4] This scalar operator D, which will not appear again, is not to be confused with the electric flux density.

3.5.3 Divergence Theorem

We close the treatment of divergence by presenting a theorem that brings the discussion full circle, the *divergence theorem*. This theorem applies to any vector field for which the appropriate partial derivatives exist, although it is easiest for us to develop it for the electric flux density. We have actually obtained it already and now have little more to do than point it out and name it, for starting from Gauss's law, we have

$$\oint_S \mathbf{D} \cdot d\mathbf{S} = Q = \int_{\text{vol}} \rho_v \, dv = \int_{\text{vol}} \nabla \cdot \mathbf{D} \, dv$$

The first and last expressions constitute the divergence theorem,

$$\boxed{\oint_S \mathbf{D} \cdot d\mathbf{S} = \int_{\text{vol}} \nabla \cdot \mathbf{D} \, dv} \qquad (17)$$

which may be stated as follows:

> *The integral of the normal component of any vector field over a closed surface is equal to the integral of the divergence of this vector field throughout the volume enclosed by the closed surface.*

The divergence theorem is also known as *Gauss's theorem*, and in fact Gauss's law as we have used it is nothing more than an application of the divergence theorem to electrostatics. Again, we emphasize that the theorem is true for any vector field, and we will have occasion later to apply it to several different fields. Its benefits derive from the fact that it relates a triple integration *throughout some volume* to a double integration *over the surface* of that volume. For example, it is much easier to look for leaks in a bottle full of some agitated liquid by inspecting the surface than by calculating the velocity at every internal point. It should also be pointed out that Eq. (17), as applied to the electric flux density, is Maxwell's first equation *in integral form*.

The divergence theorem becomes obvious physically if we consider a volume *v*, shown in cross section in Figure 3.7, which is surrounded by a closed surface *S*.

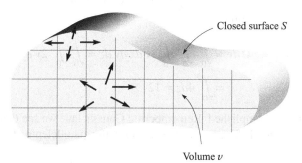

Closed surface *S*

Volume *v*

Figure 3.7 The divergence theorem states that the total flux crossing the closed surface is equal to the integral of the divergence of the flux density throughout the enclosed volume. The volume is shown here in cross section.

Division of the volume into a number of small compartments of differential size and consideration of one cell show that the flux diverging from such a cell *enters*, or *converges* on, the adjacent cells unless the cell contains a portion of the outer surface. In summary, the divergence of the flux density throughout a volume leads, then, to the same result as determining the net flux crossing the enclosing surface.

EXAMPLE 3.5

Evaluate both sides of the divergence theorem for the field $\mathbf{D} = 2xy\mathbf{a}_x + x^2\mathbf{a}_y$ C/m^2 and the rectangular parellelepiped formed by the planes $x = 0$ and 1, $y = 0$ and 2, and $z = 0$ and 3.

Solution. Evaluating the surface integral first, we note that $\mathbf{D}$ is parallel to the surfaces at $z = 0$ and $z = 3$, so $\mathbf{D} \cdot d\mathbf{S} = 0$ there. For the remaining four surfaces we have

$$\oint_S \mathbf{D} \cdot d\mathbf{S} = \int_0^3 \int_0^2 (\mathbf{D})_{x=0} \cdot (-dy\, dz\; \mathbf{a}_x) + \int_0^3 \int_0^2 (\mathbf{D})_{x=1} \cdot (dy\, dz\; \mathbf{a}_x)$$

$$+ \int_0^3 \int_0^1 (\mathbf{D})_{y=0} \cdot (-dx\, dz\; \mathbf{a}_y) + \int_0^3 \int_0^1 (\mathbf{D})_{y=2} \cdot (dx\, dz\; \mathbf{a}_y)$$

$$= -\int_0^3 \int_0^2 (D_x)_{x=0}\, dy\, dz + \int_0^3 \int_0^2 (D_x)_{x=1}\, dy\, dz$$

$$- \int_0^3 \int_0^1 (D_y)_{y=0}\, dx\, dz + \int_0^3 \int_0^1 (D_y)_{y=2}\, dx\, dz$$

However, $(D_x)_{x=0} = 0$, and $(D_y)_{y=0} = (D_y)_{y=2}$, which leaves only

$$\oint_S \mathbf{D} \cdot d\mathbf{S} = \int_0^3 \int_0^2 (D_x)_{x=1}\, dy\, dz = \int_0^3 \int_0^2 2y\, dy\, dz$$

$$= \int_0^3 4\, dz = 12$$

Since

$$\nabla \cdot \mathbf{D} = \frac{\partial}{\partial x}(2xy) + \frac{\partial}{\partial y}(x^2) = 2y$$

the volume integral becomes

$$\int_{\text{vol}} \nabla \cdot \mathbf{D}\, dv = \int_0^3 \int_0^2 \int_0^1 2y\, dx\, dy\, dz = \int_0^3 \int_0^2 2y\, dy\, dz$$

$$= \int_0^3 4\, dz = 12$$

and the check is accomplished. Remembering Gauss's law, we see that we have also determined that a total charge of 12 C lies within this parallelepiped.

> **D3.9.** Given the field $\mathbf{D} = 6\rho \sin\frac{1}{2}\phi\, \mathbf{a}_\rho + 1.5\rho \cos\frac{1}{2}\phi\, \mathbf{a}_\phi$ C/m^2, evaluate both sides of the divergence theorem for the region bounded by $\rho = 2$, $\phi = 0$, $\phi = \pi$, $z = 0$, and $z = 5$.
>
> **Ans.** 225; 225

REFERENCES

1. Kraus, J. D., and D. A. Fleisch. *Electromagnetics*. 5th ed. New York: McGraw-Hill, 1999. The static electric field in free space is introduced in Chapter 2.

2. Plonsey, R., and R. E. Collin. *Principles and Applications of Electromagnetic Fields*. New York: McGraw-Hill, 1961. The level of this text is somewhat higher than the one we are reading now, but it is an excellent text to read next. Gauss's law appears in the second chapter.

3. Plonus, M. A. *Applied Electromagnetics*. New York: McGraw-Hill, 1978. This book contains rather detailed descriptions of many practical devices that illustrate electromagnetic applications. For example, see the discussion of xerography on pp. 95–98 as an electrostatics application.

4. Skilling, H. H. *Fundamentals of Electric Waves*. 2d ed. New York: John Wiley & Sons, 1948. The operations of vector calculus are well illustrated. Divergence is discussed on pp. 22 and 38. Chapter 1 is interesting reading.

5. Thomas, G. B., Jr., and R. L. Finney. (See Suggested References for Chapter 1.) The divergence theorem is developed and illustrated from several different points of view on pp. 976–980.

CHAPTER 3 PROBLEMS

3.1 Suppose that the Faraday concentric sphere experiment is performed in free space using a central charge at the origin, Q_1, and with hemispheres of radius a. A second charge Q_2 (this time a point charge) is located at distance R from Q_1, where $R \gg a$. (*a*) What is the force on the point charge before the hemispheres are assembled around Q_1? (*b*) What is the force on the point charge after the hemispheres are assembled but before they are discharged? (*c*) What is the force on the point charge after the hemispheres are assembled and after they are discharged? (*d*) Qualitatively, describe what happens as Q_2 is moved toward the sphere assembly to the extent that the condition $R \gg a$ is no longer valid.

3.2 An electric field in free space is $\mathbf{E} = (5z^2/\epsilon_0)\hat{\mathbf{a}}_z$ V/m. Find the total charge contained within a cube, centered at the origin, of 4-m side length, in which all sides are parallel to coordinate axes (and therefore each side intersects an axis at ± 2).

3.3 Consider an electric dipole in free space, consisting of point charge q at location $z = +d/2$, and point charge $-q$ at location $z = -d/2$. The electric field intensity in the xy plane is (see Problem 2.7):

$$\mathbf{E} = \frac{-qd\,\mathbf{a}_z}{4\pi\epsilon_0 [\rho^2 + (d/2)^2]^{3/2}}$$

where ρ is the radius from the origin in cylindrical coordinates.
(*a*) Determine the net electric flux associated with this field that penetrates the xy plane. (*b*) Interpret your result as it relates to Gauss's law.

3.4 An electric field in free space is $\mathbf{E} = (5z^3/\epsilon_0)\,\hat{\mathbf{a}}_z$ V/m. Find the total charge contained within a sphere of 3-m radius, centered at the origin.

3.5 A volume charge distribution in free space is characterized by the density

$$\rho_v = \frac{q}{2Ad}\exp(-|z|/d)$$

where d is a distance along z, A is the area of a surface parallel to the xy plane, and q is a fixed charge quantity. The charge distribution exists everywhere. (a) Find the electric field intensity, $\mathbf{E}$, everywhere. (b) What is the interpretation of q?

3.6 In free space, a volume charge of constant density $\rho_v = \rho_0$ exists within the region $-\infty < x < \infty$, $-\infty < y < \infty$, and $-d/2 < z < d/2$. Find $\mathbf{D}$ and $\mathbf{E}$ everywhere.

3.7 A spherically symmetric charge distribution in free space is characterized by the charge density

$$\rho_v = \frac{qb}{r^2}\exp(-br)\quad \text{C/m}^3 \quad (0 < r < \infty)$$

(a) Find the electric field intensity, $\mathbf{E}(r)$, everywhere. (b) Find the total charge present.

3.8 Use Gauss's law in integral form to show that an inverse distance field in spherical coordinates, $\mathbf{D} = Aa_r/r$, where A is a constant, requires every spherical shell of 1 m thickness to contain $4\pi A$ coulombs of charge. Does this indicate a continuous charge distribution? If so, find the charge density variation with r.

3.9 A sphere of radius a in free space contains charge of density $\rho_v = \rho_0\, r/a$, where ρ_0 is a constant. (a) Find the electric field intensity, $\mathbf{E}_{\text{I}}$, inside the sphere. (b) Find the electric field intensity, $\mathbf{E}_{\text{II}}$, outside the sphere. (c) A spherical shell of radius b is positioned concentrically around the sphere. What surface charge density, ρ_s, must exist on the shell so that the electric field at locations $r > b$ is zero? (d) What electrostatic force per unit area is exerted by the solid sphere on the spherical shell?

3.10 An infinitely long cylindrical dielectric of radius b contains charge within its volume of density $\rho_v = a\rho^2$, where a is a constant. Find the electric field strength, $\mathbf{E}$, both inside and outside the cylinder.

3.11 Consider a cylindrical charge distribution having infinite length in z, but which has a radial dependence in charge density given by the gaussian, $\rho_v(\rho) = \rho_0\exp[-(\rho/b)^2]$. (a) Find the electric field intensity, $\mathbf{E}$, at large radii, $\rho \gg b$. This enables the enclosed charge integral in Gauss's law to be approximated using an infinite upper limit in radius. (b) Compare your result to the field outside a charged cylinder of radius b containing uniform charge density ρ_0.

3.12 The sun radiates a total power of about 3.86×10^{26} watts (W). If we imagine the sun's surface to be marked off in latitude and longitude and assume uniform radiation, (*a*) what power is radiated by the region lying between latitude 50° N and 60° N and longitude 12° W and 27° W? (*b*) What is the power density on a spherical surface 93,000,000 miles from the sun in W/m²?

3.13 Spherical surfaces at $r = 2$, 4, and 6 m carry uniform surface charge densities of 20 nC/m², −4 nC/m², and ρ_{S0}, respectively. (*a*) Find **D** at $r = 1$, 3, and 5 m. (*b*) Determine ρ_{S0} such that **D** = 0 at $r = 7$ m.

3.14 A certain light-emitting diode (LED) is centered at the origin with its surface in the *xy* plane. At far distances, the LED appears as a point, but the glowing surface geometry produces a far-field radiation pattern that follows a raised cosine law: that is, the optical power (flux) density in W/m² is given in spherical coordinates by

$$P_d = P_0 \frac{\cos^2\theta}{2\pi r^2} \mathbf{a}_r \quad \text{W/m}^2$$

where θ is the angle measured with respect to the direction that is normal to the LED surface (in this case, the *z* axis), and *r* is the radial distance from the origin at which the power is detected. (*a*) In terms of P_0, find the total power in watts emitted in the upper half-space by the LED. (*b*) Find the cone angle, θ_1, within which half the total power is radiated, that is, within the range $0 < \theta < \theta_1$. (*c*) An optical detector, having a 1-mm² cross-sectional area, is positioned at $r = 1$ m and at $\theta = 45°$, such that it faces the LED. If one nanowatt is measured by the detector, what (to a very good estimate) is the value of P_0?

3.15 Volume charge density is located as follows: $\rho_v = 0$ for $\rho < 1$ mm and for $\rho > 2$ mm, $\rho_v = 4\rho$ μC/m³ for $1 < \rho < 2$ mm. (*a*) Calculate the total charge in the region $0 < \rho < \rho_1$, $0 < z < L$, where $1 < \rho_1 < 2$ mm. (*b*) Use Gauss's law to determine D_ρ at $\rho = \rho_1$. (*c*) Evaluate D_ρ at $\rho = 0.8$ mm, 1.6 mm, and 2.4 mm.

3.16 An electric flux density is given by $\mathbf{D} = D_0\, \mathbf{a}_\rho$, where D_0 is a given constant. (*a*) What charge density generates this field? (*b*) For the specified field, what total charge is contained within a cylinder of radius *a* and height *b*, where the cylinder axis is the *z* axis?

3.17 In a region having spherical symmetry, volume charge is distributed according to:

$$\rho_v(r) = \rho_0 \frac{\sin(\pi r/a)}{r^2} \quad \text{C/m}^3$$

Find the surfaces on which **E** = 0.

3.18 State whether the divergence of the following vector fields is positive, negative, or zero: (*a*) the thermal energy flow in J/(m² − s) at any point in a freezing ice cube; (*b*) the current density in A/m² in a bus bar carrying direct current; (c) the mass flow rate in kg/(m^2 – s) below the surface of water in a basin, in which the water is circulating clockwise as viewed from above.

3.19 A spherical surface of radius 3 mm is centered at $P(4, 1, 5)$ in free space. Let $\mathbf{D} = x\mathbf{a}_x$ C/m^2. Use the results of Section 3.4 to estimate the net electric flux leaving the spherical surface.

3.20 A radial electric field distribution in free space is given in spherical coordinates as:

$$\mathbf{E}_1 = \frac{r\rho_0}{3\epsilon_0}\,\mathbf{a}_r \qquad (r \leq a)$$

$$\mathbf{E}_2 = \frac{(2a^3 - r^3)\rho_0}{3\epsilon_0 r^2}\,\mathbf{a}_r \qquad (a \leq r \leq b)$$

$$\mathbf{E}_3 = \frac{(2a^3 - b^3)\rho_0}{3\epsilon_0 r^2}\,\mathbf{a}_r \qquad (r \geq b)$$

where ρ_0, a, and b are constants. (a) Determine the volume charge density in the entire region $(0 \leq r \leq \infty)$ by the appropriate use of $\nabla \cdot \mathbf{D} = \rho_v$. (b) In terms of given parameters, find the total charge, Q, within a sphere of radius r where $r > b$.

3.21 In a region exhibiting spherical symmetry, electric flux density is found to be $\mathbf{D}_1 = \rho_0\, r/3\,\mathbf{a}_r\,(0 < r < a)$, $\mathbf{D}_2 = 0\,(a < r < b)$, and $\mathbf{D}_3 = (a^3\rho_0)/(3r^2)\,\mathbf{a}_r\,(r < b)$. (a) Find the charge configuration that would produce the given field. (b) What total charge is present?

3.22 (a) A flux density field is given as $\mathbf{F}_1 = 5\mathbf{a}_z$. Evaluate the outward flux of $\mathbf{F}_1$ through the hemispherical surface, $r = a, 0 < \theta < \pi/2, 0 < \phi < 2\pi$. (b) What simple observation would have saved a lot of work in part a? (c) Now suppose the field is given by $\mathbf{F}_2 = 5z\mathbf{a}_z$. Using the appropriate surface integrals, evaluate the net outward flux of $\mathbf{F}_2$ through the closed surface consisting of the hemisphere of part a and its circular base in the xy plane. (d) Repeat part c by using the divergence theorem and an appropriate volume integral.

3.23 (a) A point charge Q lies at the origin. Show that div $\mathbf{D}$ is zero everywhere except at the origin. (b) Replace the point charge with a uniform volume charge density ρ_{v0} for $0 < r < a$. Relate ρ_{v0} to Q and a so that the total charge is the same. Find div $\mathbf{D}$ everywhere.

3.24 In a region in free space, electric flux density is found to be

$$\mathbf{D} = \begin{cases} \rho_0(z + 2d)\,\mathbf{a}_z \text{ C/}m^2 & (-2d \leq z \leq 0) \\ -\rho_0(z - 2d)\,\mathbf{a}_z \text{ C/}m^2 & (0 \leq z \leq 2d) \end{cases}$$

Everywhere else, $\mathbf{D} = 0$. (a) Using $\nabla \cdot \mathbf{D} = \rho_v$, find the volume charge density as a function of position everywhere. (b) Determine the electric flux that passes through the surface defined by $z = 0$, $-a \leq x \leq a$, $-b \leq y \leq b$. (c) Determine the total charge contained within the region $-a \leq x \leq a$, $-b \leq y \leq b$, $-d \leq z \leq d$. (d) Determine the total charge contained within the region $-a \leq x \leq a$, $-b \leq y \leq b$, $0 \leq z \leq 2d$.

3.25 Within the spherical shell, $3 < r < 4$ m, the electric flux density is given as $\mathbf{D} = 5(r - 3)^3 \, \mathbf{a}_r$ C/m^2. (a) What is the volume charge density at $r = 4$? (b) What is the electric flux density at $r = 4$? (c) How much electric flux leaves the sphere $r = 4$? (d) How much charge is contained within the sphere $r = 4$?

3.26 If we have a perfect gas of mass density ρ_m kg/m^3, and we assign a velocity $\mathbf{U}$ m/s to each differential element, then the mass flow rate is $\rho_m \mathbf{U}$ kg/ (m^2 − s). Physical reasoning then leads to the *continuity equation*, $\nabla \cdot (\rho_m \mathbf{U}) = -\partial \rho_m / \partial t$. (a) Explain in words the physical interpretation of this equation. (b) Show that $\oint_s \rho_m \mathbf{U} \cdot d\mathbf{S} = -dM/dt$, where M is the total mass of the gas within the constant closed surface S, and explain the physical significance of the equation.

3.27 Consider a slab of material containing a volume charge distribution throughout. The slab is of length d in the z direction, and its dimensions in x and y represent a cross-sectional area of A. Free space permittivity exists throughout. The electric field in the slab is given by

$$\mathbf{E} = \frac{\rho_0}{\varepsilon_0 \alpha} \exp(-\alpha z) \, \mathbf{a}_z \quad \text{V/m}$$

where ρ_0 is a positive constant. (a) Find the volume charge density ρ_v in the slab. (b) Find the total charge in the slab. (c) Verify your result for part b by evaluating the net outward flux of $\mathbf{D}$ through the slab surfaces.

3.28 Repeat Problem 3.8, but use $\nabla \cdot \mathbf{D} = \rho_v$ and take an appropriate volume integral.

3.29 In the region of free space that includes the volume $2 < x, y, z < 3$, $\mathbf{D} = \frac{2}{z^2}\left(yz \, \mathbf{a}_x + xz \, \mathbf{a}_y - 2xy \, \mathbf{a}_z\right)$ C/m^2. (a) Evaluate the volume integral side of the divergence theorem for the volume defined here. (b) Evaluate the surface integral side for the corresponding closed surface.

3.30 (a) Use Maxwell's first equation, $\nabla \cdot \mathbf{D} = \rho_v$, to describe the variation of the electric field intensity with x in a region in which no charge density exists and in which a nonhomogeneous dielectric has a permittivity that increases exponentially with x. The field has an x component only; (b) repeat part a, but with a radially directed electric field (spherical coordinates) in which again $\rho_v = 0$, but in which the permittivity *decreases* exponentially with r.

3.31 Given the flux density $\mathbf{D} = \frac{16}{r} \cos(2\theta) \, \mathbf{a}_\theta$ C/m^2, use two different methods to find the total charge within the region $1 < r < 2$ m, $1 < \theta < 2$ rad, $1 < \phi < 2$ rad.

CHAPTER 4 Energy and Potential

术语

potential 电位，电位的值依赖于参考点，即参考点不同电位值也不同，单位是伏［特］（V）。

potential difference 电位差。

conservative field 守恒场，也叫保守场。

gradient 梯度，对一个标量函数求梯度，结果是一个矢量函数。

dipole 偶极子。

dipole moment 偶极矩。

electrostatic field 静电场，涉及的是静止电荷所产生的电场。

In Chapters 2 and 3 we became acquainted with Coulomb's law and its use in finding the electric field about several simple distributions of charge, and also with Gauss's law and its application in determining the field about some symmetrical charge arrangements. The use of Gauss's law was invariably easier for these highly symmetrical distributions because the problem of integration always disappeared when the proper closed surface was chosen.

If we had tried to find a more complicated field, such as that of two unlike point charges separated by a small distance, it would have been impossible to choose a suitable gaussian surface and obtain an answer. Coulomb's law, however, is more powerful and enables us to solve problems for which Gauss's law is not applicable. The application of Coulomb's law is laborious, detailed, and often quite complex; the reason for this is that the electric field intensity, a vector field, must be found directly from the charge distribution. Three different integrations are needed in general, one for each component, and the resolution of the vector into components usually adds to the complexity of the integrals.

It would be desirable if we could find some as-yet-undefined scalar function with a single integration and then determine the electric field from this scalar by some simple, straightforward procedure, such as differentiation.

This scalar function does exist and is known as the *potential* or *potential field*. We will find that it has a very real physical interpretation and is more familiar to most of us than the electric field it will be used to find.

We should expect, then, to be equipped soon with a third method of finding electric fields—a single scalar integration, although not always as simple as we might wish, followed by a pleasant differentiation. ■

4.1 ENERGY EXPENDED IN MOVING A POINT CHARGE IN AN ELECTRIC FIELD

The electric field intensity was defined as the force on a unit test charge at that point at which we wish to find the value of this vector field. If the test charge is moved against the electric field, we have to exert a force equal and opposite to that exerted by the field, and this requires us to expend energy or do work. If we wish to move the charge in the direction of the field, our energy expenditure turns out to be negative; we do not do the work, the field does.

Suppose we wish to move a charge Q a distance $d\mathbf{L}$ in an electric field $\mathbf{E}$. The force on Q arising from the electric field is

$$\boxed{\mathbf{F}_E = Q\mathbf{E}} \tag{1}$$

where the subscript is a reminder that this force arises from the field. The component of this force in the direction $d\mathbf{L}$ which must be overcome is

$$F_{EL} = \mathbf{F} \cdot \mathbf{a}_L = Q\mathbf{E} \cdot \mathbf{a}_L$$

where $\mathbf{a}_L$ = a unit vector in the direction of $d\mathbf{L}$.

The force that must be applied is equal and opposite to the force associated with the field,

$$F_{\text{appl}} = -Q\mathbf{E} \cdot \mathbf{a}_L$$

and the expenditure of energy is the product of the force and distance. That is, the differential work done by an external source moving charge Q is $dW = -Q\mathbf{E} \cdot \mathbf{a}_L dL$,

or

$$\boxed{dW = -Q\mathbf{E} \cdot d\mathbf{L}} \tag{2}$$

where we have replaced $\mathbf{a}_L dL$ by the simpler expression $d\mathbf{L}$.

This differential amount of work required may be zero under several conditions determined easily from Eq. (2). There are the trivial conditions for which $\mathbf{E}$, Q, or $d\mathbf{L}$ is zero, and a much more important case in which $\mathbf{E}$ and $d\mathbf{L}$ are perpendicular. Here the charge is moved always in a direction at right angles to the electric field. We can draw on a good analogy between the electric field and the gravitational field, where, again, energy must be expended to move against the field. Sliding a mass around with constant velocity on a frictionless surface is an effortless process if the mass is moved along a constant elevation contour; positive or negative work must be done in moving it to a higher or lower elevation, respectively.

Returning to the charge in the electric field, the work required to move the charge a finite distance must be determined by integrating,

$$\boxed{W = -Q\int_{\text{init}}^{\text{final}} \mathbf{E} \cdot d\mathbf{L}} \tag{3}$$

导读

1. 电场对带点粒子有作用力可以做功，表示电场具有能量属性。电场的能量可以通过电荷与电位或场强与电位移来表示。能量的单位是焦[耳]（J）。

2. 电场强度与电位的关系：电场强度等于电位的负梯度；沿指定方向上的电场强度分量等于电位函数沿该方向的方向导数。一点的电位等于从该点开始到参考点的任意路径上的电场强度的线积分。静电场中两点之间的电位差或电压定义为电场使单位正电荷从一点移动到另一点所做的功，其量值等于从一点到另一点沿任意曲线对电场强度的线积分，即该积分值只与两点的位置有关，而与积分的路径无关。但在时变电磁场中，沿不同路径对电场强度的积分值将不同，故两点电压不再唯一。

3. 一个标量函数可以求梯度，其结果为一个矢量函数；函数在一点的梯度值等于函数在该点的最大方向导数，即所有方向上的那个最大变化率，而梯度矢量的方向为函数有最大变化率的方向。无论是梯度还是散度与旋度，其实都是一种微分运算。

Interactives

where the path must be specified before the integral can be evaluated. The charge is assumed to be at rest at both its initial and final positions.

This definite integral is basic to field theory, and we shall devote the following section to its interpretation and evaluation.

> **D4.1.** Given the electric field $\mathbf{E} = \frac{1}{z^2}(8xyz\,\mathbf{a}_x + 4x^2z\,\mathbf{a}_y - 4x^2y\,\mathbf{a}_z)$V/m, find the differential amount of work done in moving a 6-nC charge a distance of 2 μm, starting at $P(2, -2, 3)$ and proceeding in the direction $\mathbf{a}_L = (a) -\frac{6}{7}\mathbf{a}_x + \frac{3}{7}\mathbf{a}_y + \frac{2}{7}\mathbf{a}_z$; $(b) \frac{6}{7}\mathbf{a}_x - \frac{3}{7}\mathbf{a}_y - \frac{2}{7}\mathbf{a}_z$; $(c) \frac{3}{7}\mathbf{a}_x + \frac{6}{7}\mathbf{a}_y$.
>
> **Ans.** (a) −149.3 fJ; (b) 149.3 fJ; (c) 0

4.2 THE LINE INTEGRAL

The integral expression for the work done in moving a point charge Q from one position to another, Eq. (3), is an example of a line integral, which in vector-analysis notation always takes the form of the integral along some prescribed path of the dot product of a vector field and a differential vector path length $d\mathbf{L}$. Without using vector analysis we write

$$W = -Q\int_{\text{init}}^{\text{final}} E_L dL$$

where E_L = component of **E** along $d\mathbf{L}$.

A line integral is like many other integrals which appear in advanced analysis, including the surface integral appearing in Gauss's law, in that it is essentially descriptive. It tells us to choose a path, break it up into a large number of very small segments, multiply the component of the field along each segment by the length of the segment, and then add the results for all the segments. This is a summation, of course, and the integral is obtained exactly only when the number of segments becomes infinite.

This procedure is indicated in Figure 4.1, where a path has been chosen from an initial position B to a final position[1] A and a *uniform electric field* is selected for simplicity. The path is divided into six segments, $\Delta\mathbf{L}_1, \Delta\mathbf{L}_2, \ldots, \Delta\mathbf{L}_6$, and the components of **E** along each segment are denoted by $E_{L1}, E_{L2}, \ldots, E_{L6}$. The work involved in moving a charge Q from B to A is then approximately

$$W = -Q(E_{L1}\Delta L_1 + E_{L2}\Delta L_2 + \cdots + E_{L6}\Delta L_6)$$

or, using vector notation,

$$W = -Q(\mathbf{E}_1 \cdot \Delta\mathbf{L}_1 + \mathbf{E}_2 \cdot \Delta\mathbf{L}_2 + \cdots + \mathbf{E}_6 \cdot \Delta\mathbf{L}_6)$$

---

[1] The final position is given the designation A to correspond with the convention for potential difference, as discussed in the following section.

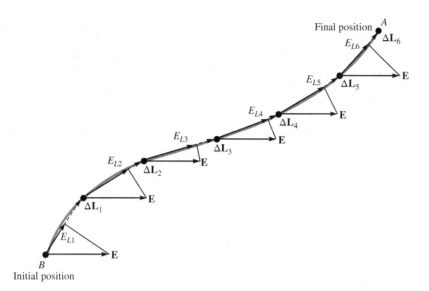

Figure 4.1 A graphical interpretation of a line integral in a uniform field. The line integral of **E** between points B and A is independent of the path selected, even in a nonuniform field; this result is not, in general, true for time-varying fields.

and because we have assumed a uniform field,

$$\mathbf{E}_1 = \mathbf{E}_2 = \cdots = \mathbf{E}_6$$
$$W = -Q\mathbf{E} \cdot (\Delta\mathbf{L}_1 + \Delta\mathbf{L}_2 + \cdots + \Delta\mathbf{L}_6)$$

What is this sum of vector segments in the preceding parentheses? Vectors add by the parallelogram law, and the sum is just the vector directed from the initial point B to the final point A, $\mathbf{L}_{BA}$. Therefore

$$W = -Q\mathbf{E} \cdot \mathbf{L}_{BA} \qquad \text{(uniform } \mathbf{E}) \tag{4}$$

Remembering the summation interpretation of the line integral, this result for the uniform field can be obtained rapidly now from the integral expression

$$W = -Q \int_B^A \mathbf{E} \cdot d\mathbf{L} \tag{5}$$

as applied to a uniform field

$$W = -Q\mathbf{E} \cdot \int_B^A d\mathbf{L}$$

where the last integral becomes $\mathbf{L}_{BA}$ and

$$W = -Q\mathbf{E} \cdot \mathbf{L}_{BA} \qquad \text{(uniform } \mathbf{E})$$

For this special case of a uniform electric field intensity, we should note that the work involved in moving the charge depends only on Q, $\mathbf{E}$, and $\mathbf{L}_{BA}$, a vector drawn from the initial to the final point of the path chosen. It does not depend on the path we choose to carry the charge. Whether we proceed from B to A on a straight line or via the Old Chisholm Trail, the answer is the same. We show in Section 4.5 that an identical statement may be made for any nonuniform (static) $\mathbf{E}$ field.

A few examples follow that illustrate the mechanics of setting up the line integral in Eq. (5).

EXAMPLE 4.1

We are given the nonuniform field

$$\mathbf{E} = y\mathbf{a}_x + x\mathbf{a}_y + 2\mathbf{a}_z$$

and we are asked to determine the work expended in carrying 2 C from $B(1, 0, 1)$ to $A(0.8, 0.6, 1)$ along the shorter arc of the circle

$$x^2 + y^2 = 1 \quad z = 1$$

Solution. Use $W = -Q\int_B^A \mathbf{E} \cdot d\mathbf{L}$, where $\mathbf{E}$ is not necessarily constant. Working in rectangular coordinates, the differential path $d\mathbf{L}$ is $dx\mathbf{a}_x + dy\mathbf{a}_y + dz\mathbf{a}_z$, and the integral becomes

$$W = -Q\int_B^A \mathbf{E} \cdot d\mathbf{L}$$

$$= -2\int_B^A (y\mathbf{a}_x + x\mathbf{a}_y + 2\mathbf{a}_z) \cdot (dx\ \mathbf{a}_x + dy\ \mathbf{a}_y + dz\ \mathbf{a}_z)$$

$$= -2\int_1^{0.8} y\,dx - 2\int_0^{0.6} x\,dy - 4\int_1^1 dz$$

where the limits on the integrals have been chosen to agree with the initial and final values of the appropriate variable of integration. Using the equation of the circular path (and selecting the sign of the radical which is correct for the quadrant involved), we have

$$W = -2\int_1^{0.8} \sqrt{1 - x^2}\,dx - 2\int_0^{0.6} \sqrt{1 - y^2}\,dy - 0$$

$$= -\left[x\sqrt{1 - x^2} + \sin^{-1}x\right]_1^{0.8} - \left[y\sqrt{1 - y^2} + \sin^{-1}y\right]_0^{0.6}$$

$$= -(0.48 + 0.927 - 0 - 1.571) - (0.48 + 0.644 - 0 - 0)$$

$$= -0.96\ \text{J}$$

EXAMPLE 4.2

Again find the work required to carry 2 C from B to A in the same field, but this time use the straight-line path from B to A.

Solution. Begin by determining the equations of the straight line. Any two of the following three equations for planes passing through the line are sufficient to define the line:

$$y - y_B = \frac{y_A - y_B}{x_A - x_B}(x - x_B)$$

$$z - z_B = \frac{z_A - z_B}{y_A - y_B}(y - y_B)$$

$$x - x_B = \frac{x_A - x_B}{z_A - z_B}(z - z_B)$$

From the first equation we have

$$y = -3(x - 1)$$

and from the second we obtain

$$z = 1$$

Thus,

$$W = -2\int_1^{0.8} y\,dx - 2\int_0^{0.6} x\,dy - 4\int_1^{1} dz$$

$$= 6\int_1^{0.8}(x - 1)\,dx - 2\int_0^{0.6}\left(1 - \frac{y}{3}\right)dy$$

$$= -0.96\,\text{J}$$

This is the same answer we found using the circular path between the same two points, and it again demonstrates the statement (unproved) that the work done is independent of the path taken in any electrostatic field.

It should be noted that the equations of the straight line show that $dy = -3dx$ and $dx = -\frac{1}{3}dy$. These substitutions may be made in the first two integrals, along with a change in limits, and the answer may be obtained by evaluating the new integrals. This method is often simpler if the integrand is a function of only one variable.

Note that the expressions for $d\mathbf{L}$ in our three coordinate systems use the differential lengths obtained in Chapter 1 (rectangular in Section 1.3, cylindrical in Section 1.8, and spherical in Section 1.9):

$d\mathbf{L} = dx\,\mathbf{a}_x + dy\,\mathbf{a}_y + dz\,\mathbf{a}_z$	(rectangular)	(6)
$d\mathbf{L} = d\rho\,\mathbf{a}_\rho + \rho\,d\phi\,\mathbf{a}_\phi + dz\,\mathbf{a}_z$	(cylindrical)	(7)
$d\mathbf{L} = dr\,\mathbf{a}_r + r\,d\theta\,\mathbf{a}_\theta + r\sin\theta\,d\phi\,\mathbf{a}_\phi$	(spherical)	(8)

The interrelationships among the several variables in each expression are determined from the specific equations for the path.

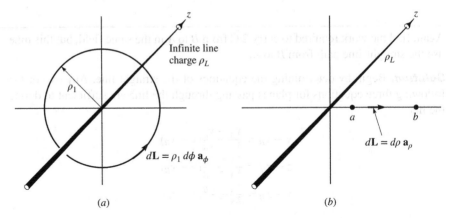

Figure 4.2   (*a*) A circular path and (*b*) a radial path along which a charge of *Q* is carried in the field of an infinite line charge. No work is expected in the former case.

As a final illustration, we investigate several paths that we might take near an infinite line charge. The field has been obtained several times and is entirely in the radial direction,

$$\mathbf{E} = E_\rho \mathbf{a}_\rho = \frac{\rho_L}{2\pi\epsilon_0\rho}\mathbf{a}_\rho$$

First we find the work done in carrying the positive charge Q about a circular path of radius ρ_b, centered at the line charge, as illustrated in Figure 4.2*a*. Without lifting a pencil, we see that the work must be nil, for the path is always perpendicular to the electric field intensity, or the force on the charge is always exerted at right angles to the direction in which we are moving it. For practice, however, we will set up the integral and obtain the answer.

The differential element $d\mathbf{L}$ is chosen in cylindrical coordinates, and the circular path selected demands that $d\rho$ and dz be zero, so $d\mathbf{L} = \rho_1\,d\phi\,\mathbf{a}_\phi$. The work is then

$$W = -Q\int_{\text{init}}^{\text{final}} \frac{\rho_L}{2\pi\epsilon_0\rho_1}\mathbf{a}_\rho \cdot \rho_1\,d\phi\,\mathbf{a}_\phi$$

$$= -Q\int_0^{2\pi} \frac{\rho_L}{2\pi\epsilon_0}d\phi\,\mathbf{a}_\rho \cdot \mathbf{a}_\phi = 0$$

We will now carry the charge from $\rho = a$ to $\rho = b$ along a radial path (Figure 4.2*b*). Here $d\mathbf{L} = d\rho\,\mathbf{a}_\rho$ and

$$W = -Q\int_{\text{init}}^{\text{final}} \frac{\rho_L}{2\pi\epsilon_0\rho}\mathbf{a}_\rho \cdot d\rho\,\mathbf{a}_\rho = -Q\int_a^b \frac{\rho_L}{2\pi\epsilon_0}\frac{d\rho}{\rho}$$

or

$$W = -\frac{Q\rho_L}{2\pi\epsilon_0}\ln\frac{b}{a}$$

Because b is larger than a, ln (b/a) is positive, and the work done is negative, indicating that the external source that is moving the charge receives energy.

One of the pitfalls in evaluating line integrals is a tendency to use too many minus signs when a charge is moved in the direction of a *decreasing* coordinate value. This is taken care of completely by the limits on the integral, and no misguided attempt should be made to change the sign of $d\mathbf{L}$. Suppose we carry Q from b to a (Figure 4.2b). We still have $d\mathbf{L} = d\rho\,\mathbf{a}_\rho$ and show the different direction by recognizing $\rho = b$ as the initial point and $\rho = a$ as the final point,

$$W = -Q\int_b^a \frac{\rho_L}{2\pi\,\epsilon_0}\frac{d\rho}{\rho} = \frac{Q\rho_L}{2\pi\,\epsilon_0}\ln\frac{b}{a}$$

This is the negative of the previous answer and is obviously correct.

D4.2. Calculate the work done in moving a 4-C charge from $B(1, 0, 0)$ to $A(0, 2, 0)$ along the path $y = 2 - 2x$, $z = 0$ in the field $\mathbf{E} = (a)\ 5\mathbf{a}_x$V/m; $(b)\ 5x\mathbf{a}_x$V/m; $(c)\ 5x\mathbf{a}_x + 5y\mathbf{a}_y$V/m.

Ans. (a) 20 J; (b) 10 J; (c) −30 J

D4.3. We will see later that a time-varying $\mathbf{E}$ field need not be conservative. (If it is not conservative, the work expressed by Eq. (3) may be a function of the path used.) Let $\mathbf{E} = y\mathbf{a}_x$V/m at a certain instant of time, and calculate the work required to move a 3-C charge from $(1, 3, 5)$ to $(2, 0, 3)$ along the straight-line segments joining: (a) $(1, 3, 5)$ to $(2, 3, 5)$ to $(2, 0, 5)$ to $(2, 0, 3)$; (b) $(1, 3, 5)$ to $(1, 3, 3)$ to $(1, 0, 3)$ to $(2, 0, 3)$.

Ans. (a) −9 J; (b) 0

4.3 DEFINITION OF POTENTIAL DIFFERENCE AND POTENTIAL

We are now ready to define a new concept from the expression for the work done by an external source in moving a charge Q from one point to another in an electric field $\mathbf{E}$, "Potential difference and work."

$$W = -Q\int_{\text{init}}^{\text{final}} \mathbf{E}\cdot d\mathbf{L}$$

In much the same way as we defined the electric field intensity as the force on a *unit* test charge, we now define *potential difference V* as the work done (by an external source) in moving a *unit* positive charge from one point to another in an electric field,

$$\boxed{\text{Potential difference} = V = -\int_{\text{init}}^{\text{final}} \mathbf{E}\cdot d\mathbf{L}}$$

(9)

We have to agree on the direction of movement, and we do this by stating that V_{AB} signifies the potential difference between points A and B and is the work done in moving the unit charge from B (last named) to A (first named). Thus, in determining V_{AB}, B is the initial point and A is the final point. The reason for this somewhat peculiar definition will become clearer shortly, when it is seen that the initial point B is often taken at infinity, whereas the final point A represents the fixed position of the charge; point A is thus inherently more significant.

Potential difference is measured in joules per coulomb, for which the *volt* is defined as a more common unit, abbreviated as V. Hence the potential difference between points A and B is

$$\boxed{V_{AB} = -\int_B^A \mathbf{E} \cdot d\mathbf{L} \ \text{V}} \tag{10}$$

and V_{AB} is positive if work is done in carrying the positive charge from B to A.

From the line-charge example of Section 4.2 we found that the work done in taking a charge Q from $\rho = b$ to $\rho = a$ was

$$W = \frac{Q\rho_L}{2\pi\epsilon_0}\ln\frac{b}{a}$$

Thus, the potential difference between points at $\rho = a$ and $\rho = b$ is

$$V_{ab} = \frac{W}{Q} = \frac{\rho_L}{2\pi\epsilon_0}\ln\frac{b}{a} \tag{11}$$

We can try out this definition by finding the potential difference between points A and B at radial distances r_A and r_B from a point charge Q. Choosing an origin at Q,

$$\mathbf{E} = E_r\mathbf{a}_r = \frac{Q}{4\pi\epsilon_0 r^2}\mathbf{a}_r$$

and

$$d\mathbf{L} = dr\,\mathbf{a}_r$$

we have

$$V_{AB} = -\int_B^A \mathbf{E}\cdot d\mathbf{L} = -\int_{r_B}^{r_A}\frac{Q}{4\pi\,\epsilon_0 r^2}\,dr = \frac{Q}{4\pi\,\epsilon_0}\left(\frac{1}{r_A} - \frac{1}{r_B}\right) \tag{12}$$

If $r_B > r_A$, the potential difference V_{AB} is positive, indicating that energy is expended by the external source in bringing the positive charge from r_B to r_A. This agrees with the physical picture showing the two like charges repelling each other.

It is often convenient to speak of the *potential*, or *absolute potential*, of a point, rather than the potential difference between two points, but this means only that we agree to measure every potential difference with respect to a specified reference point that we consider to have zero potential. Common agreement must be reached on the zero reference before a statement of the potential has any significance. A person having one hand on the deflection plates of a cathode-ray tube that are "at a potential of 50 V" and the other hand on the cathode terminal would probably be too shaken up to understand that the cathode is not the zero reference, but that all potentials in that circuit are customarily measured with respect to the metallic shield about the tube. The cathode may be several thousands of volts negative with respect to the shield.

Perhaps the most universal zero reference point in experimental or physical potential measurements is "ground," by which we mean the potential of the surface region of the earth itself. Theoretically, we usually represent this surface by an infinite plane at zero potential, although some large-scale problems, such as those involving propagation across the Atlantic Ocean, require a spherical surface at zero potential.

Another widely used reference "point" is infinity. This usually appears in theoretical problems approximating a physical situation in which the earth is relatively far removed from the region in which we are interested, such as the static field near the wing tip of an airplane that has acquired a charge in flying through a thunderhead, or the field inside an atom. Working with the *gravitational* potential field on earth, the zero reference is normally taken at sea level; for an interplanetary mission, however, the zero reference is more conveniently selected at infinity.

A cylindrical surface of some definite radius may occasionally be used as a zero reference when cylindrical symmetry is present and infinity proves inconvenient. In a coaxial cable the outer conductor is selected as the zero reference for potential. And, of course, there are many special problems, such as those for which a two-sheeted hyperboloid or an oblate spheroid must be selected as the zero-potential reference, but these need not concern us immediately.

If the potential at point A is V_A and that at B is V_B, then

$$\boxed{V_{AB} = V_A - V_B} \tag{13}$$

where we necessarily agree that V_A and V_B shall have the same zero reference point.

D4.4. An electric field is expressed in rectangular coordinates by $\mathbf{E} = 6x^2\mathbf{a}_x + 6y\mathbf{a}_y + 4\mathbf{a}_z$ V/m. Find: (a) V_{MN} if points M and N are specified by $M(2, 6, -1)$ and $N(-3, -3, 2)$; (b) V_M if $V = 0$ at $Q(4, -2, -35)$; (c) V_N if $V = 2$ at $P(1, 2, -4)$.

Ans. (a) −139.0 V; (b) −120.0 V; (c) 19.0 V

4.4 THE POTENTIAL FIELD OF A POINT CHARGE

In Section 4.3 we found an expression Eq. (12) for the potential difference between two points located at $r = r_A$ and $r = r_B$ in the field of a point charge Q placed at the origin. How might we conveniently define a zero reference for potential? The simplest possibility is to let $V = 0$ at infinity. If we let the point at $r = r_B$ recede to infinity, the potential at r_A becomes

$$V_A = \frac{Q}{4\pi\,\epsilon_0\,r_A}$$

or, as there is no reason to identify this point with the A subscript,

$$V = \frac{Q}{4\pi\epsilon_0 r} \qquad (14)$$

This expression defines the potential at any point distance r from a point charge Q at the origin, the potential at infinite radius being taken as the zero reference. Returning to a physical interpretation, we may say that $Q/4\pi\epsilon_0 r$ joules of work must be done in carrying a unit charge from infinity to any point r meters from the charge Q.

A convenient method to express the potential without selecting a specific zero reference entails identifying r_A as r once again and letting $Q/4\pi\epsilon_0 r_B$ be a constant. Then

$$V = \frac{Q}{4\pi\epsilon_0 r} + C_1 \qquad (15)$$

and C_1 may be selected so that $V = 0$ at any desired value of r. We could also select the zero reference indirectly by choosing to let V be V_0 at $r = r_0$.

It should be noted that the *potential difference* between two points is not a function of C_1.

Equations (14) and (15) represent the potential field of a point charge. The potential is a scalar field and does not involve any unit vectors.

We now define an *equipotential surface* as a surface composed of all those points having the same value of potential. All field lines would be perpendicular to such a surface at the points where they intersect it. Therefore, no work is involved in moving a unit charge around on an equipotential surface. The equipotential surfaces in the potential field of a point charge are spheres centered at the point charge.

An inspection of the form of the potential field of a point charge shows that it is an inverse-distance field, whereas the electric field intensity was found to be an inverse-square-law function. A similar result occurs for the gravitational force field of a point mass (inverse-square law) and the gravitational potential field (inverse distance). The gravitational force exerted by the earth on an object one million miles from it is four times that exerted on the same object two million miles away. The kinetic energy given to a freely falling object starting from the end of the universe with zero velocity, however, is only twice as much at one million miles as it is at two million miles.

D4.5. A 15-nC point charge is at the origin in free space. Calculate V_1 if point P_1 is located at $P_1(-2, 3, -1)$ and (a) $V = 0$ at (6, 5, 4); (b) $V = 0$ at infinity; (c) $V = 5$ V at (2, 0, 4).

Ans. (a) 20.67 V; (b) 36.0 V; (c) 10.89 V

4.5 THE POTENTIAL FIELD OF A SYSTEM OF CHARGES: CONSERVATIVE PROPERTY

The potential at a point has been defined as the work done in bringing a unit positive charge from the zero reference to the point, and we have suspected that this work, and hence the potential, is independent of the path taken. If it were not, potential would not be a very useful concept.

4.5.1 Potential Field of an Ensemble of Point Charges

Consider the potential field of the single point charge for which we showed, in Section 4.4, the independence with regard to the path, noting that the field is linear with respect to charge so that superposition is applicable. It will then follow that the potential of a system of charges has a value at any point which is independent of the path taken in carrying the test charge to that point.

Thus the potential field of a single point charge, which we shall identify as Q_1 and locate at $\mathbf{r}_1$, involves only the distance $|\mathbf{r} - \mathbf{r}_1|$ from Q_1 to the point at $\mathbf{r}$ where we are establishing the value of the potential. For a zero reference at infinity, we have

$$V(\mathbf{r}) = \frac{Q_1}{4\pi\epsilon_0|\mathbf{r} - \mathbf{r}_1|}$$

The potential arising from two charges, Q_1 at $\mathbf{r}_1$ and Q_2 at $\mathbf{r}_2$, is a function only of $|\mathbf{r} - \mathbf{r}_1|$ and $|\mathbf{r} - \mathbf{r}_2|$, the distances from Q_1 and Q_2 to the field point, respectively.

$$V(\mathbf{r}) = \frac{Q_1}{4\pi\epsilon_0|\mathbf{r} - \mathbf{r}_1|} + \frac{Q_2}{4\pi\epsilon_0|\mathbf{r} - \mathbf{r}_2|}$$

Continuing to add charges, we find that the potential arising from n point charges is

$$V(\mathbf{r}) = \sum_{m=1}^{n} \frac{Q_m}{4\pi\epsilon_0|\mathbf{r} - \mathbf{r}_m|} \tag{16}$$

4.5.2 Potential Field of a Continuous Charge Distribution

If each of the preceding point charges is now represented as a small element of a continuous volume charge distribution $\rho v \Delta v$, then

$$V(\mathbf{r}) = \frac{\rho_v(\mathbf{r}_1)\,\Delta v_1}{4\pi\epsilon_0|\mathbf{r} - \mathbf{r}_1|} + \frac{\rho_v(\mathbf{r}_2)\,\Delta v_2}{4\pi\epsilon_0|\mathbf{r} - \mathbf{r}_2|} + \cdots + \frac{\rho_v(\mathbf{r}_n)\,\Delta v_n}{4\pi\epsilon_0|\mathbf{r} - \mathbf{r}_n|}$$

As the number of elements becomes infinite, we obtain the integral expression

$$V(\mathbf{r}) = \int_{\text{vol}} \frac{\rho_v(\mathbf{r}')\,dv'}{4\pi\epsilon_0|\mathbf{r} - \mathbf{r}'|} \tag{17}$$

We have come quite a distance from the potential field of the single point charge, and it might be helpful to examine Eq. (17) and refresh ourselves as to the meaning of each term. The potential $V(\mathbf{r})$ is determined with respect to a zero reference potential at infinity and is an exact measure of the work done in bringing a unit charge from infinity to the field point at **r** where we are finding the potential. The volume charge density $\rho_v(\mathbf{r}')$ and differential volume element dv' combine to represent a differential amount of charge $\rho_v(\mathbf{r}')\,dv'$ located at $\mathbf{r}'$. The distance $|\mathbf{r} - \mathbf{r}'|$ is that distance from the source point to the field point. The integral is a multiple (volume) integral.

If the charge distribution takes the form of a line charge or a surface charge, the integration is along the line or over the surface:

$$V(\mathbf{r}) = \int \frac{\rho_L(\mathbf{r}')\,dL'}{4\pi\epsilon_0|\mathbf{r} - \mathbf{r}'|} \tag{18}$$

$$V(\mathbf{r}) = \int_S \frac{\rho_S(\mathbf{r}')\,dS'}{4\pi\epsilon_0|\mathbf{r} - \mathbf{r}'|} \tag{19}$$

The most general expression for potential is obtained by combining Eqs. (16)–(19).

These integral expressions for potential in terms of the charge distribution should be compared with similar expressions for the electric field intensity, such as Eq. (15) in Section 2.3:

$$\mathbf{E}(\mathbf{r}) = \int_{\text{vol}} \frac{\rho_v(\mathbf{r}')\,dv'}{4\pi\,\epsilon_0|\mathbf{r} - \mathbf{r}'|^2}\frac{\mathbf{r} - \mathbf{r}'}{|\mathbf{r} - \mathbf{r}'|}$$

The potential again is inverse distance, and the electric field intensity, inverse-square law. The latter, of course, is also a vector field.

EXAMPLE 4.3

To illustrate the use of one of these potential integrals, we will find V on the z axis for a uniform line charge ρ_L in the form of a ring, $\rho = a$, in the $z = 0$ plane, as shown in Figure 4.3.

Solution. Working with Eq. (18), we have $dL' = a\,d\phi'$, $\mathbf{r} = z\mathbf{a}_z$, $\mathbf{r}' = a\mathbf{a}_\rho$, $|\mathbf{r} - \mathbf{r}'| = \sqrt{a^2 + z^2}$, and

$$V = \int_0^{2\pi} \frac{\rho_L a\,d\phi'}{4\pi\,\epsilon_0\sqrt{a^2 + z^2}} = \frac{\rho_L a}{2\epsilon_0\sqrt{a^2 + z^2}}$$

4.5.3   The Conservative Nature of the Static Electric Field

For a zero reference at infinity, the potential field has the following attributes:

1. The potential arising from a single point charge is the work done in carrying a unit positive charge from infinity to the point at which we desire the potential, and the work is independent of the path chosen between those two points.

2. The potential field in the presence of a number of point charges is the sum of the individual potential fields arising from each charge.

3. The potential arising from a number of point charges or any continuous charge distribution may therefore be found by carrying a unit charge from infinity to the point in question along any path we choose.

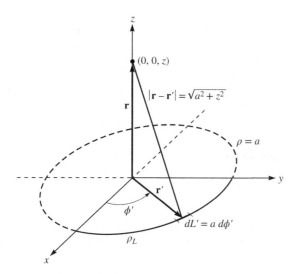

Figure 4.3 The potential field of a ring of uniform line charge density is easily obtained from $V = \int \rho_L(\mathbf{r}')dL' / (4\pi\epsilon_0|r - \mathbf{r}'|)$.

In other words, the expression for potential (zero reference at infinity),

$$V_A = -\int_\infty^A \mathbf{E} \cdot d\mathbf{L}$$

or potential difference,

$$V_{AB} = V_A - V_B = -\int_B^A \mathbf{E} \cdot d\mathbf{L}$$

is not dependent on the path chosen for the line integral, regardless of the source of the **E** field.

This result is often stated concisely by recognizing that no work is done in carrying the unit charge around any *closed path*, or

$$\boxed{\oint \mathbf{E} \cdot d\mathbf{L} = 0} \qquad (20)$$

A small circle is placed on the integral sign to indicate the closed nature of the path. This symbol also appeared in the formulation of Gauss's law, where a closed *surface* integral was used.

Equation (20) is true for *static* fields, but we will see in Chapter 9 that Faraday demonstrated it was incomplete when time-varying magnetic fields were present. One of Maxwell's greatest contributions to electromagnetic theory was in showing that a time-varying electric field produces a magnetic field, and therefore we should expect to find later that Eq. (20) is not correct when either **E** or the magnetic field varies with time.

Restricting our attention to the static case where **E** does not change with time, consider the dc circuit shown in Figure 4.4. Two points, *A* and *B*, are marked, and

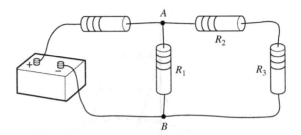

Figure 4.4 A simple dc-circuit problem that must be solved by applying $\oint \mathbf{E} \cdot d\mathbf{L} = 0$ in the form of Kirchhoff's voltage law.

(20) states that no work is involved in carrying a unit charge from A through R_2 and R_3 to B and back to A through R_1, or that the sum of the potential differences around any closed path is zero.

Equation (20) is therefore just a more general form of Kirchhoff's circuital law for voltages, more general in that we can apply it to any region where an electric field exists and we are not restricted to a conventional circuit composed of wires, resistances, and batteries. Equation (20) must be amended before we can apply it to time-varying fields.

Any field that satisfies an equation of the form of Eq. (20), (i.e., where the closed line integral of the field is zero) is said to be a *conservative field*. The name arises from the fact that no work is done (or that energy is *conserved*) around a closed path. The gravitational field is also conservative, for any energy expended in moving (raising) an object against the field is recovered exactly when the object is returned (lowered) to its original position. A nonconservative gravitational field could solve our energy problems forever.

Given a *nonconservative* field, it is of course possible that the line integral may be zero for certain closed paths. For example, consider the force field, $\mathbf{F} = \sin \pi\rho \, \mathbf{a}_\phi$. Around a circular path of radius $\rho = \rho_1$, we have $d\mathbf{L} = \rho \, d\phi \, \mathbf{a}_\phi$, and

$$\oint \mathbf{F} \cdot d\mathbf{L} = \int_0^{2\pi} \sin\pi\rho_1 \mathbf{a}_\phi \cdot \rho_1 d\phi \, \mathbf{a}_\phi = \int_0^{2\pi} \rho_1 \sin\pi\rho_1 \, d\phi$$
$$= 2\pi\rho_1 \sin\pi\rho_1$$

The integral is zero if $\rho_1 = 1, 2, 3, \ldots$, etc., but it is not zero for other values of ρ_1, or for most other closed paths, and the given field is not conservative. A conservative field must yield a zero value for the line integral around every possible closed path.

D4.6. If we take the zero reference for potential at infinity, find the potential at $(0, 0, 2)$ caused by this charge configuration in free space (*a*) 12 nC/m on the line $\rho = 2.5$ m, $z = 0$; (*b*) point charge of 18 nC at $(1, 2, -1)$; (*c*) 12 nC/m on the line $y = 2.5$, $z = 0$, $-1.0 < x < 1.0$.

Ans. (*a*) 529 V; (*b*) 43.2 V; (*c*) 66.3 V

4.6 POTENTIAL GRADIENT

We now have two methods of determining potential, one directly from the electric field intensity by means of a line integral, and another from the basic charge distribution itself by a volume integral. Neither method is very helpful in determining the fields in most practical problems, however, for as we will see later, neither the electric field intensity nor the charge distribution is very often known. Preliminary information is much more apt to consist of a description of two equipotential surfaces, such as the statement that we have two parallel conductors of circular cross section at potentials of 100 and −100 V. Perhaps we wish to find the capacitance between the conductors, or the charge and current distribution on the conductors from which losses may be calculated.

These quantities may be easily obtained from the potential field, and our immediate goal will be to acquire a simple method of finding the electric field intensity from the potential.

4.6.1 General Relation Between the Electric and Potential Fields

We already have the general line-integral relationship between these quantities,

$$V = -\int \mathbf{E} \cdot d\mathbf{L}$$ (21)

but this is much easier to use in the reverse direction: given $\mathbf{E}$, find V.

However, Eq. (21) may be applied to a very short element of length ΔL along which $\mathbf{E}$ is essentially constant, leading to an incremental potential difference ΔV,

$$\Delta V \doteq -\mathbf{E} \cdot \Delta \mathbf{L}$$ (22)

Now consider a general region of space, as shown in Figure 4.5, in which $\mathbf{E}$ and V both change as we move from point to point. Equation (22) tells us to choose an incremental vector element of length $\Delta \mathbf{L} = \Delta L \, \mathbf{a}_L$ and multiply its magnitude by the

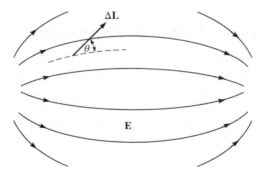

Figure 4.5 A vector incremental element of length $\Delta\mathbf{L}$ is shown making an angle of θ with an $\mathbf{E}$ field, indicated by its streamlines. The sources of the field are not shown.

component of $\mathbf{E}$ in the direction of $\mathbf{a}_L$ (one interpretation of the dot product) to obtain the small potential difference between the final and initial points of $\Delta \mathbf{L}$.

If we designate the angle between $\Delta \mathbf{L}$ and $\mathbf{E}$ as θ, then

$$\Delta V \doteq -E \Delta L \cos\theta$$

We now pass to the limit and consider the derivative dV/dL. To do this, we need to show that V may be interpreted as a *function* $V(x, y, z)$. So far, V is merely the result of the line integral (21). If we assume a specified starting point or zero reference and then let our end point be (x, y, z), we know that the result of the integration is a unique function of the end point (x, y, z) because $\mathbf{E}$ is a conservative field. Therefore V is a single-valued function $V(x, y, z)$. We may then pass to the limit and obtain

$$\frac{dV}{dL} = -E \cos\theta$$

In which direction should $\Delta \mathbf{L}$ be placed to obtain a maximum value of ΔV? Remember that $\mathbf{E}$ is a definite value at the point at which we are working and is independent of the direction of $\Delta \mathbf{L}$. The magnitude ΔL is also constant, and our variable is $\mathbf{a}_L$, the unit vector showing the direction of $\Delta \mathbf{L}$. It is obvious that the maximum positive increment of potential, ΔV_{max}, will occur when $\cos \theta$ is -1, or $\Delta \mathbf{L}$ points in the direction *opposite* to $\mathbf{E}$. For this condition,

$$\left.\frac{dV}{dL}\right|_{max} = E$$

This little exercise shows us two characteristics of the relationship between $\mathbf{E}$ and V at any point:

1. The magnitude of the electric field intensity is given by the maximum value of the rate of change of potential with distance.
2. This maximum value is obtained when the direction of the distance increment is opposite to $\mathbf{E}$ or, in other words, the direction of $\mathbf{E}$ is *opposite* to the direction in which the potential is *increasing* the most rapidly.

4.6.2   Static Electric Field as the Negative Gradient of Potential

We now illustrate the preceding relationships in terms of potential. Figure 4.6 is intended to show the information we have been given about some potential field. It does this by showing the equipotential surfaces (shown as lines in the two-dimensional sketch). We want information about the electric field intensity at point P. Starting at P, we lay off a small incremental distance $\Delta \mathbf{L}$ in various directions, hunting for that direction in which the potential is changing (increasing) the most rapidly. From the sketch, this direction appears to be left and slightly upward. From our second characteristic above, the electric field intensity is therefore oppositely directed, or to the right and slightly downward at P. Its magnitude is given by dividing the small increase in potential by the small element of length.

It seems likely that the direction in which the potential is increasing the most rapidly is perpendicular to the equipotentials (in the direction of *increasing* potential),

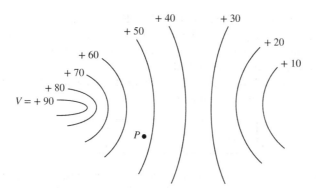

Figure 4.6 A potential field is shown by its equipotential surfaces. At any point the **E** field is normal to the equipotential surface passing through that point and is directed toward the more negative surfaces.

and this is correct, for if $\Delta\mathbf{L}$ is directed along an equipotential, $\Delta V = 0$ by our definition of an equipotential surface. But then

$$\Delta V = -\mathbf{E} \cdot \Delta\mathbf{L} = 0$$

and as neither $\mathbf{E}$ nor $\Delta\mathbf{L}$ is zero, $\mathbf{E}$ must be perpendicular to this $\Delta\mathbf{L}$ or perpendicular to the equipotentials.

Because the potential field information is more likely to be determined first, we will describe the direction of $\Delta\mathbf{L}$, which leads to a maximum increase in potential mathematically in terms of the potential field rather than the electric field intensity. We do this by letting $\mathbf{a}_N$ be a unit vector normal to the equipotential surface and directed toward the higher potentials. The electric field intensity is then expressed in terms of the potential,

$$\mathbf{E} = -\frac{dV}{dL}\bigg|_{\text{max}} \mathbf{a}_N \qquad (23)$$

which shows that the magnitude of $\mathbf{E}$ is given by the maximum space rate of change of V and the direction of $\mathbf{E}$ is *normal* to the equipotential surface (in the direction of *decreasing* potential).

Because $dV/dL|_{\text{max}}$ occurs when $\Delta\mathbf{L}$ is in the direction of $\mathbf{a}_N$, we may remind ourselves of this fact by letting

$$\frac{dV}{dL}\bigg|_{\text{max}} = \frac{dV}{dN}$$

and

$$\mathbf{E} = -\frac{dV}{dN}\mathbf{a}_N \qquad (24)$$

Either Eq. (23) or Eq. (24) provides a physical interpretation of the process of finding the electric field intensity from the potential. Both are descriptive of a general procedure, and we do not intend to use them directly to obtain quantitative information. This procedure leading from V to $\mathbf{E}$ is not unique to this pair of quantities, however, but has appeared as the relationship between a scalar and a vector field in hydraulics, thermodynamics, and magnetics, and indeed in almost every field to which vector analysis has been applied.

The operation on V by which $-\mathbf{E}$ is obtained is known as the *gradient*, and the gradient of a scalar field T is defined as

$$\boxed{\text{Gradient of } T = \text{grad } T = \frac{dT}{dN}\mathbf{a}_N} \tag{25}$$

where $\mathbf{a}_N$ is a unit vector normal to the equipotential surfaces, and that normal is chosen which points in the direction of increasing values of T.

Using this new term, we now may write the relationship between V and $\mathbf{E}$ as

$$\boxed{\mathbf{E} = -\text{grad } V} \tag{26}$$

4.6.3 Computation of the Gradient

Because we have shown that V is a unique function of x, y, and z, we may take its total differential

$$dV = \frac{\partial V}{\partial x}dx + \frac{\partial V}{\partial y}dy + \frac{\partial V}{\partial z}dz$$

But we also have

$$dV = -\mathbf{E} \cdot d\mathbf{L} = -E_x\,dx - E_y\,dy - E_z\,dz$$

Because both expressions are true for any dx, dy, and dz, then

$$E_x = -\frac{\partial V}{\partial x}$$

$$E_y = -\frac{\partial V}{\partial y}$$

$$E_z = -\frac{\partial V}{\partial z}$$

These results may be combined vectorially to yield

$$\boxed{\mathbf{E} = -\left(\frac{\partial V}{\partial x}\mathbf{a}_x + \frac{\partial V}{\partial y}\mathbf{a}_y + \frac{\partial V}{\partial z}\mathbf{a}_z \right)} \tag{27}$$

and comparing Eqs. (26) and (27) provides us with an expression which may be used to evaluate the gradient in rectangular coordinates,

$$\boxed{\text{grad } V = \frac{\partial V}{\partial x}\mathbf{a}_x + \frac{\partial V}{\partial y}\mathbf{a}_y + \frac{\partial V}{\partial z}\mathbf{a}_z} \tag{28}$$

The gradient of a scalar is a vector, and old quizzes show that the unit vectors that are often incorrectly added to the divergence expression appear to be those that were incorrectly removed from the gradient. Once the physical interpretation of the gradient, expressed by Eq. (25), is grasped as showing the maximum space rate of change of a scalar quantity and *the direction in which this maximum occurs*, the vector nature of the gradient should be self-evident.

The vector operator

$$\nabla = \frac{\partial}{\partial x}\mathbf{a}_x + \frac{\partial}{\partial y}\mathbf{a}_y + \frac{\partial}{\partial z}\mathbf{a}_z$$

may be used formally as an operator on a scalar, T, ∇T, producing

$$\nabla T = \frac{\partial T}{\partial x}\mathbf{a}_x + \frac{\partial T}{\partial y}\mathbf{a}_y + \frac{\partial T}{\partial z}\mathbf{a}_z$$

from which we see that

$$\boxed{\nabla T = \text{grad } T}$$

This allows us to use a very compact expression to relate $\mathbf{E}$ and V,

$$\boxed{\mathbf{E} = -\nabla V} \tag{29}$$

Interactives

The gradient may be expressed in terms of partial derivatives in other coordinate systems through the application of its definition Eq. (25). These expressions are derived in Appendix A and repeated here for convenience when dealing with problems having cylindrical or spherical symmetry. They also appear at the end of this book.

$$\boxed{\nabla V = \frac{\partial V}{\partial x}\mathbf{a}_x + \frac{\partial V}{\partial y}\mathbf{a}_y + \frac{\partial V}{\partial z}\mathbf{a}_z \quad \text{(rectangular)}} \tag{30}$$

$$\boxed{\nabla V = \frac{\partial V}{\partial \rho}\mathbf{a}_\rho + \frac{1}{\rho}\frac{\partial V}{\partial \phi}\mathbf{a}_\phi + \frac{\partial V}{\partial z}\mathbf{a}_z \quad \text{(cylindrical)}} \tag{31}$$

$$\boxed{\nabla V = \frac{\partial V}{\partial r}\mathbf{a}_r + \frac{1}{r}\frac{\partial V}{\partial \theta}\mathbf{a}_\theta + \frac{1}{r\sin\theta}\frac{\partial V}{\partial \phi}\mathbf{a}_\phi \quad \text{(spherical)}} \tag{32}$$

Note that the denominator of each term has the form of one of the components of $d\mathbf{L}$ in that coordinate system, except that partial differentials replace ordinary differentials; for example, $r \sin \theta \, d\phi$ becomes $r \sin \theta \, \partial\phi$.

We now illustrate the gradient concept with an example.

EXAMPLE 4.4

Given the potential field, $V = 2x^2y - 5z$, and a point $P(-4, 3, 6)$, we wish to find several numerical values at point P: the potential V, the electric field intensity $\mathbf{E}$, the direction of $\mathbf{E}$, the electric flux density $\mathbf{D}$, and the volume charge density ρ_v.

Solution. The potential at $P(-4, 5, 6)$ is

$$V_P = 2(-4)^2(3) - 5(6) = 66 \text{ V}$$

Next, we may use the gradient operation to obtain the electric field intensity,

$$\mathbf{E} = -\nabla V = -4xy\,\mathbf{a}_x - 2x^2\,\mathbf{a}_y + 5\,\mathbf{a}_z \text{ V/m}$$

The value of $\mathbf{E}$ at point P is

$$\mathbf{E}_P = 48\,\mathbf{a}_x - 32\,\mathbf{a}_y + 5\,\mathbf{a}_z \text{ V/m}$$

and

$$|\mathbf{E}_P| = \sqrt{48^2 + (-32)^2 + 5^2} = 57.9 \text{ V/m}$$

The direction of $\mathbf{E}$ at P is given by the unit vector

$$\mathbf{a}_{E,P} = (48\,\mathbf{a}_x - 32\,\mathbf{a}_y + 5\,\mathbf{a}_z)/57.9$$
$$= 0.829\,\mathbf{a}_x - 0.553\,\mathbf{a}_y + 0.086\,\mathbf{a}_z$$

If we assume these fields exist in free space, then

$$\mathbf{D} = \epsilon_0\mathbf{E} = -35.4xy\,\mathbf{a}_x - 17.71x^2\,\mathbf{a}_y + 44.3\,\mathbf{a}_z \text{ pC/m}^3$$

Finally, we may use the divergence relationship to find the volume charge density that is the source of the given potential field,

$$\rho_v = \nabla \cdot \mathbf{D} = -35.4y \text{ pC/m}^3$$

At P, $\rho_v = -106.2$ pC/m^3.

D4.7. A portion of a two-dimensional ($E_z = 0$) potential field is shown in Figure 4.7. The grid lines are 1 mm apart in the actual field. Determine approximate values for $\mathbf{E}$ in rectangular coordinates at: (a) a; (b) b; (c) c.

Ans. (a) $-1075\mathbf{a}_y$ V/m; (b) $-600\mathbf{a}_x - 700\mathbf{a}_y$ V/m; (c) $-500\mathbf{a}_x - 650\mathbf{a}_y$ V/m

---

D4.8. Given the potential field in cylindrical coordinates, $V = \dfrac{100}{z^2 + 1}\rho\cos\phi$ V, and point P at $\rho = 3$ m, $\phi = 60°$, $z = 2$ m, find values at P for (a) V; (b) $\mathbf{E}$; (c) E; (d) dV/dN; (e) $\mathbf{a}_N$; (f) ρ_v in free space.

Ans. (a) 30.0 V; (b) $-10.00\mathbf{a}_\rho + 17.3\mathbf{a}_\phi + 24.0\mathbf{a}_z$ V/m; (c) 31.2 V/m; (d) 31.2 V/m; (e) $0.32\mathbf{a}_\rho - 0.55\mathbf{a}_\phi - 0.77\mathbf{a}_z$; (f) -234 pC/m^3

4.7 THE ELECTRIC DIPOLE

The dipole fields that are developed in this section are quite important because they form the basis for the behavior of dielectric materials in electric fields, as discussed in Chapter 5. Moreover, this development will serve to illustrate the importance of the potential concept presented in this chapter.

An *electric dipole*, or simply a *dipole*, is the name given to two point charges of equal magnitude and opposite sign, separated by a distance that is small compared to the distance to the point P at which we want to know the electric and potential fields.

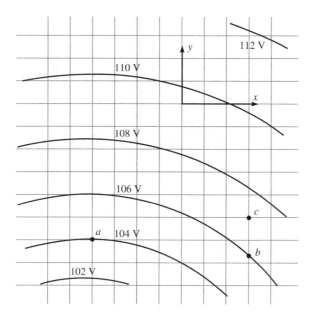

Figure 4.7   See Problem D4.7.

The dipole is shown in Figure 4.8*a*. The distant point *P* is described by the spherical coordinates *r*, θ, and $\phi = 90°$, in view of the azimuthal symmetry. The positive and negative point charges have separation *d* and rectangular coordinates $(0, 0, \frac{1}{2}d)$ and $(0, 0, -\frac{1}{2}d)$, respectively.

So much for the geometry. What would we do next? Should we find the total electric field intensity by adding the known fields of each point charge? Would it be easier to find the total potential field first? In either case, having found one, we will find the other from it before calling the problem solved.

If we choose to find **E** first, we will have two components to keep track of in spherical coordinates (symmetry shows E_ϕ is zero), and then the only way to find *V* from **E** is by use of the line integral. This last step includes establishing a suitable zero reference for potential, since the line integral gives us only the potential difference between the two points at the ends of the integral path.

On the other hand, the determination of *V* first is a much simpler problem. This is because we find the potential as a function of position by simply adding the scalar potentials from the two charges. The position-dependent vector magnitude and direction of **E** are subsequently evaluated with relative ease by taking the negative gradient of *V*.

4.7.1 Obtaining the Dipole Potential in the Far Zone

To find the potential field, we let the distances from *Q* and −*Q* to *P* be R_1 and R_2, respectively, and we write the total potential as

$$V = \frac{Q}{4\pi\epsilon_0}\left(\frac{1}{R_1} - \frac{1}{R_2}\right) = \frac{Q}{4\pi\epsilon_0}\frac{R_2 - R_1}{R_1 R_2}$$

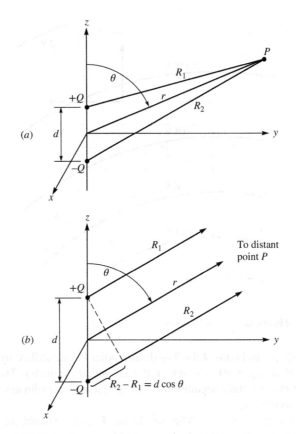

Figure 4.8 (a) The geometry of the problem of an electric dipole. The dipole moment $p = Qd$ is in the $\mathbf{a}_z$ direction. (b) For a distant point P, R_1 is essentially parallel to R_2, and we find that $R_2 - R_1 = d \cos \theta$.

Note that the plane $z = 0$, midway between the two point charges, is the locus of points for which $R_1 = R_2$, and is therefore at zero potential, as are all points at infinity.

Consider a point at distance at location r which satisfies the *far field approximation*: $r >> d$, at which the point is said to lie in the *far zone*. If this is true, then $R_1 \doteq R_2$, and the $R_1 R_2$ product in the denominator may be replaced by r^2. The approximation may not be made in the numerator, however, without obtaining the trivial answer that the potential field approaches zero as we go very far away from the dipole. Coming back a little closer to the dipole, we see from Figure 4.8b that $R_2 - R_1$ may be approximated very easily if R_1 and R_2 are parallel, a condition that is a good approximation in the far zone

Therefore:

$$R_2 - R_1 \doteq d \cos\theta$$

The final result is then

$$V = \frac{Qd\cos\theta}{4\pi\epsilon_0 r^2} \qquad r \gg d \tag{33}$$

Again, we note that the plane $z = 0$ ($\theta = 90°$) is at zero potential.

4.7.2 Obtaining the Electric Field

Using the gradient relationship in spherical coordinates,

$$\mathbf{E} = -\nabla V = -\left(\frac{\partial V}{\partial r}\mathbf{a}_r + \frac{1}{r}\frac{\partial V}{\partial \theta}\mathbf{a}_\theta + \frac{1}{r\sin\theta}\frac{\partial V}{\partial \phi}\mathbf{a}_\phi\right)$$

and applying this to Eq. (33), we find

$$\mathbf{E} = -\left(-\frac{Qd\cos\theta}{2\pi\epsilon_0 r^3}\mathbf{a}_r - \frac{Qd\sin\theta}{4\pi\epsilon_0 r^3}\mathbf{a}_\theta\right) \tag{34}$$

or

$$\mathbf{E} = \frac{Qd}{4\pi\epsilon_0 r^3}(2\cos\theta\ \mathbf{a}_r + \sin\theta\ \mathbf{a}_\theta) \tag{35}$$

These are the desired far zone fields of the dipole, obtained with a very small amount of work. Any student who has several hours to spend may try to work the problem in the reverse direction—the authors consider the process too long and detailed to include here, even for effect.

To obtain a plot of the potential field, we choose a dipole such that $Qd/(4\pi\epsilon_0) = 1$, and then $\cos\theta = Vr^2$. The colored lines in Figure 4.9 indicate equipotentials for which $V = 0$, $+0.2$, $+0.4$, $+0.6$, $+0.8$, and $+1$, as indicated. The dipole axis is vertical, with the positive charge on the top. The streamlines for the electric field are obtained by applying the methods of Section 2.6 in spherical coordinates,

$$\frac{E_\theta}{E_r} = \frac{r\,d\theta}{dr} = \frac{\sin\theta}{2\cos\theta}$$

or

$$\frac{dr}{r} = 2\cot\theta\,d\theta$$

from which we obtain

$$r = C_1\sin^2\theta$$

The black streamlines shown in Figure 4.9 are for $C_1 = 1$, 1.5, 2, and 2.5.

4.7.3 Rewriting the Potential Field

The potential field of the dipole, Eq. (33), may be simplified by making use of the dipole moment. We first identify the vector length directed from $-Q$

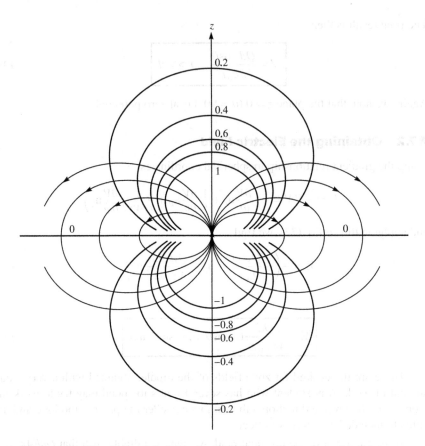

Figure 4.9 The electrostatic field of a point dipole with its moment in the $\mathbf{a}_z$ direction. Six equipotential surfaces are labeled with relative values of V.

to $+Q$ as $\mathbf{d}$ and then define the *dipole moment* as $Q\mathbf{d}$ and assign it the symbol $\mathbf{p}$. Thus

$$\boxed{\mathbf{p} = Q\mathbf{d}} \tag{36}$$

The units of $\mathbf{p}$ are C · m.

Because $\mathbf{d} \cdot \mathbf{a}_r = d \cos \theta$, we then have

$$V = \frac{\mathbf{p} \cdot \mathbf{a}_r}{4\pi \,\epsilon_0 r^2} \tag{37}$$

This result may be generalized as

$$V = \frac{1}{4\pi \,\epsilon_0 |\mathbf{r} - \mathbf{r}'|^2} \mathbf{p} \cdot \frac{\mathbf{r} - \mathbf{r}'}{|\mathbf{r} - \mathbf{r}'|} \tag{38}$$

where $\mathbf{r}$ locates the field point P, and $\mathbf{r}'$ determines the dipole center. Equation (38) is independent of any coordinate system.

The dipole moment **p** will appear again when we discuss dielectric materials. Since it is equal to the product of the charge and the separation, neither the dipole moment nor the potential will change as Q increases and **d** decreases, provided the product remains constant. The limiting case of a *point dipole* is achieved when we let **d** approach zero and Q approach infinity such that the product **p** is finite.

Turning our attention to the resultant fields, it is interesting to note that the potential field is now proportional to the inverse *square* of the distance, and the electric field intensity is proportional to the inverse *cube* of the distance from the dipole. Each field falls off faster than the corresponding field for the point charge, but this is no more than we should expect because the opposite charges appear to be closer together at greater distances and to act more like a single point charge of zero coulombs.

Symmetrical arrangements of larger numbers of point charges produce fields proportional to the inverse of higher and higher powers of r. These charge distributions are called *multipoles*, and they are used in infinite series to approximate more unwieldy charge configurations.

D4.9. An electric dipole located at the origin in free space has a moment **p** = $3\mathbf{a}_x - 2\mathbf{a}_y + \mathbf{a}_z$ nC · m. (*a*) Find V at $P_A(2, 3, 4)$. (*b*) Find V at $r = 2.5$, $\theta = 30°$, $\phi = 40°$.

Ans. (*a*) 0.23 V; (*b*) 1.97 V

D4.10. A dipole of moment **p** = $6\mathbf{a}_z$ nC · m is located at the origin in free space. (*a*) Find V at $P(r = 4, \theta = 20°, \phi = 0°)$. (*b*) Find **E** at P.

Ans. (*a*) 3.17 V; (*b*) $1.58\mathbf{a}_r + 0.29\mathbf{a}_\theta$ V/m

---

4.8 ELECTROSTATIC ENERGY

We have introduced the potential concept by considering the work done, or energy expended, in moving a point charge around in an electric field, and now we must tie up the loose ends of that discussion by tracing the energy flow one step further.

4.8.1 Stored Energy in a Distribution of Charge

Bringing a positive charge from infinity into the field of another positive charge requires work, the work being done by the external source moving the charge. Imagine that the external source carries the charge up to a point near the fixed charge and then holds it there. Energy must be conserved, and the energy expended in bringing this charge into position now represents potential energy, for if the external source released its hold on the charge, it would accelerate away from the fixed charge, acquiring kinetic energy of its own and the capability of doing work.

In order to find the potential energy present in a system of charges, we must find the work done by an external source in positioning the charges.

We may start by visualizing an empty universe. Bringing a charge Q_1 from infinity to any position requires no work, for there is no field present.[2] The positioning of Q_2 at a point in the field of Q_1 requires an amount of work given by the product of the charge Q_2 and the potential at that point due to Q_1. We represent this potential as V, where the first subscript indicates the location and the second subscript the source. That is, V is the potential at the location of Q_2 due to Q_1. Then

$$\text{Work to position } Q_2 = Q_2 V_{2,1}$$

Similarly, we may express the work required to position each additional charge in the field of all those already present:

$$\text{Work to position } Q_3 = Q_3 V_{3,1} + Q_3 V_{3,2}$$
$$\text{Work to position } Q_4 = Q_4 V_{4,1} + Q_4 V_{4,2} + Q_4 V_{4,3}$$

and so forth. The total work is obtained by adding each contribution:

$$\text{Total positioning work} = \text{potential energy of field}$$
$$= W_E = Q_2 V_{2,1} + Q_3 V_{3,1} + Q_3 V_{3,2} + Q_4 V_{4,1} \tag{39}$$
$$+ Q_4 V_{4,2} + Q_4 V_{4,3} + \cdots$$

Noting the form of a representative term in the preceding equation,

$$Q_3 V_{3,1} = Q_3 \frac{Q_1}{4\pi \epsilon_0 R_{13}} = Q_1 \frac{Q_3}{4\pi \epsilon_0 R_{31}}$$

where R_{13} and R_{31} each represent the scalar distance between Q_1 and Q_3, we see that it might equally well have been written as $Q_1 V_{1,3}$. If each term of the total energy expression is replaced by its equal, we have

$$W_E = Q_1 V_{1,2} + Q_1 V_{1,3} + Q_2 V_{2,3} + Q_1 V_{1,4} + Q_2 V_{2,4} + Q_3 V_{3,4} + \cdots \tag{40}$$

Adding the two energy expressions (39) and (40) gives us a chance to simplify the result a little:

$$2W_E = Q_1(V_{1,2} + V_{1,3} + V_{1,4} + \cdots)$$
$$+ Q_2(V_{2,1} + V_{2,3} + V_{2,4} + \cdots)$$
$$+ Q_3(V_{3,1} + V_{3,2} + V_{3,4} + \cdots)$$
$$+ \cdots$$

Each sum of potentials in parentheses is the combined potential due to all the charges except for the charge at the point where this combined potential is being found. In other words,

$$V_{1,2} + V_{1,3} + V_{1,4} + \cdots = V_1$$

---

[2] However, somebody in the workshop at infinity had to do an infinite amount of work to create the point charge in the first place! How much energy is required to bring two half-charges into coincidence to make a unit charge?

V_1 is the potential at the location of Q_1 due to the presence of Q_2, Q_3, ... We therefore have

$$W_E = \tfrac{1}{2}(Q_1 V_1 + Q_2 V_2 + Q_3 V_3 + \cdots) = \tfrac{1}{2}\sum_{m=1}^{m=N} Q_m V_m \qquad (41)$$

In order to obtain an expression for the energy stored in a region of continuous charge distribution, each charge is replaced by $\rho_v dv$, and the summation becomes an integral,

$$W_E = \tfrac{1}{2}\int_{\text{vol}} \rho_v V \, dv \qquad (42)$$

Equations (41) and (42) allow us to find the total potential energy present in a system of point charges or distributed volume charge density. Similar expressions may be easily written in terms of line or surface charge density. Usually we prefer to use Eq. (42) and let it represent all the various types of charge which may have to be considered. This may always be done by considering point charges, line charge density, or surface charge density to be continuous distributions of volume charge density over very small regions. We will illustrate such a procedure with an example shortly.

4.8.2 Stored Energy in the Electric Field

Before we undertake any interpretation of the preceding result, we should consider a few lines of more difficult vector analysis and obtain an expression equivalent to Eq. (42) but written in terms of **E** and **D**.

We begin by making the expression a little bit longer. Using Maxwell's first equation, replace ρ_v with its equal $\nabla \cdot \mathbf{D}$ and make use of a vector identity which is true for any scalar function V and any vector function **D**,

$$\nabla \cdot (V\mathbf{D}) \equiv V(\nabla \cdot \mathbf{D}) + \mathbf{D} \cdot (\nabla V) \qquad (43)$$

This may be proved readily by expansion in rectangular coordinates. We then have, successively,

$$W_E = \tfrac{1}{2}\int_{\text{vol}} \rho_v V \, dv = \tfrac{1}{2}\int_{\text{vol}} (\nabla \cdot \mathbf{D}) V \, dv$$
$$= \tfrac{1}{2}\int_{\text{vol}} [\nabla \cdot (V\mathbf{D}) - \mathbf{D} \cdot (\nabla V)] \, dv$$

Using the divergence theorem from Chapter 3, the first volume integral of the last equation is changed into a closed surface integral, where the closed surface surrounds the volume considered. This volume, first appearing in Eq. (42), must contain *every* charge, and there can then be no charges outside of the volume. We may therefore consider the volume as *infinite* in extent if we wish. We have

$$W_E = \tfrac{1}{2}\oint_S (V\mathbf{D}) \cdot d\mathbf{S} - \tfrac{1}{2}\int_{\text{vol}} \mathbf{D} \cdot (\nabla V) \, dv$$

The surface integral is equal to zero, for over this closed surface surrounding the universe we see that V is approaching zero at least as rapidly as $1/r$ (the charges look like point charges from there), and **D** is approaching zero at least as rapidly as $1/r^2$. The integrand therefore approaches zero at least as rapidly as $1/r^3$, while the differential area of the surface, looking more and more like a portion of a sphere, is increasing only as r^2. Consequently, in the limit as $r \to \infty$, the integrand and the integral both approach zero. Substituting $\mathbf{E} = -\nabla V$ in the remaining volume integral, we have our answer,

$$W_E = \frac{1}{2} \int_{\text{vol}} \mathbf{D} \cdot \mathbf{E} \, dv = \frac{1}{2} \int_{\text{vol}} \epsilon_0 E^2 \, dv \qquad (44)$$

4.8.3 Field Energy in a Coaxial Cable

We may now use Eq. (44) to calculate the energy stored in the electrostatic field of a section of a coaxial cable or capacitor of length L. We found in Section 3.3 that

$$D_\rho = \frac{a\rho_S}{\rho}$$

Hence,

$$\mathbf{E} = \frac{a\rho_S}{\epsilon_0 \rho} \mathbf{a}_\rho$$

where ρ_S is the surface charge density on the inner conductor, whose radius is a. Thus,

$$W_E = \frac{1}{2} \int_0^L \int_0^{2\pi} \int_a^b \epsilon_0 \frac{a^2 \rho_S^2}{\epsilon_0^2 \rho^2} \rho \, d\rho \, d\phi \, dz = \frac{\pi L a^2 \rho_S^2}{\epsilon_0} \ln \frac{b}{a}$$

This same result may be obtained from Eq. (42). We choose the outer conductor as our zero-potential reference, and the potential of the inner cylinder is then

$$V_a = -\int_b^a E_\rho \, d\rho = -\int_b^a \frac{a\rho_S}{\epsilon_0 \rho} \, d\rho = \frac{a\rho_S}{\epsilon_0} \ln \frac{b}{a}$$

The surface charge density ρ_S at $\rho = a$ can be interpreted as a volume charge density $\rho_v = \rho_S/t$, extending from $\rho = a - \frac{1}{2}t$ to $\rho = a + \frac{1}{2}t$, where $t \ll a$. The integrand in Eq. (42) is therefore zero everywhere between the cylinders (where the volume charge density is zero), as well as at the outer cylinder (where the potential is zero). The integration is therefore performed only within the thin cylindrical shell at $\rho = a$,

$$W_E = \frac{1}{2} \int_{\text{vol}} \rho_v \, V \, dV = \frac{1}{2} \int_0^L \int_0^{2\pi} \int_{a-t/2}^{a+t/2} \frac{\rho_S}{t} a \frac{\rho_S}{\epsilon_0} \ln \frac{b}{a} \rho \, d\rho \, d\phi \, dz$$

from which

$$W_E = \frac{a^2 \rho_S^2 \ln(b/a)}{\epsilon_0} \pi L$$

once again.

This expression takes on a more familiar form if we recognize the total charge on the inner conductor as $Q = 2\pi a L \rho_S$. Combining this with the potential difference between the cylinders, V_a, we see that

$$W_E = \frac{1}{2} Q V_a$$

which should be familiar as the energy stored in a capacitor.

4.8.4 Electric Field Energy Density

The question of where the energy is stored in an electric field has not yet been answered. Potential energy can never be pinned down precisely in terms of physical location. Someone lifts a pencil, and the pencil acquires potential energy. Is the energy stored in the molecules of the pencil, in the gravitational field between the pencil and the earth, or in some obscure place? Is the energy in a capacitor stored in the charges themselves, in the field, or where? No one can offer any proof for his or her own private opinion, and the matter of deciding may be left to the philosophers.

Electromagnetic field theory makes it easy to believe that the energy of an electric field or a charge distribution is stored in the field itself, for if we take Eq. (44), an exact and rigorously correct expression,

$$W_E = \frac{1}{2} \int_{vol} \mathbf{D} \cdot \mathbf{E} \, dv$$

and write it on a differential basis,

$$dW_E = \frac{1}{2} \mathbf{D} \cdot \mathbf{E} \, dv$$

or

$$\frac{dW_E}{dv} \equiv w_E = \frac{1}{2} \mathbf{D} \cdot \mathbf{E} \ \text{J/m}^3 \qquad (45)$$

we obtain a quantity $w_E = \frac{1}{2} \mathbf{D} \cdot \mathbf{E}$, which has the dimensions of an energy density, or joules per cubic meter. We know that if we integrate this energy density over the entire field-containing volume, the result is truly the total energy present, but we have no more justification for saying that the energy stored in each differential volume element dv is $\frac{1}{2} \mathbf{D} \cdot \mathbf{E} \, dv$ than we have for looking at Eq. (42) and saying that the stored energy is $\frac{1}{2} \rho_v V dv$. The interpretation afforded by Eq. (45), however, is a convenient one, and we will use it until proved wrong.

D4.11. Find the energy stored in free space for the region 2 mm $< r <$ 3 mm, 0 $<$ $\theta <$ 90°, 0 $< \phi <$ 90°, given the potential field $V =$: (a) $\frac{200}{r}$ V; (b) $\frac{300 \ \cos\theta}{r^2}$ V.

Ans. (a) 46.4 μJ; (b) 36.7 J

REFERENCES

1. Attwood, S. S. *Electric and Magnetic Fields.* 3d ed. New York: John Wiley & Sons, 1949. There are a large number of well-drawn field maps of various charge distributions, including the dipole field. Vector analysis is not used.

2. Skilling, H. H. *Fundamentals of Electric Waves.* 2d ed. New York: John Wiley & Sons, 1948. Gradient is described on pp. 19–21.

3. Thomas, G. B., Jr., and R. L. Finney. *Calculus and Analytic Geometry.* 6th ed. Reading, Mass.: Addison-Wesley Publishing Company, 1984. The directional derivative and the gradient are presented on pp. 823–30.

Quizzes

CHAPTER 4   PROBLEMS

4.1 Given $\mathbf{E} = E_x\mathbf{a}_x + E_y\mathbf{a}_y + E_z\mathbf{a}_z$ V/m, where E_x, E_y, and E_z are constants, determine the incremental work required to move charge q through a distance δ: (a) along the positive x axis; (b) in a direction at 45 degrees from the x axis in the first quadrant; (c) along a line in the first octant having equal x, y, and z components, and moving away from the origin.

4.2 A positive point charge of magnitude q_1 lies at the origin. Derive an expression for the incremental work done in moving a second point charge q_2 through a distance dx from the starting position (x, y, z), in the direction of $-\mathbf{a}_x$.

4.3 Given $\mathbf{E} = E_\rho\mathbf{a}_\rho + E_\phi\mathbf{a}_\phi + E_z\mathbf{a}_z$ V/m, where E_ρ, E_ϕ, and E_z are constants, (a) find the incremental work done in moving charge q through distance δ in a direction having equal ρ and ϕ components. (b) If the initial charge in part a was at radius $\rho = b$, what change in angle ϕ occurred in moving the charge?

4.4 An electric field in free space is given by $\mathbf{E} = x\mathbf{a}_x + y\mathbf{a}_y + z\mathbf{a}_z$ V/m. Find the work done in moving a 1-μC charge through this field (a) from $(1, 1, 1)$ to $(0, 0, 0)$; (b) from $(\rho = 2, \phi = 0)$ to $(\rho = 2, \phi = 90°)$; (c) from $(r = 10, \theta = \theta_0)$ to $(r = 10, \theta = \theta_0 + 180°)$.

4.5 Consider the vector field $\mathbf{G} = (A/\rho)\,\mathbf{a}_\phi$ where A is a constant: (a) evaluate the line integral of $\mathbf{G}$ over a circular path segment of radius b (with center at origin), along which the change in ϕ is α. (b) Same as part a, except the path is at radius $c > b$, and the change in ϕ is $-\alpha$. (c) Evaluate $\oint \mathbf{G} \cdot d\mathbf{L}$ over a four-segment path that includes those of parts a and b as two of the segments.

4.6 An electric field in free space is given as $\mathbf{E} = x\,\hat{\mathbf{a}}_x + 4z\,\hat{\mathbf{a}}_y + 4y\,\hat{\mathbf{a}}_z$. Given $V(1, 1, 1) = 10$ V, determine $V(3, 3, 3)$.

4.7 A rectangular waveguide (presented in Chapter 13) is a hollow conducting pipe oriented along the z axis, with an interior cross section enclosed by conducting planes at $x = 0$ and a, and $y = 0$ and b (see Figure 13.7 for an illustration). The transverse electric mode field, TE_{11}, defined for $(0 < x < a)$, $(0 < y < b)$, is given by

$$\mathbf{E}_{11} = E_0\,\frac{\pi}{b}\,\cos\left(\frac{\pi x}{a}\right)\sin\left(\frac{\pi y}{b}\right)\mathbf{a}_x - E_0\,\frac{\pi}{a}\sin\left(\frac{\pi x}{a}\right)\cos\left(\frac{\pi y}{b}\right)\mathbf{a}_y$$

where E_0 is a constant. (a) Find the potential difference between $(0, 0)$ and the center of the guide $(a/2, b/2)$. (b) Using the given field, provide a simple demonstration that the guide walls form an equipotential surface.

4.8 Given $\mathbf{E} = -x\mathbf{a}_x + y\mathbf{a}_y$, (a) find the work involved in moving a unit positive charge on a circular arc, the circle centered at the origin, from $x = a$ to $x = y = a/\sqrt{2}$; (b) verify that the work done in moving the charge around the full circle from $x = a$ is zero.

4.9 An electric field intensity in spherical coordinates is given as

$$\mathbf{E} = \frac{V_0}{a}\,e^{-r/a}\,\mathbf{a}_r\quad\text{V/m}$$

where V_0 and a are constants, and where the field exists everywhere. (a) Find the potential field $V(r)$, using a zero reference at infinity. (b) What is the significance of V_0?

4.10 A sphere of radius a carries a surface charge density of ρ_{s0} C/m². (a) Find the absolute potential at the sphere surface. (b) A grounded conducting shell of radius b where $b > a$ is now positioned around the charged sphere. What is the potential at the inner sphere surface in this case?

4.11 At large distances from a dipole antenna (to be addressed in Chapter 14), the electric field amplitude that it radiates assumes the simplified form

$$\mathbf{E} = \frac{A}{r}\sin(\theta)\mathbf{a}_\theta \quad \text{V/m}$$

where A is a constant. A second dipole antenna, receiving radiation from the first, is located at distance r from the first, has length L, and is oriented along the $\mathbf{a}_\theta$ direction, thus presenting its full length to the transmitting antenna at the origin. The angular position of the receiving antenna is $\theta = \theta_0$. As observed from the transmitting antenna, the receiving antenna subtends angle $\Delta\theta$. (a) Find the voltage amplitude across the length of the receiving antenna, and express your result in terms of A, θ_0, L, and r. (b) Specialize your result for the case in which $L \ll r$ (the usual case).

4.12 In spherical coordinates, $\mathbf{E} = 2r/(r^2 + a^2)^2\mathbf{a}_r$ V/m. Find the potential at any point, using the reference (a) $V = 0$ at infinity; (b) $V = 0$ at $r = 0$; (c) $V = 100$ V at $r = a$.

4.13 Three identical point charges of 4 pC each are located at the corners of an equilateral triangle 0.5 mm on a side in free space. How much work must be done to move one charge to a point equidistant from the other two and on the line joining them?

4.14 Given the electric field $\mathbf{E} = (y + 1)\mathbf{a}_x + (x - 1)\mathbf{a}_y + 2\mathbf{a}_z$ find the potential difference between the points (a) (2, −2, −1) and (0, 0, 0); (b) (3, 2, −1) and (−2, −3, 4).

4.15 Two uniform line charges, 8 nC/m each, are located at $x = 1$, $z = 2$, and at $x = -1$, $y = 2$ in free space. If the potential at the origin is 100 V, find V at $P(4, 1, 3)$.

4.16 A spherically symmetric charge distribution in free space (with $a < r < \infty$) is known to have a potential function $V(r) = V_0 a^2/r^2$, where V_0 and a are constants. (a) Find the electric field intensity. (b) Find the volume charge density. (c) Find the charge contained inside radius a. (d) Find the total energy stored in the charge (or equivalently, in its electric field).

4.17 Uniform surface charge densities of 6 and 2 nC/m² are present at $\rho = 2$ and 6 cm, respectively, in free space. Assume $V = 0$ at $\rho = 4$ cm, and calculate V at (a) $\rho = 5$ cm; (b) $\rho = 7$ cm.

4.18 Find the potential at the origin produced by a line charge $\rho_L = kx/(x^2 + a^2)$ extending along the x axis from $x = a$ to $+\infty$, where $a > 0$. Assume a zero reference at infinity.

4.19 Volume charge density is given as $\rho_v = \rho_0 e^{-r}/r$ C/m^3, valid everywhere in free space. (a) Find the potential at the origin, using Eq. (17). (b) (Harder) Find the potential by first finding **E** from Gauss's law, then line-integrating that result from infinity to the origin.

4.20 In a certain medium, the electric potential is given by

$$V(x) = \frac{\rho_0}{a\epsilon_0}(1 - e^{-ax})$$

where ρ_0 and a are constants. (a) Find the electric field intensity, **E**. (b) Find the potential difference between the points $x = d$ and $x = 0$. (c) If the medium permittivity is given by $\epsilon(x) = \epsilon_0 e^{ax}$, find the electric flux density, **D**, and the volume charge density, ρ_v, in the region. (d) Find the stored energy in the region $(0 < x < d)$, $(0 < y < 1)$, $(0 < z < 1)$.

4.21 A capacitor is formed in free space from two conducting spherical shells, arranged concentrically with their centers at the origin; the shell radii are a and b, where $b > a$. With the inner shell at potential V_0 and the outer shell grounded, the potential field (using methods to be discussed in Chapter 6) is found to be:

$$V(r) = V_0 \frac{\frac{1}{r} - \frac{1}{b}}{\frac{1}{a} - \frac{1}{b}} \quad (a < r < b)$$

The charge present is in the form of surface densities ρ_{sa} and ρ_{sb} on the inner and outer sphere surfaces, respectively. (a) Find **E** between spheres. (b) Find **E** for $r < a$ and for $r > b$. (c) Find the electric field energy density and the stored energy in the system. (d) Knowing that the energy can be found through $\frac{1}{2}\int_s \rho_s V\, da$, use your part c result to find the charge density ρ_{sa} on the inner sphere and thereby determine the relation between ρ_{sa} and **E**$(r = a)$. (e) What is the potential at the origin?

4.22 A line charge of infinite length lies along the z axis and carries a uniform linear charge density of ρ_ℓ C/m. A perfectly conducting cylindrical shell, whose axis is the z axis, surrounds the line charge. The cylinder (of radius b) is at ground potential. Under these conditions, the potential function inside the cylinder $(\rho < b)$ is given by

$$V(\rho) = k - \frac{\rho_\ell}{2\pi\epsilon_0}\ln(\rho)$$

where k is a constant. (a) Find k in terms of given or known parameters. (b) Find the electric field strength, **E**, for $\rho < b$. (c) Find the electric field strength, **E**, for $\rho > b$. (d) Find the stored energy in the electric field *per unit length* in the z direction within the volume defined by $\rho > a$, where $a < b$.

4.23 In free space, an electric potential is given in cylindrical coordinates by

$$V(\rho) = \frac{a^2\rho_0}{\epsilon_0}e^{-\rho/a}$$

(*a*) Find the electric field intensity, **E**. (*b*) Find the volume charge density. (*c*) Using Eq. (42), find the stored energy in the region $(0 < \rho < \infty)$, $(0 < \phi < 2\pi)$, $(0 < z < 1)$. (*d*) Repeat part *c*, but use Eq. (44).

4.24 A certain spherically symmetric charge configuration in free space produces an electric field given in spherical coordinates by

$$E(r) = \begin{cases} (\rho_0 r^2)/(100\,\epsilon_0)\ \mathbf{a}_r\ \text{V/m} & (r \le 10) \\ (100\rho_0)/(\epsilon_0 r^2)\ \mathbf{a}_r\ \text{V/m} & (r \ge 10) \end{cases}$$

where ρ_0 is a constant. (*a*) Find the charge density as a function of position. (*b*) Find the absolute potential as a function of position in the two regions, $r \le 10$ and $r \ge 10$. (*c*) Check your result for part *b* by using the gradient. (*d*) Find the stored energy in the charge by an integral of the form of Eq. (42). (*e*) Find the stored energy in the field by an integral of the form of Eq. (44).

4.25 Consider an electric field intensity in free space that exhibits a gaussian function of radius in spherical coordinates

$$\mathbf{E} = \frac{\rho_0\, a}{\varepsilon_0} e^{-r^2/a^2}\ \mathbf{a}_r$$

where ρ_0 and a are constants, and where the field exists everywhere. (*a*) What charge density would produce this field? (*b*) What total charge is present? (*c*) What is the potential at the origin? (*d*) What is the net stored energy in the field?

4.26 Let us assume that we have a very thin, square, imperfectly conducting plate 2 m on a side, located in the plane $z = 0$ with one corner at the origin such that it lies entirely within the first quadrant. The potential at any point in the plate is given as $V = -e^{-x}\sin y$. (*a*) An electron enters the plate at $x = 0$, $y = \pi/3$ with zero initial velocity; in what direction is its initial movement? (*b*) Because of collisions with the particles in the plate, the electron achieves a relatively low velocity and little acceleration (the work that the field does on it is converted largely into heat). The electron therefore moves approximately along a streamline. Where does it leave the plate and in what direction is it moving at the time?

4.27 By performing an appropriate line integral from infinity, show that Eq. (33) can be found from Eq. (35).

4.28 Use the electric field intensity of the dipole [Section 4.7, Eq. (35)] to find the difference in potential between points at θ_a and θ_b, each point having the same r and ϕ coordinates. Under what conditions does the answer agree with Eq. (33), for the potential at θ_a?

4.29 A dipole having a moment $\mathbf{p} = 3\mathbf{a}_x - 5\mathbf{a}_y + 10\mathbf{a}_z$ nC · m is located at $Q(1, 2, -4)$ in free space. Find V at $P(2, 3, 4)$.

4.30 A dipole for which $\mathbf{p} = 10\epsilon_0\mathbf{a}_z$ C · m is located at the origin. What is the equation of the surface on which $E_z = 0$ but $\mathbf{E} \neq 0$?

4.31 A potential field in free space is expressed as $V = 20/(xyz)$V. (a) Find the total energy stored within the cube $1 < x, y, z < 2$. (b) What value would be obtained by assuming a uniform energy density equal to the value at the center of the cube?

4.32 (a) Using Eq. (35), find the energy stored in the dipole field in the region $r > a$. (b) Why can we not let a approach zero as a limit?

4.33 A copper sphere of radius 4 cm carries a uniformly distributed total charge of 5 μC in free space. (a) Use Gauss's law to find **D** external to the sphere. (b) Calculate the total energy stored in the electrostatic field. (c) Use $W_E = Q^2/(2C)$ to calculate the capacitance of the isolated sphere.

4.34 A sphere of radius a contains volume charge of uniform density ρ_0 C/m^3. Find the total stored energy by applying (a) Eq. (42); (b) Eq. (44).

4.35 Four 0.8 nC point charges are located in free space at the corners of a square 4 cm on a side. (a) Find the total potential energy stored. (b) A fifth 0.8 nC charge is installed at the center of the square. Again find the total stored energy.

4.36 Surface charge of uniform density ρ_s lies on a spherical shell of radius b, centered at the origin in free space. (a) Find the absolute potential everywhere, with zero reference at infinity. (b) Find the stored energy in the sphere by considering the charge density and the potential in a two-dimensional version of Eq. (42). (c) Find the stored energy in the electric field and show that the results of parts b and c are identical.

CHAPTER 5 Conductors and Dielectrics

术语

dielectric 电介质。

conductivity 电导率，单位是西［门子］/ 米（S / m）。

Ohm's law 欧姆定律，相对于电路中的欧姆定律，此处应该叫欧姆定律的微分形式，其为导电媒质中每一点上电流密度与电场强度的关系。

resistance 电阻，单位是欧［姆］（Ω）。电阻是表征导体损耗电能或将电能转换为热能特性的元件，它属于导体中的电流场内容范畴，而与电介质中的静电场内容无关。

the method of image 镜像法。

bound charge 束缚电荷。

polarization 极化强度，其为单位体积的偶极矩。

electric susceptibility 电极化率。

perfect dielectric material 完纯介质，指电导率为零的理想介质。当然，实际材料一般都具有介质特性，因而具有一定的介电常数，同时也具有导电特性，从而具有一定的电导率。

In this chapter, the methods we have learned are applied to some of the materials with which an engineer must work. In the first part of the chapter, we consider conducting materials by describing the parameters that relate current to an applied electric field. This leads to a general definition of Ohm's law. We then develop methods of evaluating resistances of conductors in a few simple geometric forms. Conditions that must be met at a conducting boundary are obtained next, and this knowledge leads to a discussion of the method of images. The properties of semiconductors are described to conclude the discussion of conducting media.

In the second part of the chapter, we consider insulating materials, or dielectrics. Such materials differ from conductors in that ideally, there is no free charge that can be transported within them to produce conduction current. Instead, all charge is confined to molecular or lattice sites by coulomb forces. An applied electric field has the effect of displacing the charges slightly, leading to the formation of ensembles of electric dipoles. The extent to which this occurs is measured by the relative permittivity, or dielectric constant. Polarization of the medium may modify the electric field, whose magnitude and direction may differ from the values it would have in a different medium or in free space. Boundary conditions for the fields at interfaces between dielectrics are developed to evaluate these differences.

It should be noted that most materials will possess both dielectric and conductive properties; that is, a material considered a dielectric may be slightly conductive, and a material that is mostly conductive may be slightly polarizable. These departures from the ideal cases lead to some interesting behavior, particularly as to the effects on electromagnetic wave propagation, as we will see later. ∎

1. 电流密度是矢量，单位是安［培］/平方米（A/m²），导体的电流密度为单位面积上流过的电流。

2. 电介质是没有可以自由移动的电子或电荷的材料，其受外场的作用会被极化，极化后会产生电场。而导体是存在大量可以自由移动电子的材料。在静电场中的导体特性是体内电场强度等于零，这是由于累积在导体表面上的感应电荷在导体内各点所产生的电场会抵消掉外场，这种现象叫静电感应。而在分析电流的问题中，导体内的电场不为零。

3. 静电场分析的是电介质中的电场；而电流场是分析导体内的电场，主要分析电流密度与电场的关系。5.1~5.3节的电流与电阻内容插在静电场内容中容易混淆不同类型的场，应注意这两种场研究的问题截然不同。

5.1 CURRENT AND CURRENT DENSITY

Electric charges in motion constitute a *current,* which we treat as a flux quantity. The unit of current is the ampere (A), defined as a rate of movement of charge passing a given reference point (or crossing a given reference plane) of one coulomb per second. Current is symbolized by I, and therefore

$$I = \frac{dQ}{dt} \tag{1}$$

Current is thus defined as the motion of positive charges, even though conduction in metals takes place through the motion of electrons, as we will see shortly.

In field theory, we are usually interested in events occurring at a point rather than within a large region, and we find the concept of *current density,* measured in amperes per square meter (A/m²), more useful. Current density is a vector flux density[1] represented by **J**.

The increment of current ΔI crossing an incremental surface ΔS normal to the current density is

$$\Delta I = J_N \Delta S$$

and in the case where the current density is not perpendicular to the surface,

$$\Delta I = \mathbf{J} \cdot \Delta \mathbf{S}$$

If ΔS is of differential scale, the total current is obtained by integrating,

$$\boxed{I = \int_S \mathbf{J} \cdot d\mathbf{S}} \tag{2}$$

Current density may be related to the velocity of volume charge density at a point. Consider the element of charge $\Delta Q = \rho_v \Delta v = \rho_v \Delta S \Delta L$, as shown in Figure 5.1a. To simplify the explanation, assume that the charge element is oriented with its edges parallel to the coordinate axes and that it has only an x component of velocity. In the time interval Δt, the element of charge has moved a distance Δx, as indicated in Figure 5.1b. We have therefore moved a charge $\Delta Q = \rho_v \Delta S \Delta x$ through a reference plane perpendicular to the direction of motion in a time increment Δt, and the resulting current is

$$\Delta I = \frac{\Delta Q}{\Delta t} = \rho_v \Delta S \frac{\Delta x}{\Delta t}$$

As we take the limit with respect to time, we have

$$\Delta I = \rho_v \Delta S \, \mathrm{v}_x$$

---

[1] Current is not a vector, for it is easy to visualize a problem in which a total current I in a conductor of nonuniform cross section (such as a sphere) may have a different direction at each point of a given cross section. Current in an exceedingly fine wire, or a *filamentary current,* is occasionally defined as a vector, but we usually prefer to be consistent and give the direction to the filament, or path, and not to the current.

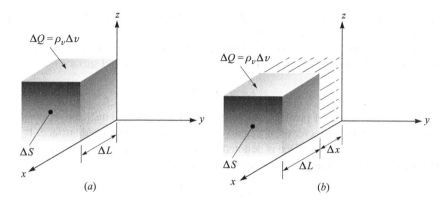

Figure 5.1 An increment of charge, $\Delta Q = \rho_v \Delta S\, \Delta L$, which moves a distance Δx in a time Δt, produces a component of current density in the limit of $J_x = \rho_v v_x$.

where v_x represents the x component of the velocity $\mathbf{v}$.[2] In terms of current density, we find

$$J_x = \rho_v\, v_x$$

and in general

$$\boxed{\mathbf{J} = \rho_v \mathbf{v}} \tag{3}$$

This last result shows clearly that charge in motion constitutes a current. We call this type of current a *convection current,* and $\mathbf{J}$ or $\rho_v \mathbf{v}$ is the *convection current density*. Note that the convection current density is related linearly to charge density as well as to velocity. The mass rate of flow of cars (cars per square foot per second) in the Holland Tunnel could be increased either by raising the density of cars per cubic foot or by going to higher speeds, if the drivers were capable of doing so.

D5.1. Given the vector current density $\mathbf{J} = 10\rho^2 z\mathbf{a}_\rho - 4\rho \cos^2 \phi\, \mathbf{a}_\phi$ mA/m²: (a) find the current density at $P(\rho = 3, \phi = 30°, z = 2)$; (b) determine the total current flowing outward through the circular band $\rho = 3, 0 < \phi < 2\pi, 2 < z < 2.8$.

Ans. (a) $180\mathbf{a}_\rho - 9\mathbf{a}_\phi$ mA/m²; (b) 3.26 A

5.2 CONTINUITY OF CURRENT

The introduction of the concept of current is logically followed by a discussion of the conservation of charge and the continuity equation. The principle of conservation of charge states simply that charges can be neither created nor destroyed, although equal

---

[2] The lowercase v is used both for volume and velocity. Note, however, that velocity always appears as a vector $\mathbf{v}$, a component v_x, or a magnitude $|\mathbf{v}|$, whereas volume appears only in differential form as dv or Δv.

amounts of positive and negative charge may be *simultaneously* created, obtained by separation, or lost by recombination.

The continuity equation follows from this principle when we consider any region bounded by a closed surface. The current through the closed surface is

$$I = \oint_S \mathbf{J} \cdot d\mathbf{S}$$

and this *outward flow* of positive charge must be balanced by a decrease of positive charge (or perhaps an increase of negative charge) within the closed surface. If the charge inside the closed surface is denoted by Q_i, then the rate of decrease is $-dQ_i/dt$ and the principle of conservation of charge requires

$$I = \oint_S \mathbf{J} \cdot d\mathbf{S} = -\frac{dQ_i}{dt} \tag{4}$$

It might be well to answer here an often-asked question. "Isn't there a sign error? I thought $I = dQ/dt$." The presence or absence of a negative sign depends on what current and charge we consider. In circuit theory we usually associate the current flow *into* one terminal of a capacitor with the time rate of increase of charge on that plate. The current of (4), however, is an *outward-flowing* current.

Equation (4) is the integral form of the continuity equation; the differential, or point, form is obtained by using the divergence theorem to change the surface integral into a volume integral:

$$\oint_S \mathbf{J} \cdot d\mathbf{S} = \int_{\text{vol}} (\nabla \cdot \mathbf{J}) dv$$

We next represent the enclosed charge Q_i by the volume integral of the charge density,

$$\int_{\text{vol}} (\nabla \cdot \mathbf{J}) dv = -\frac{d}{dt} \int_{\text{vol}} \rho_v \, dv$$

If we agree to keep the surface constant, the derivative becomes a partial derivative and may appear within the integral,

$$\int_{\text{vol}} (\nabla \cdot \mathbf{J}) dv = \int_{\text{vol}} -\frac{\partial \rho_v}{\partial t} dv$$

from which we have our point form of the continuity equation,

$$\boxed{(\nabla \cdot \mathbf{J}) = -\frac{\partial \rho_v}{\partial t}} \tag{5}$$

Remembering the physical interpretation of divergence, this equation indicates that the current, or charge per second, diverging from a small volume per unit volume is equal to the time rate of decrease of charge per unit volume at every point.

As a numerical example illustrating some of the concepts from the last two sections, let us consider a current density that is directed radially outward and decreases exponentially with time,

$$\mathbf{J} = \frac{1}{r} e^{-t} \mathbf{a}_r \ \text{A/m}^2$$

Selecting an instant of time $t = 1$ s, we may calculate the total outward current at $r = 5$ m:

$$I = J_r S = \left(\tfrac{1}{5} e^{-1}\right)(4\pi 5^2) = 23.1 \text{ A}$$

At the same instant, but for a slightly larger radius, $r = 6$ m, we have

$$I = J_r S = \left(\tfrac{1}{6} e^{-1}\right)\left(4\pi 6^2\right) = 27.7 \text{ A}$$

Thus, the total current is larger at $r = 6$ than it is at $r = 5$.

To see why this happens, we need to look at the volume charge density and the velocity. We use the continuity equation first:

$$-\frac{\partial \rho_v}{\partial t} = \nabla \cdot \mathbf{J} = \nabla \cdot \left(\tfrac{1}{r} e^{-t} \mathbf{a}_r\right) = \frac{1}{r^2} \frac{\partial}{\partial r}\left(r^2 \tfrac{1}{r} e^{-t}\right) = \frac{1}{r^2} e^{-t}$$

The volume charge density is then found by integrating with respect to t. Because ρ_v is given by a partial derivative with respect to time, the "constant" of integration may be a function of r:

$$\rho_v = -\int \frac{1}{r^2} e^{-t} \, dt + K(r) = \frac{1}{r^2} e^{-t} + K(r)$$

If we assume that $\rho_v \to 0$ as $t \to \infty$, then $K(r) = 0$, and

$$\rho_v = \frac{1}{r^2} e^{-t} \text{ C/m}^3$$

We may now use $\mathbf{J} = \rho_v \mathbf{v}$ to find the velocity,

$$v_r = \frac{J_r}{\rho_v} = \frac{\tfrac{1}{r} e^{-t}}{\tfrac{1}{r^2} e^{-t}} = r \text{ m/s}$$

The velocity is greater at $r = 6$ than it is at $r = 5$, and therefore some (unspecified) force is accelerating the charge density in an outward direction.

In summary, we have a current density that is inversely proportional to r, a charge density that is inversely proportional to r^2, and a velocity and total current that are proportional to r. All quantities vary as e^{-t}.

D5.2. Current density is given in cylindrical coordinates as $\mathbf{J} = -10^6 z^{1.5} \mathbf{a}_z$ A/m^2 in the region $0 \le \rho \le 20$ μm; for $\rho \ge 20$ μm, $\mathbf{J} = 0$. (a) Find the total current crossing the surface $z = 0.1$ m in the $\mathbf{a}_z$ direction. (b) If the charge velocity is 2×10^6 m/s at $z = 0.1$ m, find ρ_v there. (c) If the volume charge density at $z = 0.15$ m is -2000 C/m^3, find the charge velocity there.

Ans. (a) -39.7μA; (b) -15.8 mC/m^3; (c) 29.0 m/s

5.3 METALLIC CONDUCTORS

Physicists describe the behavior of the electrons surrounding the positive atomic nucleus in terms of the total energy of the electron with respect to a zero reference level for an electron at an infinite distance from the nucleus. The total energy is the sum of the kinetic and potential energies, and because energy must be given to an electron to pull it away from the nucleus, the energy of every electron in the atom is a negative quantity. Even though this picture has some limitations, it is convenient to associate these energy values with orbits surrounding the nucleus, the more negative energies corresponding to orbits of smaller radius. According to the quantum theory, only certain discrete energy levels, or energy states, are permissible in a given atom, and an electron must therefore absorb or emit discrete amounts of energy, or quanta, in passing from one level to another. A normal atom at absolute zero temperature has an electron occupying every one of the lower energy shells, starting outward from the nucleus and continuing until the supply of electrons is exhausted.

5.3.1 Influence of Material Properties on Current

In a crystalline solid, such as a metal or a diamond, atoms are packed closely together, many more electrons are present, and many more permissible energy levels are available because of the interaction forces between adjacent atoms. We find that the allowed energies of electrons are grouped into broad ranges, or "bands," each band consisting of very numerous, closely spaced, discrete levels. At a temperature of absolute zero, the normal solid also has every level occupied, starting with the lowest and proceeding in order until all the electrons are located. The electrons with the highest (least negative) energy levels, the valence electrons, are located in the *valence band*. If there are permissible higher-energy levels in the valence band, or if the valence band merges smoothly into a *conduction band*, then additional kinetic energy may be given to the valence electrons by an external field, resulting in an electron flow. The solid is called a *metallic conductor*. The filled valence band and the unfilled conduction band for a conductor at absolute zero temperature are suggested by the sketch in Figure 5.2a.

If, however, the electron with the greatest energy occupies the top level in the valence band and a gap exists between the valence band and the conduction band, then the electron cannot accept additional energy in small amounts, and the material is an insulator. This band structure is indicated in Figure 5.2b. Note that if a relatively large amount of energy can be transferred to the electron, it may be sufficiently excited to jump the gap into the next band where conduction can occur easily. Here the insulator breaks down.

An intermediate condition occurs when only a small "forbidden region" separates the two bands, as illustrated by Figure 5.2c. Small amounts of energy in the form of heat, light, or an electric field may raise the energy of the electrons at the top of the filled band and provide the basis for conduction. These materials are insulators which display many of the properties of conductors and are called *semiconductors*.

Let us first consider the conductor. Here the valence electrons, or *conduction,* or *free,* electrons, move under the influence of an electric field. With a field $\mathbf{E}$, an electron having a charge $Q = -e$ will experience a force

$$\mathbf{F} = -e\mathbf{E}$$

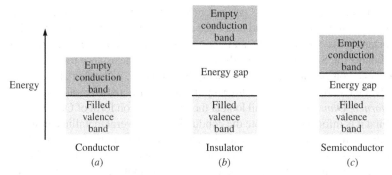

Figure 5.2 The energy-band structure in three different types of materials at 0 K. (*a*) The conductor exhibits no energy gap between the valence and conduction bands. (*b*) The insulator shows a large energy gap. (*c*) The semiconductor has only a small energy gap.

In free space, the electron would accelerate and continuously increase its velocity (and energy); in the crystalline material, the progress of the electron is impeded by continual collisions with the thermally excited crystalline lattice structure, and a constant average velocity is soon attained. This velocity $\mathbf{v}_d$ is termed the *drift velocity*, and it is linearly related to the electric field intensity by the *mobility* of the electron in the given material. We designate mobility by the symbol μ (mu), so that

$$\mathbf{v}_d = -\mu_e \mathbf{E} \tag{6}$$

where μ_e is the mobility of an electron and is positive by definition. Note that the electron velocity is in a direction opposite to the direction of $\mathbf{E}$. Equation (6) also shows that mobility is measured in the units of square meters per volt-second; typical values[3] are 0.0012 for aluminum, 0.0032 for copper, and 0.0056 for silver.

For these good conductors, a drift velocity of a few centimeters per second is sufficient to produce a noticeable temperature rise and can cause the wire to melt if the heat cannot be quickly removed by thermal conduction or radiation.

Substituting (6) into Eq. (3) of Section 5.1, we obtain

$$\boxed{\mathbf{J} = -\rho_e \mu_e \mathbf{E}} \tag{7}$$

where ρ_e is the free-electron charge density, a negative value. The total charge density ρ_v is zero because equal positive and negative charges are present in the neutral material. The negative value of ρ_e and the minus sign lead to a current density $\mathbf{J}$ that is in the same direction as the electric field intensity $\mathbf{E}$.

5.3.2 Conductivity and Ohm's Law

The relationship between $\mathbf{J}$ and $\mathbf{E}$ for a metallic conductor, however, is also specified by the conductivity σ (sigma),

$$\boxed{\mathbf{J} = \sigma \mathbf{E}} \tag{8}$$

---

[3] Wert and Thomson, p. 238, listed in the References at the end of this chapter.

where σ is measured in siemens[4] per meter (S/m). One siemens (1 S) is the basic unit of conductance in the SI system and is defined as one ampere per volt. Formerly, the unit of conductance was called the mho and was symbolized by an *inverted* Ω. Just as the siemens honors the Siemens brothers, the reciprocal unit of resistance that we call the ohm (1 Ω is one volt per ampere) honors Georg Simon Ohm, a German physicist who first described the current-voltage relationship implied by Eq. (8). We call this equation the *point form of Ohm's law;* we will look at the more common form of Ohm's law shortly.

First it is informative to note the conductivity of several metallic conductors; typical values (in siemens per meter) are 3.82×10^7 for aluminum, 5.80×10^7 for copper, and 6.17×10^7 for silver. Data for other conductors may be found in Appendix C. On seeing data such as these, it is natural to assume that we are being presented with *constant* values; this is essentially true. Metallic conductors obey Ohm's law quite faithfully, and it is a *linear* relationship; the conductivity is constant over wide ranges of current density and electric field intensity. Ohm's law and the metallic conductors are also described as *isotropic,* or having the same properties in every direction.

The conductivity is a function of temperature, however. The resistivity, which is the reciprocal of the conductivity, varies almost linearly with temperature in the region of room temperature, and for aluminum, copper, and silver it increases about 0.4 percent for a 1-degree Kelvin rise in temperature.[5] For several metals the resistivity drops abruptly to zero at a temperature of a few kelvin; this property is termed *superconductivity.* Copper and silver are not superconductors, although aluminum is (for temperatures below 1.14 K).

If we now combine Equations (7) and (8), conductivity may be expressed in terms of the charge density and the electron mobility,

$$\boxed{\sigma = -\rho_e \mu_e} \tag{9}$$

From the definition of mobility (6), it is now satisfying to note that a higher temperature infers a greater crystalline lattice vibration, more impeded electron progress for a given electric field strength, lower drift velocity, lower mobility, lower conductivity from Eq. (9), and higher resistivity as stated.

5.3.3 Resistance

The application of Ohm's law in point form to a macroscopic (visible to the naked eye) region leads to a more familiar form. Initially, assume that $\mathbf{J}$ and $\mathbf{E}$ are *uniform,* as they are in the cylindrical region shown in Figure 5.3. Because they are uniform,

$$I = \int_S \mathbf{J} \cdot d\mathbf{S} = JS \tag{10}$$

---

[4] This is the family name of two German-born brothers, Karl Wilhelm and Werner von Siemens, who were famous engineer-inventors in the nineteenth century. Karl became a British subject and was knighted, becoming Sir William Siemens.

[5] Copious temperature data for conducting materials are available in the *Standard Handbook for Electrical Engineers,* listed among the References at the end of this chapter.

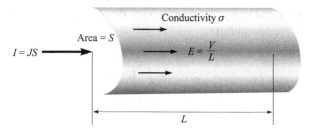

Figure 5.3 Uniform current density J and electric field intensity E in a cylindrical region of length L and cross-sectional area S. Here $V = IR$, where $R = L/\sigma S$.

and

$$V_{ab} = -\int_b^a \mathbf{E} \cdot d\mathbf{L} = -\mathbf{E} \cdot \int_b^a d\mathbf{L} = -\mathbf{E} \cdot \mathbf{L}_{ba} \tag{11}$$
$$= \mathbf{E} \cdot \mathbf{L}_{ab}$$

or

$$V = EL$$

Thus

$$J = \frac{I}{S} = \sigma E = \sigma \frac{V}{L}$$

or

$$V = \frac{L}{\sigma S} I$$

The ratio of the potential difference between the two ends of the cylinder to the current entering the more positive end, however, is recognized from elementary circuit theory as the *resistance* of the cylinder, and therefore

$$\boxed{V = IR} \tag{12}$$

where

$$\boxed{R = \frac{L}{\sigma S}} \tag{13}$$

Equation (12) is, of course, known as Ohm's law, and Eq. (13) enables us to compute the resistance R, measured in ohms (abbreviated as Ω), of conducting objects which possess uniform fields. If the fields are not uniform, the resistance may still be defined as the ratio of V to I, where V is the potential difference between two specified equipotential surfaces in the material and I is the total current crossing the more positive surface into the material. From the general integral relationships in

Eqs. (10) and (11), and using (8), we may write this general expression for resistance when the fields are nonuniform,

$$R = \frac{V_{ab}}{I} = \frac{-\int_b^a \mathbf{E} \cdot d\mathbf{L}}{\int_S \sigma \mathbf{E} \cdot d\mathbf{S}} \tag{14}$$

The line integral is taken between two equipotential surfaces in the conductor, and the surface integral is evaluated over the more positive of these two equipotentials.

EXAMPLE 5.1

As an example of the determination of the resistance of a cylinder, we find the resistance of a 1-mile length of #16 copper wire, which has a diameter of 0.0508 in.

Solution. The diameter of the wire is $0.0508 \times 0.0254 = 1.291 \times 10^{-3}$ m, the area of the cross section is $\pi(1.291 \times 10^{-3}/2)^2 = 1.308 \times 10^{-6}$ m^2, and the length is 1609 m. Using a conductivity of 5.80×10^7 S/m, the resistance of the wire is, therefore,

$$R = \frac{1609}{(5.80 \times 10^7)(1.308 \times 10^{-6})} = 21.2 \ \Omega$$

This wire can safely carry about 10 A dc, corresponding to a current density of $10/(1.308 \times 10^{-6}) = 7.65 \times 10^6$ A/m^2, or 7.65 A/mm^2. With this current, the potential difference between the two ends of the wire is 212 V, the electric field intensity is 0.312 V/m, the drift velocity is 0.000 422 m/s, or a little more than one furlong a week, and the free-electron charge density is -1.81×10^{10} C/m^3, or about one electron within a cube two angstroms on a side.

D5.3. Find the magnitude of the current density in a sample of silver for which $\sigma = 6.17 \times 10^7$ S/m and $\mu_e = 0.0056$ m^2/V · s if (a) the drift velocity is 1.5 μm/s; (b) the electric field intensity is 1 mV/m; (c) the sample is a cube 2.5 mm on a side having a voltage of 0.4 mV between opposite faces; (d) the sample is a cube 2.5 mm on a side carrying a total current of 0.5 A.

Ans. (a) 16.5 kA/m^2; (b) 61.7 kA/m^2; (c) 9.9 MA/m^2; (d) 80.0 kA/m^2

D5.4. A copper conductor has a diameter of 0.6 in. and it is 1200 ft long. Assume that it carries a total dc current of 50 A. (a) Find the total resistance of the conductor. (b) What current density exists in it? (c) What is the dc voltage between the conductor ends? (d) How much power is dissipated in the wire?

Ans. (a) 0.035 Ω; (b) 2.74×10^5 A/m^2; (c) 1.73 V; (d) 86.4 W

5.4 CONDUCTOR PROPERTIES AND BOUNDARY CONDITIONS

Once again, we must temporarily depart from our assumed static conditions and let time vary for a few microseconds to see what happens when the charge distribution is suddenly unbalanced within a conducting material. Suppose, for the sake of argument, that there suddenly appear a number of electrons in the interior of a conductor. The electric fields set up by these electrons are not counteracted by any positive charges, and the electrons therefore begin to accelerate away from each other. This continues until the electrons reach the surface of the conductor or until a number of electrons equal to the number injected have reached the surface.

Here, the outward progress of the electrons is stopped, for the material surrounding the conductor is an insulator not possessing a convenient conduction band. No charge may remain within the conductor. If it did, the resulting electric field would force the charges to the surface.

Hence the final result within a conductor is zero charge density, and a surface charge density resides on the exterior surface. This is one of the two characteristics of a good conductor.

The other characteristic, stated for static conditions in which no current may flow, follows directly from Ohm's law: the electric field intensity within the conductor is zero. Physically, we see that if an electric field were present, the conduction electrons would move and produce a current, thus leading to a nonstatic condition.

Summarizing for electrostatics, no charge and no electric field may exist at any point *within* a conducting material. Charge may, however, appear on the surface as a surface charge density, and our next investigation concerns the fields *external* to the conductor.

We wish to relate these external fields to the charge on the surface of the conductor. The problem is a simple one, and we may first talk our way to the solution with a little mathematics.

If the external electric field intensity is decomposed into two components, one tangential and one normal to the conductor surface, the tangential component is seen to be zero. If it were not zero, a tangential force would be applied to the elements of the surface charge, resulting in their motion and nonstatic conditions. Because static conditions are assumed, the tangential electric field intensity and electric flux density are zero.

Gauss's law answers our questions concerning the normal component. The electric flux leaving a small increment of surface must be equal to the charge residing on that incremental surface. The flux cannot penetrate into the conductor, for the total field there is zero. It must then leave the surface normally. Quantitatively, we may say that the electric flux density in coulombs per square meter leaving the surface normally is equal to the surface charge density in coulombs per square meter, or $D_N = \rho_S$.

If we use some of our previously derived results in making a more careful analysis (and incidentally introducing a general method which we must use later), we should set up a boundary between a conductor and free space (Figure 5.4) showing

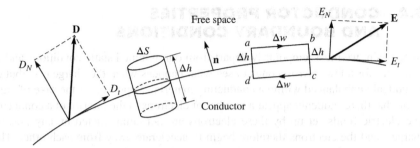

Figure 5.4 An appropriate closed path and gaussian surface are used to determine boundary conditions at a boundary between a conductor and free space; $E_t = 0$ and $D_N = \rho_S$.

tangential and normal components of **D** and **E** on the free-space side of the boundary. Both fields are zero in the conductor. The tangential field may be determined by applying Section 4.5, Eq. (21),

$$\oint \mathbf{E} \cdot d\mathbf{L} = 0$$

around the small closed path *abcda*. The integral must be broken up into four parts

$$\int_a^b + \int_b^c + \int_c^d + \int_d^a = 0$$

Remembering that **E** = 0 within the conductor, we let the length from *a* to *b* or *c* to *d* be Δw and from *b* to *c* or *d* to *a* be Δh, and we obtain

$$E_t \Delta w - E_{N,\text{at}b} \frac{1}{2} \Delta h + E_{N,\text{at}a} \frac{1}{2} \Delta h = 0$$

As we allow Δh to approach zero, keeping Δw small but finite, it makes no difference whether or not the normal fields are equal at *a* and *b*, for Δh causes these products to become negligibly small. Hence, $E_t \Delta w = 0$ and, therefore, $E_t = 0$.

The condition on the normal field is found most readily by considering D_N rather than E_N and choosing a small cylinder as the gaussian surface. Let the height be Δh and the area of the top and bottom faces be ΔS. Again, we let Δh approach zero. Using Gauss's law,

$$\oint_S \mathbf{D} \cdot d\mathbf{S} = Q$$

we integrate over the three distinct surfaces

$$\int_{\text{top}} + \int_{\text{bottom}} + \int_{\text{sides}} = Q$$

and find that the last two are zero (for different reasons). Then

$$D_N \Delta S = Q = \rho_S \Delta S$$

or

$$D_N = \rho_S$$

These are the desired *boundary conditions* for the conductor-to-free-space boundary in electrostatics,

$$D_t = E_t = 0 \qquad\qquad (15)$$

$$D_N = \epsilon_0 E_N = \rho_S \qquad\qquad (16)$$

The electric flux leaves the conductor in a direction normal to the surface, and the value of the electric flux density is numerically equal to the surface charge density. Equations (15) and (16) can be more formally expressed using the vector fields

$$\boxed{\mathbf{E} \times \mathbf{n}|_s = 0} \qquad\qquad (17)$$

$$\boxed{\mathbf{D} \cdot \mathbf{n}|_s = \rho_s} \qquad\qquad (18)$$

where **n** is the unit normal vector at the surface that points *away* from the conductor, as shown in Figure 5.4, and where both operations are evaluated at the conductor surface, *s*. Taking the cross product or the dot product of either field quantity with **n** gives the tangential or the normal component of the field, respectively.

An immediate and important consequence of a zero tangential electric field intensity is the fact that a conductor surface is an equipotential surface. The evaluation of the potential difference between any two points on the surface by the line integral leads to a zero result, because the path may be chosen on the surface itself where $\mathbf{E} \cdot d\mathbf{L} = 0$.

To summarize the principles which apply to conductors in electrostatic fields, we may state that

1. The static electric field intensity inside a conductor is zero.
2. The static electric field intensity at the surface of a conductor is everywhere directed normal to that surface.
3. The conductor surface is an equipotential surface.

Using these three principles, there are a number of quantities that may be calculated at a conductor boundary, given a knowledge of the potential field.

EXAMPLE 5.2

Given the potential,

$$V = 100(x^2 - y^2)$$

and a point $P(2, -1, 3)$ that is stipulated to lie on a conductor-to-free-space boundary, find V, **E**, **D**, and ρ_S at P, and also the equation of the conductor surface.

Solution. The potential at point P is

$$V_P = 100[2^2 - (-1)^2] = 300 \text{ V}$$

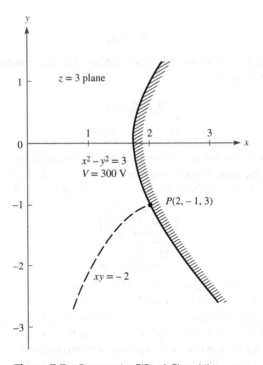

Figure 5.5 Given point $P(2, -1, 3)$ and the potential field, $V = 100(x^2 - y^2)$, we find the equipotential surface through P is $x^2 - y^2 = 3$, and the streamline through P is $xy = -2$.

Because the conductor is an equipotential surface, the potential at the entire surface must be 300 V. Moreover, if the conductor is a solid object, then the potential everywhere in and on the conductor is 300 V, for $\mathbf{E} = 0$ within the conductor.

The equation representing the locus of all points having a potential of 300 V is

$$300 = 100(x^2 - y^2)$$

or

$$x^2 - y^2 = 3$$

This is therefore the equation of the conductor surface; it happens to be a hyperbolic cylinder, as shown in Figure 5.5. Let us assume arbitrarily that the solid conductor lies above and to the right of the equipotential surface at point P, whereas free space is down and to the left.

Next, we find $\mathbf{E}$ by the gradient operation,

$$\mathbf{E} = -100\nabla(x^2 - y^2) = -200x\mathbf{a}_x + 200y\mathbf{a}_y$$

At point P,

$$\mathbf{E}_p = -400\mathbf{a}_x - 200\mathbf{a}_y \text{ V/m}$$

Because $\mathbf{D} = \epsilon_0 \mathbf{E}$, we have

$$\mathbf{D}_P = 8.854 \times 10^{-12}\,\mathbf{E}_P = -3.54\mathbf{a}_x - 1.771\mathbf{a}_y \ \text{nC/m}^2$$

The field is directed downward and to the left at P; it is normal to the equipotential surface. Therefore,

$$D_N = |\mathbf{D}_P| = 3.96 \ \text{nC/m}^2$$

Thus, the surface charge density at P is

$$\rho_{S,P} = D_N = 3.96 \ \text{nC/m}^2$$

Note that if we had taken the region to the left of the equipotential surface as the conductor, the $\mathbf{E}$ field would *terminate* on the surface charge and we would let $\rho_S = -3.96 \ \text{nC/m}^2$.

EXAMPLE 5.3

Finally, let us determine the equation of the streamline passing through P.

Solution.
We see that

$$\frac{E_y}{E_x} = \frac{200y}{-200x} = -\frac{y}{x} = \frac{dy}{dx}$$

Thus,

$$\frac{dy}{y} + \frac{dx}{x} = 0$$

and

$$\ln y + \ln x = C_1$$

Therefore,

$$xy = C_2$$

The line (or surface) through P is obtained when $C_2 = (2)(-1) = -2$. Thus, the streamline is the trace of another hyperbolic cylinder,

$$xy = -2$$

This is also shown on Figure 5.5.

D5.5. Given the potential field in free space, $V = 100 \sinh 5x \sin 5y$ V, and a point $P(0.1, 0.2, 0.3)$, find at P: (a) V; (b) $\mathbf{E}$; (c) $|\mathbf{E}|$; (d) $|\rho_S|$ if it is known that P lies on a conductor surface.

Ans. (a) 43.8 V; (b) $-474\mathbf{a}_x - 140.8\mathbf{a}_y$ V/m; (c) 495 V/m; (d) 4.38 nC/m^2

5.5   THE METHOD OF IMAGES

One important characteristic of the dipole field that we developed in Chapter 4 is the infinite plane at zero potential that exists midway between the two charges. Such a plane may be represented by a vanishingly thin conducting plane that is infinite in extent. The conductor is an equipotential surface at a potential $V = 0$, and the electric field intensity is therefore normal to the surface. Thus, if we replace the dipole configuration shown in Figure 5.6a with the single charge and conducting plane shown in Figure 5.6b, the fields in the upper half of each figure are the same. Below the conducting plane, all fields are zero, as we have not provided any charges in that region. Of course, we might also substitute a single negative charge below a conducting plane for the dipole arrangement and obtain equivalence for the fields in the lower half of each region.

If we approach this equivalence from the opposite point of view, we begin with a single charge above a perfectly conducting plane and then see that we may maintain the same fields above the plane by removing the plane and locating a negative charge at a symmetrical location below the plane. This charge is called the *image* of the original charge, and it is the negative of that value.

If we can do this once, linearity allows us to do it again and again, and thus *any* charge configuration above an infinite ground plane may be replaced by an arrangement composed of the given charge configuration, its image, and no conducting plane. This is suggested by the two illustrations of Figure 5.7. In many cases, the potential field of the new system is much easier to find since it does not contain the conducting plane with its unknown surface charge distribution.

As an example of the use of images, we find the surface charge density at $P(2, 5, 0)$ on the conducting plane $z = 0$ if there is a line charge of 30 nC/m located at $x = 0$, $z = 3$, as shown in Figure 5.8a. We remove the plane and install an image line charge of -30 nC/m at $x = 0$, $z = -3$, as illustrated in Figure 5.8b. The field at P may now be obtained by superposition of the known fields of the line charges. The radial vector

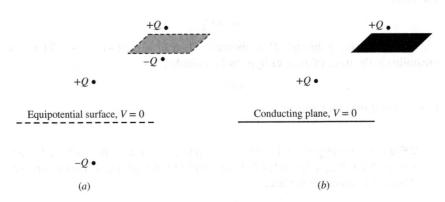

Figure 5.6   (a) Two equal but opposite charges may be replaced by (b) a single charge and a conducting plane without affecting the fields above the $V = 0$ surface.

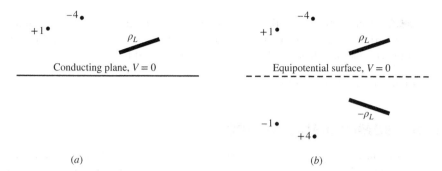

Figure 5.7 (a) A given charge configuration above an infinite conducting plane may be replaced by (b) the given charge configuration plus the image configuration, without the conducting plane.

from the positive line charge to P is $\mathbf{R}_+ = 2\mathbf{a}_x - 3\mathbf{a}_z$, while $\mathbf{R}_- = 2\mathbf{a}_x + 3\mathbf{a}_z$. Thus, the individual fields are

$$\mathbf{E}_+ = \frac{\rho_L}{2\pi\epsilon_0 R_+}\mathbf{a}_{R+} = \frac{30 \times 10^{-9}}{2\pi\epsilon_0 \sqrt{13}}\frac{2\mathbf{a}_x - 3\mathbf{a}_z}{\sqrt{13}}$$

and

$$\mathbf{E}_- = \frac{30 \times 10^{-9}}{2\pi\epsilon_0 \sqrt{13}}\frac{2\mathbf{a}_x + 3\mathbf{a}_z}{\sqrt{13}}$$

Adding these results, we have

$$\mathbf{E} = \frac{-180 \times 10^{-9}\mathbf{a}_z}{2\pi\epsilon_0(13)} = -249\mathbf{a}_z \text{ V/m}$$

This then is the field at (or just above) P in both the configurations of Figure 5.8, and it is certainly satisfying to note that the field is normal to the conducting plane, as it must be. Thus, $\mathbf{D} = \epsilon_0\mathbf{E} = -2.20\mathbf{a}_z \text{ nC/m}^2$, and because this is directed *toward* the conducting plane, ρ_S is negative and has a value of -2.20 nC/m^2 at P.

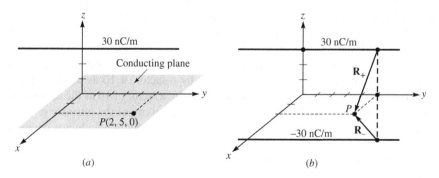

Figure 5.8 (a) A line charge above a conducting plane. (b) The conductor is removed, and the image of the line charge is added.

> **D5.6.** A perfectly conducting plane is located in free space at $x = 4$, and a uniform infinite line charge of 40 nC/m lies along the line $x = 6$, $y = 3$. Let $V = 0$ at the conducting plane. At $P(7, -1, 5)$ find: (a) V; (b) $\mathbf{E}$.
>
> **Ans.** (a) 317 V; (b) $-45.3\mathbf{a}_x - 99.2\mathbf{a}_y$ V/m

5.6   SEMICONDUCTORS

If we now turn our attention to an intrinsic semiconductor material, such as pure germanium or silicon, two types of current carriers are present, electrons and holes. The electrons are those from the top of the filled valence band that have received sufficient energy (usually thermal) to cross the relatively small forbidden band into the conduction band. The forbidden-band energy gap in typical semiconductors is of the order of one electronvolt. The vacancies left by these electrons represent unfilled energy states in the valence band which may also move from atom to atom in the crystal. The vacancy is called a hole, and many semiconductor properties may be described by treating the hole as if it had a positive charge of e, a mobility, μ_h, and an effective mass comparable to that of the electron. Both carriers move in an electric field, and they move in opposite directions; hence each contributes a component of the total current which is in the same direction as that provided by the other. The conductivity is therefore a function of both hole and electron concentrations and mobilities,

$$\sigma = -\rho_e\mu_e + \rho_h\mu_h \qquad (19)$$

For pure, or *intrinsic,* silicon, the electron and hole mobilities are 0.12 and 0.025, respectively, whereas for germanium, the mobilities are, respectively, 0.36 and 0.17. These values are given in square meters per volt-second and range from 10 to 100 times as large as those for aluminum, copper, silver, and other metallic conductors.[6] These mobilities are given for a temperature of 300 K.

The electron and hole concentrations depend strongly on temperature. At 300 K the electron and hole volume charge densities are both 0.0024 C/m^3 in magnitude in intrinsic silicon and 3.0 C/m^3 in intrinsic germanium. These values lead to conductivities of 0.000 35 S/m in silicon and 1.6 S/m in germanium. As temperature increases, the mobilities decrease, but the charge densities increase very rapidly. As a result, the conductivity of silicon increases by a factor of 10 as the temperature increases from 300 to about 330 K and decreases by a factor of 10 as the temperature drops from 300 to about 275 K. Note that the conductivity of the intrinsic semiconductor increases with temperature, whereas that of a metallic conductor decreases with temperature; this is one of the characteristic differences between the metallic conductors and the intrinsic semiconductors.

---

[6] Mobility values for semiconductors are given in References 2, 3, and 5 listed at the end of this chapter.

Intrinsic semiconductors also satisfy the point form of Ohm's law; that is, the conductivity is reasonably constant with current density and with the direction of the current density.

The number of charge carriers and the conductivity may both be increased dramatically by adding very small amounts of impurities. *Donor* materials provide additional electrons and form *n-type* semiconductors, whereas *acceptors* furnish extra holes and form *p-type* materials. The process is known as *doping,* and a donor concentration in silicon as low as one part in 10^7 causes an increase in conductivity by a factor of 10^5.

The range of value of the conductivity is extreme as we go from the best insulating materials to semiconductors and the finest conductors. In siemens per meter, σ ranges from 10^{-17} for fused quartz, 10^{-7} for poor plastic insulators, and roughly unity for semiconductors to almost 10^8 for metallic conductors at room temperature. These values cover the remarkably large range of some 25 orders of magnitude.

D5.7. Using the values given in this section for the electron and hole mobilities in silicon at 300 K, and assuming hole and electron charge densities are 0.0029 C/m^3 and −0.0029 C/m^3, respectively, find: (*a*) the component of the conductivity due to holes; (*b*) the component of the conductivity due to electrons; (*c*) the conductivity.

Ans. (*a*) 72.5 μS/m; (*b*) 348 μS/m; (*c*) 421 μS/m

Quizzes

5.7 THE NATURE OF DIELECTRIC MATERIALS

A dielectric in an electric field can be viewed as a free-space arrangement of microscopic electric dipoles, each of which is composed of a positive and a negative charge whose centers do not quite coincide. These are not free charges, and they cannot contribute to the conduction process. Rather, they are bound in place by atomic and molecular forces and can only shift positions slightly in response to external fields. They are called *bound* charges, in contrast to the free charges that determine conductivity. The bound charges can be treated as any other sources of the electrostatic field. Therefore, we would not need to introduce the dielectric constant as a new parameter or to deal with permittivities different from the permittivity of free space; however, the alternative would be to consider *every charge within a piece of dielectric material.* This is too great a price to pay for using all our previous equations in an unmodified form, and we shall therefore spend some time theorizing about dielectrics in a qualitative way; introducing polarization **P**, permittivity ϵ, and relative permittivity ϵ_r; and developing some quantitative relationships involving these new parameters.

Interactives

5.7.1 Bound Charge and the Polarization Field

The characteristic that all dielectric materials have in common, whether they are solid, liquid, or gas, and whether or not they are crystalline in nature, is their ability to store electric energy. This storage takes place by means of a shift in the relative positions of the internal, bound positive and negative charges against the normal molecular and atomic forces.

This displacement against a restraining force is analogous to lifting a weight or stretching a spring, and it represents potential energy. The source of the energy is the external field, the motion of the shifting charges resulting perhaps in a transient current through a battery that is producing the field.

The actual mechanism of the charge displacement differs in the various dielectric materials. Some molecules, termed *polar* molecules, have a permanent displacement existing between the centers of "gravity" of the positive and negative charges, and each pair of charges acts as a dipole. Normally the dipoles are oriented in a random way throughout the interior of the material, and the action of the external field is to align these molecules, to some extent, in the same direction. A sufficiently strong field may even produce an additional displacement between the positive and negative charges.

A *nonpolar* molecule does not have this dipole arrangement until after a field is applied. The negative and positive charges shift in opposite directions against their mutual attraction and produce a dipole that is aligned with the electric field.

Either type of dipole may be described by its dipole moment $\mathbf{p}$, as developed in Section 4.7, Eq. (36),

$$\mathbf{p} = Q\mathbf{d} \tag{20}$$

where Q is the positive one of the two bound charges composing the dipole, and $\mathbf{d}$ is the vector from the negative to the positive charge. We note again that the units of $\mathbf{p}$ are coulomb-meters.

If there are n dipoles per unit volume and we deal with a volume Δv, then there are $n \, \Delta v$ dipoles, and the total dipole moment is obtained by the vector sum,

$$\mathbf{p}_{\text{total}} = \sum_{i=1}^{n \Delta v} \mathbf{p}_i$$

If the dipoles are aligned in the same general direction, $\mathbf{p}_{\text{total}}$ may have a significant value. However, a random orientation may cause $\mathbf{p}_{\text{total}}$ to be essentially zero.

We now define the *polarization field*, $\mathbf{P}$ (a flux density), as the dipole moment per unit volume,

$$\mathbf{P} = \lim_{\Delta v \to 0} \frac{1}{\Delta v} \sum_{i=1}^{n \Delta v} \mathbf{p}_i \tag{21}$$

with units of coulombs per square meter. We will treat $\mathbf{P}$ as a typical continuous field, even though it is obvious that it is essentially undefined at points within an atom or molecule. Instead, we should think of its value at any point as an average value taken over a sample volume Δv—large enough to contain many molecules ($n \, \Delta v$ in number), but yet sufficiently small to be considered incremental in concept.

Our immediate goal is to show that the bound-volume charge density acts like the free-volume charge density in producing an external field; we will obtain a result similar to Gauss's law.

To be specific, assume that we have a dielectric containing nonpolar molecules. No molecule has a dipole moment, and $\mathbf{P} = 0$ throughout the material. Somewhere in the interior of the dielectric we select an incremental surface element $\Delta \mathbf{S}$, as shown in

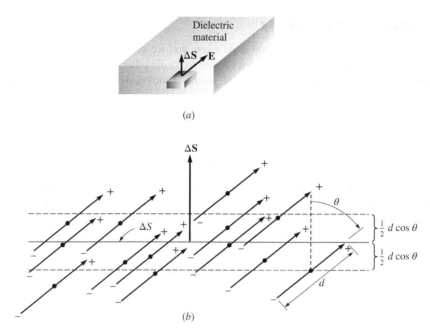

Figure 5.9 (a) An incremental surface element ΔS is shown in the interior of a dielectric in which an electric field **E** is present. (b) The nonpolar molecules form dipole moments **p** and a polarization **P**. There is a net transfer of bound charge across ΔS.

Figure 5.9a, and apply an electric field **E**. The electric field produces a moment $\mathbf{p} = Q\mathbf{d}$ in each molecule, such that **p** and **d** make an angle θ with ΔS, as indicated in Figure 5.9b.

The bound charges will now move across ΔS. Each of the charges associated with the creation of a dipole must have moved a distance $\frac{1}{2}d\cos\theta$ in the direction perpendicular to ΔS. Thus, any positive charges initially lying below the surface ΔS and within the distance $\frac{1}{2}d\cos\theta$ of the surface must have crossed ΔS going upward. Also, any negative charges initially lying above the surface and within that distance $(\frac{1}{2}d\cos\theta)$ from ΔS must have crossed ΔS going downward. Therefore, because there are n molecules/m³, the net total charge that crosses the elemental surface in an upward direction is equal to $nQd\cos\theta\Delta S$, or

$$\Delta Q_b = n\,Q\mathbf{d}\cdot\Delta\mathbf{S}$$

where the subscript on Q_b reminds us that we are dealing with a bound charge and not a free charge. In terms of the polarization, we have

$$\Delta Q_b = \mathbf{P}\cdot\Delta\mathbf{S}$$

If we interpret ΔS as an element of a *closed* surface inside the dielectric material, then the direction of $\Delta\mathbf{S}$ is outward, and the net increase in the bound charge *within* the closed surface is obtained through the integral

$$Q_b = -\oint_S \mathbf{P}\cdot d\mathbf{S} \tag{22}$$

5.7.2 Electric Flux Density

Eq. (22) has some resemblance to Gauss's law, and we may now generalize our definition of electric flux density so that it applies to media other than free space. We first write Gauss's law in terms of $\epsilon_0\mathbf{E}$ and QT, the *total* enclosed charge, bound plus free:

$$Q_T = \oint_S \epsilon_0 \mathbf{E} \cdot d\mathbf{S} \tag{23}$$

where

$$Q_T = Q_b + Q$$

and Q is the total *free* charge enclosed by the surface S. Note that the free charge appears without a subscript because it is the most important type of charge and will appear in Maxwell's equations.

Combining these last three equations, we obtain an expression for the free charge enclosed,

$$Q = Q_T - Q_b = \oint_S (\epsilon_0 \mathbf{E} + \mathbf{P}) \cdot d\mathbf{S} \tag{24}$$

Electric flux density, $\mathbf{D}$, is now defined in more general terms than was done in Chapter 3,

$$\boxed{\mathbf{D} = \epsilon_0 \mathbf{E} + \mathbf{P}} \tag{25}$$

There is thus an added term to $\mathbf{D}$ that appears when polarizable material is present. Thus,

$$Q = \oint_S \mathbf{D} \cdot d\mathbf{S} \tag{26}$$

where Q is the free charge enclosed.

Utilizing the several volume charge densities, we have

$$Q_b = \int_v \rho_b \, dv$$

$$Q = \int_v \rho_v \, dv$$

$$Q_T = \int_v \rho_T \, dv$$

With the help of the divergence theorem, we may therefore transform Eqs. (22), (23), and (26) into the equivalent divergence relationships,

$$\nabla \cdot \mathbf{P} = -\rho_b$$

$$\nabla \cdot \epsilon_0 \mathbf{E} = \rho_T$$

$$\boxed{\nabla \cdot \mathbf{D} = \rho_v} \tag{27}$$

We will emphasize only Eq. (26) and (27), the two expressions involving the free charge, in the work that follows.

5.7.3 Material Susceptibility and Dielectric Constant

In order to make any real use of these new concepts, it is necessary to know the relationship between the electric field intensity $\mathbf{E}$ and the polarization $\mathbf{P}$ that results. This relationship will, of course, be a function of the type of material, and we will essentially limit our discussion to those isotropic materials for which $\mathbf{E}$ and $\mathbf{P}$ are linearly related. In an isotropic material, the vectors $\mathbf{E}$ and $\mathbf{P}$ are always parallel, regardless of the orientation of the field. Although most engineering dielectrics are linear for moderate-to-large field strengths and are also isotropic, single crystals may be anisotropic. The periodic nature of crystalline materials causes dipole moments to be formed most easily along the crystal axes, and not necessarily in the direction of the applied field.

In *ferroelectric* materials, the relationship between $\mathbf{P}$ and $\mathbf{E}$ not only is nonlinear, but also shows hysteresis effects; that is, the polarization produced by a given electric field intensity depends on the past history of the sample. Important examples of this type of dielectric are barium titanate, often used in ceramic capacitors, and Rochelle salt.

The linear relationship between $\mathbf{P}$ and $\mathbf{E}$ is

$$\boxed{\mathbf{P} = \chi_e \epsilon_0 \mathbf{E}} \qquad (28)$$

where χ_e (chi) is a dimensionless quantity called the *electric susceptibility* of the material.

Using this relationship in Eq. (25), we have

$$\mathbf{D} = \epsilon_0 \mathbf{E} + \chi_e \epsilon_0 \mathbf{E} = (\chi_e + 1)\epsilon_0 \mathbf{E}$$

The expression within the parentheses is now defined as

$$\epsilon_r = \chi_e + 1 \qquad (29)$$

This is another dimensionless quantity, and it is known as the *relative permittivity*, or *dielectric constant* of the material. Thus,

$$\mathbf{D} = \epsilon_0 \epsilon_r \mathbf{E} = \epsilon \mathbf{E} \qquad (30)$$

where

$$\boxed{\epsilon = \epsilon_0 \epsilon_r} \qquad (31)$$

and ϵ is the *permittivity*. The dielectric constants are given for some representative materials in Appendix C.

Anisotropic dielectric materials cannot be described in terms of a simple susceptibility or permittivity parameter. Instead, we find that each component of $\mathbf{D}$ may be a function of every component of $\mathbf{E}$, and $\mathbf{D} = \epsilon \mathbf{E}$ becomes a matrix equation where $\mathbf{D}$ and $\mathbf{E}$ are each 3×1 column matrices and ϵ is a 3×3 square matrix. Expanding the matrix equation gives

$$D_x = \epsilon_{xx} E_x + \epsilon_{xy} E_y + \epsilon_{xz} E_z$$
$$D_y = \epsilon_{yx} E_x + \epsilon_{yy} E_y + \epsilon_{yz} E_z$$
$$D_z = \epsilon_{zx} E_x + \epsilon_{zy} E_y + \epsilon_{zz} E_z$$

Note that the elements of the matrix depend on the selection of the coordinate axes in the anisotropic material. Certain choices of axis directions lead to simpler matrices.[7]

D and **E** (and **P**) are no longer parallel, and although $\mathbf{D} = \epsilon_0\mathbf{E} + \mathbf{P}$ remains a valid equation for anisotropic materials, we may continue to use $\mathbf{D} = \epsilon\mathbf{E}$ only by interpreting it as a matrix equation. We will concentrate our attention on linear isotropic materials and reserve the general case for a more advanced text.

In summary, then, we now have a relationship between **D** and **E** that depends on the dielectric material present,

$$\boxed{\mathbf{D} = \epsilon\mathbf{E}} \tag{30}$$

where

$$\boxed{\epsilon = \epsilon_0\epsilon_r} \tag{31}$$

This electric flux density is still related to the free charge by either the point or integral form of Gauss's law:

$$\boxed{\nabla \cdot \mathbf{D} = \rho_v} \tag{27}$$

$$\boxed{\oint_S \mathbf{D} \cdot d\mathbf{S} = Q} \tag{26}$$

Use of the relative permittivity, as indicated by Eq. (31), makes consideration of the polarization, dipole moments, and bound charge unnecessary. However, when anisotropic or nonlinear materials must be considered, the relative permittivity, in the simple scalar form that we have discussed, is no longer applicable.

EXAMPLE 5.4

We locate a slab of Teflon in the region $0 \le x \le a$, and assume free space where $x < 0$ and $x > a$. Outside the Teflon there is a uniform field $\mathbf{E}_{\text{out}} = E_0\mathbf{a}_x$ V/m. We seek values for **D**, **E**, and **P** everywhere.

Solution. The dielectric constant of the Teflon is 2.1, and thus the electric susceptibility is 1.1.

Outside the slab, we have immediately $\mathbf{D}_{\text{out}} = \epsilon_0 E_0\mathbf{a}_x$. Also, as there is no dielectric material there, $\mathbf{P}_{\text{out}} = 0$. Now, any of the last four or five equations will enable us to relate the several fields inside the material to each other. Thus

$$\mathbf{D}_{\text{in}} = 2.1\,\epsilon_0\,\mathbf{E}_{\text{in}} \qquad (0 \le x \le a)$$
$$\mathbf{P}_{\text{in}} = 1.1\,\epsilon_0\,\mathbf{E}_{\text{in}} \qquad (0 \le x \le a)$$

---

[7] A more complete discussion of this matrix may be found in the Ramo, Whinnery, and Van Duzer reference listed at the end of this chapter.

As soon as we establish a value for any of these three fields within the dielectric, the other two can be found immediately. The difficulty lies in crossing over the boundary from the known fields external to the dielectric to the unknown ones within it. To do this we need a boundary condition, and this is the subject of the next section. We will complete this example then.

In the remainder of this text we will describe polarizable materials in terms of **D** and ϵ rather than **P** and χ_e. We will limit our discussion to isotropic materials.

D5.8. A slab of dielectric material has a relative dielectric constant of 3.8 and contains a uniform electric flux density of 8 nC/m². If the material is lossless, find: (a) E; (b) P; (c) the average number of dipoles per cubic meter if the average dipole moment is 10^{-29} C · m.

Ans. (a) 238 V/m; (b) 5.89 nC/m²; (c) 5.89×10^{20} m^{-3}

5.8 BOUNDARY CONDITIONS FOR PERFECT DIELECTRIC MATERIALS

How do we attack a problem in which there are two different dielectrics, or a dielectric and a conductor? This is another example of a *boundary condition*, such as the condition at the surface of a conductor whereby the tangential fields are zero and the normal electric flux density is equal to the surface charge density on the conductor. Now we take the first step in solving a two-dielectric problem, or a dielectric-conductor problem, by determining the behavior of the fields at the dielectric interface.

Illustrations

5.8.1 Tangential Electric Field at a Boundary

We first consider the interface between two dielectrics having permittivities ϵ_1 and ϵ_2 and occupying regions 1 and 2, as shown in Figure 5.10. We first examine the tangential components by using

$$\oint \mathbf{E} \cdot d\mathbf{L} = 0$$

around the small closed path on the left, obtaining

$$E_{\tan 1} \Delta w - E_{\tan 2} \Delta w = 0$$

The small contribution to the line integral by the normal component of **E** along the sections of length Δh becomes negligible as Δh decreases and the closed path crowds the surface. Immediately, then,

$$\boxed{E_{\tan 1} = E_{\tan 2}} \tag{32}$$

Evidently, Kirchhoff's voltage law is still applicable to this case. Certainly we have shown that the potential difference between any two points on the boundary that are separated by a distance Δw is the same immediately above or below the boundary.

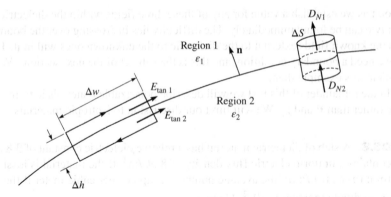

Figure 5.10 The boundary between perfect dielectrics of permittivities ϵ_1 and ϵ_2. The continuity of D_N is shown by the gaussian surface on the right, and the continuity of E_{tan} is shown by the line integral about the closed path at the left.

If the tangential electric field intensity is continuous across the boundary, then tangential **D** is discontinuous, for

$$\frac{D_{\tan 1}}{\epsilon_1} = E_{\tan 1} = E_{\tan 2} = \frac{D_{\tan 2}}{\epsilon_2}$$

or

$$\frac{D_{\tan 1}}{D_{\tan 2}} = \frac{\epsilon_1}{\epsilon_2} \qquad (33)$$

5.8.2 Normal Electric Flux Density at a Boundary

The boundary conditions on the normal components are found by applying Gauss's law to the small "pillbox" shown at the right in Figure 5.10. The sides are again very short, and the flux leaving the top and bottom surfaces is the difference

$$D_{N1}\,\Delta S - D_{N2}\,\Delta S = \Delta Q = \rho_S \Delta S$$

from which

$$\boxed{D_{N1} - D_{N2} = \rho_S} \qquad (34)$$

What is this surface charge density? It cannot be a *bound* surface charge density, because we are taking the polarization of the dielectric into effect by using a dielectric constant different from unity; that is, instead of considering bound charges in free space, we are using an increased permittivity. Also, it is extremely unlikely that any *free* charge is on the interface, for no free charge is available in the perfect dielectrics we are considering. This charge must then have been placed there deliberately, thus unbalancing the total charge in and on this dielectric body. Except for this special case, then, we may assume ρ_S is zero on the interface and

$$\boxed{D_{N1} = D_{N2}} \qquad (35)$$

or the normal component of **D** is continuous. It follows that

$$\epsilon_1 E_{N1} = \epsilon_2 E_{N2} \tag{36}$$

and normal **E** is discontinuous.

5.8.3 General Boundary Conditions and an Application

Equations (32) and (34) can be written in terms of field vectors in any direction, along with the unit normal to the surface as shown in Figure 5.10. Formally stated, the boundary conditions for the electric flux density and the electric field strength at the surface of a perfect dielectric are

$$\boxed{(\mathbf{D}_1 - \mathbf{D}_2) \cdot \mathbf{n} = \rho_s} \tag{37}$$

which is the general statement of Eq. (32), and

$$\boxed{(\mathbf{E}_1 - \mathbf{E}_2) \times \mathbf{n} = 0} \tag{38}$$

generally states Eq. (34). This construction was used previously in Eqs. (17) and (18) for a conducting surface, in which the normal or tangential components of the fields are obtained through the dot product or cross product with the normal, respectively.

These conditions may be used to show the change in the vectors **D** and **E** at the surface. Let $\mathbf{D}_1$ (and $\mathbf{E}_1$) make an angle θ_1 with a normal to the surface (Figure 5.11). Because the normal components of **D** are continuous,

$$D_{N1} = D_1 \cos\theta_1 = D_2 \cos\theta_2 = D_{N2} \tag{39}$$

The ratio of the tangential components is given by (33) as

$$\frac{D_{\tan 1}}{D_{\tan 2}} = \frac{D_1 \sin\theta_1}{D_2 \sin\theta_2} = \frac{\epsilon_1}{\epsilon_2}$$

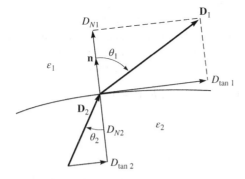

Figure 5.11 The refraction of **D** at a dielectric interface. For the case shown, $\epsilon_1 > \epsilon_2$; **E**$_1$ and **E**$_2$ are directed along **D**$_1$ and **D**$_2$, with $D_1 > D_2$ and $E_1 < E_2$.

or

$$\epsilon_2 D_1 \sin \theta_1 = \epsilon_1 D_2 \sin \theta_2 \qquad (40)$$

and the division of this equation by (39) gives

$$\frac{\tan \theta_1}{\tan \theta_2} = \frac{\epsilon_1}{\epsilon_2} \qquad (41)$$

In Figure 5.11 we have assumed that $\epsilon_1 > \epsilon_2$, and therefore $\theta_1 > \theta_2$.

The direction of $\mathbf{E}$ on each side of the boundary is identical with the direction of $\mathbf{D}$, because $\mathbf{D} = \epsilon \mathbf{E}$.

The magnitude of $\mathbf{D}$ in region 2 may be found from Eq. (39) and (40),

$$D_2 = D_1 \sqrt{\cos^2 \theta_1 + \left(\frac{\epsilon_2}{\epsilon_1}\right)^2 \sin^2 \theta_1} \qquad (42)$$

and the magnitude of $\mathbf{E}_2$ is

$$E_2 = E_1 \sqrt{\sin^2 \theta_1 + \left(\frac{\epsilon_1}{\epsilon_2}\right)^2 \cos^2 \theta_1} \qquad (43)$$

An inspection of these equations shows that D is larger in the region of larger permittivity (unless $\theta_1 = \theta_2 = 0°$ where the magnitude is unchanged) and that E is larger in the region of smaller permittivity (unless $\theta_1 = \theta_2 = 90°$, where its magnitude is unchanged).

EXAMPLE 5.5

Complete Example 5.4 by finding the fields within the Teflon ($\epsilon_r = 2.1$), given the uniform external field $\mathbf{E}_{\text{out}} = E_0 \mathbf{a}_x$ in free space.

Solution. We recall that we had a slab of Teflon extending from $x = 0$ to $x = a$, as shown in Figure 5.12, with free space on both sides of it and an external field $\mathbf{E}_{\text{out}} = E_0 \mathbf{a}_x$. We also have $\mathbf{D}_{\text{out}} = \epsilon_0 E_0 \mathbf{a}_x$ and $\mathbf{P}_{\text{out}} = 0$.

Inside, the continuity of D_N at the boundary allows us to find that $\mathbf{D}_{\text{in}} = \mathbf{D}_{\text{out}} = \epsilon_0 E_0 \mathbf{a}_x$. This gives us $\mathbf{E}_{\text{in}} = \mathbf{D}_{\text{in}}/\epsilon = \epsilon_0 E_0 \mathbf{a}_x/(\epsilon_r \epsilon_0) = 0.476 E_0 \mathbf{a}_x$. To get the polarization field in the dielectric, we use $\mathbf{D} = \epsilon_0 \mathbf{E} + \mathbf{P}$ and obtain

$$\mathbf{P}_{\text{in}} = \mathbf{D}_{\text{in}} - \epsilon_0 \mathbf{E}_{\text{in}} = \epsilon_0 E_0 \mathbf{a}_x - 0.476 \epsilon_0 E_0 \mathbf{a}_x = 0.524 \epsilon_0 E_0 \mathbf{a}_x$$

Summarizing then gives

$$\mathbf{D}_{\text{in}} = \epsilon_0 E_0 \mathbf{a}_x \qquad (0 \leq x \leq a)$$
$$\mathbf{E}_{\text{in}} = 0.476 E_0 \mathbf{a}_x \qquad (0 \leq x \leq a)$$
$$\mathbf{P}_{\text{in}} = 0.524 \epsilon_0 E_0 \mathbf{a}_x \qquad (0 \leq x \leq a)$$

A practical problem most often does not provide us with a direct knowledge of the field on either side of the boundary. The boundary conditions must be used to help us determine the fields on both sides of the boundary from the other information that is given.

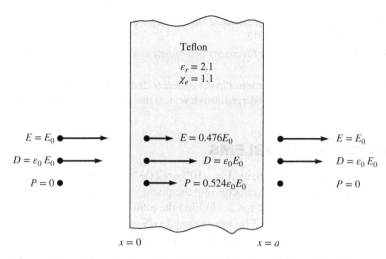

Figure 5.12 A knowledge of the electric field external to the dielectric enables us to find the remaining external fields first and then to use the continuity of normal **D** to begin finding the internal fields.

D5.9. Let Region 1 ($z < 0$) be composed of a uniform dielectric material for which $\epsilon_r = 3.2$, while Region 2 ($z > 0$) is characterized by $\epsilon_r = 2$. Let $\mathbf{D}_1 = -30\mathbf{a}_x + 50\mathbf{a}_y + 70\mathbf{a}_z$ nC/m² and find: (a) D_{N1}; (b) $\mathbf{D}_{t1}$; (c) D_{t1}; (d) D_1; (e) θ_1; (f) $\mathbf{P}_1$.

Ans. (a) 70 nC/m²; (b) $-30\mathbf{a}_x + 50\mathbf{a}_y$ nC/m²; (c) 58.3 nC/m²; (d) 91.1 nC/m²; (e) 39.8°; (f) $-20.6\mathbf{a}_x + 34.4\mathbf{a}_y + 48.1\mathbf{a}_z$ nC/m²

D5.10. Continue Problem D5.9 by finding: (a) $\mathbf{D}_{N2}$; (b) $\mathbf{D}_{t2}$; (c) $\mathbf{D}_2$; (d) $\mathbf{P}_2$; (e) θ_2.

Ans. (a) $70\mathbf{a}_z$ nC/m²; (b) $-18.75\mathbf{a}_x + 31.25\mathbf{a}_y$ nC/m²; (c) $-18.75\mathbf{a}_x + 31.25\mathbf{a}_y + 70\mathbf{a}_z$ nC/m²; (d) $-9.38\mathbf{a}_x + 15.63\mathbf{a}_y + 35\mathbf{a}_z$ nC/m²; (e) 27.5°

REFERENCES

1. Fano, R. M., L. J. Chu, and R. B. Adler. *Electromagnetic Fields, Energy, and Forces.* Cambridge, Mass.: MIT Press, 1968. Polarization in dielectrics is discussed in Chapter 5. This junior-level text presupposes a full-term physics course in electricity and magnetism, and it is therefore slightly more advanced in level. The introduction beginning on p. 1 should be read.

2. Dekker, A. J. *Electrical Engineering Materials.* Englewood Cliffs, NJ: Prentice-Hall, 1963. This admirable little book covers dielectrics, conductors, semiconductors, and magnetic materials.

3. Fink, D. G., and H. W. Beaty. *Standard Handbook for Electrical Engineers.* 16th ed. New York: McGraw-Hill, 2013.

4. Maxwell, J. C. *A Treatise on Electricity and Magnetism.* New York: Cambridge University Press, 2010.

5. Wert, C. A., and R. M. Thomson. *Physics of Solids.* 2nd ed. New York: McGraw-Hill, 1970. This is an advanced undergraduate-level text that covers metals, semiconductors, and dielectrics.

CHAPTER 5   PROBLEMS

5.1 Given the current density $\mathbf{J} = -10^4[\sin(2x)e^{-2y}\mathbf{a}_x + \cos(2x)e^{-2y}\mathbf{a}_y]$ kA/m^2 (*a*) Find the total current crossing the plane $y = 1$ in the $\mathbf{a}_y$ direction in the region $0 < x < 1$, $0 < z < 2$. (*b*) Find the total current leaving the region $0 < x, y < 1$, $2 < z < 3$ by integrating $\mathbf{J} \cdot d\mathbf{S}$ over the surface of the cube. (*c*) Repeat part *b*, but use the divergence theorem.

5.2 Given $\mathbf{J} = -10^{-4}(y\mathbf{a}_x + x\mathbf{a}_y)$ A/m^2, find the current crossing the $y = 0$ plane in the $-\mathbf{a}_y$ direction between $z = 0$ and 1, and $x = 0$ and 2.

5.3 A solid sphere of radius b contains charge Q, uniformly distributed throughout the sphere volume. The sphere rotates about the z axis at angular velocity Ω rad/s. What total current is flowing?

5.4 If volume charge density is given as $\rho_v = (\cos \omega t)/r^2$ C/m^3 in spherical coordinates, find $\mathbf{J}$. It is reasonable to assume that $\mathbf{J}$ is not a function of θ or ϕ.

5.5 Consider the following time-varying current density, which is in the form of a propagating wave at frequency f, traveling at velocity v (more about this in Chapter 10):

$$\mathbf{J}(z,t) = J_0 \cos(\omega t - \beta z)\, \mathbf{a}_z \, \text{A/m}^2$$

where J_0 is a constant amplitude, $\omega = 2\pi f$ is the radian frequency, and β is a constant to be found. Determine β as a function of ω and v such that the given waveform satisfies both Eqs. (3) and (5). Begin with Eq. (5).

5.6 In spherical coordinates, a current density $\mathbf{J} = -k/(r \sin \theta)\, \mathbf{a}_\theta$ A/m^2 exists in a conducting medium, where k is a constant. Determine the total current in the $\mathbf{a}_z$ direction that crosses a circular disk of radius R, centered on the z axis and located at (*a*) $z = 0$; (*b*) $z = h$.

5.7 Assuming that there is no transformation of mass to energy or vice versa, it is possible to write a continuity equation for mass. (*a*) If we use the continuity equation for charge as our model, what quantities correspond to $\mathbf{J}$ and ρ_v? (*b*) Given a cube 1 cm on a side, experimental data show that the rates at which mass is leaving each of the six faces are 10.25, −9.85, 1.75, −2.00, −4.05, and 4.45 mg/s. If we assume that the cube is an incremental volume element, determine an approximate value for the time rate of change of density at its center.

5.8 A truncated cone has a height of 16 cm. The circular faces on the top and bottom have radii of 2 mm and 0.1 mm, respectively. If the material from which this solid cone is constructed has a conductivity of 2×10^6 S/m, use some good approximations to determine the resistance between the two circular faces.

5.9 (a) Using data tabulated in Appendix C, calculate the required diameter for a 2-m-long nichrome wire that will dissipate an average power of 450 W when 120 V rms at 60 Hz is applied to it. (b) Calculate the rms current density in the wire.

5.10 A large brass washer has a 2-cm inside diameter, a 5-cm outside diameter, and is 0.5 cm thick. Its conductivity is $\sigma = 1.5 \times 10^7$ S/m. The washer is cut in half along a diameter, and a voltage is applied between the two rectangular faces of one part. The resultant electric field in the interior of the half-washer is $\mathbf{E} = (0.5/\rho)\, \mathbf{a}_\phi$ V/m in cylindrical coordinates, where the z axis is the axis of the washer. (a) What potential difference exists between the two rectangular faces? (b) What total current is flowing? (c) What is the resistance between the two faces?

5.11 Two perfectly conducting cylindrical surfaces of length ℓ are located at $\rho = 3$ and $\rho = 5$ cm. The total current passing radially outward through the medium between the cylinders is 3 A dc. (a) Find the voltage and resistance between the cylinders, and $\mathbf{E}$ in the region between the cylinders, if a conducting material having $\sigma = 0.05$ S/m is present for $3 < \rho < 5$ cm. (b) Show that integrating the power dissipated per unit volume over the volume gives the total dissipated power.

5.12 Two identical conducting plates, each having area A, are located at $z = 0$ and $z = d$. The region between plates is filled with a material having z-dependent conductivity, $\sigma(z) = \sigma_0 e^{-z/d}$, where σ_0 is a constant. Voltage V_0 is applied to the plate at $z = d$; the plate at $z = 0$ is at zero potential. Find, in terms of the given parameters, (a) the resistance of the material; (b) the total current flowing between plates; (c) the electric field intensity $\mathbf{E}$ within the material.

5.13 A hollow cylindrical tube with a rectangular cross section has external dimensions of 0.5 in. by 1 in. and a wall thickness of 0.05 in. Assume that the material is brass, for which $\sigma = 1.5 \times 10^7$ S/m. A current of 200 A dc is flowing down the tube. (a) What voltage drop is present across a 1 m length of the tube? (b) Find the voltage drop if the interior of the tube is filled with a conducting material for which $\sigma = 1.5 \times 10^5$ S/m.

5.14 A rectangular conducting plate lies in the xy plane, occupying the region $0 < x < a$, $0 < y < b$. An identical conducting plate is positioned directly above and parallel to the first, at $z = d$. The region between plates is filled with material having conductivity $\sigma(x) = \sigma_0 e^{-x/a}$, where σ_0 is a constant.

Voltage V_0 is applied to the plate at $z = d$; the plate at $z = 0$ is at zero potential. Find, in terms of the given parameters, (a) the electric field intensity $\mathbf{E}$ within the material; (b) the total current flowing between plates; (c) the resistance of the material.

5.15 A conducting medium is in the shape of a hemispherical shell, having inner and outer radii, a and b, respectively. The conductivity varies radially as $\sigma(r) = \sigma_0\, a/r$ S/m, where σ_0 is a constant. The surfaces at $r = a$ and b are coated with silver (essentially infinite conductivity for this problem). The inner surface is raised to potential V_0; the outer surface is grounded. A radial current, I_0, flows between surfaces. (a) Find the current density, $\mathbf{J}$, in terms of I_0. (b) Find $\mathbf{E}$ between surfaces in terms of I_0. (c) Find V_0 in terms of I_0. (d) Find the resistance R.

5.16 A coaxial transmission line has inner and outer conductor radii a and b. Between conductors $(a < \rho < b)$ lies a conductive medium whose conductivity is $\sigma(\rho) = \sigma_0/\rho$, where σ_0 is a constant. The inner conductor is charged to potential V_0, and the outer conductor is grounded. (a) Assuming dc radial current I per unit length in z, determine the radial current density field $\mathbf{J}$ in A/m². (b) Determine the electric field intensity $\mathbf{E}$ in terms of I and other parameters, given or known. (c) By taking an appropriate line integral of $\mathbf{E}$ as found in part b, find an expression that relates V_0 to I. (d) Find an expression for the conductance of the line per unit length, G.

5.17 Consider the setup as in Problem 5.15, except find R by considering the medium as made up of a stack of hemispherical shells, each of thickness dr, and where the resistance of each shell will be of the form of Eq. (13). The overall resistance will be the series combination of the shell resistances.

5.18 Two parallel circular plates of radius a are located at $z = 0$ and $z = d$. The top plate $(z = d)$ is raised to potential V_0; the bottom plate is grounded. Between the plates is a conducting material having radial-dependent conductivity, $\sigma(\rho) = \sigma_0\rho$, where σ_0 is a constant. (a) Find the ρ-independent electric field strength, $\mathbf{E}$, between plates. (b) Find the current density, $\mathbf{J}$ between plates. (c) Find the total current, I, in the structure. (d) Find the resistance between plates.

5.19 Consider the setup as in Problem 5.18, except find R by considering the medium as made up of concentric cylindrical shells, each of thickness $d\rho$. The resistance will be the parallel combination of the shell resistances.

5.20 Consider the basic image problem of a point charge q at $z = d$, suspended over an infinite conducting plane at $z = 0$. (a) Apply Eq. (10) in Chapter 2 to find $\mathbf{E}$ and $\mathbf{D}$ everywhere at the conductor surface as functions of cylindrical radius ρ. (b) Use your result from part a to find the charge density, ρ_s, and the total induced charge on the conductor.

5.21 Let the surface $y = 0$ be a perfect conductor in free space. Two uniform infinite line charges of 30 nC/m each are located at $x = 0$, $y = 1$, and $x = 0$,

$y = 2$. (a) Let $V = 0$ at the plane $y = 0$, and find V at $P(1, 2, 0)$. (b) Find **E** at P.

5.22 The line segment $x = 0$, $-1 \leq y \leq 1$, $z = 1$, carries a linear charge density $\rho_L = \pi|y| \, \mu\text{C/m}$. Let $z = 0$ be a conducting plane, and determine the surface charge density at: (a) $(0, 0, 0)$; (b) $(0, 1, 0)$.

5.23 A dipole with $\mathbf{p} = 0.1\mathbf{a}_z \, \mu\text{C} \cdot \text{m}$ is located at $A(1, 0, 0)$ in free space, and the $x = 0$ plane is perfectly conducting. (a) Find V at $P(2, 0, 1)$. (b) Find the equation of the 200 V equipotential surface in rectangular coordinates.

5.24 At a certain temperature, the electron and hole mobilities in intrinsic germanium are given as 0.43 and 0.21 $\text{m}^2/\text{V} \cdot \text{s}$, respectively. If the electron and hole concentrations are both $2.3 \times 10^{19} \, \text{m}^{-3}$, find the conductivity at this temperature.

5.25 Electron and hole concentrations increase with temperature. For pure silicon, suitable expressions are $\rho_h = -\rho_e = 6200T^{1.5}e^{-7000/T} \, \text{C/m}^3$. The functional dependence of the mobilities on temperature is given by $\mu_h = 2.3 \times 10^5 T^{-2.7} \, \text{m}^2/\text{V} \cdot \text{s}$ and $\mu_e = 2.1 \times 10^5 T^{-2.5} \, \text{m}^2/\text{V} \cdot \text{s}$, where the temperature, T, is in degrees Kelvin. Find σ at: (a) 0°C; (b) 40°C; (c) 80°C.

5.26 A semiconductor sample has a rectangular cross section 1.5 by 2.0 mm, and a length of 11.0 mm. The material has electron and hole densities of 1.8×10^{18} and $3.0 \times 10^{15} \, \text{m}^{-3}$, respectively. If $\mu_e = 0.082 \, \text{m}^2/\text{V} \cdot \text{s}$ and $\mu_h = 0.0021 \, \text{m}^2/\text{V} \cdot \text{s}$, find the resistance offered between the end faces of the sample.

5.27 Atomic hydrogen contains 5.5×10^{25} atoms/m^3 at a certain temperature and pressure. When an electric field of 4 kV/m is applied, each dipole formed by the electron and positive nucleus has an effective length of 7.1×10^{-19} m. (a) Find P. (b) Find ϵ_r.

5.28 Find the dielectric constant of a material in which the electric flux density is four times the polarization.

5.29 A coaxial conductor has radii $a = 0.8$ mm and $b = 3$ mm and a polystyrene dielectric for which $\epsilon_r = 2.56$. If $\mathbf{P} = (2/\rho)\mathbf{a}_\rho \, \text{nC/m}^2$ in the dielectric, find (a) **D** and **E** as functions of ρ; (b) V_{ab} and χ_e. (c) If there are 4×10^{19} molecules per cubic meter in the dielectric, find $\mathbf{p}(\rho)$.

5.30 Consider a composite material made up of two species, having number densities N_1 and N_2 molecules/m^3, respectively. The two materials are uniformly mixed, yielding a total number density of $N = N_1 + N_2$. The presence of an electric field **E** induces molecular dipole moments $\mathbf{p}_1$ and $\mathbf{p}_2$ within the individual species, whether mixed or not. Show that the dielectric constant of the composite material is given by $\epsilon_r = f\epsilon_{r1} + (1 - f)\epsilon_{r2}$, where f is the number fraction of species 1 dipoles in the composite, and where ϵ_{r1} and ϵ_{r2} are the dielectric constants that the unmixed species would have if each had number density N.

5.31 A circular cylinder of radius a and dielectric constant ϵ_r is positioned with its axis on the z axis. An electric field $\mathbf{E} = E_0 \, \mathbf{a}_x$ is incident from free space onto the cylinder surface. E_0 is a constant. (a) Find the electric field in the cylinder interior, and express the result in rectangular components. (b) Consider a field line incident at $\phi = 135°$. Find ϵ_r so that the interior field line will be oriented at $22.5°$ from the x axis.

5.32 Two equal but opposite-sign point charges of $3 \, \mu C$ are held x meters apart by a spring that provides a repulsive force given by $F_{sp} = 12(0.5 - x)$ N. Without any force of attraction, the spring would be fully extended to 0.5 m. (a) Determine the charge separation. (b) What is the dipole moment?

5.33 Two perfect dielectrics have relative permittivities $\epsilon_{r1} = 2$ and $\epsilon_{r2} = 8$. The planar interface between them is the surface $x - y + 2z = 5$. The origin lies in region 1. If $\mathbf{E}_1 = 100\mathbf{a}_x + 200\mathbf{a}_y - 50\mathbf{a}_z$ V/m, find $\mathbf{E}_2$.

5.34 A sphere of radius b and dielectric constant ϵ_r is centered at the origin in free space. An electric field $\mathbf{E} = E_0 \, \mathbf{a}_x$ is incident from free space onto the sphere surface. E_0 is a constant. (a) Find the electric field in the sphere interior and express the result in rectangular components. (b) Specialize your result from part a for the case in which $\phi = \pi$. (c) Specialize your result from part a for the case in which $\theta = \pi/2$. Compare both results (parts b and c) to the answer to Problem 5.31.

5.35 Let the cylindrical surfaces $\rho = 4$ cm and $\rho = 9$ cm enclose two wedges of perfect dielectrics, $\epsilon_{r1} = 2$ for $0 < \phi < \pi/2$ and $\epsilon_{r2} = 5$ for $\pi/2 < \phi < 2\pi$. If $\mathbf{E}_1 = (2000/\rho)\mathbf{a}_\rho$ V/m, find (a) $\mathbf{E}_2$; (b) the total electrostatic energy stored in a 1 m length of each region.

CHAPTER 6 Capacitance

术语

    capacitance   电容，单位是法［拉］，简称法（F）；电容是储存电荷或电场能的元件，属于电介质中的静电场内容范畴。

    capacitance of a two-wire line   双导线电容，指作为电压正负极的两根导线间的电容，是将每根导线都视为等位体。

    two-dimensional problem   二维问题，指场的分布只是空间两个变量的函数，不随另一个变量而变化；在此指二维平行平面场，如双平行长直导线的电容问题，若设导线方向为z轴方向，则在垂直于z轴的所有平面上场的分布相同，只分析一个二维平面上的场即可。

    Poisson's and Laplace's equations   泊松方程和拉普拉斯方程。

Capacitance measures the capability of energy storage in electrical devices. It can be designed for a specific purpose, or it may exist as an unavoidable by-product of the device structure that one must live with. Understanding capacitance and its impact on device or system operation is critical in every aspect of electrical engineering.

    A capacitor is a device that stores energy; energy thus stored can either be associated with accumulated charge or it can be related to the stored electric field, as was discussed in Section 4.8. In fact, one can think of a capacitor as a device that stores electric *flux*, in a similar way that an inductor—an analogous device—stores magnetic flux (or ultimately magnetic field energy). We will explore this in Chapter 8. A primary goal in this chapter is to present the methods for calculating capacitance for a number of cases, including transmission line geometries, and to be able to make judgments on how capacitance will be altered by changes in materials or their configuration. ∎

6.1 CAPACITANCE DEFINED

Consider two conductors embedded in a homogeneous dielectric (Figure 6.1). Conductor M_2 carries a total positive charge Q, and M_1 carries an equal negative charge. There are no other charges present, and the *total* charge of the system is zero.

    We now know that the charge is carried on the surface as a surface charge density and also that the electric field is normal to the conductor surface. Each conductor is, moreover, an equipotential surface. Because M_2 carries the positive charge, the electric flux is directed from M_2 to M_1, and M_2 is at the more positive potential. In other words, work must be done to carry a positive charge from M_1 to M_2.

    We designate the potential difference between M_2 and M_1 as V_0. The *capacitance* of this two-conductor system is defined as the ratio of the magnitude of the total

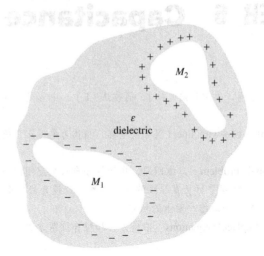

Figure 6.1 Two oppositely charged conductors M_1 and M_2 surrounded by a uniform dielectric. The ratio of the magnitude of the charge on either conductor to the magnitude of the potential difference between them is the capacitance C.

charge on either conductor to the magnitude of the potential difference between conductors,

$$C = \frac{Q}{V_0} \qquad (1)$$

In general terms, we determine Q by a surface integral over the positive conductors, and we find V_0 by carrying a unit positive charge from the negative to the positive surface,

$$C = \frac{\oint_S \epsilon \mathbf{E} \cdot d\mathbf{S}}{-\int_-^+ \mathbf{E} \cdot d\mathbf{L}} \qquad (2)$$

The capacitance is independent of the potential and total charge, for their ratio is constant. If the charge density is increased by a factor of N, Gauss's law indicates that the electric flux density or electric field intensity also increases by N, as does the potential difference. The capacitance is a function only of the physical dimensions of the system of conductors and of the permittivity of the homogeneous dielectric.

Capacitance is measured in *farads* (F), where a farad is defined as one coulomb per volt. Common values of capacitance are apt to be very small fractions of a farad, and consequently more practical units are the microfarad (μF), the nanofarad (nF), and the picofarad (pF).

6.2 PARALLEL-PLATE CAPACITOR

We can apply the definition of capacitance to a simple two-conductor system in which the conductors are identical, infinite parallel planes with separation d (Figure 6.2). Choosing the lower conducting plane at $z = 0$ and the upper one at $z = d$, a uniform sheet of surface charge $\pm \rho_S$ on each conductor leads to the uniform field [Section 2.5, Eq. (18)]

$$\mathbf{E} = \frac{\rho_S}{\epsilon}\mathbf{a}_z$$

where the permittivity of the homogeneous dielectric is ϵ, and

$$\mathbf{D} = \rho_S \mathbf{a}_z$$

Note that this result could be obtained by applying the boundary condition at a conducting surface (Eq. (18), Chapter 5) at either *one* of the plate surfaces. Referring to the surfaces and their unit normal vectors in Fig. 6.2, where $\mathbf{n}_\ell = \mathbf{a}_z$ and $\mathbf{n}_u = -\mathbf{a}_z$, we find on the lower plane:

$$\mathbf{D} \cdot \mathbf{n}_\ell|_{z=0} = \mathbf{D} \cdot \mathbf{a}_z = \rho_s \Rightarrow \mathbf{D} = \rho_s \mathbf{a}_z$$

On the upper plane, we get the same result

$$\mathbf{D} \cdot \mathbf{n}_u|_{z=d} = \mathbf{D} \cdot (-\mathbf{a}_z) = -\rho_s \Rightarrow \mathbf{D} = \rho_s \mathbf{a}_z$$

This is a key advantage of the conductor boundary condition, in that we need to apply it only to a single boundary to obtain the *total* field there (arising from all other sources).

The potential difference between lower and upper planes is

$$V_0 = -\int_{\text{upper}}^{\text{lower}} \mathbf{E} \cdot d\mathbf{L} = -\int_d^0 \frac{\rho_S}{\epsilon}dz = \frac{\rho_S}{\epsilon}d$$

Since the total charge on either plane is infinite, the capacitance is infinite. A more practical answer is obtained by considering planes, each of area S, whose linear dimensions are much greater than their separation d. The electric field and charge

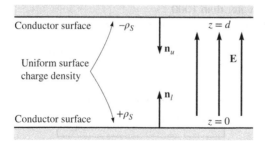

Figure 6.2 The problem of the parallel-plate capacitor. The capacitance per square meter of surface area is ϵ/d.

distribution are then almost uniform at all points not adjacent to the edges, and this latter region contributes only a small percentage of the total capacitance, allowing us to write the familiar result

$$Q = \rho_S S$$
$$V_0 = \frac{\rho_S}{\epsilon} d$$

$$\boxed{C = \frac{Q}{V_0} = \frac{\epsilon S}{d}}$$ (3)

More rigorously, we might consider Eq. (3) as the capacitance of a portion of the infinite-plane arrangement having a surface area S. Methods of calculating the effect of the unknown and nonuniform distribution near the edges must wait until we are able to solve more complicated potential problems.

EXAMPLE 6.1

Calculate the capacitance of a parallel-plate capacitor having a mica dielectric, $\epsilon_r = 6$, a plate area of 10 in.^{2}, and a separation of 0.01 in.

Solution. We find that

$$S = 10 \times 0.0254^2 = 6.45 \times 10^{-3} \text{ m}^2$$
$$d = 0.01 \times 0.0254 = 2.54 \times 10^{-4} \text{ m}$$

and therefore

$$C = \frac{6 \times 8.854 \times 10^{-12} \times 6.45 \times 10^{-3}}{2.54 \times 10^{-4}} = 1.349 \text{ nF}$$

A large plate area is obtained in capacitors of small physical dimensions by stacking smaller plates in 50- or 100-decker sandwiches, or by rolling up foil plates separated by a flexible dielectric.

Table C.1 in Appendix C also indicates that materials are available having dielectric constants greater than 1000.

Finally, the total energy stored in the capacitor is

$$W_E = \frac{1}{2}\int_{\text{vol}} \epsilon E^2 \, dv = \frac{1}{2}\int_0^S \int_0^d \frac{\epsilon \rho_S^2}{\epsilon^2} dz \, dS = \frac{1}{2}\frac{\rho_S^2}{\epsilon} Sd = \frac{1}{2}\frac{\epsilon S}{d}\frac{\rho_S^2 d^2}{\epsilon^2}$$

or

$$\boxed{W_E = \frac{1}{2}C \, V_0^2 = \frac{1}{2}Q \, V_0 = \frac{1}{2}\frac{Q^2}{C}}$$ (4)

which are all familiar expressions. Equation (4) also indicates that the energy stored in a capacitor with a fixed potential difference across it increases as the dielectric constant of the medium increases.

D6.1. Find the relative permittivity of the dielectric material present in a parallel-plate capacitor if: (a) $S = 0.12 \text{ m}^2$, $d = 80 \, \mu\text{m}$, $V_0 = 12$ V, and the capacitor contains 1 μJ of energy; (b) the stored energy density is 100 J/m³, $V_0 = 200$ V, and $d = 45 \, \mu\text{m}$; (c) $E = 200$ kV/m and $\rho_S = 20 \, \mu\text{C/m}^2$.

Ans. (a) 1.05; (b) 1.14; (c) 11.3

6.3 SEVERAL CAPACITANCE EXAMPLES

The methods just presented can be applied without much difficulty to other geometries in the other coordinate systems. A few examples follow.

6.3.1 Coaxial Cable

As a first brief example, we choose a coaxial cable or coaxial capacitor of inner radius a, outer radius b, and length L. No great derivational struggle is required, because the potential difference is given as Eq. (11) in Section 4.3, and we find the capacitance very simply by dividing this by the total charge $\rho_L L$ in the length L. Thus,

$$C = \frac{2\pi \epsilon L}{\ln(b/a)} \tag{5}$$

6.3.2 Spherical Capacitor

Next we consider a spherical capacitor formed of two concentric spherical conducting shells of radius a and b, $b > a$. The expression for the electric field was obtained previously by Gauss's law,

$$E_r = \frac{Q}{4\pi \epsilon r^2}$$

where the region between the spheres is a dielectric with permittivity ϵ. The expression for potential difference was found from this by the line integral [Section 4.3, Eq. (12)]. Thus,

$$V_{ab} = \frac{Q}{4\pi \epsilon} \left(\frac{1}{a} - \frac{1}{b} \right)$$

Here Q represents the total charge on the inner sphere, and the capacitance becomes

$$C = \frac{Q}{V_{ab}} = \frac{4\pi \epsilon}{\dfrac{1}{a} - \dfrac{1}{b}} \tag{6}$$

If we allow the outer sphere to become infinitely large, we obtain the capacitance of an isolated spherical conductor,

$$C = 4\pi \epsilon a \tag{7}$$

For a diameter of 1 cm, or a sphere about the size of a marble, $C = 0.556$ pF in free space.

6.3.3 Capacitors with Multiple Dielectrics

Suppose the sphere in the previous example were to be coated with a different dielectric layer, for which $\epsilon = \epsilon_1$, extending from $r = a$ to $r = r_1$. As the charge is still Q, the electric flux density is unaffected by the dielectric layer, but the electric field will evaluate differently in the two media. We have:

$$D_r = \frac{Q}{4\pi r^2} \qquad (a < r < \infty)$$

$$E_r = \frac{Q}{4\pi \epsilon_1 r^2} \qquad (a < r < r_1)$$

$$= \frac{Q}{4\pi \epsilon_0 r^2} \qquad (r_1 < r)$$

The potential difference between the conductor and infinity is now:

$$V_0 = V_a - V_\infty = -\int_{r_1}^{a} \frac{Q\,dr}{4\pi \epsilon_1 r^2} - \int_{\infty}^{r_1} \frac{Q\,dr}{4\pi \epsilon_0 r^2}$$

$$= \frac{Q}{4\pi}\left[\frac{1}{\epsilon_1}\left(\frac{1}{a} - \frac{1}{r_1}\right) + \frac{1}{\epsilon_0 r_1}\right]$$

Therefore, the capacitance is

$$C = \frac{Q}{V_0} = \frac{4\pi}{\frac{1}{\epsilon_1}\left(\frac{1}{a} - \frac{1}{r_1}\right) + \frac{1}{\epsilon_0 r_1}} \qquad (8)$$

In order to look at the problem of multiple dielectrics a little more thoroughly, consider a parallel-plate capacitor of area S and spacing d, with the usual assumption that d is small compared to the linear dimensions of the plates. The capacitance is $\epsilon_1 S/d$, using a dielectric of permittivity ϵ_1. Now replace a part of this dielectric with another of permittivity ϵ_2, placing the boundary between the two dielectrics parallel to the plates (Figure 6.3).

Some of us may immediately suspect that this combination is effectively two capacitors in series, yielding a total capacitance of

$$C = \frac{1}{\frac{1}{C_1} + \frac{1}{C_2}}$$

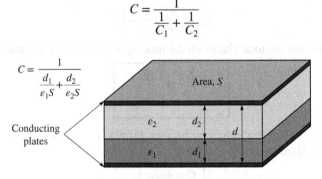

$$C = \frac{1}{\frac{d_1}{\epsilon_1 S} + \frac{d_2}{\epsilon_2 S}}$$

Figure 6.3 A parallel-plate capacitor containing two dielectrics with the dielectric interface parallel to the conducting plates.

where $C_1 = \epsilon_1 S/d_1$ and $C_2 = \epsilon_2 S/d_2$. This is the correct result, but we can obtain it using less intuition and a more basic approach.

Because the capacitance definition, $C = Q/V_0$, involves a charge and a voltage, we may assume either and then find the other in terms of it. The capacitance is not a function of either, but only of the dielectrics and the geometry. Suppose we assume a potential difference V_0 between the plates. The electric field intensities in the two regions, E_2 and E_1, are both uniform, and $V_0 = E_1 d_1 + E_2 d_2$. At the dielectric interface, E is normal, and our boundary condition, Eq. (35) in Chapter 5, tells us that $D_{N1} = D_{N2}$, or $\epsilon_1 E_1 = \epsilon_2 E_2$. This assumes (correctly) that there is no surface charge at the interface. Eliminating E_2 in our V_0 relation, we have

$$E_1 = \frac{V_0}{d_1 + d_2(\epsilon_1/\epsilon_2)}$$

and the surface charge density on the lower plate therefore has the magnitude

$$\rho_{S1} = D_1 = \epsilon_1 E_1 = \frac{V_0}{\dfrac{d_1}{\epsilon_1} + \dfrac{d_2}{\epsilon_2}}$$

Because $D_1 = D_2$, the magnitude of the surface charge is the same on each plate. The capacitance is then

$$C = \frac{Q}{V_0} = \frac{\rho_S S}{V_0} = \frac{1}{\dfrac{d_1}{\epsilon_1 S} + \dfrac{d_2}{\epsilon_2 S}} = \frac{1}{\dfrac{1}{C_1} + \dfrac{1}{C_2}}$$

As an alternate (and slightly simpler) solution, we might assume a charge Q on one plate, leading to a charge density Q/S and a value of D that is also Q/S. This is true in both regions, as $D_{N1} = D_{N2}$ and D is normal. Then $E_1 = D/\epsilon_1 = Q/(\epsilon_1 S)$, $E_2 = D/\epsilon_2 = Q/(\epsilon_2 S)$, and the potential differences across the regions are $V_1 = E_1 d_1 = Qd_1/(\epsilon_1 S)$, and $V_2 = E_2 d_2 = Qd_2/(\epsilon_2 S)$. The capacitance is

$$C = \frac{Q}{V_0} = \frac{Q}{V_1 + V_2} = \frac{1}{\dfrac{d_1}{\epsilon_1 S} + \dfrac{d_2}{\epsilon_2 S}} \tag{9}$$

How would the method of solution or the answer change if there were a third conducting plane along the interface? We would expect to find surface charge on each side of this conductor, and the magnitudes of these charges should be equal. In other words, we think of the electric lines not as passing directly from one outer plate to the other, but as terminating on one side of this interior plane and then continuing on the other side. The capacitance is unchanged, provided, of course, that the added conductor is of negligible thickness. The addition of a thick conducting plate will increase the capacitance if the separation of the outer plates is kept constant, and this is an example of a more general theorem which states that the replacement of any portion of the dielectric by a conducting body will cause an increase in the capacitance.

If the dielectric boundary were placed *normal* to the two conducting plates and the dielectrics occupied areas of S_1 and S_2, then an assumed potential difference V_0 would produce field strengths $E_1 = E_2 = V_0/d$. These are tangential fields at the interface, and they must be equal. Then we may find in succession D_1, D_2, ρS_1, ρS_2, and Q, obtaining a capacitance

$$C = \frac{\epsilon_1 S_1 + \epsilon_2 S_2}{d} = C_1 + C_2 \tag{10}$$

as we expect.

D6.2. Determine the capacitance of: (*a*) a 1-ft length of 35B/U coaxial cable, which has an inner conductor 0.1045 in. in diameter, a polyethylene dielectric ($\epsilon_r = 2.26$ from Table C.1), and an outer conductor that has an inner diameter of 0.680 in.; (*b*) a conducting sphere of radius 2.5 mm, covered with a polyethylene layer 2 mm thick, surrounded by a conducting sphere of radius 4.5 mm; (*c*) two rectangular conducting plates, 1 cm by 4 cm, with negligible thickness, between which are three sheets of dielectric, each 1 cm by 4 cm, and 0.1 mm thick, having dielectric constants of 1.5, 2.5, and 6.

Ans. (*a*) 20.5 pF; (*b*) 1.41 pF; (*c*) 28.7 pF

6.4 CAPACITANCE OF A TWO-WIRE LINE

We next consider the problem of the two-wire line. The configuration consists of two parallel conducting cylinders, each of circular cross section, and we will find complete information about the electric field intensity, the potential field, the surface-charge-density distribution, and the capacitance. This arrangement is an important type of transmission line, as is the coaxial cable.

We begin by investigating the potential field of two infinite line charges. Figure 6.4 shows a positive line charge in the *xz* plane at $x = a$ and a negative line charge at $x = -a$. The potential of a single line charge with zero reference at a radius of R_0 is

$$V = \frac{\rho_L}{2\pi\epsilon} \ln \frac{R_0}{R}$$

We now write the expression for the combined potential field in terms of the radial distances from the positive and negative lines, R_1 and R_2, respectively,

$$V = \frac{\rho_L}{2\pi\epsilon}\left(\ln\frac{R_{10}}{R_1} - \ln\frac{R_{20}}{R_2}\right) = \frac{\rho_L}{2\pi\epsilon}\ln\frac{R_{10}R_2}{R_{20}R_1}$$

We choose $R_{10} = R_{20}$, thus placing the zero reference at equal distances from each line. This surface is the $x = 0$ plane. Expressing R_1 and R_2 in terms of x and y,

$$V = \frac{\rho_L}{2\pi\epsilon}\ln\sqrt{\frac{(x+a)^2 + y^2}{(x-a)^2 + y^2}} = \frac{\rho_L}{4\pi\epsilon}\ln\frac{(x+a)^2 + y^2}{(x-a)^2 + y^2} \tag{11}$$

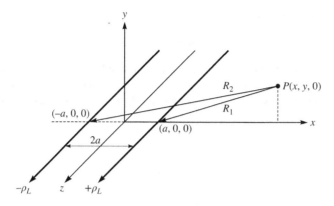

Figure 6.4 Two parallel infinite line charges carrying opposite charge. The positive line is at $x = a$, $y = 0$, and the negative line is at $x = -a$, $y = 0$. A general point $P(x, y, 0)$ in the xy plane is radially distant R_1 and R_2 from the positive and negative lines, respectively. The equipotential surfaces are circular cylinders.

In order to recognize the equipotential surfaces and adequately understand the problem we are going to solve, some algebraic manipulations are necessary. Choosing an equipotential surface $V = V_1$, we define K_1 as a dimensionless parameter that is a function of the potential V_1,

$$K_1 = e^{4\pi\epsilon V_1/\rho_L} \qquad (12)$$

so that

$$K_1 = \frac{(x + a)^2 + y^2}{(x - a)^2 + y^2}$$

After multiplying and collecting like powers, we obtain

$$x^2 - 2ax\frac{K_1 + 1}{K_1 - 1} + y^2 + a^2 = 0$$

We next work through a couple of lines of algebra and complete the square,

$$\left(x - a\frac{K_1 + 1}{K_1 - 1}\right)^2 + y^2 = \left(\frac{2a\sqrt{K_1}}{K_1 - 1}\right)^2$$

This shows that the $V = V_1$ equipotential surface is independent of z (or is a cylinder) and intersects the xy plane in a circle of radius b,

$$b = \frac{2a\sqrt{K_1}}{K_1 - 1}$$

which is centered at $x = h$, $y = 0$, where

$$h = a\frac{K_1 + 1}{K_1 - 1}$$

Now consider a zero-potential conducting plane located at $x = 0$, and a conducting cylinder of radius b and potential V_0 with its axis located a distance h from the plane. We solve the last two equations for a and K_1 in terms of the dimensions b and h,

$$a = \sqrt{h^2 - b^2} \qquad (13)$$

and

$$\sqrt{K_1} = \frac{h + \sqrt{h^2 - b^2}}{b} \qquad (14)$$

But the potential of the cylinder is V_0, so Eq. (12) leads to

$$\sqrt{K_1} = e^{2\pi\epsilon V_0/\rho_L}$$

Therefore,

$$\rho_L = \frac{4\pi\epsilon V_0}{\ln K_1} \qquad (15)$$

Thus, given h, b, and V_0, we may determine a, ρ_L, and the parameter K_1. The capacitance between the cylinder and plane is now available. For a length L in the z direction, we have

$$C = \frac{\rho_L L}{V_0} = \frac{4\pi\epsilon L}{\ln K_1} = \frac{2\pi\epsilon L}{\ln \sqrt{K_1}}$$

or

$$C = \frac{2\pi\epsilon L}{\ln[(h + \sqrt{h^2 - b^2})/b]} = \frac{2\pi\epsilon L}{\cosh^{-1}(h/b)} \qquad (16)$$

The solid line in Figure 6.5 shows the cross section of a cylinder of 5 m radius at a potential of 100 V in free space, with its axis 13 m distant from a plane at zero potential. Thus, $b = 5$, $h = 13$, $V_0 = 100$, and we rapidly find the location of the equivalent line charge from Eq. (13),

$$a = \sqrt{h^2 - b^2} = \sqrt{13^2 - 5^2} = 12 \text{ m}$$

the value of the potential parameter K_1 from Eq. (14),

$$\sqrt{K_1} = \frac{h + \sqrt{h^2 - b^2}}{b} = \frac{13 + 12}{5} = 5 \qquad K_1 = 25$$

the strength of the equivalent line charge from Eq. (15),

$$\rho_L = \frac{4\pi\epsilon V_0}{\ln K_1} = \frac{4\pi \times 8.854 \times 10^{-12} \times 100}{\ln 25} = 3.46 \text{ nC/m}$$

and the capacitance between cylinder and plane from Eq. (16),

$$C = \frac{2\pi\epsilon}{\cosh^{-1}(h/b)} = \frac{2\pi \times 8.854 \times 10^{-12}}{\cosh^{-1}(13/5)} = 34.6 \text{ pF/m}$$

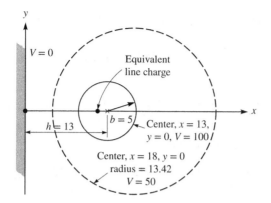

Figure 6.5 A numerical example of the capacitance, linear charge density, position of an equivalent line charge, and characteristics of the mid-equipotential surface for a cylindrical conductor of 5 m radius at a potential of 100 V, parallel to and 13 m from a conducting plane at zero potential.

We may also identify the cylinder representing the 50 V equipotential surface by finding new values for K_1, h, and b. We first use Eq. (12) to obtain

$$K_1 = e^{4\pi\epsilon V_1/\rho_L} = e^{4\pi\times8.854\times10^{-12}\times50/3.46\times10^{-9}} = 5.00$$

Then the new radius is

$$b = \frac{2a\sqrt{K_1}}{K_1 - 1} = \frac{2\times12\sqrt{5}}{5-1} = 13.42 \text{ m}$$

and the corresponding value of h becomes

$$h = a\frac{K_1 + 1}{K_1 - 1} = 12\frac{5+1}{5-1} = 18 \text{ m}$$

This cylinder is shown in color in Figure 6.5.

The electric field intensity can be found by taking the gradient of the potential field, as given by Eq. (11),

$$\mathbf{E} = -\nabla\left[\frac{\rho_L}{4\pi\epsilon}\ln\frac{(x+a)^2 + y^2}{(x-a)^2 + y^2}\right]$$

Thus,

$$\mathbf{E} = -\frac{\rho_L}{4\pi\epsilon}\left[\frac{2(x+a)\mathbf{a}_x + 2y\,\mathbf{a}_y}{(x+a)^2 + y^2} - \frac{2(x-a)\mathbf{a}_x + 2y\,\mathbf{a}_y}{(x-a)^2 + y^2}\right]$$

and

$$\mathbf{D} = \epsilon\mathbf{E} = -\frac{\rho_L}{2\pi}\left[\frac{(x+a)\mathbf{a}_x + y\,\mathbf{a}_y}{(x+a)^2 + y^2} - \frac{(x-a)\mathbf{a}_x + y\,\mathbf{a}_y}{(x-a)^2 + y^2}\right]$$

If we evaluate D_x at $x = h - b$, $y = 0$, we may obtain $\rho_{S,\text{max}}$

$$\rho_{S,\text{max}} = -D_{x,x=h-b,y=0} = \frac{\rho_L}{2\pi}\left[\frac{h-b+a}{(h-b+a)^2} - \frac{h-b-a}{(h-b-a)^2}\right]$$

For our example,

$$\rho_{S,\text{max}} = \frac{3.46 \times 10^{-9}}{2\pi}\left[\frac{13-5+12}{(13-5+12)^2} - \frac{13-5-12}{(13-5-12)^2}\right] = 0.165 \ \text{nC/m}^2$$

Similarly, $\rho_{S,\text{min}} = D_{x,x=h+b,y=0}$, and

$$\rho_{S,\text{min}} = \frac{3.46 \times 10^{-9}}{2\pi}\left[\frac{13+5+12}{30^2} - \frac{13+5-12}{6^2}\right] = 0.073 \ \text{nC/m}^2$$

Thus,

$$\rho_{S,\text{max}} = 2.25\,\rho_{S,\text{min}}$$

If we apply Eq. (16) to the case of a conductor for which $b \ll h$, then

$$\ln\left[(h + \sqrt{h^2 - b^2})/b\right] \doteq \ln\left[(h + h)/b\right] \doteq \ln(2h/b)$$

and

$$C = \frac{2\pi\epsilon L}{\ln(2h/b)} \qquad (b \ll h) \tag{17}$$

The capacitance between two circular conductors separated by a distance $2h$ is one-half the capacitance given by Eqs. (16) or (17). This last answer is of interest because it gives us an expression for the capacitance of a section of two-wire transmission line, one of the types of transmission lines studied later in Chapter 13.

D6.3. A conducting cylinder with a radius of 1 cm and at a potential of 20 V is parallel to a conducting plane which is at zero potential. The plane is 5 cm distant from the cylinder axis. If the conductors are embedded in a perfect dielectric for which $\epsilon_r = 4.5$, find: (a) the capacitance per unit length between cylinder and plane; (b) $\rho_{S,\text{max}}$ on the cylinder.

Ans. (a) 109.2 pF/m; (b) 42.6 nC/m^2

6.5 USING FIELD SKETCHES TO ESTIMATE CAPACITANCE IN TWO-DIMENSIONAL PROBLEMS

In capacitance problems in which the conductor configurations cannot be easily described using a single coordinate system, other analysis techniques are usually applied. Such methods typically involve a numerical determination of field or potential values over a grid within the region of interest. In this section, another approach is described that involves making sketches of field lines and equipotential surfaces in a manner that follows a few simple rules. This approach, although lacking the accuracy of more elegant methods, allows fairly quick estimates of capacitance while providing a useful visualization of the field configuration.

The method, requiring only pencil and paper, is capable of yielding good accuracy if used skillfully and patiently. Fair accuracy (5 to 10 percent on a capacitance determination) may be obtained by a beginner who does no more than follow the few rules and hints of the art. The method to be described is applicable only to fields in which no variation exists in the direction normal to the plane of the sketch. The procedure is based on several facts that we have already demonstrated:

1. A conductor boundary is an equipotential surface.
2. The electric field intensity and electric flux density are both perpendicular to the equipotential surfaces.
3. **E** and **D** are therefore perpendicular to the conductor boundaries and possess zero tangential values.
4. The lines of electric flux, or streamlines, begin and terminate on charge and hence, in a charge-free, homogeneous dielectric, begin and terminate only on the conductor boundaries.

We consider the implications of these statements by drawing the streamlines on a sketch that already shows the equipotential surfaces. In Figure 6.6a, two conductor boundaries are shown, and equipotentials are drawn with a constant potential difference between lines. We should remember that these lines are only the cross sections of the equipotential surfaces, which are cylinders (although not circular). No variation in the direction normal to the surface of the paper is permitted. We arbitrarily choose to begin a streamline, or flux line, at A on the surface of the more positive conductor. It leaves the surface normally and must cross at right angles the undrawn but very real equipotential surfaces between the conductor and the first surface shown. The line is continued to the other conductor, obeying the single rule that the intersection with each equipotential must be square.

In a similar manner, we may start at B and sketch another streamline ending at B'. We need to understand the meaning of this pair of streamlines. The streamline, by

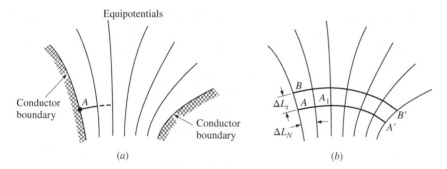

Figure 6.6 (a) Sketch of the equipotential surfaces between two conductors. The increment of potential between each of the two adjacent equipotentials is the same. (b) One flux line has been drawn from A to A', and a second from B to B'.

definition, is everywhere tangent to the electric field intensity or to the electric flux density. Because the streamline is tangent to the electric flux density, the flux density is tangent to the streamline, and no electric flux may cross any streamline. In other words, if there is a charge of 5 μC on the surface between A and B (and extending 1 m into the paper), then 5 μC of flux begins in this region, and all must terminate between A' and B'. Such a pair of lines is sometimes called a flux *tube* because it physically seems to carry flux from one conductor to another without losing any.

We next construct a third streamline, and both the mathematical and visual interpretations we may make from the sketch will be greatly simplified if we draw this line starting from some point C chosen so that the same amount of flux is carried in the tube BC as is contained in AB. How do we choose the position of C?

The electric field intensity at the midpoint of the line joining A to B may be found approximately by assuming a value for the flux in the tube AB, say $\Delta\Psi$, which allows us to express the electric flux density by $\Delta\Psi/\Delta L_t$, where the depth of the tube into the paper is 1 m and ΔL_t is the length of the line joining A to B. The magnitude of E is then

$$E = \frac{1}{\epsilon} \frac{\Delta\Psi}{\Delta L_t}$$

We may also find the magnitude of the electric field intensity by dividing the potential difference between points A and A_1, lying on two adjacent equipotential surfaces, by the distance from A to A_1. If this distance is designated ΔL_N and an increment of potential between equipotentials of ΔV is assumed, then

$$E = \frac{\Delta V}{\Delta L_N}$$

This value applies most accurately to the point at the middle of the line segment from A to A_1, while the previous value was most accurate at the midpoint of the line segment from A to B. If, however, the equipotentials are close together (ΔV small) and the two streamlines are close together ($\Delta\Psi$ small), the two values found for the electric field intensity must be approximately equal,

$$\frac{1}{\epsilon} \frac{\Delta\Psi}{\Delta L_t} = \frac{\Delta V}{\Delta L_N} \qquad (18)$$

Throughout our sketch we have assumed a homogeneous medium (ϵ constant), a constant increment of potential between equipotentials (ΔV constant), and a constant amount of flux per tube ($\Delta\Psi$ constant). To satisfy all these conditions, Eq. (18) shows that

$$\boxed{\frac{\Delta L_t}{\Delta L_N} = \text{constant} = \frac{1}{\epsilon} \frac{\Delta\Psi}{\Delta V}} \qquad (19)$$

A similar argument might be made at any point in our sketch, and we are therefore led to the conclusion that a constant ratio must be maintained between the distance between streamlines as measured along an equipotential, and the distance between equipotentials as measured along a streamline. It is this *ratio* that must have the same value at every point, not the individual lengths. Each length must decrease in regions of greater field strength, because ΔV is constant.

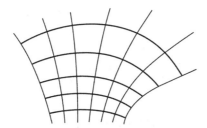

Figure 6.7 The remaining streamlines have been added to Fig. 6.6*b* by beginning each new line normally to the conductor and maintaining curvilinear squares throughout the sketch.

The simplest ratio we can use is unity, and the streamline from B to B' shown in Figure 6.6*b* was started at a point for which $\Delta L_t = \Delta L_N$. Because the ratio of these distances is kept at unity, the streamlines and equipotentials divide the field-containing region into curvilinear squares, a term implying a planar geometric figure that differs from a true square in having slightly curved and slightly unequal sides but which approaches a square as its dimensions decrease. Those incremental surface elements in our three coordinate systems which are planar may also be drawn as curvilinear squares.

We may now sketch in the remainder of the streamlines by keeping each small box as square as possible. One streamline is begun, an equipotential line is roughed in, another streamline is added, forming a curvilinear square, and the map is gradually extended throughout the desired region. The complete sketch is shown in Figure 6.7.

The construction of a useful field map is an art; the science merely furnishes the rules. Proficiency in any art requires practice. A good problem for beginners is the coaxial cable or coaxial capacitor, since all the equipotentials are circles while the flux lines are straight lines. The next sketch attempted should be two parallel circular conductors, where the equipotentials are again circles but with different centers. Each of these is given as a problem at the end of the chapter.

Figure 6.8 shows a completed map for a cable containing a square inner conductor surrounded by a circular conductor. The capacitance is found from $C = Q/V_0$ by replacing Q with $N_Q\,\Delta Q = N_Q\Delta\Psi$, where N_Q is the number of flux tubes joining the two conductors, and letting $V_0 = N_V\,\Delta V$, where N_V is the number of potential increments between conductors,

$$C = \frac{N_Q\,\Delta Q}{N_V\,\Delta V}$$

and then using Eq. (19),

$$\boxed{C = \frac{N_Q}{N_V}\epsilon\frac{\Delta L_t}{\Delta L_N} = \epsilon\frac{N_Q}{N_V}} \tag{20}$$

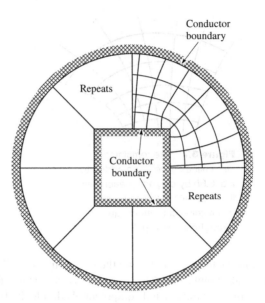

Figure 6.8   An example of a curvilinear-square field map. The side of the square is two-thirds the radius of the circle. $N_V = 4$ and $N_Q = 8 \times 3.25 = 26$, and therefore $C = \varepsilon_0 \, N_Q/N_V = 57.6$ pF/m.

since $\Delta L_t/\Delta L_N = 1$. The determination of the capacitance from a flux plot merely consists of counting squares in two directions, between conductors and around either conductor. From Figure 6.8 we obtain

$$C = \varepsilon_0 \frac{8 \times 3.25}{4} = 57.6 \text{ pF/m}$$

Ramo, Whinnery, and Van Duzer have an excellent discussion with examples of the construction of field maps by curvilinear squares. They offer the following suggestions:[1]

1. Plan on making a number of rough sketches, taking only a minute or so apiece, before starting any plot to be made with care. The use of transparent paper over the basic boundary will speed up this preliminary sketching.

2. Divide the known potential difference between electrodes into an equal number of divisions, say four or eight to begin with.

3. Begin the sketch of equipotentials in the region where the field is known best, for example, in some region where it approaches a uniform field. Extend the equipotentials according to your best guess throughout the plot. Note that they should tend to hug acute angles of the conducting boundary and be spread out in the vicinity of obtuse angles of the boundary.

---

[1] By permission from S. Ramo, J. R. Whinnery, and T. Van Duzer, pp. 51–52. See References at the end of this chapter. Curvilinear maps are discussed on pp. 50–52.

4. Draw in the orthogonal set of field lines. As these are started, they should form curvilinear squares, but as they are extended the condition of orthogonality should be kept paramount, even though this will result in some rectangles with ratios other than unity.

5. Look at the regions with poor side ratios and try to see what was wrong with the first guess of equipotentials. Correct them and repeat the procedure until reasonable curvilinear squares exist throughout the plot.

6. In regions of low field intensity, there will be large figures, often of five or six sides. To judge the correctness of the plot in this region, these large units should be subdivided. The subdivisions should be started back away from the region needing subdivision, and each time a flux tube is divided in half, the potential divisions in this region must be divided by the same factor.

D6.4. Figure 6.9 shows the cross section of two circular cylinders at potentials of 0 and 60 V. The axes are parallel and the region between the cylinders is air-filled. Equipotentials at 20 V and 40 V are also shown. Prepare a curvilinear-square map on the figure and use it to establish suitable values for: (*a*) the capacitance per meter length; (*b*) E at the left side of the 60 V conductor if its true radius is 2 mm; (*c*) ρ_S at that point.

Ans. (*a*) 69 pF/m; (*b*) 60 kV/m; (*c*) 550 nC/m^2

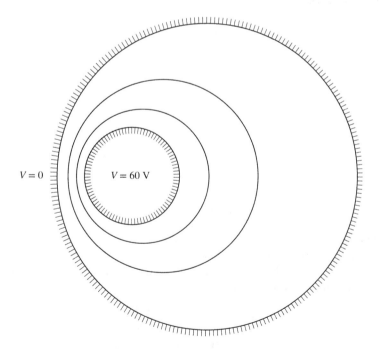

$V = 0$ $V = 60$ V

Figure 6.9 See Problem D6.4.

6.6   POISSON'S AND LAPLACE'S EQUATIONS

In preceding sections, we have found capacitance by first assuming a known charge distribution on the conductors and then finding the potential difference in terms of the assumed charge. An alternate approach would be to start with known potentials on each conductor, and then work backward to find the charge in terms of the known potential difference. The capacitance in either case is found by the ratio Q/V.

The first objective in the latter approach is thus to find the potential function between conductors, given values of potential on the boundaries, along with possible volume charge densities in the region of interest. The mathematical tools that enable this to happen are Poisson's and Laplace's equations, to be explored in the remainder of this chapter. Problems involving one to three dimensions can be solved either analytically or numerically. Laplace's and Poisson's equations, when compared to other methods, are probably the most widely useful because many problems in engineering practice involve devices in which applied potential differences are known, and in which constant potentials occur at the boundaries.

Obtaining Poisson's equation is exceedingly simple, for from the point form of Gauss's law,

$$\nabla \cdot \mathbf{D} = \rho_v \tag{21}$$

the definition of $\mathbf{D}$,

$$\mathbf{D} = \epsilon \mathbf{E} \tag{22}$$

and the gradient relationship,

$$\mathbf{E} = -\nabla V \tag{23}$$

by substitution we have

$$\nabla \cdot \mathbf{D} = \nabla \cdot (\epsilon \mathbf{E}) = -\nabla \cdot (\epsilon \nabla V) = \rho_v$$

or

$$\nabla \cdot \nabla V = -\frac{\rho_v}{\epsilon} \tag{24}$$

for a homogeneous region in which ϵ is constant.

Equation (24) is *Poisson's equation*, but the "double ∇" operation must be interpreted and expanded, at least in rectangular coordinates, before the equation can be useful. In rectangular coordinates,

$$\nabla \cdot \mathbf{A} = \frac{\partial A_x}{\partial x} + \frac{\partial A_y}{\partial y} + \frac{\partial A_z}{\partial z}$$

$$\nabla V = \frac{\partial V}{\partial x}\mathbf{a}_x + \frac{\partial V}{\partial y}\mathbf{a}_y + \frac{\partial V}{\partial z}\mathbf{a}_z$$

and therefore

$$\nabla \cdot \nabla V = \frac{\partial}{\partial x}\left(\frac{\partial V}{\partial x}\right) + \frac{\partial}{\partial y}\left(\frac{\partial V}{\partial y}\right) + \frac{\partial}{\partial z}\left(\frac{\partial V}{\partial z}\right)$$
$$= \frac{\partial^2 V}{\partial x^2} + \frac{\partial^2 V}{\partial y^2} + \frac{\partial^2 V}{\partial z^2} \tag{25}$$

Usually the operation $\nabla \cdot \nabla$ is abbreviated ∇^2 (and pronounced "del squared"), a good reminder of the second-order partial derivatives appearing in Eq. (25), and so Poisson's equation becomes

$$\boxed{\nabla^2 V = \frac{\partial^2 V}{\partial x^2} + \frac{\partial^2 V}{\partial y^2} + \frac{\partial^2 V}{\partial z^2} = -\frac{\rho_v}{\epsilon}} \tag{26}$$

in rectangular coordinates.

If $\rho_v = 0$, indicating zero *volume* charge density, but allowing point charges, line charge, and surface charge density to exist at singular locations as sources of the field, then

$$\boxed{\nabla^2 V = 0} \tag{27}$$

which is *Laplace's equation*. The ∇^2 operation is called the *Laplacian of V*.

In rectangular coordinates Laplace's equation is

$$\boxed{\nabla^2 V = \frac{\partial^2 V}{\partial x^2} + \frac{\partial^2 V}{\partial y^2} + \frac{\partial^2 V}{\partial z^2} = 0 \quad \text{(rectangular)}} \tag{28}$$

and the form of $\nabla^2 V$ in cylindrical and spherical coordinates may be obtained by using the expressions for the divergence and gradient already obtained in those coordinate systems. For reference, the Laplacian in cylindrical coordinates is

$$\boxed{\nabla^2 V = \frac{1}{\rho}\frac{\partial}{\partial \rho}\left(\rho\frac{\partial V}{\partial \rho}\right) + \frac{1}{\rho^2}\left(\frac{\partial^2 V}{\partial \phi^2}\right) + \frac{\partial^2 V}{\partial z^2} \quad \text{(cylindrical)}} \tag{29}$$

and in spherical coordinates is

$$\boxed{\nabla^2 V = \frac{1}{r^2}\frac{\partial}{\partial r}\left(r^2\frac{\partial V}{\partial r}\right) + \frac{1}{r^2\sin\theta}\frac{\partial}{\partial \theta}\left(\sin\theta\frac{\partial V}{\partial \theta}\right) + \frac{1}{r^2\sin^2\theta}\frac{\partial^2 V}{\partial \phi^2} \quad \text{(spherical)}} \tag{30}$$

These equations may be expanded by taking the indicated partial derivatives, but it is usually more helpful to have them in the forms just given; furthermore, it is much easier to expand them later if necessary than it is to put the broken pieces back together again.

Laplace's equation is all-embracing, for, applying as it does wherever volume charge density is zero, it states that every conceivable configuration of electrodes or conductors produces a field for which $\nabla^2 V = 0$. All these fields are different, with different potential values and different spatial rates of change, yet for each of them $\nabla^2 V = 0$. Because *every* field (if $\rho_v = 0$) satisfies Laplace's equation, how can we expect to reverse the procedure and use Laplace's equation to find one specific field in which we happen to have an interest? Obviously, more information is required, and we will find that Laplace's equation must be solved subject to certain *boundary conditions*.

Every physical problem must contain at least one conducting boundary and usually contains two or more. The potentials on these boundaries are assigned values, perhaps V_0, V_1, ..., or perhaps numerical values. These definite equipotential surfaces will provide the boundary conditions for the type of problem to be solved. In other types of problems, the boundary conditions take the form of specified values of E (alternatively, a surface charge density, ρ_S) on an enclosing surface, or a mixture of known values of V and E.

Before using Laplace's equation or Poisson's equation in several examples, it must be stated that if our answer satisfies Laplace's equation and also satisfies the boundary conditions, then it is the only possible answer. This is a statement of the Uniqueness Theorem, the proof of which is presented in Appendix D.

D6.5. Calculate numerical values for V and ρ_v at point P in free space if:
(a) $V = \frac{4yz}{x^2 + 1}$, at $P(1, 2, 3)$; (b) $V = 5\rho^2 \cos 2\phi$, at $P(\rho = 3, \phi = \frac{\pi}{3}, z = 2)$;
(c) $V = \frac{2 \cos \phi}{r^2}$, at $P(r = 0.5, \theta = 45°, \phi = 60°)$.

Ans. (a) 12 V, -106.2 pC/m^3; (b) -22.5 V, 0; (c) 4 V, 0

6.7 EXAMPLES OF THE SOLUTION OF LAPLACE'S EQUATION

Several methods have been developed for solving Laplace's equation. The simplest method is that of direct integration. We will use this technique to work several examples involving one-dimensional potential variation in various coordinate systems in this section.

The method of direct integration is applicable only to problems that are "one-dimensional," or in which the potential field is a function of only one of the three coordinates. Since we are working with only three coordinate systems, it might seem, then, that there are nine problems to be solved, but a little reflection will show that a field that varies only with x is fundamentally the same as a field that varies only with y. Rotating the physical problem a quarter turn is no change. Actually, there are only five problems to be solved, one in rectangular coordinates, two in cylindrical, and two in spherical. We will solve them all.

First, let us assume that V is a function only of x and worry later about which physical problem we are solving when we have a need for boundary conditions. Laplace's equation reduces to

$$\frac{\partial^2 V}{\partial x^2} = 0$$

and the partial derivative may be replaced by an ordinary derivative, since V is not a function of y or z,

$$\frac{d^2 V}{dx^2} = 0$$

We integrate twice, obtaining

$$\frac{dV}{dx} = A$$

and

$$V = Ax + B \tag{31}$$

where A and B are constants of integration. Equation (31) contains two such constants, as we would expect for a second-order differential equation. These constants can be determined only from the boundary conditions.

Since the field varies only with x and is not a function of y and z, then V is a constant if x is a constant or, in other words, the equipotential surfaces are parallel planes normal to the x axis. The field is thus that of a parallel-plate capacitor, and as soon as we specify the potential on any two planes, we may evaluate our constants of integration.

EXAMPLE 6.2

Start with the potential function, Eq. (31), and find the capacitance of a parallel-plate capacitor of plate area S, plate separation d, and potential difference V_0 between plates.

Solution. Take $V = 0$ at $x = 0$ and $V = V_0$ at $x = d$. Then from Eq. (31),

$$A = \frac{V_0}{d} \quad B = 0$$

and

$$\boxed{V = \frac{V_0 x}{d}} \tag{32}$$

We still need the total charge on either plate before the capacitance can be found. We should remember that when we first solved this capacitor problem, the sheet of charge provided our starting point. We did not have to work very hard to find the charge, for all the fields were expressed in terms of it. The work then was spent in finding potential difference. Now the problem is reversed (and simplified).

The necessary steps are these, after the choice of boundary conditions has been made:

1. Given V, use $\mathbf{E} = -\nabla V$ to find $\mathbf{E}$.
2. Use $\mathbf{D} = \epsilon \mathbf{E}$ to find $\mathbf{D}$.
3. Evaluate $\mathbf{D}$ at either capacitor plate, $\mathbf{D} = \mathbf{D}_S = D_N \mathbf{a}_N$.
4. Recognize that $\rho_S = D_N$.
5. Find Q by a surface integration over the capacitor plate, $Q = \int_S \rho_S \, dS$.

Here we have

$$V = V_0 \frac{x}{d}$$

$$\mathbf{E} = -\frac{V_0}{d}\mathbf{a}_x$$

$$\mathbf{D} = -\epsilon\frac{V_0}{d}\mathbf{a}_x$$

$$\mathbf{D}_S = \mathbf{D}|_{x=0} = -\epsilon\frac{V_0}{d}\mathbf{a}_x$$

$$\mathbf{a}_N = \mathbf{a}_x$$

$$D_N = -\epsilon\frac{V_0}{d} = \rho_S$$

$$Q = \int_S \frac{-\epsilon V_0}{d}dS = -\epsilon\frac{V_0 S}{d}$$

and the capacitance is

$$\boxed{C = \frac{|Q|}{V_0} = \frac{\epsilon S}{d}}$$
(33)

We will use this procedure several times in the examples to follow.

EXAMPLE 6.3

Because no new problems are solved by choosing fields which vary only with y or with z in rectangular coordinates, we pass on to cylindrical coordinates for our next example. Variations with respect to z are again nothing new, and we next assume variation with respect to ρ only. Laplace's equation becomes

$$\frac{1}{\rho}\frac{\partial}{\partial\rho}\left(\rho\frac{\partial V}{\partial\rho}\right) = 0$$

Noting the ρ in the denominator, we exclude $\rho = 0$ from our solution and then multiply by ρ and integrate,

$$\rho\frac{dV}{d\rho} = A$$

where a total derivative replaces the partial derivative because V varies only with ρ. Next, rearrange, and integrate again,

$$V = A\ln\rho + B$$
(34)

The equipotential surfaces are given by $\rho = $ constant and are cylinders, and the problem is that of the coaxial capacitor or coaxial transmission line. We choose a potential difference of V_0 by letting $V = V_0$ at $\rho = a$, $V = 0$ at $\rho = b$, $b > a$, and obtain

$$V = V_0 \frac{\ln(b/\rho)}{\ln(b/a)}$$

(35)

from which

$$\mathbf{E} = \frac{V_0}{\rho} \frac{1}{\ln(b/a)} \mathbf{a}_\rho$$

$$D_{N(\rho=a)} = \frac{\epsilon V_0}{a \ln(b/a)}$$

$$Q = \frac{\epsilon V_0 2\pi a L}{a \ln(b/a)}$$

$$C = \frac{2\pi\epsilon L}{\ln(b/a)}$$

(36)

which agrees with our result in Section 6.3 (Eq. (5)).

EXAMPLE 6.4

Now assume that V is a function only of ϕ in cylindrical coordinates. We might look at the physical problem first for a change and see that equipotential surfaces are given by ϕ = constant. These are radial planes. Boundary conditions might be $V = 0$ at $\phi = 0$ and $V = V_0$ at $\phi = \alpha$, leading to the physical problem detailed in Figure 6.10.

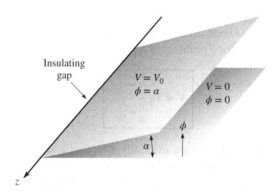

Figure 6.10 Two infinite radial planes with an interior angle α. An infinitesimal insulating gap exists at $\rho = 0$. The potential field may be found by applying Laplace's equation in cylindrical coordinates.

Laplace's equation is now

$$\frac{1}{\rho^2}\frac{\partial^2 V}{\partial \phi^2} = 0$$

We exclude $\rho = 0$ and have

$$\frac{d^2 V}{d\phi^2} = 0$$

The solution is

$$V = A\phi + B$$

The boundary conditions determine A and B, and

$$\boxed{V = V_0 \frac{\phi}{\alpha}} \tag{37}$$

Taking the gradient of Eq. (37) produces the electric field intensity,

$$\boxed{\mathbf{E} = -\frac{V_0 \mathbf{a}_\phi}{\alpha \rho}} \tag{38}$$

and it is interesting to note that E is a function of ρ and not of ϕ. This does not contradict our original assumptions, which were restrictions only on the potential field. Note, however, that the *vector* field $\mathbf{E}$ is in the ϕ direction.

A problem involving the capacitance of these two radial planes is included at the end of the chapter.

EXAMPLE 6.5

We now turn to spherical coordinates, dispose immediately of variations with respect to ϕ only as having just been solved, and treat first $V = V(r)$.

The details are left for a problem later, but the final potential field is given by

$$\boxed{V = V_0 \frac{\frac{1}{r} - \frac{1}{b}}{\frac{1}{a} - \frac{1}{b}}} \tag{39}$$

where the boundary conditions are evidently $V = 0$ at $r = b$ and $V = V_0$ at $r = a$, $b > a$. The problem is that of concentric spheres. The capacitance was found previously in Section 6.3 (by a somewhat different method) and is

$$\boxed{C = \frac{4\pi\epsilon}{\frac{1}{a} - \frac{1}{b}}} \tag{40}$$

EXAMPLE 6.6

In spherical coordinates we now restrict the potential function to $V = V(\theta)$, obtaining

$$\frac{1}{r^2 \sin\theta} \frac{d}{d\theta}\left(\sin\theta \frac{dV}{d\theta}\right) = 0$$

We exclude $r = 0$ and $\theta = 0$ or π and have

$$\sin\theta \frac{dV}{d\theta} = A$$

The second integral is then

$$V = \int \frac{A\,d\theta}{\sin\theta} + B$$

which is not as obvious as the previous ones. From integral tables (or a good memory) we have

$$V = A\ln\left(\tan\frac{\theta}{2}\right) + B \tag{41}$$

The equipotential surfaces of Eq. (41) are cones. Figure 6.11 illustrates the case where $V = 0$ at $\theta = \pi/2$ and $V = V_0$ at $\theta = \alpha$, $\alpha < \pi/2$. We obtain

$$\boxed{V = V_0 \frac{\ln\left(\tan\dfrac{\theta}{2}\right)}{\ln\left(\tan\dfrac{\alpha}{2}\right)}} \tag{42}$$

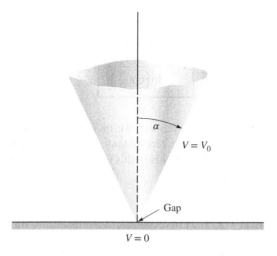

Figure 6.11 For the cone $\theta = \alpha$ at V_0 and the plane $\theta = \pi/2$ at $V = 0$, the potential field is given by $V = V_0[\ln(\tan\theta/2)]/[\ln(\tan\alpha/2)]$.

In order to find the capacitance between a conducting cone with its vertex separated from a conducting plane by an infinitesimal insulating gap and its axis normal to the plane, we first find the field strength:

$$E = -\nabla V = \frac{-1}{r}\frac{\partial V}{\partial \theta}\mathbf{a}_\theta = -\frac{V_0}{r \sin\theta \ln\left(\tan\frac{\alpha}{2}\right)}\mathbf{a}_\theta$$

The surface charge density on the cone is then

$$\rho_S = \frac{-\epsilon V_0}{r \sin\alpha \ln\left(\tan\frac{\alpha}{2}\right)}$$

producing a total charge Q,

$$Q = \frac{-\epsilon V_0}{\sin\alpha \ln\left(\tan\frac{\alpha}{2}\right)}\int_0^\infty \int_0^{2\pi} \frac{r \sin\alpha\, d\phi\, dr}{r}$$

$$= \frac{-2\pi\epsilon_0 V_0}{\ln\left(\tan\frac{\alpha}{2}\right)}\int_0^\infty dr$$

This leads to an infinite value of charge and capacitance, and it becomes necessary to consider a cone of finite size. Our answer will now be only an approximation because the theoretical equipotential surface is $\theta = \alpha$, a conical surface extending from $r = 0$ to $r = \infty$, whereas our physical conical surface extends only from $r = 0$ to, say, $r = r_1$. The approximate capacitance is

$$\boxed{C \doteq \frac{2\pi\epsilon r_1}{\ln\left(\cot\frac{\alpha}{2}\right)}}$$
(43)

If we desire a more accurate answer, we may make an estimate of the capacitance of the base of the cone to the zero-potential plane and add this amount to our answer. Fringing, or nonuniform, fields in this region have been neglected and introduce an additional source of error.

D6.6. Find $|E|$ at $P(3, 1, 2)$ in rectangular coordinates for the field of: (a) two coaxial conducting cylinders, $V = 50$ V at $\rho = 2$ m, and $V = 20$ V at $\rho = 3$ m; (b) two radial conducting planes, $V = 50$ V at $\phi = 10°$, and $V = 20$ V at $\phi = 30°$.

Ans. (a) 23.4 V/m; (b) 27.2 V/m

6.8 EXAMPLE OF THE SOLUTION OF POISSON'S EQUATION: THE P-N JUNCTION CAPACITANCE

To select a reasonably simple problem that might illustrate the application of Poisson's equation, we must assume that the volume charge density is specified. This is not usually the case, however; in fact, it is often the quantity about which we are seeking further information. The type of problem which we might encounter later would begin with a knowledge only of the boundary values of the potential, the electric field intensity, and the current density. From these we would have to apply Poisson's equation, the continuity equation, and some relationship expressing the forces on the charged particles, such as the Lorentz force equation or the diffusion equation, and solve the whole system of equations simultaneously. Such an ordeal is beyond the scope of this text, and we will therefore assume a reasonably large amount of information.

As an example, consider a *pn* junction between two halves of a semiconductor bar extending in the *x* direction. We will assume that the region for $x < 0$ is doped *p* type and that the region for $x > 0$ is *n* type. The degree of doping is identical on each side of the junction. To review some of the facts about the semiconductor junction, we note that initially there are excess holes to the left of the junction and excess electrons to the right. Each diffuses across the junction until an electric field is built up in such a direction that the diffusion current drops to zero. Thus, to prevent more holes from moving to the right, the electric field in the neighborhood of the junction must be directed to the left; E_x is negative there. This field must be produced by a net positive charge to the right of the junction and a net negative charge to the left. Note that the layer of positive charge consists of two parts—the holes which have crossed the junction and the positive donor ions from which the electrons have departed. The negative layer of charge is constituted in the opposite manner by electrons and negative acceptor ions.

The type of charge distribution that results is shown in Figure 6.12*a*, and the negative field which it produces is shown in Figure 6.12*b*. After looking at these two figures, one might profitably read the previous paragraph again.

A charge distribution of this form may be approximated by many different expressions. One of the simpler expressions is

$$\rho_v = 2\rho_{v0} \operatorname{sech} \frac{x}{a} \tanh \frac{x}{a} \qquad (44)$$

which has a maximum charge density $\rho_{v,max} = \rho_{v0}$ that occurs at $x = 0.881a$. The maximum charge density ρ_{v0} is related to the acceptor and donor concentrations N_a and N_d by noting that all the donor and acceptor ions in this region (the *depletion* layer) have been stripped of an electron or a hole, and thus

$$\rho_{v0} = e N_a = e N_d$$

We now solve Poisson's equation,

$$\nabla^2 V = -\frac{\rho_v}{\epsilon}$$

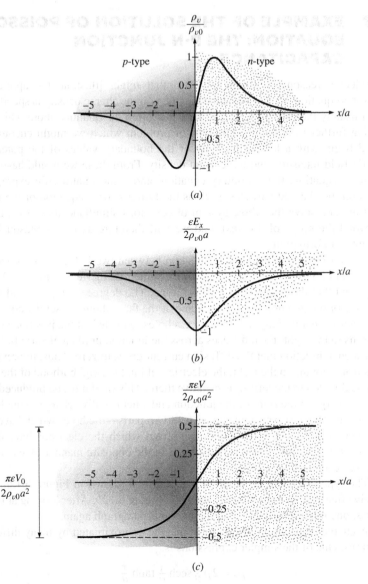

Figure 6.12 (a) The charge density, (b) the electric field intensity, and (c) the potential are plotted for a *pn* junction as functions of distance from the center of the junction. The *p*-type material is on the left, and the *n*-type is on the right.

subject to the charge distribution assumed above,

$$\frac{d^2 V}{dx^2} = -\frac{2\rho_{v0}}{\epsilon}\operatorname{sech}\frac{x}{a}\tanh\frac{x}{a}$$

in this one-dimensional problem in which variations with y and z are not present. We integrate once,

$$\frac{dV}{dx} = \frac{2\rho_{v0}a}{\epsilon}\operatorname{sech}\frac{x}{a} + C_1$$

and obtain the electric field intensity,

$$E_x = -\frac{2\rho_{v0}a}{\epsilon}\operatorname{sech}\frac{x}{a} - C_1$$

To evaluate the constant of integration C_1, we note that no net charge density and no fields can exist *far* from the junction. Thus, as $x \to \pm\infty$, E_x must approach zero. Therefore $C_1 = 0$, and

$$E_x = -\frac{2\rho_{v0}a}{\epsilon}\operatorname{sech}\frac{x}{a} \tag{45}$$

Integrating again,

$$V = \frac{4\rho_{v0}a^2}{\epsilon}\tan^{-1} e^{x/a} + C_2$$

The zero reference of potential is arbitrarily set at the center of the junction, $x = 0$,

$$0 = \frac{4\rho_{v0}a^2}{\epsilon}\frac{\pi}{4} + C_2$$

and finally,

$$V = \frac{4\rho_{v0}a^2}{\epsilon}\left(\tan^{-1} e^{x/a} - \frac{\pi}{4}\right) \tag{46}$$

Figure 6.12 shows the charge distribution (a), electric field intensity (b), and the potential (c), as given by Eqs. (44), (45), and (46), respectively.

The potential is constant once we are a distance of about $4a$ or $5a$ from the junction. The total potential difference V_0 across the junction is obtained from Eq. (46),

$$V_0 = V_{x\to\infty} - V_{x\to-\infty} = \frac{2\pi\rho_{v0}a^2}{\epsilon} \tag{47}$$

This expression suggests the possibility of determining the total charge on one side of the junction and then using Eq. (47) to find a junction capacitance. The total positive charge is

$$Q = S\int_0^\infty 2\rho_{v0}\operatorname{sech}\frac{x}{a}\tanh\frac{x}{a}dx = 2\rho_{v0}aS$$

where S is the area of the junction cross section. If we make use of Eq. (47) to eliminate the distance parameter a, the charge becomes

$$Q = S\sqrt{\frac{2\rho_{v0}\epsilon V_0}{\pi}} \tag{48}$$

Because the total charge is a function of the potential difference, we have to be careful in defining a capacitance. Thinking in "circuit" terms for a moment,

$$I = \frac{dQ}{dt} = C\frac{dV_0}{dt}$$

and thus

$$C = \frac{dQ}{dV_0}$$

By differentiating Eq. (48), we therefore have the capacitance

$$C = \sqrt{\frac{\rho_{v0}\epsilon}{2\pi V_0}}S = \frac{\epsilon S}{2\pi a} \tag{49}$$

The first form of Eq. (49) shows that the capacitance varies inversely as the square root of the voltage. That is, a higher voltage causes a greater separation of the charge layers and a smaller capacitance. The second form is interesting in that it indicates that we may think of the junction as a parallel-plate capacitor with a "plate" separation of $2\pi a$. In view of the dimensions of the region in which the charge is concentrated, this is a logical result.

D6.7. In the neighborhood of a certain semiconductor junction, the volume charge density is given by $\rho_v = 750 \text{ sech } 10^6\pi x \text{ tanh } 10^6\pi x$ C/m^3. The dielectric constant of the semiconductor material is 10 and the junction area is 2×10^{-7} m^2. Find: (a) V_0; (b) C; (c) E at the junction.

Ans. (a) 2.70 V; (b) 8.85 pF; (c) 2.70 MV/m

D6.8. Given the volume charge density $\rho_v = -2 \times 10^7\epsilon_0\sqrt{x}$ C/m^3 in free space, let $V = 0$ at $x = 0$ and let $V = 2$ V at $x = 2.5$ mm. At $x = 1$ mm, find: (a) V; (b) E_x.

Ans. (a) 0.302 V; (b) −555 V/m

REFERENCES

1. Matsch, L. W. *Capacitors, Magnetic Circuits, and Transformers*. Englewood Cliffs, NJ: Prentice-Hall, 1964. Many of the practical aspects of capacitors are discussed in Chapter 2.

2. Ramo, S., J. R. Whinnery, and T. Van Duzer. *Fields and Waves in Communication Electronics*. 3rd ed. New York: John Wiley and Sons, 1994. This classic text is primarily directed toward beginning graduate students, but it may be read by anyone familiar with basic electromagnetics concepts. Curvilinear square plotting is described on pp. 50–52. A more advanced discussion of methods of solving Laplace's equation is given in Chapter 7.

3. Dekker, A. J. See references for Chapter 5.

4. Hayt, W. H. Jr., and J. E. Kemmerly. *Engineering Circuit Analysis*. 5th ed. New York: McGraw-Hill, 1993.

5. Collin, R. E., and R. E. Plonsey. *Principles and Applications of Electromagnetic Fields.* New York: McGraw-Hill, 1961. Provides an excellent treatment of methods of solving Laplace's and Poisson's equations.

6. Smythe, W. R. *Static and Dynamic Electricity.* 3rd ed. Taylor and Francis, 1989. An advanced treatment of potential theory is given in Chapter 4.

CHAPTER 6 PROBLEMS

Quizzes

6.1 Consider a coaxial capacitor having inner radius a, outer radius b, unit length, and filled with a material with dielectric constant, ϵ_r. Compare this to a parallel-plate capacitor having plate width w, plate separation d, filled with the same dielectric, and having unit length. Express the ratio b/a in terms of the ratio d/w, such that the two structures will store the same energy for a given applied voltage.

6.2 Let $S = 100$ mm^2, $d = 3$ mm, and $\epsilon_r = 12$ for a parallel-plate capacitor. (*a*) Calculate the capacitance. (*b*) After connecting a 6-V battery across the capacitor, calculate E, D, Q, and the total stored electrostatic energy. (*c*) With the source still connected, the dielectric is carefully withdrawn from between the plates. With the dielectric gone, recalculate E, D, Q, and the energy stored in the capacitor. (*d*) If the charge and energy found in part c are less than the values found in part b (which you should have discovered), what became of the missing charge and energy? Assume in both cases that the dielectric itself carries no charge on removal.

6.3 Capacitors tend to be more expensive as their capacitance and maximum voltage V_{max} increase. The voltage V_{max} is limited by the field strength at which the dielectric breaks down, E_{BD}. Which of these dielectrics will give the largest CV_{max} product for equal plate areas? (*a*) Air: $\epsilon_r = 1$, $E_{BD} = 3$ MV/m. (*b*) Barium titanate: $\epsilon_r = 1200$, $E_{BD} = 3$ MV/m. (*c*) Silicon dioxide: $\epsilon_r = 3.78$, $E_{BD} = 16$ MV/m. (*d*) Polyethylene: $\epsilon_r = 2.26$, $E_{BD} = 4.7$ MV/m.

6.4 An air-filled parallel-plate capacitor with plate separation d and plate area A is connected to a battery that applies a voltage V_0 between plates. With the battery left connected, the plates are moved apart to a distance of $10d$. Determine by what factor each of the following quantities changes: (*a*) V_0; (*b*) C; (*c*) E; (*d*) D; (*e*) Q; (*f*) ρ_S; (*g*) W_E. Repeat the exercise for the case in which the battery is disconnected before the plates are separated.

6.5 A parallel-plate capacitor having plate area A and plate spacing d is partially filled with dielectric of relative permittivity ϵ_r. The material fills the capacitor volume from the lower plate (at $z = 0$) up to location $z = b$, where $b < d$. The remaining volume is free space. The dielectric has the effect of increasing the capacitance by a certain factor over the value found when the capacitor is completely air-filled. Find this factor in terms of b, ϵ_r, and other known parameters.

6.6 A parallel-plate capacitor is made using two circular plates of radius a, with the bottom plate on the xy plane, centered at the origin. The top plate is located at $z = d$, with its center on the z axis. Charge Q is on the top plate; $-Q$ is on the bottom plate. Dielectric having z-*dependent* permittivity fills the region between plates. The permittivity is given by $\epsilon(z) = \epsilon_0(1 + z^2/d^2)$. Find (a) **D**; (b) **E**; (c) V_0; (d) C.

6.7 For the capacitor of Problem 6.6, consider the dielectric as made up of a stack of layers, each having differential thickness dz, and where each layer (at location z) has dielectric constant $\epsilon_r = (1 + z^2/d^2)$. Evaluate the capacitance by considering the structure as a series combination of the layer capacitances and evaluating an appropriate integral.

6.8 A parallel-plate capacitor is made using two circular plates of radius a, with the bottom plate on the xy plane, centered at the origin. The top plate is located at $z = d$, with its center on the z axis. Potential V_0 is on the top plate; the bottom plate is grounded. Dielectric having *radially dependent* permittivity fills the region between plates. The permittivity is given by $\epsilon(\rho) = \epsilon_0(1 + \rho^2/a^2)$. Find (a) **E**; (b) **D**; (c) Q; (d) C.

6.9 For the capacitor of Problem 6.8, consider the dielectric as made up of concentric cylindrical shells, each of thickness $d\rho$, and where the dielectric constant of each shell (at radius ρ) is $\epsilon_r = (1 + \rho^2/a^2)$. Evaluate the capacitance by considering the structure as a parallel combination of the shell capacitances and by evaluating an appropriate integral.

6.10 A coaxial cable has conductor dimensions of $a = 1.0$ mm and $b = 2.7$ mm. The inner conductor is supported by dielectric spacers ($\epsilon_r = 5$) in the form of washers with a hole radius of 1 mm and an outer radius of 2.7 mm, and with a thickness of 3.0 mm. The spacers are located every 2 cm down the cable. (a) By what factor do the spacers increase the capacitance per unit length? (b) If 100 V is maintained across the cable, find **E** at all points.

6.11 A coaxial transmission line has inner and outer conductors of radii a and b. The volume between conductors is filled with two dielectrics having relative permittivities ϵ_{r1} ($a < \rho < c$) and ϵ_{r2} ($c < \rho < b$). (a) If $c = (a + b)/2$, find the ratio of ϵ_{r1} to ϵ_{r2} such that the maximum electric field intensities in the two regions are equal. (b) Find the capacitance per unit length under the condition of part a.

6.12 (a) Determine the capacitance of an isolated conducting sphere of radius a in free space (consider an outer conductor existing at $r \rightarrow \infty$). (b) The sphere is to be covered with a dielectric layer of thickness d and dielectric contant ϵ_r. If $\epsilon_r = 3$, find d in terms of a such that the capacitance is twice that of part a.

6.13 With reference to Figure 6.5, let $b = 6$ m, $h = 15$ m, and the conductor potential be 250 V. Take $\epsilon = \epsilon_0$. Find values for K_1, ρ_L, a, and C.

6.14 Two #16 copper conductors (1.29 mm diameter) are parallel with a separation d between axes. Determine d so that the capacitance between wires in air is 30 pF/m.

6.15 A 2-cm-diameter conductor is suspended in air with its axis 5 cm from a conducting plane. Let the potential of the cylinder be 100 V and that of the plane be 0 V. (a) Find the surface charge density on the cylinder at a point nearest the plane. (b) Find the surface charge density on the plane at a point nearest the cylinder. (c) Find the capacitance per unit length.

6.16 Consider an arrangement of two isolated conducting surfaces of any shape that form a capacitor. Use the definitions of capacitance (Eq. (2) in this chapter) and resistance (Eq. (14) in Chapter 5) to show that when the region between the conductors is filled with either conductive material (conductivity σ) or a perfect dielectric (permittivity ϵ), the resulting resistance and capacitance of the structures are related through the simple formula $RC = \epsilon/\sigma$. What basic properties must be true about both the dielectric and the conducting medium for this condition to hold for certain?

6.17 Construct a curvilinear-square map for a coaxial capacitor of 3 cm inner radius and 8 cm outer radius. These dimensions are suitable for the drawing. (a) Use your sketch to calculate the capacitance per meter length, assuming $\epsilon_r = 1$. (b) Calculate an exact value for the capacitance per unit length.

6.18 Construct a curvilinear-square map of the potential field about two parallel circular cylinders, each of 2.5 cm radius, separated by a center-to-center distance of 13 cm. These dimensions are suitable for the actual sketch if symmetry is considered. As a check, compute the capacitance per meter both from your sketch and from the exact formula. Assume $\epsilon_r = 1$.

6.19 Construct a curvilinear-square map of the potential field between two parallel circular cylinders, one of 4 cm radius inside another of 8 cm radius. The two axes are displaced by 2.5 cm. These dimensions are suitable for the drawing. As a check on the accuracy, compute the capacitance per meter from the sketch and from the exact expression:

$$C = \frac{2\pi\epsilon}{\cosh^{-1}[(a^2 + b^2 - D^2)/(2ab)]}$$

where a and b are the conductor radii and D is the axis separation.

6.20 A solid conducting cylinder of 4 cm radius is centered within a rectangular conducting cylinder with a 12 cm by 20 cm cross section. (a) Make a full-size sketch of one quadrant of this configuration and construct a curvilinear-square map for its interior. (b) Assume $\epsilon = \epsilon_0$ and estimate C per meter length.

6.21 The inner conductor of the transmission line shown in Figure 6.13 has a square cross section $2a \times 2a$, whereas the outer square is $4a \times 5a$. The axes are displaced as shown. (a) Construct a good-sized drawing of this transmission line, say with $a = 2.5$ cm, and then prepare a curvilinear-square plot of the electrostatic field between the conductors. (b) Use the map to calculate the capacitance per meter length if $\epsilon = 1.6\epsilon_0$. (c) How would your result to part b change if $a = 0.6$ cm?

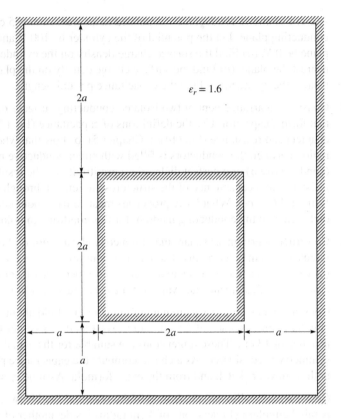

Figure 6.13   See Problem 6.21.

6.22 A parallel-plate capacitor is air filled and has plate area A and plate separation d. Sufficient dielectric material of relative permittivity ϵ_r is available to fill half the capacitor volume. How should the material be used to maximize the capacitance, and by what factor will the capacitance increase over the air-filled case?

6.23 A two-wire transmission line consists of two parallel perfectly conducting cylinders, each having a radius of 0.2 mm, separated by a center-to-center distance of 2 mm. The medium surrounding the wires has $\epsilon_r = 3$ and $\sigma = 1.5$ mS/m. A 100-V battery is connected between the wires. (a) Calculate the magnitude of the charge per meter length on each wire. (b) Using the result of Problem 6.16, find the battery current.

6.24 A potential field in free space is given in spherical coordinates as

$$V(r) = \begin{cases} [\rho_0/(6\epsilon_0)][3a^2 - r^2] & (r \le a) \\ (a^3\rho_0)/(3\epsilon_0 r) & (r \ge a) \end{cases}$$

where ρ_0 and a are constants. (a) Use Poisson's equation to find the volume charge density everywhere. (b) Find the total charge present.

6.25 A capacitor is formed from concentric spherical conductors having radii a and b, where $b > a$. The inner conductor is raised to potential V_0; the outer conductor is grounded. Under these conditions, derive Eq. (39) using Laplace's equation.

6.26 Given the spherically symmetric potential field in free space, $V = V_0 e^{-r/a}$, find. (a) ρ_v at $r = a$; (b) the electric field at $r = a$; (c) the total charge.

6.27 Let $V(x, y) = 4e^{2x} + f(x) - 3y^2$ in a region of free space where $\rho_v = 0$. It is known that both E_x and V are zero at the origin. Find $f(x)$ and $V(x, y)$.

6.28 Show that in a homogeneous medium of conductivity σ, the potential field V satisfies Laplace's equation if any volume charge density present does not vary with time.

6.29 What total charge must be located within a unit sphere centered at the origin in free space in order to produce the potential field $V(r) = -6r^5/\epsilon_0$ for $r \le 1$?

6.30 A parallel-plate capacitor has plates located at $z = 0$ and $z = d$. The region between plates is filled with a material that contains volume charge of uniform density ρ_0 C/m^3 and has permittivity ϵ. Both plates are held at ground potential. (a) Determine the potential field between plates. (b) Determine the electric field intensity **E** between plates. (c) Repeat parts a and b for the case of the plate at $z = d$ raised to potential V_0, with the $z = 0$ plate grounded.

6.31 For the parallel-plate capacitor shown in Figure 6.3, find the potential field in the interior if the upper plate (at $z = d$) is raised to potential V_0, while the lower plate (at $z = 0$) is grounded. Do this by solving Laplace's equation separately in each of the two dielectrics. These solutions, as well as the electric flux density, must be continuous across the dielectric interface. Take the interface to lie at $z = b$.

6.32 A uniform volume charge has constant density $\rho_v = \rho_0$ C/m^3 and fills the region $r < a$, in which permittivity ϵ is assumed. A conducting spherical shell is located at $r = a$ and is held at ground potential. Find (a) the potential everywhere; (b) the electric field intensity, **E**, everywhere.

6.33 The functions $V_1(\rho, \phi, z)$ and $V_2(\rho, \phi, z)$ both satisfy Laplace's equation in the region $a < \rho < b$, $0 \le \phi < 2\pi$, $-L < z < L$; each is zero on the surfaces $\rho = b$ for $-L < z < L$; $z = -L$ for $a < \rho < b$; and $z = L$ for $a < \rho < b$; and each is 100 V on the surface $\rho = a$ for $-L < z < L$. (a) In the region specified, is Laplace's equation satisfied by the functions $V_1 + V_2$, $V_1 - V_2$, $V_1 + 3$, and $V_1 V_2$? (b) On the boundary surfaces specified, are the potential values given in this problem obtained from the functions $V_1 + V_2$, $V_1 - V_2$, $V_1 + 3$, and V_1V_2? (c) Are the functions $V_1 + V_2$, $V_1 - V_2$, $V_1 + 3$, and $V_1 V_2$ identical with V_1?

6.34 Consider the parallel-plate capacitor of Problem 6.30, but this time the charged dielectric exists only between $z = 0$ and $z = b$, where $b < d$. Free space fills the region $b < z < d$. Both plates are at ground potential. By solving Laplace's *and* Poisson's equations, find (*a*) $V(z)$ for $0 < z < d$; (*b*) the electric field intensity for $0 < z < d$. No surface charge exists at $z = b$, so both V and $\mathbf{D}$ are continuous there.

6.35 In spherical coordinates, a potential is known to be a function of θ only. (*a*) Find the function $V(\theta)$ if $V = 10$ V at $\theta = 90°$ and $\mathbf{E} = -500\ \mathbf{a}_\theta$ V/m at $\theta = 30°$, $r = 0.4$ m; (*b*) find the electric field intensity in rectangular coordinates at $\theta = 90°$, $r = 1$ m.

6.36 The derivation of Laplace's and Poisson's equations assumed constant permittivity, but there are cases of spatially varying permittivity in which the equations will still apply. Consider the vector identity, $\nabla \cdot (\psi \mathbf{G}) = \mathbf{G} \cdot \nabla\psi + \psi\nabla \cdot \mathbf{G}$, where ψ and $\mathbf{G}$ are scalar and vector functions, respectively. Determine a general rule on the allowed *directions* in which ϵ may vary with respect to the local electric field.

6.37 Coaxial conducting cylinders are located at $\rho = 0.5$ cm and $\rho = 1.2$ cm. The region between the cylinders is filled with a homogeneous perfect dielectric. If the inner cylinder is at 100 V and the outer at 0 V, find (*a*) the location of the 20 V equipotential surface; (*b*) $E_{\rho\ \max}$; (*c*) ϵ_r if the charge per meter length on the inner cylinder is 20 nC/m.

6.38 Repeat Problem 6.37, but with the dielectric only partially filling the volume, within $0 < \phi < \pi$, and with free space in the remaining volume.

6.39 The two conducting planes illustrated in Figure 6.14 are defined by $0.001 < \rho < 0.120$ m, $0 < z < 0.1$ m, $\phi = 0.179$ and 0.188 rad. The medium surrounding the planes is air. For Region 1, $0.179 < \phi < 0.188$; neglect fringing and find (*a*) $V(\phi)$; (*b*) $\mathbf{E}(\rho)$; (*c*) $\mathbf{D}(\rho)$; (*d*) ρ_s on the upper surface of the lower plane; (*e*) Q on the upper surface of the lower plane. (*f*) Repeat parts *a* through *c* for Region 2 by letting the location of the upper plane be $\phi = .188 - 2\pi$, and then find ρ_s and Q on the lower surface of the lower plane. (*g*) Find the total charge on the lower plane and the capacitance between the planes.

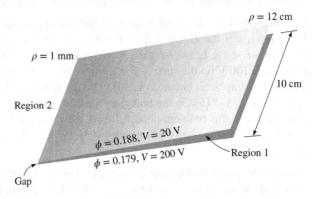

Figure 6.14   See Problem 6.39.

6.40 A parallel-plate capacitor is made using two circular plates of radius a, with the bottom plate on the xy plane, centered at the origin. The top plate is located at $z = d$, with its center on the z axis. Potential V_0 is on the top plate; the bottom plate is grounded. Dielectric having *radially dependent* permittivity fills the region between plates. The permittivity is given by $\epsilon(\rho) = \epsilon_0(1 + \rho^2/a^2)$. Find (a) $V(z)$; (b) **E**; (c) Q; (d) C. This is a reprise of Problem 6.8, but it starts with Laplace's equation.

6.41 Concentric conducting spheres are located at $r = 5$ mm and $r = 20$ mm. The region between the spheres is filled with a perfect dielectric. If the inner sphere is at 100 V and the outer sphere is at 0 V, (a) find the location of the 20 V equipotential surface. (b) Find $E_{r,max}$. (c) Find ϵ_r if the surface charge density on the inner sphere is 1.0 μC/m^2.

6.42 The hemisphere $0 < r < a$, $0 < \theta < \pi/2$, is composed of homogeneous conducting material of conductivity σ. The flat side of the hemisphere rests on a perfectly conducting plane. Now, the material within the conical region $0 < \theta < \alpha$, $0 < r < a$ is drilled out and replaced with material that is perfectly conducting. An air gap is maintained between the $r = 0$ tip of this new material and the plane. What resistance is measured between the two perfect conductors? Neglect fringing fields.

6.43 Two concentric spherical conducting surfaces exist at $r = a$ and $r = b$, where $a < b$. Between conductors are two dielectrics having relative permittivities ϵ_{r1} ($a < r < c$) and ϵ_{r2} ($c < r < b$). The inner conductor is raised to potential V_0; the outer conductor is grounded. Solve Laplace's equation for the potential everywhere between conductors. This is the spherical coordinate version of Problem 6.31, and the procedure is the same. As a test of your final result, the potential should reduce to that given in Eq. (39) when $\epsilon_{r1} = \epsilon_{r2}$.

6.44 A potential field in free space is given as $V = 100 \ln \tan(\theta/2) + 50$ V. (a) Find the maximum value of $|\mathbf{E}_\theta|$ on the surface $\theta = 40°$ for $0.1 < r < 0.8$ m, $60° < \phi < 90°$. (b) Describe the surface $V = 80$ V.

6.45 In free space, let $\rho_v = 200\epsilon_0/r^{2.4}$. (a) Use Poisson's equation to find $V(r)$ if it is assumed that $r^2 E_r \to 0$ when $r \to 0$, and also that $V \to 0$ as $r \to \infty$. (b) Now find $V(r)$ by using Gauss's law and a line integral.

6.46 By appropriate solution of Laplace's *and* Poisson's equations, determine the absolute potential at the center of a sphere of radius a, containing uniform volume charge of density ρ_0. Assume permittivity ϵ_0 everywhere. HINT: What must be true about the potential and the electric field at $r = 0$ and at $r = a$?

CHAPTER 7　The Steady Magnetic Field

术语

steady magnetic field　恒定磁场，直流电流产生的磁场。
magnetic field intensity　磁场强度，单位是安［培］／米（A／m）
Biot-Savart law　毕奥-萨伐定律，已知无限大均匀磁媒质中分布的电流，其磁场可以通过对电流的积分求得。
Ampere's circuital law　安培环路定律。
curl　旋度。
magnetic flux density　磁通密度，简称磁密；国内教材中多称之为磁感应强度，单位是韦［伯］／平方米（Wb／m²），但国标已规定采用磁通密度，而非磁感应强度；国际单位制中的单位是特［斯拉］（T）。
permeability　磁导率，单位是亨［利］／米（H／m）；亨［利］是电感的单位，所以磁导率有电感率的含义。
scalar magnetic potential　标量磁位，单位是安［培］（A）。
vector magnetic potential　矢量磁位，单位是韦［伯］／米（Wb／m）。

At this point, the concept of a field should be a familiar one. Since we first accepted the experimental law of forces existing between two point charges and defined electric field intensity as the force per unit charge on a test charge in the presence of a second charge, we have discussed numerous fields. These fields possess no real physical basis, for physical measurements must always be in terms of the forces on the charges in the detection equipment. Those charges that are the source cause measurable forces to be exerted on other charges, which we may think of as detector charges. The fact that we attribute a field to the source charges and then determine the effect of this field on the detector charges amounts merely to a division of the basic problem into two parts for convenience.

We will begin our study of the magnetic field with a definition of the magnetic field itself and show how it arises from a current distribution. The effect of this field on other currents, or the second half of the physical problem, will be discussed in Chapter 8. As we did with the electric field, we confine our initial discussion to free-space conditions, and the effect of material media will also be saved for discussion in Chapter 8.

The relation of the steady magnetic field to its source is more complicated than is the relation of the electrostatic field to its source. We will find it necessary to accept several laws temporarily on faith alone. Their derivations are presented in Sec. 7.7. ∎

7.1　BIOT-SAVART LAW

The source of the steady magnetic field may be a permanent magnet, an electric field changing linearly with time, or a direct current. We will largely ignore the permanent magnet and save the time-varying electric field for a later discussion. Our present study will concern the magnetic field produced by a differential dc element in free space.

7.1.1　Differential Form of the Biot-Savart Law

We may think of a differential current element as a vanishingly small section of a current-carrying filamentary conductor, where a filamentary conductor is the limiting

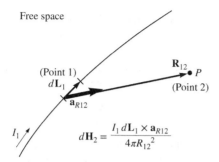

Free space

(Point 1)
$d\mathbf{L}_1$

$\mathbf{R}_{12}$ P
(Point 2)

$\mathbf{a}_{R12}$

I_1

$$d\mathbf{H}_2 = \frac{I_1\, d\mathbf{L}_1 \times \mathbf{a}_{R12}}{4\pi R_{12}{}^2}$$

Figure 7.1 The law of Biot-Savart expresses the magnetic field intensity $d\mathbf{H}_2$ produced by a differential current element $I_1 d\mathbf{L}_1$. The direction of $d\mathbf{H}_2$ is into the page.

case of a cylindrical conductor of circular cross section as the radius approaches zero. We assume a current I flowing in a differential vector length of the filament $d\mathbf{L}$. The law of Biot-Savart[1] then states that at any point P the magnitude of the magnetic field intensity produced by the differential element is proportional to the product of the current, the magnitude of the differential length, and the sine of the angle lying between the filament and a line connecting the filament to the point P at which the field is desired; also, the magnitude of the magnetic field intensity is inversely proportional to the square of the distance from the differential element to the point P. The direction of the magnetic field intensity is normal to the plane containing the differential filament and the line drawn from the filament to the point P. Of the two possible normals, that one to be chosen is the one which is in the direction of progress of a right-handed screw turned from $d\mathbf{L}$ through the smaller angle to the line from the filament to P. Using rationalized mks units, the constant of proportionality is $1/4\pi$.

The *Biot-Savart law,* just described in some 150 words, may be written concisely using vector notation as

$$d\mathbf{H} = \frac{I d\mathbf{L} \times \mathbf{a}_R}{4\pi R^2} = \frac{I d\mathbf{L} \times \mathbf{R}}{4\pi R^3} \tag{1}$$

The units of the *magnetic field intensity* $\mathbf{H}$ are evidently amperes per meter (A/m). The geometry is illustrated in Figure 7.1. Subscripts may be used to indicate the point to which each of the quantities in (1) refers. If we locate the current element at point 1 and describe the point P at which the field is to be determined as point 2, then

$$\boxed{d\mathbf{H}_2 = \frac{I_1\, d\mathbf{L}_1 \times \mathbf{a}_{R12}}{4\pi R_{12}^2}} \tag{2}$$

---

[1] Biot and Savart were colleagues of Ampère, and all three were professors of physics at the Collège de France at one time or another. The Biot-Savart law was proposed in 1820.

导读

1. 恒定磁场的源是恒定电流。

2. 求解恒定磁场的4种方法为：（1）基于毕奥-萨伐定律的积分法，该方法适用于无限大均匀媒质空间中已知电流的所有问题，但有时积分不易求解，需要采用数值积分方法。（2）利用安培环路定律的积分形式法，该方法仅能求解一些具有对称结构的问题，其关键是选择合适的环路。（3）求解标量磁位的边值问题法，即标量磁位满足的拉普拉斯方程［书中式（44）］加上其边界条件与交界面条件构成的边值问题；由于标量磁位仅适用于电流为零的区域，故该方法的应用受到很大限制。（4）求解矢量磁位的边值问题法，即矢量磁位满足的泊松方程［书中式（64）］加上其边界条件与交界面条件构成的边值问题。

3. 注意安培环路定律积分形式中的电流是指与积分环路相交链的电流，就是说电流必须是闭合的（或为无限长电流在无限远处闭合），若为一段电流等式就不成立了。

4. 安培环路定律的微分形式，即磁场强度的旋度等于电流密度［书中式（28）］和磁通连续性定理的微分形式［即式（36）］构成了描述恒定磁场特性的基本方程组。

The law of Biot-Savart is sometimes called *Ampère's law for the current element,* but we will retain the former name because of possible confusion with Ampère's circuital law, to be discussed later.

In some aspects, the Biot-Savart law is reminiscent of Coulomb's law when that law is written for a differential element of charge,

$$d\mathbf{E}_2 = \frac{dQ_1\, \mathbf{a}_{R12}}{4\pi\epsilon_0 R_{12}^2}$$

Both show an inverse-square-law dependence on distance, and both show a linear relationship between source and field. The chief difference appears in the direction of the field.

7.1.2 Integral Form of the Biot-Savart Law

It is impossible to check experimentally the law of Biot-Savart as expressed by (1) or (2) because the differential current element cannot be isolated. We have restricted our attention to direct currents only, so the charge density is not a function of time. The continuity equation in Section 5.2, Eq. (5),

$$\nabla \cdot \mathbf{J} = -\frac{\partial \rho_v}{\partial t}$$

therefore shows that

$$\nabla \cdot \mathbf{J} = 0$$

or upon applying the divergence theorem,

$$\oint_s \mathbf{J} \cdot d\mathbf{S} = 0$$

The total current crossing any closed surface is zero, and this condition may be satisfied only by assuming a current flow around a closed path. It is this current flowing in a closed circuit that must be our experimental source, not the differential element.

It follows that only the integral form of the Biot-Savart law can be verified experimentally,

$$\boxed{\mathbf{H} = \oint \frac{Id\mathbf{L} \times \mathbf{a}_R}{4\pi R^2}} \tag{3}$$

Equation (1) or (2), of course, leads directly to the integral form (3), but other differential expressions also yield the same integral formulation. Any term may be added to (1) whose integral around a closed path is zero. That is, any conservative field could be added to (1). The gradient of any scalar field always yields a conservative field, and we could therefore add a term ∇G to (1), where G is a general scalar field, without changing (3) in the slightest. This qualification on (1) or (2) is mentioned to show that if we later ask some foolish questions, not subject to any experimental check, concerning the force exerted by one *differential* current element on another, we should expect foolish answers.

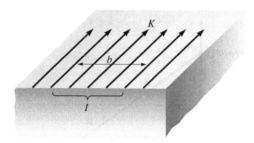

Figure 7.2 The total current I within a transverse width b, in which there is a *uniform* surface current density K, is Kb.

The Biot-Savart law may also be expressed in terms of distributed sources, such as current density $\mathbf{J}$ and *surface current density* $\mathbf{K}$. Surface current flows in a sheet of vanishingly small thickness, and the current density $\mathbf{J}$, measured in amperes per square meter, is therefore infinite. Surface current density, however, is measured in amperes per meter width and designated by $\mathbf{K}$. If the surface current density is uniform, the total current I in any width b is

$$I = Kb$$

where we assume that the width b is measured perpendicularly to the direction in which the current is flowing. The geometry is illustrated by Figure 7.2. For a nonuniform surface current density, integration is necessary:

$$\boxed{I = \int K \, dN} \tag{4}$$

where dN is a differential element of the path *across* which the current is flowing. Thus the differential current element $I \, d\mathbf{L}$, where $d\mathbf{L}$ is in the direction of the current, may be expressed in terms of surface current density $\mathbf{K}$ or current density $\mathbf{J}$,

$$I \, d\mathbf{L} = \mathbf{K} \, dS = \mathbf{J} \, dv \tag{5}$$

and alternate forms of the Biot-Savart law obtained,

$$\boxed{\mathbf{H} = \int_s \frac{\mathbf{K} \times \mathbf{a}_R \, dS}{4\pi R^2}} \tag{6}$$

and

$$\boxed{\mathbf{H} = \int_{\text{vol}} \frac{\mathbf{J} \times \mathbf{a}_R \, dv}{4\pi R^2}} \tag{7}$$

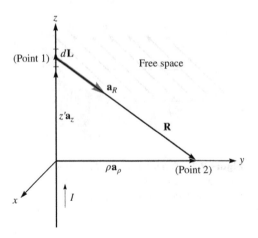

Figure 7.3 An infinitely long, straight filament carrying a direct current I. The field at point 2 is $H = (I/2\pi\rho)\mathbf{a}_\phi$.

7.1.3 Magnetic Field of a Current Filament

We illustrate an application of the Biot-Savart law by considering an infinitely long straight filament. We apply (2) first and then integrate. This, of course, is the same as using the integral form (3) in the first place.[2]

Referring to Figure 7.3, we should recognize the symmetry of this field. No variation with z or with ϕ can exist. Point 2, at which we will determine the field, is therefore chosen in the $z = 0$ plane. The field point $\mathbf{r}$ is therefore $r = \rho\mathbf{a}_\rho$. The source point $\mathbf{r}'$ is given by $\mathbf{r}' = z'\mathbf{a}_z$, and therefore

$$\mathbf{R}_{12} = \mathbf{r} - \mathbf{r}' = \rho\mathbf{a}_\rho - z'\mathbf{a}_z$$

so that

$$\mathbf{a}_{R12} = \frac{\rho\mathbf{a}_\rho - z'\mathbf{a}_z}{\sqrt{\rho^2 + z'^2}}$$

We take $d\mathbf{L} = dz'\mathbf{a}_z$ and (2) becomes

$$d\mathbf{H}_2 = \frac{I\,dz'\mathbf{a}_z \times (\rho\mathbf{a}_\rho - z'\mathbf{a}_z)}{4\pi(\rho^2 + z'^2)^{3/2}}$$

Because the current is directed toward increasing values of z', the limits are $-\infty$ and ∞ on the integral, and we have

$$\mathbf{H}_2 = \int_{-\infty}^{\infty} \frac{I\,dz'\mathbf{a}_z \times (\rho\mathbf{a}_\rho - z'\mathbf{a}_z)}{4\pi(\rho^2 + z'^2)^{3/2}}$$

$$= \frac{I}{4\pi}\int_{-\infty}^{\infty} \frac{\rho\,dz'\mathbf{a}_\phi}{(\rho^2 + z'^2)^{3/2}}$$

---

[2] The closed path for the current may be considered to include a return filament parallel to the first filament and infinitely far removed. An outer coaxial conductor of infinite radius is another theoretical possibility. Practically, the problem is an impossible one, but we should realize that our answer will be quite accurate near a very long, straight wire having a distant return path for the current.

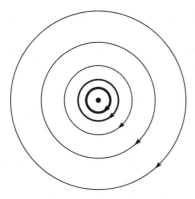

Figure 7.4 The streamlines of the magnetic field intensity about an infinitely long straight filament carrying a direct current *I*. The direction of *I* is into the page.

At this point the unit vector $\mathbf{a}_\phi$ under the integral sign should be investigated, for it is not always a constant, as are the unit vectors of the rectangular coordinate system. A vector is constant when its magnitude and direction are both constant. The unit vector certainly has constant magnitude, but its direction may change. Here $\mathbf{a}_\phi$ changes with the coordinate ϕ but not with ρ or z. Fortunately, the integration here is with respect to z', and $\mathbf{a}_\phi$ is a constant and may be removed from under the integral sign,

$$\mathbf{H}_2 = \frac{I\rho\mathbf{a}_\phi}{4\pi} \int_{-\infty}^{\infty} \frac{dz'}{(\rho^2 + z'^2)^{3/2}}$$

$$= \frac{I\rho\mathbf{a}_\phi}{4\pi} \frac{z'}{\rho^2 \sqrt{\rho^2 + z'^2}} \Bigg|_{-\infty}^{\infty}$$

and

$$\boxed{\mathbf{H}_2 = \frac{I}{2\pi\rho}\mathbf{a}_\phi} \tag{8}$$

The magnitude of the field is not a function of ϕ or z, and it varies inversely with the distance from the filament. The direction of the magnetic-field-intensity vector is circumferential. The streamlines are therefore circles about the filament, and the field may be mapped in cross section as in Figure 7.4.

The separation of the streamlines is proportional to the radius, or inversely proportional to the magnitude of **H**. To be specific, the streamlines have been drawn with curvilinear squares in mind. As yet, we have no name for the family of lines[3] that are perpendicular to these circular streamlines, but the spacing of the streamlines

---

[3] If you can't wait, see Section 7.6.

has been adjusted so that the addition of this second set of lines will produce an array of curvilinear squares.

A comparison of Figure 7.4 with the map of the *electric* field about an infinite line *charge* shows that the streamlines of the magnetic field correspond exactly to the equipotentials of the electric field, and the unnamed (and undrawn) perpendicular family of lines in the magnetic field corresponds to the streamlines of the electric field. This correspondence is not an accident, but there are several other concepts which must be mastered before the analogy between electric and magnetic fields can be explored more thoroughly.

Using the Biot-Savart law to find **H** is in many respects similar to the use of Coulomb's law to find **E**. Each requires the determination of a moderately complicated integrand containing vector quantities, followed by an integration. When we were concerned with Coulomb's law we solved a number of examples, including the fields of the point charge, line charge, and sheet of charge. The law of Biot-Savart can be used to solve analogous problems in magnetic fields, and some of these problems appear as exercises at the end of the chapter rather than as examples here.

One useful result is the field of the finite-length current element, shown in Figure 7.5. It turns out (see Problem 7.8 at the end of the chapter) that **H** is most easily expressed in terms of the angles α_1 and α_2, as identified in the figure. The result is

$$\mathbf{H} = \frac{I}{4\pi\rho}(\sin\alpha_2 - \sin\alpha_1)\mathbf{a}_\phi \tag{9}$$

If one or both ends are below point 2, then α_1 is or both α_1 and α_2 are negative.

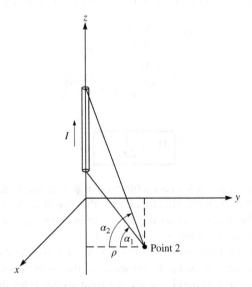

Figure 7.5 The magnetic field intensity caused by a finite-length current filament on the z axis is $(I/4\pi\rho)(\sin\alpha_2 - \sin\alpha_1)\mathbf{a}_\phi$.

Equation (9) may be used to find the magnetic field intensity caused by current filaments arranged as a sequence of straight-line segments.

EXAMPLE 7.1

As a numerical example illustrating the use of (9), we determine $\mathbf{H}$ at $P_2(0.4, 0.3, 0)$ in the field of an 8-ampere filamentary current. The current is directed inward from infinity to the origin on the positive x axis, and then outward to infinity along the y axis. This arrangement is shown in Figure 7.6.

Solution. We first consider the semi-infinite current on the x axis, identifying the two angles, $\alpha_{1x} = -90°$ and $\alpha_{2x} = \tan^{-1}(0.4/0.3) = 53.1°$. The radial distance ρ is measured from the x axis, and we have $\rho_x = 0.3$. Thus, this contribution to $\mathbf{H}_2$ is

$$\mathbf{H}_{2(x)} = \frac{8}{4\pi(0.3)}(\sin 53.1° + 1)\mathbf{a}_\phi = \frac{2}{0.3\pi}(1.8)\mathbf{a}_\phi = \frac{12}{\pi}\mathbf{a}_\phi$$

The unit vector $\mathbf{a}_\phi$ must also be referred to the x axis. We see that it becomes $-\mathbf{a}_z$. Therefore,

$$\mathbf{H}_{2(x)} = -\frac{12}{\pi}\mathbf{a}_z \text{ A/m}$$

For the current on the y axis, we have $\alpha_{1y} = -\tan^{-1}(0.3/0.4) = -36.9°$, $\alpha_{2y} = 90°$, and $\rho_y = 0.4$. It follows that

$$\mathbf{H}_{2(y)} = \frac{8}{4\pi(0.4)}(1 + \sin 36.9°)(-\mathbf{a}_z) = -\frac{8}{\pi}\mathbf{a}_z \text{ A/m}$$

Adding these results, we have

$$\mathbf{H}_2 = \mathbf{H}_{2(x)} + \mathbf{H}_{2(y)} = -\frac{20}{\pi}\mathbf{a}_z = -6.37\mathbf{a}_z \text{ A/m}$$

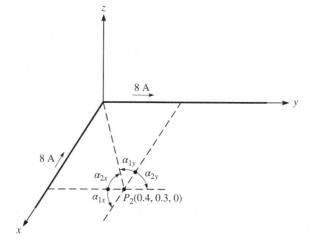

Figure 7.6 The individual fields of two semi-infinite current segments are found by (9) and added to obtain $\mathbf{H}_2$ at P_2.

D7.1. Given the following values for P_1, P_2, and $I_1\Delta L_1$, calculate $\Delta \mathbf{H}_2$:
(a) $P_1(0, 0, 2)$, $P_2(4, 2, 0)$, $2\pi\,\mathbf{a}_z\mu A \cdot m$; (b) $P_1(0, 2, 0)$, $P_2(4, 2, 3)$, $2\pi\,\mathbf{a}_z\mu A \cdot m$;
(c) $P_1(1, 2, 3)$, $P_2(-3, -1, 2)$, $2\pi(-\mathbf{a}_x + \mathbf{a}_y + 2\mathbf{a}_z)\mu A \cdot m$.

Ans. (a) $-8.51\mathbf{a}_x + 17.01\mathbf{a}_y$ nA/m; (b) $16\mathbf{a}_y$ nA/m; (c) $18.9\mathbf{a}_x - 33.9\mathbf{a}_y + 26.4\mathbf{a}_z$ nA/m

D7.2. A current filament carrying 15 A in the $\mathbf{a}_z$ direction lies along the entire z axis. Find $\mathbf{H}$ in rectangular coordinates at: (a) $P_A(\sqrt{20}, 0, 4)$; (b) $P_B(2, -4, 4)$.

Ans. (a) $0.534\mathbf{a}_y$ A/m; (b) $0.477\mathbf{a}_x + 0.239\mathbf{a}_y$ A/m

7.2 AMPÈRE'S CIRCUITAL LAW

After solving a number of simple electrostatic problems with Coulomb's law, we found that the same problems could be solved much more easily by using Gauss's law whenever a high degree of symmetry was present. Again, an analogous procedure exists in magnetic fields. Here, the law that helps us solve problems more easily is known as *Ampère's circuital[4] law,* sometimes called Ampère's work law. This law may be derived from the Biot-Savart law (see Section 7.7).

7.2.1 Definition of Ampere's Law

Ampère's circuital law states that the line integral of $\mathbf{H}$ about any *closed* path is exactly equal to the direct current enclosed by that path,

$$\oint \mathbf{H} \cdot d\mathbf{L} = I \tag{10}$$

We define positive current as flowing in the direction of advance of a right-handed screw turned in the direction in which the closed path is traversed.

Referring to Figure 7.7, which shows a circular wire carrying a direct current I, the line integral of $\mathbf{H}$ about the closed paths lettered a and b results in an answer of I; the integral about the closed path c which passes through the conductor gives an answer less than I and is exactly that portion of the total current that is enclosed by the path c. Although paths a and b give the same answer, the integrands are, of course, different. The line integral directs us to multiply the component of $\mathbf{H}$ in the direction of the path by a small increment of path length at one point of the path, move along the path to the next incremental length, and repeat the process, continuing until the path is completely traversed. Because $\mathbf{H}$ will generally vary from point to point, and because paths a and b are not alike, the contributions to the integral made by, say, each micrometer of path length are quite different. Only the final answers are the same.

---
[4] The preferred pronunciation puts the accent on "circ-."

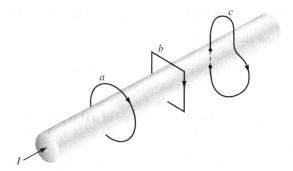

Figure 7.7 A conductor has a total current *I*. The line integral of **H** about the closed paths *a* and *b* is equal to *I*, and the integral around path *c* is less than *I*, since the entire current is not enclosed by the path.

We should also consider exactly what is meant by the expression "current enclosed by the path." Suppose we solder a circuit together after passing the conductor once through a rubber band, which we use to represent the closed path. Some strange and formidable paths can be constructed by twisting and knotting the rubber band, but if neither the rubber band nor the conducting circuit is broken, the current enclosed by the path is that carried by the conductor. Now replace the rubber band with a circular ring of spring steel across which is stretched a rubber sheet. The steel loop forms the closed path, and the current-carrying conductor must pierce the rubber sheet if the current is to be enclosed by the path. Again, we may twist the steel loop, and we may also deform the rubber sheet by pushing our fist into it or folding it in any way we wish. A single current-carrying conductor still pierces the sheet once, and this is the true measure of the current enclosed by the path. If we should thread the conductor once through the sheet from front to back and once from back to front, the total current enclosed by the path is the algebraic sum, which is zero.

In more general language, given a closed path, we recognize this path as the perimeter of an infinite number of surfaces (not closed surfaces). Any current-carrying conductor enclosed by the path must pass through every one of these surfaces once. Certainly some of the surfaces may be chosen in such a way that the conductor pierces them twice in one direction and once in the other direction, but the algebraic total current is still the same.

We will find that the nature of the closed path is usually extremely simple and can be drawn on a plane. The simplest surface is, then, that portion of the plane enclosed by the path. We need merely find the total current passing through this region of the plane.

The application of Gauss's law involves finding the total charge enclosed by a closed surface; the application of Ampère's circuital law involves finding the total current enclosed by a closed path.

7.2.2 Application of Ampere's Law to a Filament Current

Here we again find the magnetic field intensity produced by an infinitely long filament carrying a current I. The filament lies on the z axis in free space (as in Figure 7.3), and the current flows in the direction given by $\mathbf{a}_z$. Symmetry inspection comes first, showing that there is no variation with z or ϕ. Next we determine which components of $\mathbf{H}$ are present by using the Biot-Savart law. Without specifically using the cross product, we may say that the direction of $d\mathbf{H}$ is perpendicular to the plane conaining $d\mathbf{L}$ and $\mathbf{R}$ and therefore is in the direction of $\mathbf{a}_\phi$. Hence the only component of $\mathbf{H}$ is H_ϕ, and it is a function only of ρ.

We therefore choose a path, to any section of which $\mathbf{H}$ is either perpendicular or tangential, and along which H is constant. The first requirement (perpendicularity or tangency) allows us to replace the dot product of Ampère's circuital law with the product of the scalar magnitudes, except along that portion of the path where $\mathbf{H}$ is normal to the path and the dot product is zero; the second requirement (constancy) then permits us to remove the magnetic field intensity from the integral sign. The integration required is usually trivial and consists of finding the length of that portion of the path to which $\mathbf{H}$ is parallel.

In our example, the path must be a circle of radius ρ, and Ampère's circuital law becomes

$$\oint \mathbf{H} \cdot d\mathbf{L} = \int_0^{2\pi} H_\phi \rho \, d\phi = H_\phi \rho \int_0^{2\pi} d\phi = H_\phi 2\pi\rho = I$$

or

$$H_\phi = \frac{I}{2\pi\rho}$$

as before.

7.2.3 Magnetic Field Within a Coaxial Cable

As a second example of the application of Ampère's circuital law, consider an infinitely long coaxial transmission line carrying a uniformly distributed total current I in the center conductor and $-I$ in the outer conductor. The line is shown in Figure 7.8a. Symmetry shows that H is not a function of ϕ or z. In order to determine the components present, we may use the results of the previous example by considering the solid conductors as being composed of a large number of filaments. No filament has a z component of $\mathbf{H}$. Furthermore, the H_ρ component at $\phi = 0°$, produced by one filament located at $\rho = \rho_1$, $\phi = \phi_1$, is canceled by the H_ρ component produced by a symmetrically located filament at $\rho = \rho_1$, $\phi = -\phi_1$. This symmetry is illustrated by Figure 7.8b. Again we find only an H_ϕ component which varies with ρ.

A circular path of radius ρ, where ρ is larger than the radius of the inner conductor but less than the inner radius of the outer conductor, then leads immediately to

$$H_\phi = \frac{I}{2\pi\rho} \quad (a < \rho < b)$$

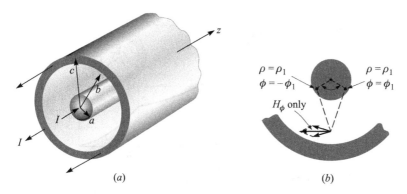

(a) (b)

Figure 7.8 (a) Cross section of a coaxial cable carrying a uniformly distributed current I in the inner conductor and $-I$ in the outer conductor. The magnetic field at any point is most easily determined by applying Ampère's circuital law about a circular path. (b) Current filaments at $\rho = \rho_1$, $\phi = \pm \phi_1$, produces $\mathbf{H}_\rho$ components which cancel. For the total field, $\mathbf{H} = \mathbf{H}_\phi \mathbf{a}_\phi$.

If we choose ρ smaller than the radius of the inner conductor, the current enclosed is

$$I_{\text{encl}} = I \frac{\rho^2}{a^2}$$

and

$$2\pi\rho H_\phi = I \frac{\rho^2}{a^2}$$

or

$$H_\phi = \frac{I\rho}{2\pi a^2} \quad (\rho < a)$$

If the radius ρ is larger than the outer radius of the outer conductor, no current is enclosed and

$$H_\phi = 0 \quad (\rho > c)$$

Finally, if the path lies within the outer conductor, we have

$$2\pi\rho H_\phi = I - I\left(\frac{\rho^2 - b^2}{c^2 - b^2}\right)$$

$$H_\phi = \frac{I}{2\pi\rho} \frac{c^2 - \rho^2}{c^2 - b^2} \quad (b < \rho < c)$$

The magnetic-field-strength variation with radius is shown in Figure 7.9 for a coaxial cable in which $b = 3a$, $c = 4a$. It should be noted that the magnetic field intensity $\mathbf{H}$ is continuous at all the conductor boundaries. In other words, a slight increase in the radius of the closed path does not result in the enclosure of a tremendously different current. The value of H_ϕ shows no sudden jumps.

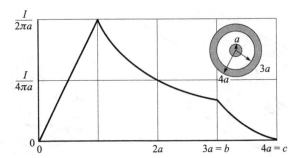

Figure 7.9 The magnetic field intensity as a function of radius in an infinitely long coaxial transmission line with the dimensions shown.

The external field is zero. This, we see, results from equal positive and negative currents enclosed by the path. Each produces an external field of magnitude $I/2\pi\rho$, but complete cancellation occurs. This is another example of "shielding"; such a coaxial cable carrying large currents would, in principle, not produce any noticeable effect in an adjacent circuit.

7.2.4 Magnetic Field of a Surface Current

As a final example, consider a sheet of current flowing in the positive y direction and located in the $z = 0$ plane. We may think of the return current as equally divided between two distant sheets on either side of the sheet we are considering. A sheet of uniform surface current density $\mathbf{K} = K_y \mathbf{a}_y$ is shown in Figure 7.10. $\mathbf{H}$ cannot vary with x or y. If the sheet is subdivided into a number of filaments, it is evident that no filament can produce an H_y component. Moreover, the Biot-Savart law shows that the contributions to H_z produced by a symmetrically located pair of filaments cancel. Thus, H_z is zero also; only an H_x component is present. We therefore choose the path 1-1'-2'-2-1 composed of straight-line segments that are either parallel or perpendicular to H_x. Ampère's circuital law gives

$$H_{x1}L + H_{x2}(-L) = K_y L$$

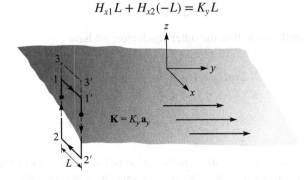

Figure 7.10 A uniform sheet of surface current $\mathbf{K} = K_y \mathbf{a}_y$ in the $z = 0$ plane. $\mathbf{H}$ may be found by applying Ampère's circuital law about the paths 1-1'-2'-2-1 and 3-3'-2'-2-3.

or

$$H_{x1} - H_{x2} = K_y$$

If the path 3-3′-2′-2-3 is now chosen, the same current is enclosed, and

$$H_{x3} - H_{x2} = K_y$$

and therefore

$$H_{x3} = H_{x1}$$

It follows that H_x is the same for all positive z. Similarly, H_x is the same for all negative z. Because of the symmetry, then, the magnetic field intensity on one side of the current sheet is the negative of that on the other. Above the sheet,

$$H_x = \tfrac{1}{2} K_y \quad (z > 0)$$

while below it

$$H_x = -\tfrac{1}{2} K_y \quad (z < 0)$$

Letting $\mathbf{a}_N$ be a unit vector normal (outward) to the current sheet, the result may be written in a form correct for all z as

$$\boxed{\mathbf{H} = \tfrac{1}{2} \mathbf{K} \times \mathbf{a}_N} \tag{11}$$

If a second sheet of current flowing in the opposite direction, $\mathbf{K} = -K_y \mathbf{a}_y$, is placed at $z = h$, (11) shows that the field in the region between the current sheets is

$$\boxed{\mathbf{H} = \mathbf{K} \times \mathbf{a}_N \quad (0 < z < h)} \tag{12}$$

and is zero elsewhere,

$$\boxed{\mathbf{H} = 0 \quad (z < 0, z > h)} \tag{13}$$

The most difficult part of the application of Ampère's circuital law is the determination of the components of the field that are present. The surest method is the logical application of the Biot-Savart law and a knowledge of the magnetic fields of simple form.

7.2.5 Magnetic Fields Within Solenoids and Toroids

Problem 7.13 at the end of this chapter outlines the steps involved in applying Ampère's circuital law to an infinitely long solenoid of radius a and uniform current density $K_a \mathbf{a}_\phi$, as shown in Figure 7.11a. For reference, the result is

$$\mathbf{H} = K_a \mathbf{a}_z \quad (\rho < a) \tag{14a}$$
$$\mathbf{H} = 0 \quad (\rho > a) \tag{14b}$$

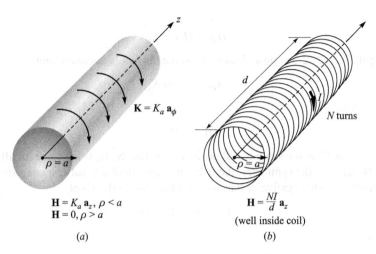

$\mathbf{K} = K_a\,\mathbf{a}_\phi$

$\mathbf{H} = K_a\,\mathbf{a}_z,\ \rho < a$
$\mathbf{H} = 0,\ \rho > a$

(a)

N turns

$\mathbf{H} = \dfrac{NI}{d}\,\mathbf{a}_z$
(well inside coil)

(b)

Figure 7.11   (a) An ideal solenoid of infinite length with a circular current sheet $\mathbf{K} = K_a\mathbf{a}_\phi$. (b) An N-turn solenoid of finite length d.

If the solenoid has a finite length d and consists of N closely wound turns of a filament that carries a current I (Figure 7.11b), then the field at points well within the solenoid is given closely by

$$\mathbf{H} = \frac{NI}{d}\mathbf{a}_z \quad \text{(well within the solenoid)} \tag{15}$$

The approximation is useful if it is not applied closer than two radii to the open ends, nor closer to the solenoid surface than twice the separation between turns.

For the toroids shown in Figure 7.12, it can be shown that the magnetic field intensity for the ideal case, Figure 7.12a, is

$$\mathbf{H} = K_a \frac{\rho_0 - a}{\rho}\mathbf{a}_\phi \quad \text{(inside toroid)} \tag{16a}$$

$$\mathbf{H} = 0 \qquad \text{(outside)} \tag{16b}$$

For the N-turn toroid of Figure 7.12b, we have the good approximations,

$$\mathbf{H} = \frac{NI}{2\pi\rho}\,\mathbf{a}_\phi \quad \text{(inside toroid)} \tag{17a}$$

$$\mathbf{H} = 0 \qquad \text{(outside)} \tag{17b}$$

as long as we consider points removed from the toroidal surface by several times the separation between turns.

Toroids having rectangular cross sections are also treated quite readily, as you can see for yourself by trying Problem 7.14.

Accurate formulas for solenoids, toroids, and coils of other shapes are available in Section 2 of the *Standard Handbook for Electrical Engineers* (see References for Chapter 5).

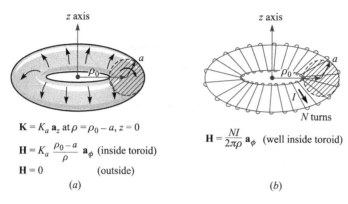

$\mathbf{K} = K_a \, \mathbf{a}_z$ at $\rho = \rho_0 - a$, $z = 0$

$\mathbf{H} = K_a \dfrac{\rho_0 - a}{\rho} \, \mathbf{a}_\phi$ (inside toroid)

$\mathbf{H} = 0$ (outside)

(a)

$\mathbf{H} = \dfrac{NI}{2\pi\rho} \, \mathbf{a}_\phi$ (well inside toroid)

(b)

Figure 7.12 (a) An ideal toroid carrying a surface current **K** in the direction shown. (b) An N-turn toroid carrying a filamentary current I.

D7.3. Express the value of **H** in rectangular components at $P(0, 0.2, 0)$ in the field of: (a) a current filament, 2.5 A in the $\mathbf{a}_z$ direction at $x = 0.1$, $y = 0.3$; (b) a coax, centered on the z axis, with $a = 0.3$, $b = 0.5$, $c = 0.6$, $I = 2.5$ A in the $\mathbf{a}_z$ direction in the center conductor; (c) three current sheets, $2.7\mathbf{a}_x$ A/m at $y = 0.1$, $-1.4\mathbf{a}_x$ A/m at $y = 0.15$, and $-1.3\mathbf{a}_x$ A/m at $y = 0.25$.

Ans. (a) $1.989\mathbf{a}_x - 1.989\mathbf{a}_y$ A/m; (b) $-0.884\mathbf{a}_x$ A/m; (c) $1.300\mathbf{a}_z$ A/m

7.3 CURL

We completed our study of Gauss's law by applying it to a differential volume element and were led to the concept of divergence. We now apply Ampère's circuital law to the perimeter of a differential surface element and discuss the third and last of the special derivatives of vector analysis, the curl. Our objective is to obtain the point form of Ampère's circuital law.

7.3.1 Development and Definition of Curl

Again we choose rectangular coordinates, and an incremental closed path of sides Δx and Δy is selected (Figure 7.13). We assume that some current, as yet unspecified, produces a reference value for **H** at the *center* of this small rectangle,

$$\mathbf{H}_0 = H_{x0}\mathbf{a}_x + H_{y0}\mathbf{a}_y + H_{z0}\mathbf{a}_z$$

The closed line integral of **H** about this path is then approximately the sum of the four values of $\mathbf{H} \cdot \Delta\mathbf{L}$ on each side. We choose the direction of traverse as 1-2-3-4-1, which corresponds to a current in the $\mathbf{a}_z$ direction, and the first contribution is therefore

$$(\mathbf{H} \cdot \Delta\mathbf{L})_{1-2} = H_{y,1-2}\,\Delta y$$

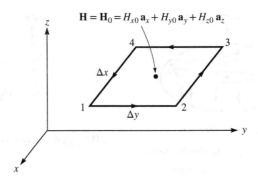

Figure 7.13 An incremental closed path in rectangular coordinates is selected for the application of Ampère's circuital law to determine the spatial rate of change of **H**.

The value of H_y *on this section* of the path may be given in terms of the reference value H_{y0} at the center of the rectangle, the rate of change of H_y with x, and the distance $\Delta x/2$ from the center to the midpoint of side 1–2:

$$H_{y,1-2} \doteq H_{y0} + \frac{\partial H_y}{\partial x}\left(\frac{1}{2}\Delta x\right)$$

Thus

$$(H \cdot \Delta L)_{1-2} \doteq \left(H_{y0} + \frac{1}{2}\frac{\partial H_y}{\partial x}\Delta x\right)\Delta y$$

Along the next section of the path we have

$$(\mathbf{H} \cdot \Delta \mathbf{L})_{2-3} \doteq H_{x,2-3}(-\Delta x) \doteq -\left(H_{x0} + \frac{1}{2}\frac{\partial H_x}{\partial y}\Delta y\right)\Delta x$$

Continuing for the remaining two segments and adding the results,

$$\oint \mathbf{H} \cdot d\mathbf{L} \doteq \left(\frac{\partial H_y}{\partial x} - \frac{\partial H_x}{\partial y}\right)\Delta x \Delta y$$

By Ampère's circuital law, this result must be equal to the current enclosed by the path, or the current crossing any surface bounded by the path. If we assume a general current density **J**, the enclosed current is then $\Delta I \doteq J_z \Delta x \Delta y$, and

$$\oint \mathbf{H} \cdot d\mathbf{L} \doteq \left(\frac{\partial H_y}{\partial x} - \frac{\partial H_x}{\partial y}\right)\Delta x \Delta y \doteq J_z \Delta x \Delta y$$

or

$$\frac{\oint \mathbf{H} \cdot d\mathbf{L}}{\Delta x \Delta y} \doteq \frac{\partial H_y}{\partial x} - \frac{\partial H_x}{\partial y} \doteq J_z$$

As we cause the closed path to shrink, the preceding expression becomes more nearly exact, and in the limit we have the equality

$$\lim_{\Delta x, \Delta y \to 0} \frac{\oint \mathbf{H} \cdot d\mathbf{L}}{\Delta x \Delta y} = \frac{\partial H_y}{\partial x} - \frac{\partial H_x}{\partial y} = J_z \qquad (18)$$

After beginning with Ampère's circuital law equating the closed line integral of **H** to the current enclosed, we have now arrived at a relationship involving the closed line integral of **H** *per unit area* enclosed and the current *per unit area* enclosed, or current density. We performed a similar analysis in passing from the integral form of Gauss's law, involving flux through a closed surface and charge enclosed, to the point form, relating flux through a closed surface *per unit volume* enclosed and charge *per unit volume* enclosed, or volume charge density. In each case a limit is necessary to produce an equality.

If we choose closed paths that are oriented perpendicularly to each of the remaining two coordinate axes, analogous processes lead to expressions for the x and y components of the current density,

$$\lim_{\Delta y, \Delta z \to 0} \frac{\oint \mathbf{H} \cdot d\mathbf{L}}{\Delta y \Delta z} = \frac{\partial H_z}{\partial y} - \frac{\partial H_y}{\partial z} = J_x \tag{19}$$

and

$$\lim_{\Delta z, \Delta x \to 0} \frac{\oint \mathbf{H} \cdot d\mathbf{L}}{\Delta z \Delta x} = \frac{\partial H_x}{\partial z} - \frac{\partial H_z}{\partial x} = J_y \tag{20}$$

Comparing (18)–(20), we see that a component of the current density is given by the limit of the quotient of the closed line integral of **H** about a small path in a plane normal to that component and of the area enclosed as the path shrinks to zero. This limit has its counterpart in other fields of science and long ago received the name of *curl*. The curl of any vector is a vector, and any component of the curl is given by the limit of the quotient of the closed line integral of the vector about a small path in a plane normal to that component desired and the area enclosed, as the path shrinks to zero. It should be noted that this definition of curl does not refer specifically to a particular coordinate system. The mathematical form of the definition is

$$(\text{curl } \mathbf{H})_N = \lim_{\Delta S_N \to 0} \frac{\oint \mathbf{H} \cdot d\mathbf{L}}{\Delta S_N} \tag{21}$$

where ΔS_N is the planar area enclosed by the closed line integral. The N subscript indicates that the component of the curl is that component which is *normal* to the surface enclosed by the closed path. It may represent any component in any coordinate system.

In rectangular coordinates, the definition (21) shows that the x, y, and z components of the curl **H** are given by (18)–(20), and therefore

$$\text{curl } \mathbf{H} = \left(\frac{\partial H_z}{\partial y} - \frac{\partial H_y}{\partial z}\right)\mathbf{a}_x + \left(\frac{\partial H_x}{\partial z} - \frac{\partial H_z}{\partial x}\right)\mathbf{a}_y + \left(\frac{\partial H_y}{\partial x} - \frac{\partial H_x}{\partial y}\right)\mathbf{a}_z \tag{22}$$

This result may be written in the form of a determinant,

$$\text{curl } \mathbf{H} = \begin{vmatrix} \mathbf{a}_x & \mathbf{a}_y & \mathbf{a}_z \\ \dfrac{\partial}{\partial x} & \dfrac{\partial}{\partial y} & \dfrac{\partial}{\partial z} \\ H_x & H_y & H_z \end{vmatrix} \tag{23}$$

and may also be written in terms of the vector operator,

$$\text{curl } \mathbf{H} = \nabla \times \mathbf{H} \tag{24}$$

Equation (22) is the result of applying the definition (21) to the rectangular coordinate system. We obtained the z component of this expression by evaluating Ampère's circuital law about an incremental path of sides Δx and Δy, and we could have obtained the other two components just as easily by choosing the appropriate paths. Equation (23) is a neat method of storing the rectangular coordinate expression for curl; the form is symmetrical and easily remembered. Equation (24) is even more concise and leads to (22) upon applying the definitions of the cross product and vector operator.

$$\nabla \times \mathbf{H} = \left(\frac{1}{\rho} \frac{\partial H_z}{\partial \phi} - \frac{\partial H_\phi}{\partial z} \right) \mathbf{a}_\rho + \left(\frac{\partial H_\rho}{\partial z} - \frac{\partial H_z}{\partial \rho} \right) \mathbf{a}_\phi$$
$$+ \left(\frac{1}{\rho} \frac{\partial (\rho H_\phi)}{\partial \rho} - \frac{1}{\rho} \frac{\partial H_\rho}{\partial \phi} \right) \mathbf{a}_z \quad \text{(cylindrical)} \tag{25}$$

$$\nabla \times \mathbf{H} = \frac{1}{r \sin\theta} \left(\frac{\partial (H_\phi \sin\theta)}{\partial \theta} - \frac{\partial H_\theta}{\partial \phi} \right) \mathbf{a}_r + \frac{1}{r} \left(\frac{1}{\sin\theta} \frac{\partial H_r}{\partial \phi} - \frac{\partial (r H_\phi)}{\partial r} \right) \mathbf{a}_\theta$$
$$+ \frac{1}{r} \left(\frac{\partial (r H_\theta)}{\partial r} - \frac{\partial H_r}{\partial \theta} \right) \mathbf{a}_\phi \quad \text{(spherical)} \tag{26}$$

7.3.2 Physical Meaning of Curl

Although we have described curl as a line integral per unit area, this does not provide everyone with a satisfactory physical picture of the nature of the curl operation, for the closed line integral itself requires physical interpretation. This integral was first met in the electrostatic field, where we saw that $\oint \mathbf{E} \cdot d\mathbf{L} = 0$. Inasmuch as the integral was zero, we did not belabor the physical picture. More recently we have discussed the closed line integral of $\mathbf{H}$, $\oint \mathbf{H} \cdot d\mathbf{L} = I$. Either of these closed line integrals is also known by the name of *circulation*, a term borrowed from the field of fluid dynamics.

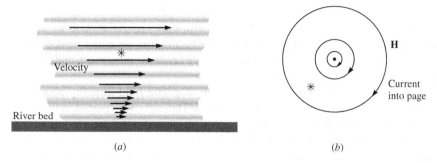

Figure 7.14 (*a*) The curl meter shows a component of the curl of the water velocity into the page. (*b*) The curl of the magnetic field intensity about an infinitely long filament is shown.

The circulation of **H**, or $\oint \mathbf{H} \cdot d\mathbf{L}$, is obtained by multiplying the component of **H** parallel to the specified closed path at each point along it by the differential path length and summing the results as the differential lengths approach zero and as their number becomes infinite. We do not require a vanishingly small path. Ampère's circuital law tells us that if **H** does possess circulation about a given path, then current passes through this path. In electrostatics we see that the circulation of **E** is zero about every path, a direct consequence of the fact that zero work is required to carry a charge around a closed path.

We may describe curl as *circulation per unit area*. The closed path is vanishingly small, and curl is defined at a point. The curl of **E** must be zero, for the circulation is zero. The curl of **H** is not zero, however; the circulation of **H** per unit area is the current density by Ampère's circuital law [or (18), (19), and (20)].

Skilling[5] suggests the use of a very small paddle wheel as a "curl meter." Our vector quantity, then, must be thought of as capable of applying a force to each blade of the paddle wheel, the force being proportional to the component of the field normal to the surface of that blade. To test a field for curl, we dip our paddle wheel into the field, with the axis of the paddle wheel lined up with the direction of the component of curl desired, and note the action of the field on the paddle. No rotation means no curl; larger angular velocities mean greater values of the curl; a reversal in the direction of spin means a reversal in the sign of the curl. To find the direction of the vector curl and not merely to establish the presence of any particular component, we should place our paddle wheel in the field and hunt around for the orientation which produces the greatest torque. The direction of the curl is then along the axis of the paddle wheel, as given by the right-hand rule.

As an example, consider the flow of water in a river. Figure 7.14*a* shows the longitudinal section of a wide river taken at the middle of the river. The water velocity is zero at the bottom and increases linearly as the surface is approached. A paddle wheel placed in the position shown, with its axis perpendicular to the paper, will turn in a clockwise direction, showing the presence of a component of curl in the

---

[5] See the References at the end of the chapter.

direction of an inward normal to the surface of the page. If the velocity of water does not change as we go up- or downstream and also shows no variation as we go across the river (or even if it decreases in the same fashion toward either bank), then this component is the only component present at the center of the stream, and the curl of the water velocity has a direction into the page.

In Figure 7.14b, the streamlines of the magnetic field intensity about an infinitely long filamentary conductor are shown. The curl meter placed in this field of curved lines shows that a larger number of blades have a clockwise force exerted on them but that this force is in general smaller than the counterclockwise force exerted on the smaller number of blades closer to the wire. It seems possible that if the curvature of the streamlines is correct and also if the variation of the field strength is just right, the net torque on the paddle wheel may be zero. Actually, the paddle wheel does not rotate in this case, for since $\mathbf{H} = (I/2\pi\rho)\mathbf{a}_\phi$, we may substitute into (25) obtaining

$$\text{curl } \mathbf{H} = -\frac{\partial H_\phi}{\partial z}\mathbf{a}_\rho + \frac{1}{\rho}\frac{\partial(\rho H_\phi)}{\partial\rho}\mathbf{a}_z = 0$$

EXAMPLE 7.2

As an example of the evaluation of curl $\mathbf{H}$ from the definition and of the evaluation of another line integral, suppose that $\mathbf{H} = 0.2z^2\mathbf{a}_x$ for $z > 0$, and $\mathbf{H} = 0$ elsewhere, as shown in Figure 7.15. Calculate $\oint \mathbf{H} \cdot d\mathbf{L}$ about a square path with side d, centered at $(0, 0, z_1)$ in the $y = 0$ plane where $z_1 > d/2$.

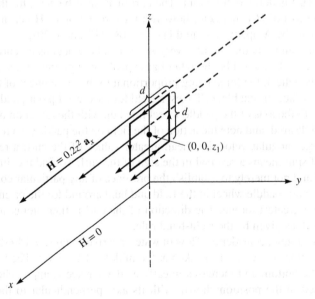

Figure 7.15 A square path of side d with its center on the z axis at $z = z_1$ is used to evaluate $\oint \mathbf{H} \cdot d\mathbf{L}$ and find curl $\mathbf{H}$.

Solution. We evaluate the line integral of $\mathbf{H}$ along the four segments, beginning at the top:

$$\oint \mathbf{H} \cdot d\mathbf{L} = 0.2 \left(z_1 + \tfrac{1}{2}d\right)^2 d + 0 - 0.2 \left(z_1 - \tfrac{1}{2}d\right)^2 d + 0$$
$$= 0.4 z_1 d^2$$

In the limit as the area approaches zero, we find

$$(\nabla \times \mathbf{H})_y = \lim_{d \to 0} \frac{\oint \mathbf{H} \cdot d\mathbf{L}}{d^2} = \lim_{d \to 0} \frac{0.4 z_1 d^2}{d^2} = 0.4 z_1$$

The other components are zero, so $\nabla \times \mathbf{H} = 0.4 z_1 \mathbf{a}_y$.

To evaluate the curl without trying to illustrate the definition or the evaluation of a line integral, we simply take the partial derivative indicated by (23):

$$\nabla \times \mathbf{H} = \begin{vmatrix} \mathbf{a}_x & \mathbf{a}_y & \mathbf{a}_z \\ \dfrac{\partial}{\partial x} & \dfrac{\partial}{\partial y} & \dfrac{\partial}{\partial z} \\ 0.2 z^2 & 0 & 0 \end{vmatrix} = \frac{\partial}{\partial z}(0.2 z^2) \mathbf{a}_y = 0.4 z \mathbf{a}_y$$

which checks with the preceding result when $z = z_1$.

Returning now to complete our original examination of the application of Ampère's circuital law to a differential-sized path, we may combine (18)–(20), (22), and (24),

$$\text{curl } \mathbf{H} = \nabla \times \mathbf{H} = \left(\frac{\partial H_z}{\partial y} - \frac{\partial H_y}{\partial z} \right) \mathbf{a}_x + \left(\frac{\partial H_x}{\partial z} - \frac{\partial H_z}{\partial x} \right) \mathbf{a}_y$$
$$+ \left(\frac{\partial H_y}{\partial x} - \frac{\partial H_x}{\partial y} \right) \mathbf{a}_z = \mathbf{J} \tag{27}$$

and write the *point form of Ampère's circuital law,*

$$\boxed{\nabla \times \mathbf{H} = \mathbf{J}} \tag{28}$$

This is the second of Maxwell's four equations as they apply to non-time-varying conditions. We may also write the third of these equations at this time; it is the point form of $\oint \mathbf{E} \cdot d\mathbf{L} = 0$, or

$$\boxed{\nabla \times \mathbf{E} = 0} \tag{29}$$

The fourth equation appears in Section 7.5.

D7.4. (a) Evaluate the closed line integral of **H** about the rectangular path $P_1(2, 3, 4)$ to $P_2(4, 3, 4)$ to $P_3(4, 3, 1)$ to $P_4(2, 3, 1)$ to P_1, given $\mathbf{H} = 3z\mathbf{a}_x - 2x^3\mathbf{a}_z$ A/m. (b) Determine the quotient of the closed line integral and the area enclosed by the path as an approximation to $(\nabla \times \mathbf{H})_y$. (c) Determine $(\nabla \times \mathbf{H})_y$ at the center of the area.

Ans. (a) 354 A; (b) 59 A/m^2; (c) 57 A/m^2

D7.5. Calculate the value of the vector current density: (a) in rectangular coordinates at $P_A(2, 3, 4)$ if $\mathbf{H} = x^2 z \mathbf{a}_y - y^2 x \mathbf{a}_z$; (b) in cylindrical coordinates at $P_B(1.5, 90°, 0.5)$ if $\mathbf{H} = \dfrac{2}{\rho}(\cos 0.2\phi)\mathbf{a}_\rho$; (c) in spherical coordinates at $P_C(2, 30°, 20°)$ if $\mathbf{H} = \dfrac{1}{\sin\theta}\mathbf{a}_\theta$.

Ans. (a) $-16\mathbf{a}_x + 9\mathbf{a}_y + 16\mathbf{a}_z$ A/m^2; (b) $0.055\mathbf{a}_z$ A/m^2; (c) $\mathbf{a}_\phi$ A/m^2

7.4 STOKES' THEOREM

Although Section 7.3 was devoted primarily to a discussion of the curl operation, the contribution to the subject of magnetic fields should not be overlooked. From Ampère's circuital law we derived one of Maxwell's equations, $\nabla \times \mathbf{H} = \mathbf{J}$. This latter equation should be considered the point form of Ampère's circuital law and applies on a "per-unit-area" basis. In this section we shall again devote a major share of the material to the mathematical theorem known as Stokes' theorem, but in the process we will show that we may obtain Ampère's circuital law from $\nabla \times \mathbf{H} = \mathbf{J}$. In other words, we are then prepared to obtain the integral form from the point form or to obtain the point form from the integral form.

Consider the surface S of Figure 7.16, which is broken up into incremental surfaces of area ΔS. If we apply the definition of the curl to one of these incremental surfaces, then

$$\frac{\oint \mathbf{H} \cdot d\mathbf{L}_{\Delta S}}{\Delta S} \doteq (\nabla \times \mathbf{H})_N$$

where the N subscript again indicates the right-hand normal to the surface. The subscript on $d\mathbf{L}_{\Delta S}$ indicates that the closed path is the perimeter of an incremental area ΔS. This result may also be written

$$\frac{\oint \mathbf{H} \cdot d\mathbf{L}_{\Delta S}}{\Delta S} \doteq (\nabla \times \mathbf{H}) \cdot \mathbf{a}_N$$

or

$$\oint \mathbf{H} \cdot d\mathbf{L}_{\Delta S} \doteq (\nabla \times \mathbf{H}) \cdot \mathbf{a}_N \Delta S = (\nabla \times \mathbf{H}) \cdot \Delta \mathbf{S}$$

where $\mathbf{a}_N$ is a unit vector in the direction of the right-hand normal to ΔS.

Now let us determine this circulation for every ΔS comprising S and sum the results. As we evaluate the closed line integral for each ΔS, some cancellation

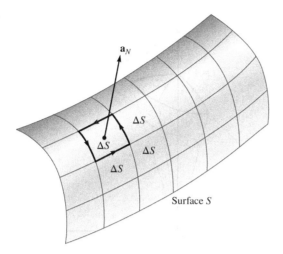

Figure 7.16 The sum of the closed line integrals about the perimeter of every ΔS is the same as the closed line integral about the perimeter of S because of cancellation on every interior path.

will occur because every *interior* wall is covered once in each direction. The only boundaries on which cancellation cannot occur form the outside boundary, the path enclosing S. Therefore we have

$$\oint \mathbf{H} \cdot d\mathbf{L} \equiv \int_S (\nabla \times \mathbf{H}) \cdot d\mathbf{S} \qquad (30)$$

where $d\mathbf{L}$ is taken only on the perimeter of S.

Equation (30) is an identity, holding for any vector field, and is known as *Stokes' theorem.*

EXAMPLE 7.3

A numerical example may help to illustrate the geometry involved in Stokes' theorem. Consider the portion of a sphere shown in Figure 7.17. The surface is specified by $r = 4$, $0 \leq \theta \leq 0.1\pi$, $0 \leq \phi \leq 0.3\pi$, and the closed path forming its perimeter is composed of three circular arcs. We are given the field $\mathbf{H} = 6r \sin \phi \mathbf{a}_r + 18r \sin \theta \cos \phi \mathbf{a}_\phi$ and are asked to evaluate each side of Stokes' theorem.

Solution. The first path segment is described in spherical coordinates by $r = 4$, $0 \leq \theta \leq 0.1\pi$, $\phi = 0$; the second one by $r = 4$, $\theta = 0.1\pi$, $0 \leq \phi \leq 0.3\pi$; and the third by $r = 4$, $0 \leq \theta \leq 0.1\pi$, $\phi = 0.3\pi$. The differential path element $d\mathbf{L}$ is the vector sum of the three differential lengths of the spherical coordinate system first discussed in Section 1.9,

$$d\mathbf{L} = dr\mathbf{a}_r + r \, d\theta \mathbf{a}_\theta + r \sin\theta \, d\phi \mathbf{a}_\phi$$

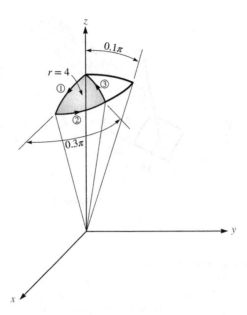

Figure 7.17 A portion of a spherical cap is used as a surface and a closed path to illustrate Stokes' theorem.

The first term is zero on all three segments of the path since $r = 4$ and $dr = 0$, the second is zero on segment 2 as θ is constant, and the third term is zero on both segments 1 and 3. Thus,

$$\oint \mathbf{H} \cdot d\mathbf{L} = \int_1 H_\theta r \, d\theta + \int_2 H_\phi r \, \sin\theta \, d\phi + \int_3 H_\theta r \, d\theta$$

Because $H\theta = 0$, we have only the second integral to evaluate,

$$\oint \mathbf{H} \cdot d\mathbf{L} = \int_0^{0.3\pi} [18(4) \sin 0.1\pi \, \cos\phi] 4 \sin 0.1\pi \, d\phi$$

$$= 288 \sin^2 0.1\pi \, \sin 0.3\pi = 22.2 \text{ A}$$

We next attack the surface integral. First, we use (26) to find

$$\nabla \times \mathbf{H} = \frac{1}{r \sin\theta}(36r \sin\theta \cos\theta \cos\phi)\mathbf{a}_r + \frac{1}{r}\left(\frac{1}{\sin\theta} 6r \cos\phi - 36r \sin\theta \cos\phi\right)\mathbf{a}_\theta$$

Because $d\mathbf{S} = r^2 \sin\theta \, d\theta \, d\phi \, \mathbf{a}_r$, the integral is

$$\int_S (\nabla \times \mathbf{H}) \cdot d\mathbf{S} = \int_0^{0.3\pi} \int_0^{0.1\pi} (36 \cos\theta \cos\phi)16 \sin\theta \, d\theta \, d\phi$$

$$= \int_0^{0.3\pi} 576\left(\frac{1}{2}\sin^2\theta\right)\Big|_0^{0.1\pi} \cos\phi \, d\phi$$

$$= 288 \, \sin^2 0.1\pi \, \sin 0.3\pi = 22.2 \text{ A}$$

Thus, the results check Stokes' theorem, and we note in passing that a current of 22.2 A is flowing upward through this section of a spherical cap.

Next, let us see how easy it is to obtain Ampère's circuital law from $\nabla \times \mathbf{H} = \mathbf{J}$. We merely have to dot each side by $d\mathbf{S}$, integrate each side over the same (open) surface S, and apply Stokes' theorem:

$$\int_S (\nabla \times \mathbf{H}) \cdot d\mathbf{S} = \int_S \mathbf{J} \cdot d\mathbf{S} = \oint \mathbf{H} \cdot d\mathbf{L}$$

The integral of the current density over the surface S is the total current I passing through the surface, and therefore

$$\oint \mathbf{H} \cdot d\mathbf{L} = I$$

This short derivation shows clearly that the current I, described as being "enclosed by the closed path," is also the current passing through any of the infinite number of surfaces that have the closed path as a perimeter.

Stokes' theorem relates a surface integral to a closed line integral. It should be recalled that the divergence theorem relates a volume integral to a closed surface integral. Both theorems find their greatest use in general vector proofs. As an example, let us find another expression for $\nabla \cdot \nabla \times \mathbf{A}$, where $\mathbf{A}$ represents any vector field. The result must be a scalar (why?), and we may let this scalar be T, or

$$\nabla \cdot \nabla \times \mathbf{A} = T$$

Multiplying by dv and integrating throughout any volume v,

$$\int_{vol} (\nabla \cdot \nabla \times \mathbf{A})dv = \int_{vol} T\, dv$$

we first apply the divergence theorem to the left side, obtaining

$$\oint_S (\nabla \times \mathbf{A}) \cdot d\mathbf{S} = \int_{vol} T\, dv$$

The left side is the surface integral of the curl of $\mathbf{A}$ over the *closed* surface surrounding the volume v. Stokes' theorem relates the surface integral of the curl of $\mathbf{A}$ over the *open* surface enclosed by a given closed path. If we think of the path as the opening of a laundry bag and the open surface as the surface of the bag itself, we see that as we gradually approach a closed surface by pulling on the drawstrings, the closed path becomes smaller and smaller and finally disappears as the surface becomes closed. Hence, the application of Stokes' theorem to a *closed* surface produces a zero result, and we have

$$\int_{vol} T\, dv = 0$$

Because this is true for any volume, it is true for the differential volume dv,

$$T \, dv = 0$$

and therefore

$$T = 0$$

or

$$\boxed{\nabla \cdot \nabla \times \mathbf{A} \equiv 0} \qquad (31)$$

Equation (31) is a useful identity of vector calculus.[6] Of course, it may also be proven easily by direct expansion in rectangular coordinates.

We next apply the identity to the non-time-varying magnetic field for which

$$\nabla \times \mathbf{H} = \mathbf{J}$$

This shows quickly that

$$\nabla \cdot \mathbf{J} = 0$$

which is the same result we obtained earlier in the chapter by using the continuity equation.

Before introducing several new magnetic field quantities in the following section, we may review our accomplishments at this point. We initially accepted the Biot-Savart law as an experimental result,

$$\mathbf{H} = \oint \frac{I \, d\mathbf{L} \times \mathbf{a}_R}{4\pi R^2}$$

and tentatively accepted Ampère's circuital law, subject to later proof,

$$\oint \mathbf{H} \cdot d\mathbf{L} = I$$

From Ampère's circuital law the definition of curl led to the point form of this same law,

$$\nabla \times \mathbf{H} = \mathbf{J}$$

We now see that Stokes' theorem enables us to obtain the integral form of Ampère's circuital law from the point form.

D7.6. Evaluate both sides of Stokes' theorem for the field $\mathbf{H} = 6xy\mathbf{a}_x - 3y^2\mathbf{a}_y$ A/m and the rectangular path around the region, $2 \leq x \leq 5, -1 \leq y \leq 1, z = 0$. Let the positive direction of $d\mathbf{S}$ be $\mathbf{a}_z$.

Ans. -126 A; -126 A

---

[6] This and other vector identities are tabulated in Appendix A.3.

7.5 MAGNETIC FLUX AND MAGNETIC FLUX DENSITY

In free space, we define the *magnetic flux density* **B** as

$$\boxed{\mathbf{B} = \mu_0 \mathbf{H}} \quad \text{(free space only)} \tag{32}$$

where **B** is measured in webers per square meter (Wb/m^2) or in a newer unit adopted in the International System of Units, tesla (T). An older unit that is often used for magnetic flux density is the gauss (G), where 1 T or 1Wb/m^2 is the same as 10,000 G. The constant μ_0 is not dimensionless and has the *defined value* for free space, in henrys per meter (H/m), of

$$\boxed{\mu_0 = 4\pi \times 10^{-7} \text{ H/m}} \tag{33}$$

The name given to μ_0 is the *permeability* of free space.

We should note that since **H** is measured in amperes per meter, the weber is dimensionally equal to the product of henrys and amperes. Considering the henry as a new unit, the weber is merely a convenient abbreviation for the product of henrys and amperes. When time-varying fields are introduced, it will be shown that a weber is also equivalent to the product of volts and seconds.

The magnetic-flux-density vector **B**, as the name weber per square meter implies, is a member of the flux-density family of vector fields. One of the possible analogies between electric and magnetic fields[7] compares the laws of Biot-Savart and Coulomb, thus establishing an analogy between **H** and **E**. The relations $\mathbf{B} = \mu_0\mathbf{H}$ and $\mathbf{D} = \epsilon_0\mathbf{E}$ then lead to an analogy between **B** and **D**. If **B** is measured in teslas or webers per square meter, then magnetic flux should be measured in webers. Magnetic flux is represented by Φ, where Φ is defined as the flux passing through any designated area,

$$\boxed{\Phi = \int_S \mathbf{B} \cdot d\mathbf{S} \text{ Wb}} \tag{34}$$

Our analogy should now remind us of the electric flux Ψ, measured in coulombs, and of Gauss's law, which states that the total flux passing through any closed surface is equal to the charge enclosed,

$$\Psi = \oint_S \mathbf{D} \cdot d\mathbf{S} = Q$$

The charge Q is the source of the lines of electric flux, and these lines begin and terminate on positive and negative charges, respectively.

---

[7] An alternate analogy is presented in Section 9.2.

No such source has ever been discovered for the lines of magnetic flux. In the example of the infinitely long straight filament carrying a direct current I, the **H** field formed concentric circles about the filament. Because $\mathbf{B} = \mu_0\mathbf{H}$, the **B** field is of the same form. The magnetic flux lines are closed and do not terminate on a "magnetic charge." For this reason Gauss's law for the magnetic field is

$$\oint_S \mathbf{B} \cdot d\mathbf{S} = 0 \tag{35}$$

and application of the divergence theorem shows us that

$$\nabla \cdot \mathbf{B} = 0 \tag{36}$$

Equation (36) is the last of Maxwell's four equations as they apply to static electric fields and steady magnetic fields. Collecting these equations, we then have for static electric fields and steady magnetic fields

$$\begin{aligned} \nabla \cdot \mathbf{D} &= \rho_v \\ \nabla \times \mathbf{E} &= 0 \\ \nabla \times \mathbf{H} &= \mathbf{J} \\ \nabla \cdot \mathbf{B} &= 0 \end{aligned} \tag{37}$$

To these equations we may add the two expressions relating **D** to **E** and **B** to **H** in free space,

$$\mathbf{D} = \epsilon_0 \mathbf{E} \tag{38}$$

$$\mathbf{B} = \mu_0 \mathbf{H} \tag{39}$$

We have also found it helpful to define an electrostatic potential,

$$\mathbf{E} = -\nabla V \tag{40}$$

and we will discuss a potential for the steady magnetic field in Section 7.6. In addition, we extended our coverage of electric fields to include conducting materials and dielectrics, and we introduced the polarization **P**. A similar treatment will be applied to magnetic fields in Chapter 8.

Returning to (37), it may be noted that these four equations specify the divergence and curl of an electric and a magnetic field. The corresponding set of four integral equations that apply to static electric fields and steady magnetic fields is

$$
\begin{aligned}
&\oint_S \mathbf{D} \cdot d\mathbf{S} = Q = \int_{vol} \rho_v dv \\
&\oint \mathbf{E} \cdot d\mathbf{L} = 0 \\
&\oint \mathbf{H} \cdot d\mathbf{L} = I = \int_S \mathbf{J} \cdot d\mathbf{S} \\
&\oint \mathbf{B} \cdot d\mathbf{S} = 0
\end{aligned}
\tag{41}
$$

Our study of electric and magnetic fields would have been much simpler if we could have begun with either set of equations, (37) or (41). With a good knowledge of vector analysis, such as we should now have, either set may be readily obtained from the other by applying the divergence theorem or Stokes' theorem. The various experimental laws can be obtained easily from these equations.

As an example of the use of flux and flux density in magnetic fields, let us find the flux between the conductors of the coaxial line of Figure 7.8a. The magnetic field intensity was found to be

$$
H_\phi = \frac{I}{2\pi\rho} \quad (a < \rho < b)
$$

and therefore

$$
\mathbf{B} = \mu_0 \mathbf{H} = \frac{\mu_0 I}{2\pi\rho}\mathbf{a}_\phi
$$

The magnetic flux contained between the conductors in a length d is the flux crossing any radial plane extending from $\rho = a$ to $\rho = b$ and from, say, $z = 0$ to $z = d$

$$
\Phi = \int_S \mathbf{B} \cdot d\mathbf{S} = \int_0^d \int_a^b \frac{\mu_0 I}{2\pi\rho}\mathbf{a}_\phi \cdot d\rho \, dz \, \mathbf{a}_\phi
$$

or

$$
\Phi = \frac{\mu_0 I d}{2\pi} \ln \frac{b}{a}
\tag{42}
$$

This expression will be used later to obtain the inductance of the coaxial transmission line.

D7.7. A solid conductor of circular cross section is made of a homogeneous nonmagnetic material. If the radius $a = 1$ mm, the conductor axis lies on the z axis, and the total current in the $\mathbf{a}_z$ direction is 20 A, find: (a) H_ϕ at $\rho = 0.5$ mm; (b) B_ϕ at $\rho = 0.8$ mm; (c) the total magnetic flux per unit length inside the conductor; (d) the total flux for $\rho < 0.5$ mm; (e) the total magnetic flux outside the conductor.

Ans. (a) 1592 A/m; (b) 3.2 mT; (c) 2 μWb/m; (d) 0.5μWb; (e) ∞

7.6   THE SCALAR AND VECTOR MAGNETIC POTENTIALS

The solution of electrostatic field problems is greatly simplified by the use of the scalar electrostatic potential V. Although this potential possesses a very real physical significance for us, it is mathematically no more than a stepping-stone which allows us to solve a problem by several smaller steps. Given a charge configuration, we may first find the potential and then from it the electric field intensity.

We should question whether or not such assistance is available in magnetic fields. Can we define a potential function which may be found from the current distribution and from which the magnetic fields may be easily determined? Can a scalar magnetic potential be defined, similar to the scalar electrostatic potential? We will show in the next few pages that the answer to the first question is yes, but the second must be answered "sometimes."

7.6.1   Scalar Potential

We begin by assuming the existence of a scalar magnetic potential, which we designate V_m, whose negative gradient gives the magnetic field intensity

$$\mathbf{H} = -\nabla V_m$$

The selection of the negative gradient provides a closer analogy to the electric potential and to problems which we have already solved.

This definition must not conflict with our previous results for the magnetic field, and therefore

$$\nabla \times \mathbf{H} = \mathbf{J} = \nabla \times (-\nabla V_m)$$

However, the curl of the gradient of any scalar is identically zero, a vector identity the proof of which is left for a leisure moment. Therefore, we see that if $\mathbf{H}$ is to be defined as the gradient of a scalar magnetic potential, then current density must be zero throughout the region in which the scalar magnetic potential is so defined. We then have

$$\boxed{\mathbf{H} = -\nabla V_m \quad (\mathbf{J} = 0)} \tag{43}$$

Because many magnetic problems involve geometries in which the current-carrying conductors occupy a relatively small fraction of the total region of interest, it is evident that a scalar magnetic potential can be useful. The scalar magnetic potential is also applicable in the case of permanent magnets. The dimensions of V_m are obviously amperes.

This scalar potential also satisfies Laplace's equation. In free space,

$$\nabla \cdot \mathbf{B} = \mu_0 \nabla \cdot \mathbf{H} = 0$$

and hence

$$\mu_0 \, \nabla \cdot (-\nabla V_m) = 0$$

or

$$\boxed{\nabla^2 V_m = 0 \quad (\mathbf{J} = 0)} \tag{44}$$

We will see later that V_m continues to satisfy Laplace's equation in homogeneous magnetic materials; it is not defined in any region in which current density is present.

Although we will consider the scalar magnetic potential to a much greater extent in Chapter 8, when we introduce magnetic materials and discuss the magnetic circuit, one difference between V and V_m should be pointed out now: V_m is not a single-valued function of position. The electric potential V is single-valued; once a zero reference is assigned, there is only one value of V associated with each point in space. Such is not the case with V_m. Consider the cross section of the coaxial line shown in Figure 7.18. In the region $a < \rho < b$, $\mathbf{J} = 0$, and we may establish a scalar magnetic potential. The value of $\mathbf{H}$ is

$$\mathbf{H} = \frac{I}{2\pi\rho}\mathbf{a}_\phi$$

where I is the total current flowing in the $\mathbf{a}_z$ direction in the inner conductor. We find V_m by integrating the appropriate component of the gradient. Applying (43),

$$\frac{I}{2\pi\rho} = -\nabla V_m\big|_\phi = -\frac{1}{\rho}\frac{\partial V_m}{\partial \phi}$$

or

$$\frac{\partial V_m}{\partial \phi} = -\frac{I}{2\pi}$$

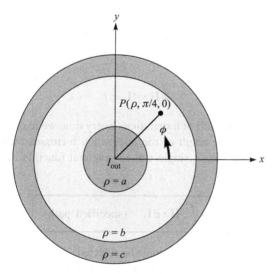

Figure 7.18 The scalar magnetic potential V_m is a multivalued function of ϕ in the region $a < \rho < b$. The electrostatic potential is always single valued.

Thus,

$$V_m = -\frac{I}{2\pi}\phi$$

where the constant of integration has been set equal to zero. What value of potential do we associate with point P, where $\phi = \pi/4$? If we let V_m be zero at $\phi = 0$ and proceed counterclockwise around the circle, the magnetic potential goes negative linearly. When we have made one circuit, the potential is $-I$, but that was the point at which we said the potential was zero a moment ago. At P, then, $\phi = \pi/4, 9\pi/4, 17\pi/4, \ldots$, or $-7\pi/4, -15\pi/4, -23\pi/4, \ldots$, or

$$V_{mP} = \frac{I}{2\pi}\left(2n - \tfrac{1}{4}\right)\pi \quad (n = 0, \pm 1, \pm 2, \ldots)$$

or

$$V_{mP} = I\left(n - \tfrac{1}{8}\right) \quad (n = 0, \pm 1, \pm 2, \ldots)$$

The reason for this multivaluedness may be shown by a comparison with the electrostatic case. There, we know that

$$\nabla \times \mathbf{E} = 0$$
$$\oint \mathbf{E} \cdot d\mathbf{L} = 0$$

and therefore the line integral

$$V_{ab} = -\int_b^a \mathbf{E} \cdot d\mathbf{L}$$

is independent of the path. In the magnetostatic case, however,

$$\nabla \times \mathbf{H} = 0 \quad (\text{wherever } \mathbf{J} = 0)$$

but

$$\oint \mathbf{H} \cdot d\mathbf{L} = I$$

even if $\mathbf{J}$ is zero along the path of integration. Every time we make another complete lap around the current, the result of the integration increases by I. If no current I is enclosed by the path, then a single-valued potential function may be defined. In general, however,

$$V_{m,ab} = -\int_b^a \mathbf{H} \cdot d\mathbf{L} \quad (\text{specified path}) \tag{45}$$

where a specific path or type of path must be selected. We should remember that the electrostatic potential V arises from a conservative $\mathbf{E}$ field; the magnetic field $\mathbf{H}$ is generally not conservative. In our coaxial problem, let us erect a barrier[8] at

---

[8] This corresponds to the more precise mathematical term "branch cut."

$\phi = \pi$; we agree not to select a path that crosses this plane. Therefore, we cannot encircle I, and a single-valued potential is possible. The result is seen to be

$$V_m = -\frac{I}{2\pi}\phi \quad (-\pi < \phi < \pi)$$

and

$$V_{mP} = -\frac{I}{8} \quad \left(\phi = \frac{\pi}{4}\right)$$

The scalar magnetic potential is evidently the quantity whose equipotential surfaces will form curvilinear squares with the streamlines of **H** in Figure 7.4. This is one more facet of the analogy between electric and magnetic fields about which we will have more to say in Chapter 8.

7.6.2 Vector Potential

We now temporarily leave the scalar magnetic potential and investigate a vector magnetic potential. This vector field is one which is extremely useful in studying radiation from antennas (as we will find in Chapter 14) as well as radiation leakage from transmission lines, waveguides, and microwave ovens. The vector magnetic potential may be used in regions where the current density is zero or nonzero, and we will also be able to extend it to the time-varying case later.

Our choice of a vector magnetic potential is indicated by noting that

$$\nabla \cdot \mathbf{B} = 0$$

Next, a vector identity that we proved in Section 7.4 shows that the divergence of the curl of any vector field is zero. Therefore, we select

$$\boxed{\mathbf{B} = \nabla \times \mathbf{A}} \tag{46}$$

where A signifies a *vector magnetic potential,* and we automatically satisfy the condition that the magnetic flux density shall have zero divergence. The **H** field is

$$\mathbf{H} = \frac{1}{\mu_0} \nabla \times \mathbf{A}$$

and

$$\nabla \times \mathbf{H} = \mathbf{J} = \frac{1}{\mu_0} \nabla \times \nabla \times \mathbf{A}$$

The curl of the curl of a vector field is not zero and is given by a fairly complicated expression,[9] which we need not know now in general form. In specific cases for which the form of **A** is known, the curl operation may be applied twice to determine the current density.

---

[9] $\nabla \times \nabla \times \mathbf{A} \equiv \nabla (\nabla \cdot \mathbf{A}) - \nabla^2\mathbf{A}$. In rectangular coordinates, it may be shown that $\nabla^2\mathbf{A} \equiv \nabla^2A_x\mathbf{a}_x + \nabla^2A_y\mathbf{a}_y + \nabla^2A_z\mathbf{a}_z$. In other coordinate systems, $\nabla^2\mathbf{A}$ may be found by evaluating the second-order partial derivatives in $\nabla^2\mathbf{A} = \nabla(\nabla \cdot \mathbf{A}) - \nabla \times \nabla \times \mathbf{A}$.

Equation (46) serves as a useful definition of the *vector magnetic potential* **A**. Because the curl operation implies differentiation with respect to a length, the units of **A** are webers per meter.

As yet we have seen only that the definition for **A** does not conflict with any previous results. It still remains to show that this particular definition can help us to determine magnetic fields more easily. We certainly cannot identify **A** with any easily measured quantity or history-making experiment.

We will show in Section 7.7 that, given the Biot-Savart law, the definition of **B**, and the definition of **A**, **A** may be determined from the differential current elements by

$$\mathbf{A} = \oint \frac{\mu_0 I \, d\mathbf{L}}{4\pi R} \qquad (47)$$

The significance of the terms in (47) is the same as in the Biot-Savart law; a direct current I flows along a filamentary conductor of which any differential length $d\mathbf{L}$ is at distance R from the point at which **A** is to be found. Because we have defined **A** only through specification of its curl, it is possible to add the gradient of any scalar field to (47) without changing **B** or **H**, for the curl of the gradient is identically zero. In steady magnetic fields, it is customary to set this added term equal to zero.

The fact that **A** is a vector magnetic *potential* is more apparent when (47) is compared with the similar expression for the electrostatic potential,

$$V = \int \frac{\rho_L dL}{4\pi \epsilon_0 R}$$

Each expression is the integral along a line source, in one case line charge and in the other case line current; each integrand is inversely proportional to the distance from the source to the point of interest; and each involves a characteristic of the medium (here free space), the permeability or the permittivity.

Equation (47) may be written in differential form,

$$d\mathbf{A} = \frac{\mu_0 I \, d\mathbf{L}}{4\pi R} \qquad (48)$$

if we again agree not to attribute any physical significance to any magnetic fields we obtain from (48) until the *entire closed path in which the current flows is considered*.

With this reservation, let us go right ahead and consider the vector magnetic potential field about a differential filament. We locate the filament at the origin in free space, as shown in Figure 7.19, and allow it to extend in the positive z direction so that $d\mathbf{L} = dz \, \mathbf{a}_z$. We use cylindrical coordinates to find $d\mathbf{A}$ at the point (ρ, ϕ, z):

$$d\mathbf{A} = \frac{\mu_0 I \, dz \, \mathbf{a}_z}{4\pi \sqrt{\rho^2 + z^2}}$$

or

$$d\mathbf{A}_z = \frac{\mu_0 I \, dz}{4\pi \sqrt{\rho^2 + z^2}} \quad dA_\phi = 0 \quad dA_\rho = 0 \qquad (49)$$

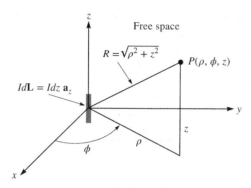

Figure 7.19 The differential current element $I\,dz\mathbf{a}_z$ at the origin establishes the differential vector magnetic potential field, $d\mathbf{A} = \dfrac{\mu_0 I dz \mathbf{a}_z}{4\pi\sqrt{\rho^2 + z^2}}$ at $P(\rho, \phi, z)$.

Interactives

We note that the direction of $d\mathbf{A}$ is the same as that of $I\,d\mathbf{L}$. Each small section of a current-carrying conductor produces a contribution to the total vector magnetic potential which is in the same direction as the current flow in the conductor. The magnitude of the vector magnetic potential varies inversely with the distance to the current element, being strongest in the neighborhood of the current and gradually falling off to zero at distant points. Skilling[10] describes the vector magnetic potential field as "like the current distribution but fuzzy around the edges, or like a picture of the current out of focus."

In order to find the magnetic field intensity, we must take the curl of (49) in cylindrical coordinates, leading to

$$dH = \frac{1}{\mu_0}\nabla \times d\mathbf{A} = \frac{1}{\mu_0}\left(-\frac{\partial dA_z}{\partial \rho}\right)\mathbf{a}_\phi$$

or

$$d\mathbf{H} = \frac{I\,dz}{4\pi}\frac{\rho}{(\rho^2 + z^2)^{3/2}}\mathbf{a}_\phi$$

which is easily shown to be the same as the value given by the Biot-Savart law.

Expressions for the vector magnetic potential $\mathbf{A}$ can also be obtained for a current source which is distributed. For a current sheet $\mathbf{K}$, the differential current element becomes

$$I\,d\mathbf{L} = \mathbf{K}\,d\mathbf{S}$$

In the case of current flow throughout a volume with a density $\mathbf{J}$, we have

$$I\,d\mathbf{L} = \mathbf{J}\,dv$$

---

[10] See the References at the end of the chapter.

In each of these two expressions the vector character is given to the current. For the filamentary element it is customary, although not necessary, to use $I\,d\mathbf{L}$ instead of $\mathbf{I}dL$. Since the magnitude of the filamentary element is constant, we have chosen the form which allows us to remove one quantity from the integral. The alternative expressions for $\mathbf{A}$ are then

$$\mathbf{A} = \int_S \frac{\mu_0 \mathbf{K}\, dS}{4\pi R} \qquad (50)$$

and

$$\mathbf{A} = \int_{vol} \frac{\mu_0 \mathbf{J}\, dv}{4\pi R} \qquad (51)$$

Equations (47), (50), and (51) express the vector magnetic potential as an integration over all of its sources. From a comparison of the form of these integrals with those which yield the electrostatic potential, it is evident that once again the zero reference for $\mathbf{A}$ is at infinity, for no finite current element can produce any contribution as $R \to \infty$. We should remember that we very seldom used the similar expressions for V; too often our theoretical problems included charge distributions that extended to infinity, and the result would be an infinite potential everywhere. Actually, we calculated very few potential fields until the differential form of the potential equation was obtained, $\nabla^2 V = -\rho_v/\epsilon$, or better yet, $\nabla^2 V = 0$. We were then at liberty to select our own zero reference.

The analogous expressions for $\mathbf{A}$ will be derived in the next section, and an example of the calculation of a vector magnetic potential field will be completed.

D7.8. A current sheet, $\mathbf{K} = 2.4\mathbf{a}_z$ A/m, is present at the surface $\rho = 1.2$ in free space. (a) Find $\mathbf{H}$ for $\rho > 1.2$. Find V_m at $P(\rho = 1.5, \phi = 0.6\pi, z = 1)$ if: (b) $V_m = 0$ at $\phi = 0$ and there is a barrier at $\phi = \pi$; (c) $V_m = 0$ at $\phi = 0$ and there is a barrier at $\phi = \pi/2$; (d) $V_m = 0$ at $\phi = \pi$ and there is a barrier at $\phi = 0$; (e) $V_m = 5$ V at $\phi = \pi$ and there is a barrier at $\phi = 0.8\,\pi$.

Ans. (a) $\frac{2.88}{\rho}\,\mathbf{a}_\phi$; (b) −5.43 V; (c) 12.7 V; (d) 3.62 V; (e) −9.48 V

D7.9. The value of $\mathbf{A}$ within a solid nonmagnetic conductor of radius a carrying a total current I in the $\mathbf{a}_z$ direction may be found easily. Using the known value of $\mathbf{H}$ or $\mathbf{B}$ for $\rho < a$, then (46) may be solved for $\mathbf{A}$. Select $A = (\mu_0 I \ln 5)/2\pi$ at $\rho = a$ (to correspond with an example in the next section) and find $\mathbf{A}$ at $\rho =$: (a) 0; (b) 0.25a; (c) 0.75a; (d) a.

Ans. (a) $0.422I\,\mathbf{a}_z\,\mu$Wb/m; (b) $0.416I\,\mathbf{a}_z\,\mu$Wb/m; (c) $0.366I\,\mathbf{a}_z\,\mu$Wb/m; (d) $0.322I\,\mathbf{a}_z\,\mu$Wb/m

7.7 DERIVATION OF THE STEADY-MAGNETIC-FIELD LAWS

We will now supply the promised proofs of the several relationships between the magnetic field quantities. All these relationships may be obtained from the definitions of **H**,

$$\mathbf{H} = \oint \frac{I\, d\mathbf{L} \times \mathbf{a}_R}{4\pi R^2} \tag{3}$$

of **B** (in free space),

$$\mathbf{B} = \mu_0 \mathbf{H} \tag{32}$$

and of **A**,

$$\mathbf{B} = \nabla \times \mathbf{A} \tag{46}$$

We first assume that we may express **A** by the last equation of Section 7.6,

$$\mathbf{A} = \int_{\text{vol}} \frac{\mu_0 \mathbf{J}\, dv}{4\pi R} \tag{51}$$

and then demonstrate the correctness of (51) by showing that (3) follows. First, we should add subscripts to indicate the point at which the current element is located (x_1, y_1, z_1) and the point at which **A** is given (x_2, y_2, z_2). The differential volume element dv is then written dv_1 and in rectangular coordinates would be $dx_1\, dy_1\, dz_1$. The variables of integration are x_1, y_1, and z_1. Using these subscripts, then,

$$\mathbf{A}_2 = \int_{\text{vol}} \frac{\mu_0 \mathbf{J}_1\, dv_1}{4\pi R_{12}} \tag{52}$$

From (32) and (46) we have

$$\mathbf{H} = \frac{\mathbf{B}}{\mu_0} = \frac{\nabla \times \mathbf{A}}{\mu_0} \tag{53}$$

To show that (3) follows from (52), it is necessary to substitute (52) into (53). This step involves taking the curl of $\mathbf{A}_2$, a quantity expressed in terms of the variables x_2, y_2, and z_2, and the curl therefore involves partial derivatives with respect to x_2, y_2, and z_2. We do this, placing a subscript on the del operator to remind us of the variables involved in the partial differentiation process,

$$\mathbf{H}_2 = \frac{\nabla_2 \times \mathbf{A}_2}{\mu_0} = \frac{1}{\mu_0}\nabla_2 \times \int_{\text{vol}} \frac{\mu_0 \mathbf{J}_1\, dv_1}{4\pi R_{12}}$$

The order of partial differentiation and integration is immaterial, and $\mu_0/4\pi$ is constant, allowing us to write

$$\mathbf{H}_2 = \frac{1}{4\pi}\int_{\text{vol}} \nabla_2 \times \frac{\mathbf{J}_1\, dv_1}{R_{12}}$$

The curl operation within the integrand represents partial differentiation with respect to x_2, y_2, and z_2. The differential volume element dv_1 is a scalar and a function only of x_1, y_1, and z_1. Consequently, it may be factored out of the curl operation as any other constant, leaving

$$\mathbf{H}_2 = \frac{1}{4\pi} \int_{\text{vol}} \left(\nabla_2 \times \frac{\mathbf{J}_1}{R_{12}}\right) dv_1 \qquad (54)$$

The curl of the product of a scalar and a vector is given by an identity which may be checked by expansion in rectangular coordinates or obtained from Appendix A.3,

$$\nabla \times (S\mathbf{V}) \equiv (\nabla S) \times \mathbf{V} + S(\nabla \times \mathbf{V}) \qquad (55)$$

This identity is used to expand the integrand of (54),

$$\mathbf{H}_2 = \frac{1}{4\pi} \int_{\text{vol}} \left[\left(\nabla_2 \frac{1}{R_{12}}\right) \times \mathbf{J}_1 + \frac{1}{R_{12}}(\nabla_2 \times \mathbf{J}_1)\right] dv_1 \qquad (56)$$

The second term of this integrand is zero because $\nabla_2 \times \mathbf{J}_1$ indicates partial derivatives of a function of x_1, y_1, and z_1, taken with respect to the variables x_2, y_2, and z_2; the first set of variables is not a function of the second set, and all partial derivatives are zero.

The first term of the integrand may be determined by expressing R_{12} in terms of the coordinate values,

$$R_{12} = \sqrt{(x_2 - x_1)^2 + (y_2 - y_1)^2 + (z_2 - z_1)^2}$$

and taking the gradient of its reciprocal. Problem 7.42 shows that the result is

$$\nabla_2 \frac{1}{R_{12}} = -\frac{\mathbf{R}_{12}}{R_{12}^3} = -\frac{\mathbf{a}_{R12}}{R_{12}^2}$$

Substituting this result into (56), we have

$$\mathbf{H}_2 = -\frac{1}{4\pi} \int_{\text{vol}} \frac{\mathbf{a}_{R12} \times \mathbf{J}_1}{R_{12}^2} dv_1$$

or

$$\mathbf{H}_2 = \int_{\text{vol}} \frac{\mathbf{J}_1 \times \mathbf{a}_{R12}}{4\pi R_{12}^2} dv_1$$

which is the equivalent of (3) in terms of current density. Replacing $\mathbf{J}_1 \, dv_1$ by $I_1 \, d\mathbf{L}_1$, we may rewrite the volume integral as a closed line integral,

$$\mathbf{H}_2 = \oint \frac{I_1 \, d\mathbf{L}_1 \times \mathbf{a}_{R12}}{4\pi R_{12}^2}$$

Equation (51) is therefore correct and agrees with the three definitions (3), (32), and (46).

Next we will prove Ampère's circuital law in point form,

$$\nabla \times \mathbf{H} = \mathbf{J} \qquad (28)$$

Combining (28), (32), and (46), we obtain

$$\nabla \times \mathbf{H} = \nabla \times \frac{\mathbf{B}}{\mu_0} = \frac{1}{\mu_0} \nabla \times \nabla \times \mathbf{A} \qquad (57)$$

We now need the expansion in rectangular coordinates for $\nabla \times \nabla \times \mathbf{A}$. Performing the indicated partial differentiations and collecting the resulting terms, we may write the result as

$$\boxed{\nabla \times \nabla \times \mathbf{A} \equiv \nabla(\nabla \cdot \mathbf{A}) - \nabla^2 \mathbf{A}} \tag{58}$$

where

$$\boxed{\nabla^2 \mathbf{A} \equiv \nabla^2 A_x \mathbf{a}_x + \nabla^2 A_y \mathbf{a}_y + \nabla^2 A_z \mathbf{a}_z} \tag{59}$$

Equation (59) is the definition (in rectangular coordinates) of the *Laplacian of a vector*.

Substituting (58) into (57), we have

$$\nabla \times \mathbf{H} = \frac{1}{\mu_0}[\nabla(\nabla \cdot \mathbf{A}) - \nabla^2 \mathbf{A}] \tag{60}$$

and now require expressions for the divergence and the Laplacian of $\mathbf{A}$.

We may find the divergence of $\mathbf{A}$ by applying the divergence operation to (52),

$$\nabla_2 \cdot \mathbf{A}_2 = \frac{\mu_0}{4\pi}\int_{\text{vol}} \nabla_2 \cdot \frac{\mathbf{J}_1}{R_{12}}dv_1 \tag{61}$$

and using the vector identity (44) of Section 4.8,

$$\nabla \cdot (S\mathbf{V}) \equiv \mathbf{V} \cdot (\nabla S) + S(\nabla \cdot \mathbf{V})$$

Thus,

$$\nabla_2 \cdot \mathbf{A}_2 = \frac{\mu_0}{4\pi}\int_{\text{vol}}\left[\mathbf{J}_1 \cdot \left(\nabla_2 \frac{1}{R_{12}}\right) + \frac{1}{R_{12}}(\nabla_2 \cdot \mathbf{J}_1)\right]dv_1 \tag{62}$$

The second part of the integrand is zero because $\mathbf{J}_1$ is not a function of x_2, y_2, and z_2.

We have already used the result that $\nabla_2(1/R_{12}) = -\mathbf{R}_{12}/R_{12}^3$, and it is just as easily shown that

$$\nabla_1 \frac{1}{R_{12}} = \frac{\mathbf{R}_{12}}{R_{12}^3}$$

or that

$$\nabla_1 \frac{1}{R_{12}} = -\nabla_2 \frac{1}{R_{12}}$$

Equation (62) can therefore be written as

$$\nabla_2 \cdot \mathbf{A}_2 = \frac{\mu_0}{4\pi}\int_{\text{vol}}\left[-\mathbf{J}_1 \cdot \left(\nabla_1 \frac{1}{R_{12}}\right)\right]dv_1$$

and the vector identity applied again,

$$\nabla_2 \cdot \mathbf{A}_2 = \frac{\mu_0}{4\pi}\int_{\text{vol}}\left[\frac{1}{R_{12}}(\nabla_1 \cdot \mathbf{J}_1) - \nabla_1 \cdot \left(\frac{\mathbf{J}_1}{R_{12}}\right)\right]dv_1 \tag{63}$$

Because we are concerned only with steady magnetic fields, the continuity equation shows that the first term of (63) is zero. Application of the divergence theorem to the second term gives

$$\nabla_2 \cdot \mathbf{A}_2 = -\frac{\mu_0}{4\pi} \oint_{S_1} \frac{\mathbf{J}_1}{R_{12}} \cdot d\mathbf{S}_1$$

where the surface S_1 encloses the volume throughout which we are integrating. This volume must include all the current, for the original integral expression for $\mathbf{A}$ was an integration such as to include the effect of all the current. Because there is no current outside this volume (otherwise we should have had to increase the volume to include it), we may integrate over a slightly larger volume or a slightly larger enclosing surface without changing $\mathbf{A}$. On this larger surface the current density $\mathbf{J}_1$ must be zero, and therefore the closed surface integral is zero, since the integrand is zero. Hence the divergence of $\mathbf{A}$ is zero.

In order to find the Laplacian of the vector $\mathbf{A}$, let us compare the x component of (51) with the similar expression for electrostatic potential,

$$A_x = \int_{\text{vol}} \frac{\mu_0 J_x \, dv}{4\pi R} \qquad V = \int_{\text{vol}} \frac{\rho_v \, dv}{4\pi \epsilon_0 R}$$

We note that one expression can be obtained from the other by a straightforward change of variable, J_x for ρ_v, μ_0 for $1/\epsilon_0$, and A_x for V. However, we have derived some additional information about the electrostatic potential which we shall not have to repeat now for the x component of the vector magnetic potential. This takes the form of Poisson's equation,

$$\nabla^2 V = -\frac{\rho_v}{\epsilon_0}$$

which becomes, after the change of variables,

$$\nabla^2 A_x = -\mu_0 J_x$$

Similarly, we have

$$\nabla^2 A_y = -\mu_0 J_y$$

and

$$\nabla^2 A_z = -\mu_0 J_z$$

or

$$\boxed{\nabla^2 \mathbf{A} = -\mu_0 \mathbf{J}} \qquad (64)$$

Returning to (60), we can now substitute for the divergence and Laplacian of $\mathbf{A}$ and obtain the desired answer,

$$\nabla \times \mathbf{H} = \mathbf{J} \qquad (28)$$

We have already shown the use of Stokes' theorem in obtaining the integral form of Ampère's circuital law from (28) and need not repeat that labor here.

We thus have succeeded in showing that every result we have essentially pulled from thin air[11] for magnetic fields follows from the basic definitions of **H**, **B**, and **A**. The derivations are not simple, but they should be understandable on a step-by-step basis.

Finally, let us return to (64) and make use of this formidable second-order vector partial differential equation to find the vector magnetic potential in one simple example. We select the field between conductors of a coaxial cable, with radii of a and b as usual, and current I in the $\mathbf{a}_z$ direction in the inner conductor. Between the conductors, $\mathbf{J} = 0$, and therefore

$$\nabla^2 \mathbf{A} = 0$$

We have already been told (and Problem 7.44 gives us the opportunity to check the results for ourselves) that the vector Laplacian may be expanded as the vector sum of the scalar Laplacians of the three components in rectangular coordinates,

$$\nabla^2 \mathbf{A} = \nabla^2 A_x \mathbf{a}_x + \nabla^2 A_y \mathbf{a}_y + \nabla^2 A_z \mathbf{a}_z$$

but such a relatively simple result is not possible in other coordinate systems. That is, in cylindrical coordinates, for example,

$$\nabla^2 \mathbf{A} \neq \nabla^2 A_\rho \mathbf{a}_\rho + \nabla^2 A_\phi \mathbf{a}_\phi + \nabla^2 A_z \mathbf{a}_z$$

However, it is not difficult to show for cylindrical coordinates that the z component of the vector Laplacian is the scalar Laplacian of the z component of **A**, or

$$\nabla^2 \mathbf{A}\big|_z = \nabla^2 A_z \tag{65}$$

and because the current is entirely in the z direction in this problem, **A** has only a z component. Therefore,

$$\nabla^2 A_z = 0$$

or

$$\frac{1}{\rho}\frac{\partial}{\partial \rho}\left(\rho \frac{\partial A_z}{\partial \rho}\right) + \frac{1}{\rho^2}\frac{\partial^2 A_z}{\partial \phi^2} + \frac{\partial^2 A_z}{\partial z^2} = 0$$

Thinking symmetrical thoughts about (51) shows us that A_z is a function only of ρ, and thus

$$\frac{1}{\rho}\frac{d}{d\rho}\left(\rho \frac{dA_z}{d\rho}\right) = 0$$

We have solved this equation before, and the result is

$$A_z = C_1 \ln \rho + C_2$$

If we choose a zero reference at $\rho = b$, then

$$A_z = C_1 \ln \frac{\rho}{b}$$

---

[11] Free space.

In order to relate C_1 to the sources in our problem, we may take the curl of $\mathbf{A}$,

$$\nabla \times \mathbf{A} = -\frac{\partial A_z}{\partial \rho}\mathbf{a}_\phi = -\frac{C_1}{\rho}\mathbf{a}_\phi = \mathbf{B}$$

obtain $\mathbf{H}$,

$$\mathbf{H} = -\frac{C_1}{\mu_0 \rho}\mathbf{a}_\phi$$

and evaluate the line integral,

$$\oint \mathbf{H} \cdot d\mathbf{L} = I = \int_0^{2\pi} -\frac{C_1}{\mu_0 \rho}\mathbf{a}_\phi \cdot \rho \, d\phi \, \mathbf{a}_\phi = -\frac{2\pi C_1}{\mu_0}$$

Thus

$$C_1 = -\frac{\mu_0 I}{2\pi}$$

or

$$A_z = \frac{\mu_0 I}{2\pi}\ln\frac{b}{\rho} \tag{66}$$

and

$$H_\phi = \frac{I}{2\pi\rho}$$

as before. A plot of A_z versus ρ for $b = 5a$ is shown in Figure 7.20; the decrease of $|\mathbf{A}|$ with distance from the concentrated current source that the inner conductor represents is evident. The results of Problem D7.9 have also been added to Figure 7.20. The extension of the curve into the outer conductor is left as Problem 7.43.

It is also possible to find A_z between conductors by applying a process some of us informally call "uncurling." That is, we know $\mathbf{H}$ or $\mathbf{B}$ for the coax, and we may

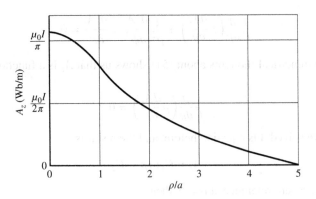

Figure 7.20 The vector magnetic potential is shown within the inner conductor and in the region between conductors for a coaxial cable with $b = 5a$ carrying I in the $\mathbf{a}_z$ direction. $A_z = 0$ is arbitrarily selected at $\rho = b$.

therefore select the ϕ component of $\nabla \times \mathbf{A} = \mathbf{B}$ and integrate to obtain A_z. Try it, you'll like it!

D7.10. Equation (66) is obviously also applicable to the exterior of any conductor of circular cross section carrying a current I in the $\mathbf{a}_z$ direction in free space. The zero reference is arbitrarily set at $\rho = b$. Now consider two conductors, each of 1 cm radius, parallel to the z axis with their axes lying in the $x = 0$ plane. One conductor whose axis is at (0, 4 cm, z) carries 12 A in the $\mathbf{a}_z$ direction; the other axis is at (0, −4 cm, z) and carries 12 A in the $-\mathbf{a}_z$ direction. Each current has its zero reference for $\mathbf{A}$ located 4 cm from its axis. Find the total $\mathbf{A}$ field at: (a) (0, 0, z); (b) (0, 8 cm, z); (c) (4 cm, 4 cm, z); (d) (2 cm, 4 cm, z).

Ans. (a) 0; (b) 2.64μWb/m; (c) 1.93μWb/m; (d) 3.40μWb/m

REFERENCES

1. Boast, W. B. (See References for Chapter 2.) The scalar magnetic potential is defined on p. 220, and its use in mapping magnetic fields is discussed on p. 444.

2. Jordan, E. C., and K. G. Balmain. *Electromagnetic Waves and Radiating Systems.* 2d ed. Englewood Cliffs, N.J.: Prentice-Hall, 1968. Vector magnetic potential is discussed on pp. 90–96.

3. Paul, C. R., K. W. Whites, and S. Y. Nasar. *Introduction to Electromagnetic Fields.* 3d ed. New York: McGraw-Hill, 1998. The vector magnetic potential is presented on pp. 216–20.

4. Skilling, H. H. (See References for Chapter 3.) The "paddle wheel" is introduced on pp. 23–25.

CHAPTER 7 PROBLEMS

Quizzes

7.1 (a) Find $\mathbf{H}$ in rectangular components at $P(2, 3, 4)$ if there is a current filament on the z axis carrying 8 mA in the $\mathbf{a}_z$ direction. (b) Repeat if the filament is located at $x = -1$, $y = 2$. (c) Find $\mathbf{H}$ if both filaments are present.

7.2 A filamentary conductor is formed into an equilateral triangle with sides of length ℓ carrying current I. Find the magnetic field intensity at the center of the triangle.

7.3 A circular current filament lies in the xy plane with its center at the origin. The loop carries current I in the positive $\mathbf{a}_\phi$ direction and is of radius a. Find $\mathbf{H}$ everywhere on the z axis.

7.4 Two circular current loops are centered on the z axis at $z = \pm h$. Each loop has radius a and carries current I in the $\mathbf{a}_\phi$ direction. (a) Find $\mathbf{H}$ on the z axis over the range $-h < z < h$. Take $I = 1$ A and plot $|\mathbf{H}|$ as a function of z/a if (b) $h = a/4$; (c) $h = a/2$; (d) $h = a$. Which choice for h gives the most

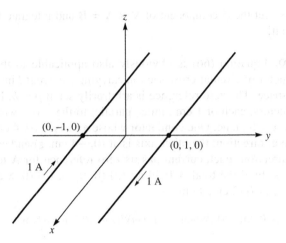

Figure 7.21 See Problem 7.5.

uniform field? These are called Helmholtz coils (of a single turn each in this case) and are used in providing uniform fields.

7.5 The parallel filamentary conductors shown in Figure 7.21 lie in free space. Plot $|\mathbf{H}|$ versus y, $-4 < y < 4$, along the line $x = 0$, $z = 2$.

7.6 A disk of radius a lies in the xy plane, with the z axis through its center. Surface charge of uniform density ρ_s lies on the disk, which rotates about the z axis at angular velocity Ω rad/s. Find $\mathbf{H}$ at any point on the z axis.

7.7 A filamentary conductor carrying current I in the $\mathbf{a}_z$ direction extends along the entire negative z axis. At $z = 0$ it connects to a copper sheet that fills the $x > 0$, $y > 0$ quadrant of the xy plane. (a) Set up the Biot-Savart law and find $\mathbf{H}$ everywhere on the z axis; (b) repeat part (a), but with the copper sheet occupying the *entire* xy plane (Hint: express $\mathbf{a}_\phi$ in terms of $\mathbf{a}_x$ and $\mathbf{a}_y$ and angle ϕ in the integral).

7.8 For the finite-length current element on the z axis, as shown in Figure 7.5, use the Biot-Savart law to derive Eq. (9) of Section 7.1.

7.9 A uniform surface charge of density $\rho_s = \rho_0$ C/m^2 is in the shape of a sphere of radius a. The sphere rotates about the z axis at angular velocity Ω rad/s. Find (a) the total current flowing, and (b) $\mathbf{H}$ at the origin. Make appropriate use of Eq. (6).

7.10 A uniform volume charge of density $\rho_v = \rho_0$ C/m^3 is in the shape of a sphere of radius a. The sphere rotates about the z axis at angular velocity Ω rad/s. Find a) the total current flowing, and b) $\mathbf{H}$ at the origin. Make appropriate use of Eq. (7).

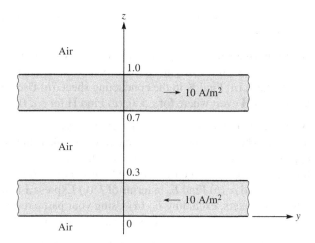

Figure 7.22   See Problem 7.12.

7.11 A solenoid of radius a is oriented with its axis along the z axis over the range $-d/2 < z < d/2$. The coil is tightly wound with wire having density n turns per meter of length, where each turn carries current I. (a) Find $\mathbf{H}$ at the origin. Begin with the current loop field (solution to Problem 7.3) modified by replacing the current in that result with $I_s = nI\,dz$. I_s is the current in a circular band of length dz on the coil; the solenoid is made up of these bands of current over the entire length. (b) Show that the midpoint field of part a reduces to $nI\,\mathbf{a}_z$ as the solenoid length approaches infinity.

7.12 In Figure 7.22, let the regions $0 < z < 0.3$ m and $0.7 < z < 1.0$ m be conducting slabs carrying uniform current densities of 10 A/m² in opposite directions as shown. Find $\mathbf{H}$ at $z =$: (a) -0.2; (b) 0.2; (c) 0.4; (d) 0.75; (e) 1.2 m.

7.13 A hollow cylindrical shell of radius a is centered on the z axis and carries a uniform surface current density of $K_a\mathbf{a}_\phi$. Using Ampere's circuital law: (a) Show that $\mathbf{H}$ is not a function of ϕ or z. (b) Show that H_ϕ and H_ρ are everywhere zero. (c) Show that $H_z = 0$ for $\rho > a$. (d) Show that $H_z = K_a$ for $\rho < a$. (e) A second shell, $\rho = b$, carries a current $K_b\mathbf{a}_\phi$. Find $\mathbf{H}$ everywhere.

7.14 A toroid having a cross section of rectangular shape is defined by the following surfaces: the cylinders $\rho = 2$ and $\rho = 3$ cm, and the planes $z = 1$ and $z = 2.5$ cm. The toroid carries a surface current density of $-50\mathbf{a}_z$ A/m on the surface $\rho = 3$ cm. Find $\mathbf{H}$ at the point $P(\rho, \phi, z)$: (a) $P_A(1.5$ cm, 0, 2 cm); (b) $P_B(2.1$ cm, 0, 2 cm); (c) $P_C(2.7$ cm, $\pi/2$, 2 cm); (d) P_D (3.5 cm, $\pi/2$, 2 cm).

7.15 A hollow spherical conducting shell of radius a has filamentary connections made at the top ($r = a$, $\theta = 0$) and bottom ($r = a$, $\theta = \pi$). A direct current I flows down the upper filament, down the spherical surface, and out the lower filament. Find $\mathbf{H}$ in spherical coordinates (a) inside and (b) outside the sphere.

7.16 A current filament carrying I in the $-\mathbf{a}_z$ direction lies along the entire positive z axis. At the origin, it connects to a conducting sheet that forms the xy plane. (a) Find $\mathbf{K}$ in the conducting sheet. (b) Use Ampere's circuital law to find $\mathbf{H}$ everywhere for $z > 0$. (c) Find $\mathbf{H}$ for $z < 0$.

7.17 A solid cylindrical wire of radius a carries current I in the positive z direction. The wire conductivity is radially dependent and is given by $\sigma(\rho) = \sigma_0 \rho$ S/m, where σ_0 is a constant. The electric field intensity, constant throughout the wire volume, is given by $\mathbf{E} = E_0 \, \mathbf{a}_z$ V/m. (a) Find $\mathbf{J}$ in terms of E_0, σ_0, and ρ. (b) Find E_0 in terms of I. (c) Express $\mathbf{J}$ in terms of I and other parameters, excluding E_0. (d) Using your part c result, find $\mathbf{H}$ inside and outside the wire. (e) Verify your part d result by applying $\nabla \times \mathbf{H} = \mathbf{J}$.

7.18 A wire of 3 mm radius is made up of an inner material ($0 < \rho < 2$ mm) for which $\sigma = 10^7$ S/m, and an outer material (2 mm $< \rho < 3$ mm) for which $\sigma = 4 \times 10^7$ S/m. If the wire carries a total current of 100 mA dc, determine $\mathbf{H}$ everywhere as a function of ρ.

7.19 In spherical coordinates, the surface of a solid conducting cone is described by $\theta = \pi/4$ and a conducting plane by $\theta = \pi/2$. Each carries a total current I. The current flows as a surface current radially inward on the plane to the vertex of the cone, and then flows radially outward throughout the cross section of the conical conductor. (a) Express the surface current density as a function of r. (b) Express the volume current density inside the cone as a function of r. (c) Determine $\mathbf{H}$ as a function of r and θ in the region between the cone and the plane. (d) Determine $\mathbf{H}$ as a function of r and θ inside the cone.

7.20 A solid conductor of circular cross section with a radius of 5 mm has a conductivity that varies with radius. The conductor is 20 m long, and there is a potential difference of 0.1 V dc between its two ends. Within the conductor, $\mathbf{H} = 10^5 \rho^2 \, \mathbf{a}_\phi$ A/m. (a) Find σ as a function of ρ. (b) What is the resistance between the two ends?

7.21 A cylindrical wire of radius a is oriented with the z axis down its center line. The wire carries a nonuniform current down its length of density $\mathbf{J} = b\rho$ $\mathbf{a}_z$ A/m^2 where b is a constant. (a) What total current flows in the wire? (b) Find $\mathbf{H}_{\text{in}}$ ($0 < \rho < a$) as a function of ρ; (c) find $\mathbf{H}_{\text{out}}$ ($\rho > a$), as a function of ρ; (d) verify your results of parts (b) and (c) by using $\nabla \times \mathbf{H} = \mathbf{J}$.

7.22 A solid cylinder of radius a and length L, where $L \gg a$, contains volume charge of uniform density ρ_0 C/m^3. The cylinder rotates about its axis (the z axis) at angular velocity Ω rad/s. (a) Determine the current density $\mathbf{J}$ as a function of position within the rotating cylinder. (b) Determine $\mathbf{H}$ on-axis by applying the results of Problem 7.6. (c) Determine the magnetic field intensity $\mathbf{H}$ inside and outside. (d) Check your result for part c by taking the curl of $\mathbf{H}$.

7.23 A solid conductor is in the shape of a circular cylinder with its axis along the z axis. A uniform electric field $\mathbf{E} = E_0\,\mathbf{a}_z$ V/m is applied, resulting in a *uniform* magnetic field inside the conductor, $\mathbf{H} = H_0\,\mathbf{a}_\phi$, where E_0 and H_0 are constants. (*a*) Using the cylindrical coordinate form of $\nabla \times \mathbf{H} = \mathbf{J}$, determine the required functional form of the medium conductivity, σ, in order for the preceding condition to be met. (*b*) Repeat part *a*, but use the spherical coordinate form of $\nabla \times \mathbf{H} = \mathbf{J}$.

7.24 Infinitely long filamentary conductors are located in the $y = 0$ plane at $x = n$ meters where $n = 0, \pm1, \pm2, \dots$ Each carries 1 A in the $\mathbf{a}_z$ direction. (*a*) Find $\mathbf{H}$ on the y axis. As a help,

$$\sum_{n=1}^{\infty} \frac{y}{y^2 + n^2} = \frac{\pi}{2} - \frac{1}{2y} + \frac{\pi}{e^{2\pi y} - 1}$$

(*b*) Compare your result for part *a* to that obtained if the filaments are replaced by a current sheet in the $y = 0$ plane that carries surface current density $\mathbf{K} = 1\mathbf{a}_z$ A/m.

7.25 When x, y, and z are positive and less than 5, a certain magnetic field intensity may be expressed as $\mathbf{H} = [x^2yz/(y + 1)]\mathbf{a}_x + 3x^2 z^2 \mathbf{a}_y - [xyz^2/(y + 1)]\mathbf{a}_z$. Find the total current in the $\mathbf{a}_x$ direction that crosses the strip $x = 2$, $1 \le y \le 4$, $3 \le z \le 4$, by a method utilizing: (*a*) a surface integral; (*b*) a closed line integral.

7.26 Consider a sphere of radius $r = 4$ centered at $(0, 0, 3)$. Let S_1 be that portion of the spherical surface that lies above the xy plane. Find $\int_{S_1} (\nabla \times \mathbf{H}) \cdot d\mathbf{S}$ if $\mathbf{H} = 3\rho\,\mathbf{a}_\phi$ in cylindrical coordinates.

7.27 The magnetic field intensity is given in a certain region of space as $\mathbf{H} = [(x + 2y)/z^2]\mathbf{a}_y + (2/z)\mathbf{a}_z$ A/m. (*a*) Find $\nabla \times \mathbf{H}$. (*b*) Find $\mathbf{J}$. (*c*) Use $\mathbf{J}$ to find the total current passing through the surface $z = 4$, $1 \le x \le 2$, $3 \le z \le 5$, in the $\mathbf{a}_z$ direction. (*d*) Show that the same result is obtained using the other side of Stokes' theorem.

7.28 Given $\mathbf{H} = (3r^2/\sin\theta)\mathbf{a}_\theta + 54r\cos\theta\mathbf{a}_\phi$ A/m in free space: (*a*) Find the total current in the $\mathbf{a}_\theta$ direction through the conical surface $\theta = 20°$, $0 \le \phi \le 2\pi$, $0 \le r \le 5$, by whatever side of Stokes' theorem you like the best. (*b*) Check the result by using the other side of Stokes' theorem.

7.29 A long, straight, nonmagnetic conductor of 0.2 mm radius carries a uniformly distributed current of 2 A dc. (*a*) Find $\mathbf{J}$ within the conductor. (*b*) Use Ampère's circuital law to find $\mathbf{H}$ and $\mathbf{B}$ within the conductor. (*c*) Show that $\nabla \times \mathbf{H} = \mathbf{J}$ within the conductor. (*d*) Find $\mathbf{H}$ and $\mathbf{B}$ outside the conductor. (*e*) Show that $\nabla \times \mathbf{H} = \mathbf{J}$ outside the conductor.

7.30 (An inversion of Problem 7.20.) A solid, nonmagnetic conductor of circular cross section has a radius of 2 mm. The conductor is inhomogeneous, with $\sigma = 10^6 (1 + 10^6 \rho^2)$ S/m. If the conductor is 1 m in length and has a voltage of 1 mV between its ends, find: (a) **H** inside; (b) the total magnetic flux inside the conductor.

7.31 The cylindrical shell defined by 1 cm $< \rho <$ 1.4 cm consists of a nonmagnetic conducting material and carries a total current of 50 A in the $\mathbf{a}_z$ direction. Find the total magnetic flux crossing the plane $\phi = 0$, $0 < z < 1$: (a) $0 < \rho <$ 1.2 cm; (b) 1.0 cm $< \rho <$ 1.4 cm; (c) 1.4 cm $< \rho <$ 20 cm.

7.32 The free space region defined by $1 < z < 4$ cm and $2 < \rho < 3$ cm is a toroid of rectangular cross section. Let the surface at $\rho = 3$ cm carry a surface current $\mathbf{K} = 2\mathbf{a}_z$ kA/m. (a) Specify the current densities on the surfaces at $\rho = 2$ cm, $z = 1$ cm, and $z = 4$ cm. (b) Find **H** everywhere. (c) Calculate the total flux within the toroid.

7.33 Use an expansion in rectangular coordinates to show that the curl of the gradient of any scalar field G is identically equal to zero.

7.34 A filamentary conductor on the z axis carries a current of 16 A in the $\mathbf{a}_z$ direction, a conducting shell at $\rho = 6$ carries a total current of 12 A in the $-\mathbf{a}_z$ direction, and another shell at $\rho = 10$ carries a total current of 4 A in the $-\mathbf{a}_z$ direction. (a) Find **H** for $0 < \rho < 12$. (b) Plot H_ϕ versus ρ. (c) Find the total flux Φ crossing the surface $1 < \rho < 7$, $0 < z < 1$, at fixed ϕ.

7.35 A current sheet, $\mathbf{K} = 20\,\mathbf{a}_z$ A/m, is located at $\rho = 2$, and a second sheet, $\mathbf{K} = -10\mathbf{a}_z$ A/m, is located at $\rho = 4$. (a) Let $V_m = 0$ at $P(\rho = 3, \phi = 0, z = 5)$ and place a barrier at $\phi = \pi$. Find $V_m(\rho, \phi, z)$ for $-\pi < \phi < \pi$. (b) Let $\mathbf{A} = 0$ at P and find $\mathbf{A}(\rho, \phi, z)$ for $2 < \rho < 4$.

7.36 Let $\mathbf{A} = (3y - z)\mathbf{a}_x + 2xz\mathbf{a}_y$ Wb/m in a certain region of free space. (a) Show that $\nabla \cdot \mathbf{A} = 0$. (b) At $P(2, -1, 3)$, find **A**, **B**, **H**, and **J**.

7.37 Let $N = 1000$, $I = 0.8$ A, $\rho_0 = 2$ cm, and $a = 0.8$ cm for the toroid shown in Figure 7.12b. Find V_m in the interior of the toroid if $V_m = 0$ at $\rho = 2.5$ cm, $\phi = 0.3\pi$. Keep ϕ within the range $0 < \phi < 2\pi$.

7.38 A square filamentary differential current loop, dL on a side, is centered at the origin in the $z = 0$ plane in free space. The current I flows generally in the $\mathbf{a}_\phi$ direction. (a) Assuming that $r \gg dL$, and following a method similar to that in Section 4.7, show that

$$d\mathbf{A} = \frac{\mu_0 I (dL)^2 \sin\theta}{4\pi r^2} \mathbf{a}_\phi$$

(b) Show that

$$d\mathbf{H} = \frac{I(dL)^2}{4\pi r^3}(2 \cos\theta\, \mathbf{a}_r + \sin\theta\, \mathbf{a}_\theta)$$

The square loop is one form of a *magnetic dipole*.

7.39 Planar current sheets of $\mathbf{K} = 30\mathbf{a}_z$ A/m and $-30\mathbf{a}_z$ A/m are located in free space at $x = 0.2$ and $x = -0.2$, respectively. For the region $-0.2 < x < 0.2$

(*a*) find **H**; (*b*) obtain an expression for V_m if $V_m = 0$ at $P(0.1, 0.2, 0.3)$; (*c*) find **B**; (*d*) obtain an expression for **A** if **A** $= 0$ at P.

7.40 Show that the line integral of the vector potential **A** about any closed path is equal to the magnetic flux enclosed by the path, or $\oint \mathbf{A} \cdot d\mathbf{L} = \int \mathbf{B} \cdot d\mathbf{S}$.

7.41 (*a*) Demonstrate the principle of Problem 7.40 using the vector potential in a coaxial cable, as given in Eq. (66). (*b*) Use the principle to find the vector magnetic potential in a long solenoid of radius a, having n turns per meter of length, in which each turn carries current I. Assume a long coil so that **B** is approximately $\mu_0 nI\, \mathbf{a}_z$ Wb/m², uniform over the coil cross section.

7.42 Show that $\nabla_2(1/R_{12}) = -\nabla_1(1/R_{12}) = \mathbf{R}_{21}/R_{12}^3$.

7.43 Compute the vector magnetic potential within the outer conductor for the coaxial line whose vector magnetic potential is shown in Figure 7.20 if the outer radius of the outer conductor is $7a$. Select the proper zero reference and sketch the results on the figure.

7.44 By expanding Eq. (58), Section 7.7 in rectangular coordinates, show that (59) is correct.

CHAPTER 8 Magnetic Forces, Materials, and Inductance

术语

inductance　电感，单位是亨［利］，简称亨（H）；电感是储存磁场能的元件，属于磁场内容范畴。

magnetization　磁化强度。

isotropic　各向同性的，（磁导率）参数在各个方向上呈现出来的值相同。

anisotropic　各向异性的。

magnetic susceptibility　磁化率。

ferromagnetic material　铁磁材料，相对磁导率很大（如电工软铁、用作变压器铁心的硅钢）。

magnetization curve　磁化曲线，表明材料中的磁通密度与磁场强度的关系为非线性关系。

hysteresis loop　磁滞回线，表明磁性材料具有历史记忆性。

saturation　饱和，这里是指当材料中的磁场强度 **H** 增加到一定程度时，**H** 仍继续增加而磁通密度 **H** 增加越来越缓慢的现象。

mutual inductance　互感，两个线圈之间存在互感；一个线圈存在自感（self-inductance），也称为电感。

flux linkages　磁链（意为磁通链）。

internal inductance　内自感，对应导体（导线）内部的磁链。

external inductance　外自感，对应导体外部的磁链。

We are now ready to undertake the second half of the magnetic field problem, that of determining the forces and torques exerted by magnetic fields on other charges. An electric field causes a force to be exerted on a charge that may be either stationary or in motion; we will see that a steady magnetic field is capable of exerting a force only on a *moving* charge. This result appears reasonable; a magnetic field may be produced by moving charges and may exert forces on moving charges; a magnetic field cannot arise from stationary charges and cannot exert any force on a stationary charge.

This chapter initially considers the forces and torques on current-carrying conductors that may either be of a filamentary nature or possess a finite cross section with a known current density distribution. The problems associated with the motion of particles in a vacuum are largely avoided.

With an understanding of the fundamental effects produced by the magnetic field, we may then consider the varied types of magnetic materials, the analysis of elementary magnetic circuits, the forces on magnetic materials, and finally, the important electrical circuit concepts of self-inductance and mutual inductance. ∎

8.1 FORCE ON A MOVING CHARGE

In an electric field, the definition of the electric field intensity shows us that the force on a charged particle is

$$\boxed{\mathbf{F} = Q\mathbf{E}} \tag{1}$$

The force is in the same direction as the electric field intensity (for a positive charge) and is directly proportional to both **E** and Q. If the charge is in motion, the force at any point in its trajectory is then given by (1).

A charged particle in motion in a magnetic field of flux density **B** is found experimentally to experience a force whose magnitude is proportional to the product of the magnitudes of the charge Q, its velocity **v**, and the flux density **B**, and to the sine of the angle between the vectors **v** and **B**. The direction of the force is perpendicular to both **v** and **B** and is given by a unit vector in the direction of $\mathbf{v} \times \mathbf{B}$. The force may therefore be expressed as

$$\boxed{\mathbf{F} = Q\mathbf{v} \times \mathbf{B}} \tag{2}$$

A fundamental difference in the effect of the electric and magnetic fields on charged particles is now apparent, for a force which is always applied in a direction at right angles to the direction in which the particle is proceeding can never change the magnitude of the particle velocity. In other words, the *acceleration* vector is always normal to the velocity vector. The kinetic energy of the particle remains unchanged, and it follows that the steady magnetic field is incapable of transferring energy to the moving charge. The electric field, on the other hand, exerts a force on the particle which is independent of the direction in which the particle is progressing and therefore effects an energy transfer between field and particle in general.

The first two problems at the end of this chapter illustrate the different effects of electric and magnetic fields on the kinetic energy of a charged particle moving in free space.

The force on a moving particle arising from combined electric and magnetic fields is obtained easily by superposition,

$$\boxed{\mathbf{F} = Q(\mathbf{E} + \mathbf{v} \times \mathbf{B})} \tag{3}$$

This equation is known as the *Lorentz force equation,* and its solution is required in determining electron orbits in the magnetron, proton paths in the cyclotron, plasma characteristics in a magnetohydrodynamic (MHD) generator, or, in general, charged-particle motion in combined electric and magnetic fields.

D8.1. The point charge $Q = 18$ nC has a velocity of 5×10^6 m/s in the direction $\mathbf{a}_v = 0.60\mathbf{a}_x + 0.75\mathbf{a}_y + 0.30\mathbf{a}_z$. Calculate the magnitude of the force exerted on the charge by the field: (a) $\mathbf{B} = -3\mathbf{a}_x + 4\mathbf{a}_y + 6\mathbf{a}_z$ mT; (b) $\mathbf{E} = -3\mathbf{a}_x + 4\mathbf{a}_y + 6\mathbf{a}_z$ kV/m; (c) **B** and **E** acting together.

Ans. (a) 660 μN; (b) 140 μN; (c) 670 μN

导读

1. 铁磁材料的一大特性是其非线性，是指铁磁材料内的磁通密度**B**与磁场强度**H**之间的关系呈非线性关系。表现为当**H**大到一定程度时，**H**仍继续增加而**B**增加越来越缓慢，即出现饱和现象。

2. 由于铁磁材料的**B-H**呈非线性关系，所以其磁导率不是常数，而是磁场强度**H**的函数，即施加在材料上的**H**不同则其磁导率就不同。特别是进入饱和区后，**H**继续增加而**B**增加越来越缓慢，所以等于**B / H**的磁导率会逐渐变小，铁磁材料会逐步失去高导磁特性。

3. 磁滞回线表明磁性材料具有历史记忆性，材料的**B-H**关系会因磁场是增加还是减小过程而不同。对于交流激励下的磁性材料，其工作磁特性是磁滞回线，而不是磁化曲线。磁化曲线只是为了避免考虑复杂的磁滞回线而对磁特性的近似。

4. 一个空心线圈的电感基本上是一个常数，仅与其结构有关，而与施加的电流大小无关。但是，对于铁心电感，由于铁磁材料的非线性，电感值与施加的电流有关，一般电流越大电感越小；仅在磁化曲线的线性范围内，可认为电感不随电流大小而改变。

5. 虽然电感的定义来自磁链，但应注意从磁场能量的角度理解电感与求解电感。

8.2 FORCE ON A DIFFERENTIAL CURRENT ELEMENT

The force on a charged particle moving through a steady magnetic field may be written as the differential force exerted on a differential element of charge,

$$d\mathbf{F} = dQ\,\mathbf{v} \times \mathbf{B} \tag{4}$$

Physically, the differential element of charge consists of a large number of very small, discrete charges occupying a volume which, although small, is much larger than the average separation between the charges. The differential force expressed by (4) is thus merely the sum of the forces on the individual charges. This sum, or resultant force, is not a force applied to a single object. In an analogous way, we might consider the differential gravitational force experienced by a small volume taken in a shower of falling sand. The small volume contains a large number of sand grains, and the differential force is the sum of the forces on the individual grains within the small volume.

If our charges are electrons in motion in a conductor, however, we can show that the force is transferred to the conductor and that the sum of this extremely large number of extremely small forces is of practical importance. Within the conductor, electrons are in motion throughout a region of immobile positive ions which form a crystalline array, giving the conductor its solid properties. A magnetic field which exerts forces on the electrons tends to cause them to shift position slightly and produces a small displacement between the centers of "gravity" of the positive and negative charges. The Coulomb forces between electrons and positive ions, however, tend to resist such a displacement. Any attempt to move the electrons, therefore, results in an attractive force between electrons and the positive ions of the crystalline lattice. The magnetic force is thus transferred to the crystalline lattice, or to the conductor itself. The Coulomb forces are so much greater than the magnetic forces in good conductors that the actual displacement of the electrons is almost immeasurable. The charge separation that does result, however, is disclosed by the presence of a slight potential difference across the conductor sample in a direction perpendicular to both the magnetic field and the velocity of the charges. The voltage is known as the *Hall voltage,* and the effect itself is called the *Hall effect.*

Figure 8.1 illustrates the direction of the Hall voltage for both positive and negative charges in motion. In Figure 8.1*a*, $\mathbf{v}$ is in the $-\mathbf{a}_x$ direction, $\mathbf{v} \times \mathbf{B}$ is in the $\mathbf{a}_y$ direction, and Q is positive, causing $\mathbf{F}_Q$ to be in the $\mathbf{a}_y$ direction; thus, the positive charges move to the right. In Figure 8.1*b*, $\mathbf{v}$ is now in the $+\mathbf{a}_x$ direction, $\mathbf{B}$ is still in the $\mathbf{a}_z$ direction, $\mathbf{v} \times \mathbf{B}$ is in the $-\mathbf{a}_y$ direction, and Q is negative; thus, $\mathbf{F}_Q$ is again in the $\mathbf{a}_y$ direction. Hence, the negative charges end up at the right edge. Equal currents provided by holes and electrons in semiconductors can therefore be differentiated by their Hall voltages. This is one method of determining whether a given semiconductor is *n*-type or *p*-type.

Devices employ the Hall effect to measure the magnetic flux density and, in some applications where the current through the device can be made proportional

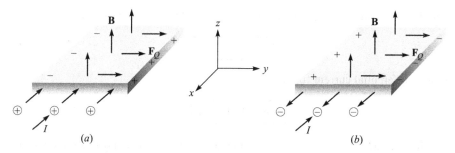

Figure 8.1 Equal currents directed into the material are provided by positive charges moving inward in (a) and negative charges moving outward in (b). The two cases can be distinguished by oppositely directed Hall voltages, as shown.

to the magnetic field across it, to serve as electronic wattmeters, squaring elements, and so forth.

Returning to (4), we may therefore say that if we are considering an element of moving charge in an electron beam, the force is merely the sum of the forces on the individual electrons in that small volume element, but if we are considering an element of moving charge within a conductor, the total force is applied to the solid conductor itself. We will now limit our attention to the forces on current-carrying conductors.

In Chapter 5 we defined convection current density in terms of the velocity of the volume charge density,

$$\boxed{\mathbf{J} = \rho_v \mathbf{v}}$$

The differential element of charge in (4) may also be expressed in terms of volume charge density,[1]

$$dQ = \rho_v \, dv$$

Thus

$$d\mathbf{F} = \rho_v \, dv \, \mathbf{v} \times \mathbf{B}$$

or

$$\boxed{d\mathbf{F} = \mathbf{J} \times \mathbf{B} \, dv} \tag{5}$$

We saw in Chapter 7 that $\mathbf{J} \, dv$ may be interpreted as a differential current element; that is,

$$\mathbf{J} \, dv = \mathbf{K} \, dS = I \, d\mathbf{L}$$

---

[1] Remember that dv is a differential volume element and not a differential increase in velocity.

and thus the Lorentz force equation may be applied to surface current density,

$$\boxed{d\mathbf{F} = \mathbf{K} \times \mathbf{B}\, dS}$$

(6)

or to a differential current filament,

$$\boxed{d\mathbf{F} = I\, d\mathbf{L} \times \mathbf{B}}$$

(7)

Integrating (5), (6), or (7) over a volume, a surface which may be either open or closed (why?), or a closed path, respectively, leads to the integral formulations

$$\mathbf{F} = \int_{vol} \mathbf{J} \times \mathbf{B}\, dv$$

(8)

$$\mathbf{F} = \int_{S} \mathbf{K} \times \mathbf{B}\, dS$$

(9)

and

$$\boxed{\mathbf{F} = \oint I\, d\mathbf{L} \times \mathbf{B} = -I \oint \mathbf{B} \times d\mathbf{L}}$$

(10)

One simple result is obtained by applying (7) or (10) to a straight conductor in a uniform magnetic field,

$$\boxed{\mathbf{F} = I\mathbf{L} \times \mathbf{B}}$$

(11)

The magnitude of the force is given by the familiar equation

$$F = BIL \sin\theta$$

(12)

where θ is the angle between the vectors representing the direction of the current flow and the direction of the magnetic flux density. Equation (11) or (12) applies only to a portion of the closed circuit, and the remainder of the circuit must be considered in any practical problem.

EXAMPLE 8.1

As a numerical example of these equations, consider Figure 8.2. We have a square loop of wire in the $z = 0$ plane carrying 2 mA in the field of an infinite filament on the y axis, as shown. We wish to find the total force on the loop.

Solution. The field produced in the plane of the loop by the straight filament is

$$\mathbf{H} = \frac{I}{2\pi x}\mathbf{a}_z = \frac{15}{2\pi x}\mathbf{a}_z \text{ A/m}$$

Therefore,

$$\mathbf{B} = \mu_0 \mathbf{H} = 4\pi \times 10^{-7}\mathbf{H} = \frac{3 \times 10^{-6}}{x}\mathbf{a}_z \text{ T}$$

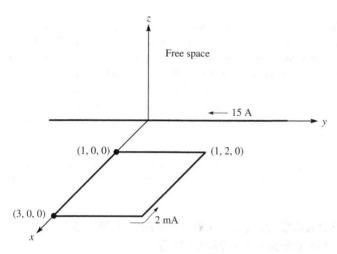

Figure 8.2 A square loop of wire in the *xy* plane carrying 2 mA is subjected to a nonuniform **B** field.

We use the integral form (10),

$$\mathbf{F} = -I \oint \mathbf{B} \times d\mathbf{L}$$

Let us assume a rigid loop so that the total force is the sum of the forces on the four sides. Beginning with the left side:

$$\mathbf{F} = -2 \times 10^{-3} \times 3 \times 10^{-6} \left[\int_{x=1}^{3} \frac{\mathbf{a}_z}{x} \times dx\, \mathbf{a}_x + \int_{y=0}^{2} \frac{\mathbf{a}_z}{3} \times dy\, \mathbf{a}_y \right.$$

$$\left. + \int_{x=3}^{1} \frac{\mathbf{a}_z}{x} \times dx\, \mathbf{a}_x + \int_{y=2}^{0} \frac{\mathbf{a}_z}{1} \times dy\, \mathbf{a}_y \right]$$

$$= -6 \times 10^{-9} \left[\ln x \Big|_1^3 \mathbf{a}_y + \frac{1}{3} y \Big|_0^2 (-\mathbf{a}_x) + \ln x \Big|_3^1 \mathbf{a}_y + y \Big|_2^0 (-\mathbf{a}_x) \right]$$

$$= -6 \times 10^{-9} \left[(\ln 3)\, \mathbf{a}_y - \frac{2}{3} \mathbf{a}_x + \left(\ln \frac{1}{3} \right) \mathbf{a}_y + 2\mathbf{a}_x \right]$$

$$= -8\mathbf{a}_x \text{ nN}$$

Thus, the net force on the loop is in the $-\mathbf{a}_x$ direction.

D8.2. The field $\mathbf{B} = -2\mathbf{a}_x + 3\mathbf{a}_y + 4\mathbf{a}_z$ mT is present in free space. Find the vector force exerted on a straight wire carrying 12 A in the $\mathbf{a}_{AB}$ direction, given $A(1, 1, 1)$ and: (*a*) $B(2, 1, 1)$; (*b*) $B(3, 5, 6)$.

Ans. (*a*) $-48\mathbf{a}_y + 36\mathbf{a}_z$ mN; (*b*) $12\mathbf{a}_x - 216\mathbf{a}_y + 168\mathbf{a}_z$ mN

> **D8.3.** The semiconductor sample shown in Figure 8.1 is n-type silicon, having a rectangular cross section of 0.9 mm by 1.1 cm and a length of 1.3 cm. Assume the electron and hole mobilities are 0.13 and 0.03 $m^2/V \cdot s$, respectively, at the operating temperature. Let $B = 0.07$ T and the electric field intensity in the direction of the current flow be 800 V/m. Find the magnitude of: (a) the voltage across the sample length; (b) the drift velocity; (c) the transverse force per coulomb of moving charge caused by B; (d) the transverse electric field intensity; (e) the Hall voltage.
>
> **Ans.** (a) 10.40 V; (b) 104.0 m/s; (c) 7.28 N/C; (d) 7.28 V/m; (e) 80.1 mV

8.3 FORCE BETWEEN DIFFERENTIAL CURRENT ELEMENTS

The concept of the magnetic field was introduced to break into two parts the problem of finding the interaction of one current distribution on a second current distribution. It is possible to express the force on one current element directly in terms of a second current element without finding the magnetic field. Because we claimed that the magnetic-field concept simplifies our work, it then behooves us to show that avoidance of this intermediate step leads to more complicated expressions.

The magnetic field at point 2 due to a current element at point 1 was found to be

$$d\mathbf{H}_2 = \frac{I_1 d\mathbf{L}_1 \times \mathbf{a}_{R12}}{4\pi R_{12}^2}$$

Now, the differential force on a differential current element is

$$d\mathbf{F} = I\,d\mathbf{L} \times \mathbf{B}$$

and we apply this to our problem by letting $\mathbf{B}$ be $d\mathbf{B}_2$ (the differential flux density at point 2 caused by current element 1), by identifying $I\,d\mathbf{L}$ as $I_2 d\mathbf{L}_2$, and by symbolizing the differential amount of our differential force on element 2 as $d(d\mathbf{F}_2)$:

$$d(d\mathbf{F}_2) = I_2 d\mathbf{L}_2 \times d\mathbf{B}_2$$

Because $d\mathbf{B}_2 = \mu_0 d\mathbf{H}_2$, we obtain the force between two differential current elements,

$$d(d\mathbf{F}_2) = \mu_0 \frac{I_1 I_2}{4\pi R_{12}^2} d\mathbf{L}_2 \times (d\mathbf{L}_1 \times \mathbf{a}_{R12}) \tag{13}$$

EXAMPLE 8.2

As an example that illustrates the use (and misuse) of these results, consider the two differential current elements shown in Figure 8.3. We seek the differential force on $d\mathbf{L}_2$.

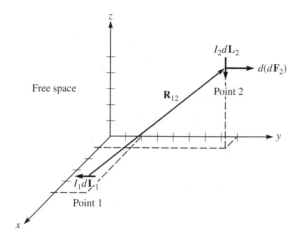

Figure 8.3 Given $P_1(5, 2, 1)$, $P_2(1, 8, 5)$, $I_1\,d\mathbf{L}_1 = -3\mathbf{a}_y$ A · m, and $I_2\,d\mathbf{L}_2 = -4\mathbf{a}_z$ A · m, the force on $I_2\,d\mathbf{L}_2$ is 8.56 nN in the $\mathbf{a}_y$ direction.

Solution. We have $I_1 d\mathbf{L}_1 = -3\mathbf{a}_y$A · m at $P_1(5, 2, 1)$, and $I_2 d\mathbf{L}_2 = -4\mathbf{a}_z$A · m at $P_2(1, 8, 5)$. Thus, $\mathbf{R}_{12} = -4\mathbf{a}_x + 6\mathbf{a}_y + 4\mathbf{a}_z$, and we may substitute these data into (13),

$$d(d\mathbf{F}_2) = \frac{4\pi\ 10^{-7}}{4\pi}\frac{(-4\mathbf{a}_z) \times [(-3\mathbf{a}_y) \times (-4\mathbf{a}_x + 6\mathbf{a}_y + 4\mathbf{a}_z)]}{(16 + 36 + 16)^{1.5}}$$

$$= 8.56\mathbf{a}_y\,\text{nN}$$

Many chapters ago, when we discussed the force exerted by one point charge on another point charge, we found that the force on the first charge was the negative of that on the second. That is, the total force on the system was zero. This is not the case with the differential current elements, and $d(d\mathbf{F}_1) = -12.84\mathbf{a}_z$ nN in Example 8.2. The reason for this different behavior lies with the nonphysical nature of the current element. Whereas point charges may be approximated quite well by small charges, the continuity of current demands that a complete circuit be considered. This we shall now do.

The total force between two filamentary circuits is obtained by integrating twice:

$$\mathbf{F}_2 = \mu_0\frac{I_1 I_2}{4\pi} \oint \left[d\mathbf{L}_2 \times \oint \frac{d\mathbf{L}_1 \times \mathbf{a}_{R12}}{R_{12}^2} \right]$$

$$= \mu_0\frac{I_1 I_2}{4\pi} \oint \left[\oint \frac{\mathbf{a}_{R12} \times d\mathbf{L}_1}{R_{12}^2} \right] \times d\mathbf{L}_2 \tag{14}$$

Equation (14) is quite formidable, but the familiarity gained in Chapter 7 with the magnetic field should enable us to recognize the inner integral as the integral necessary to find the magnetic field at point 2 due to the current element at point 1.

Although we shall only give the result, it is not very difficult to use (14) to find the force of repulsion between two infinitely long, straight, parallel, filamentary

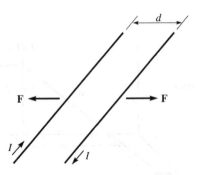

Figure 8.4 Two infinite parallel filaments with separation d and equal but opposite currents I experience a repulsive force of $\mu_0 I^2/(2\pi d)$ N/m.

conductors with separation d, and carrying equal but opposite currents I, as shown in Figure 8.4. The integrations are simple, and most errors are made in determining suitable expressions for $\mathbf{a}_{R12}$, $d\mathbf{L}_1$, and $d\mathbf{L}_2$. However, since the magnetic field intensity at either wire caused by the other is already known to be $I/(2\pi d)$, it is readily apparent that the answer is a force of $\mu_0 I^2/(2\pi d)$ newtons per meter length.

D8.4. Two differential current elements, $I_1 \Delta \mathbf{L}_1 = 3 \times 10^{-6} \mathbf{a}_y$ A · m at $P_1(1, 0, 0)$ and $I_2 \Delta \mathbf{L}_2 = 3 \times 10^{-6}(-0.5\mathbf{a}_x + 0.4\mathbf{a}_y + 0.3\mathbf{a}_z)$ A · m at $P_2(2, 2, 2)$, are located in free space. Find the vector force exerted on: (a) $I_2 \Delta \mathbf{L}_2$ by $I_1 \Delta \mathbf{L}_1$; (b) $I_1 \Delta \mathbf{L}_1$ by $I_2 \Delta \mathbf{L}_2$.

Ans. (a) $(-1.333\mathbf{a}_x + 0.333\mathbf{a}_y - 2.67\mathbf{a}_z) 10^{-20}$ N; (b) $(4.67\mathbf{a}_x + 0.667\mathbf{a}_z)10^{-20}$ N

8.4 FORCE AND TORQUE ON A CLOSED CIRCUIT

We have already obtained general expressions for the forces exerted on current systems. One special case is easily disposed of, for if we take our relationship for the force on a filamentary closed circuit, as given by Eq. (10), Section 8.2,

$$\mathbf{F} = -I \oint \mathbf{B} \times d\mathbf{L}$$

and assume a *uniform* magnetic flux density, then $\mathbf{B}$ may be removed from the integral:

$$\mathbf{F} = -I \mathbf{B} \times \oint d\mathbf{L}$$

However, we discovered during our investigation of closed line integrals in an electrostatic potential field that $\oint d\mathbf{L} = 0$, and therefore the force on a closed filamentary circuit in a uniform magnetic field is zero.

If the field is not uniform, the total force need not be zero.

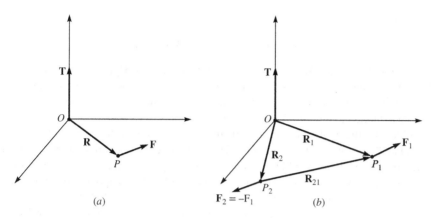

Figure 8.5 (a) Given a lever arm **R** extending from an origin O to a point P where force **F** is applied, the torque about O is **T = R × F**. (b) If **F**$_2$ = −**F**$_1$, then the torque **T = R**$_{21}$ × **F**$_1$ is independent of the choice of origin for **R**$_1$ and **R**$_2$.

This result for uniform fields does not have to be restricted to filamentary circuits only. The circuit may contain surface currents or volume current density as well. If the total current is divided into filaments, the force on each one is zero, as we have shown, and the total force is again zero. Therefore, any real closed circuit carrying direct currents experiences a total vector force of zero in a uniform magnetic field.

Although the force is zero, the torque is generally not equal to zero.

In defining the *torque,* or *moment,* of a force, it is necessary to consider both an origin at or about which the torque is to be calculated, and the point at which the force is applied. In Figure 8.5a, we apply a force **F** at point P, and we establish an origin at O with a rigid lever arm **R** extending from O to P. The torque about point O is a vector whose magnitude is the product of the magnitudes of **R**, of **F**, and of the sine of the angle between these two vectors. The direction of the vector torque **T** is normal to both the force **F** and the lever arm **R** and is in the direction of progress of a right-handed screw as the lever arm is rotated into the force vector through the smaller angle. The torque is expressible as a cross product,

$$\mathbf{T} = \mathbf{R} \times \mathbf{F}$$

Now assume that two forces, **F**$_1$ at P_1 and **F**$_2$ at P_2, having lever arms **R**$_1$ and **R**$_2$ extending from a common origin O, as shown in Figure 8.5b, are applied to an object of fixed shape and that the object does not undergo any translation. Then the torque about the origin is

$$\mathbf{T} = \mathbf{R}_1 \times \mathbf{F}_1 + \mathbf{R}_2 \times \mathbf{F}_2$$

where

$$\mathbf{F}_1 + \mathbf{F}_2 = 0$$

and therefore

$$\mathbf{T} = (\mathbf{R}_1 - \mathbf{R}_2) \times \mathbf{F}_1 = \mathbf{R}_{21} \times \mathbf{F}_1$$

The vector $\mathbf{R}_{21} = \mathbf{R}_1 - \mathbf{R}_2$ joins the point of application of $\mathbf{F}_2$ to that of $\mathbf{F}_1$ and is independent of the choice of origin for the two vectors $\mathbf{R}_1$ and $\mathbf{R}_2$. Therefore, the torque is also independent of the choice of origin, provided that the total force is zero. This may be extended to any number of forces.

Consider the application of a vertically upward force at the end of a horizontal crank handle on an elderly automobile. This cannot be the only applied force, for if it were, the entire handle would be accelerated in an upward direction. A second force, equal in magnitude to that exerted at the end of the handle, is applied in a downward direction by the bearing surface at the axis of rotation. For a 40-N force on a crank handle 0.3 m in length, the torque is 12 N · m. This figure is obtained regardless of whether the origin is considered to be on the axis of rotation (leading to 12 N · m plus 0 N · m), at the midpoint of the handle (leading to 6 N · m plus 6 N · m), or at some point not even on the handle or an extension of the handle.

We may therefore choose the most convenient origin, and this is usually on the axis of rotation and in the plane containing the applied forces if the several forces are coplanar.

With this introduction to the concept of torque, we now consider the torque on a differential current loop in a magnetic field $\mathbf{B}$. The loop lies in the xy plane (Figure 8.6); the sides of the loop are parallel to the x and y axes and are of length dx and dy. The value of the magnetic field at the center of the loop is taken as $\mathbf{B}_0$.

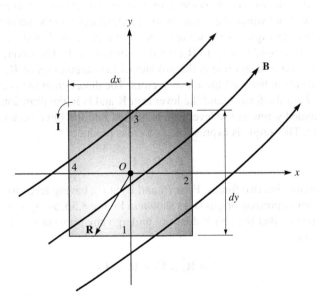

Figure 8.6   A differential current loop in a magnetic field $\mathbf{B}$. The torque on the loop is $d\mathbf{T} = I(dx\, dy\mathbf{a}_z) \times \mathbf{B}_0 = I\, d\mathbf{S} \times \mathbf{B}$.

Since the loop is of differential size, the value of **B** at all points on the loop may be taken as $\mathbf{B}_0$. (Why was this not possible in the discussion of curl and divergence?) The total force on the loop is therefore zero, and we are free to choose the origin for the torque at the center of the loop.

The vector force on side 1 is

$$d\mathbf{F}_1 = I \, dx \, \mathbf{a}_x \times \mathbf{B}_0$$

or

$$d\mathbf{F}_1 = I \, dx (B_{0y} \mathbf{a}_z - B_{0z} \mathbf{a}_y)$$

For this side of the loop the lever arm **R** extends from the origin to the midpoint of the side, $\mathbf{R}_1 = -\frac{1}{2} dy \, \mathbf{a}_y$, and the contribution to the total torque is

$$\begin{aligned}
d\mathbf{T}_1 &= \mathbf{R}_1 \times d\mathbf{F}_1 \\
&= -\tfrac{1}{2} dy \, \mathbf{a}_y \times I \, dx (B_{0y} \mathbf{a}_z - B_{0z} \mathbf{a}_y) \\
&= -\tfrac{1}{2} dx \, dy \, I B_{0y} \mathbf{a}_x
\end{aligned}$$

The torque contribution on side 3 is found to be the same,

$$\begin{aligned}
d\mathbf{T}_3 = \mathbf{R}_3 \times d\mathbf{F}_3 &= \tfrac{1}{2} dy \, \mathbf{a}_y \times (-I \, dx \, \mathbf{a}_x \times \mathbf{B}_0) \\
&= -\tfrac{1}{2} dx \, dy \, I B_{0y} \mathbf{a}_x = d\mathbf{T}_1
\end{aligned}$$

and

$$d\mathbf{T}_1 + d\mathbf{T}_3 = -dx \, dy \, I B_{0y} \mathbf{a}_x$$

Evaluating the torque on sides 2 and 4, we find

$$d\mathbf{T}_2 + d\mathbf{T}_4 = dx \, dy \, I B_{0x} \mathbf{a}_y$$

and the total torque is then

$$d\mathbf{T} = I \, dx \, dy (B_{0x} \mathbf{a}_y - B_{0y} \mathbf{a}_x)$$

The quantity within the parentheses may be represented by a cross product,

$$d\mathbf{T} = I \, dx \, dy (\mathbf{a}_z \times \mathbf{B}_0)$$

or

$$\boxed{d\mathbf{T} = I \, d\mathbf{S} \times \mathbf{B}} \tag{15}$$

where $d\mathbf{S}$ is the vector area of the differential current loop and the subscript on $\mathbf{B}_0$ has been dropped.

We now define the product of the loop current and the vector area of the loop as the differential *magnetic dipole moment* $d\mathbf{m}$, with units of $\text{A} \cdot \text{m}^2$. Thus

$$\boxed{d\mathbf{m} = I \, d\mathbf{S}} \tag{16}$$

and

$$dT = d\mathbf{m} \times \mathbf{B} \qquad (17)$$

If we extend the results we obtained in Section 4.7 for the differential *electric* dipole by determining the torque produced on it by an *electric* field, we see a similar result,

$$d\mathbf{T} = d\mathbf{p} \times \mathbf{E}$$

Equations (15) and (17) are general results that hold for differential loops of any shape, not just rectangular ones. The torque on a circular or triangular loop is also given in terms of the vector surface or the moment by (15) or (17).

Because we selected a differential current loop so that we might assume **B** was constant throughout it, it follows that the torque on a *planar* loop of any size or shape in a *uniform* magnetic field is given by the same expression,

$$\mathbf{T} = I\mathbf{S} \times \mathbf{B} = \mathbf{m} \times \mathbf{B} \qquad (18)$$

We should note that the torque on the current loop always tends to turn the loop so as to align the magnetic field produced by the loop with the applied magnetic field that is causing the torque. This is perhaps the easiest way to determine the direction of the torque.

EXAMPLE 8.3

To illustrate some force and torque calculations, consider the rectangular loop shown in Figure 8.7. Calculate the torque by using $\mathbf{T} = I\mathbf{S} \times \mathbf{B}$.

Solution. The loop has dimensions of 1 m by 2 m and lies in the uniform field $\mathbf{B}_0 = -0.6\mathbf{a}_y + 0.8\mathbf{a}_z$ T. The loop current is 4 mA, a value that is sufficiently small to avoid causing any magnetic field that might affect $\mathbf{B}_0$.

We have

$$\mathbf{T} = 4 \times 10^{-3}[(1)(2)\mathbf{a}_z] \times (-0.6\mathbf{a}_y + 0.8\mathbf{a}_z) = 4.8\mathbf{a}_x \text{ mN} \cdot \text{m}$$

Thus, the loop tends to rotate about an axis parallel to the positive x axis. The small magnetic field produced by the 4 mA loop current tends to line up with $\mathbf{B}_0$.

EXAMPLE 8.4

Now let us find the torque once more, this time by calculating the total force and torque contribution for each side.

Solution. On side 1 we have

$$\mathbf{F}_1 = I\mathbf{L}_1 \times \mathbf{B}_0 = 4 \times 10^{-3}(1\mathbf{a}_x) \times (-0.6\mathbf{a}_y + 0.8\mathbf{a}_z)$$
$$= -3.2\mathbf{a}_y - 2.4\mathbf{a}_z \text{ mN}$$

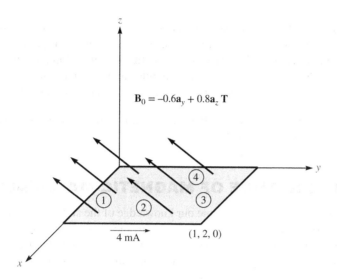

Figure 8.7 A rectangular loop is located in a uniform magnetic flux density $\mathbf{B}_0$.

On side 3 we obtain the negative of this result,

$$\mathbf{F}_3 = 3.2\mathbf{a}_y + 2.4\mathbf{a}_z \text{ mN}$$

Next, we attack side 2:

$$\mathbf{F}_2 = I\mathbf{L}_2 \times \mathbf{B}_0 = 4 \times 10^{-3}(2\mathbf{a}_y) \times (-0.6\mathbf{a}_y + 0.8\mathbf{a}_z)$$
$$= 6.4\,\mathbf{a}_x \text{ mN}$$

with side 4 again providing the negative of this result,

$$\mathbf{F}_4 = -6.4\mathbf{a}_x \text{ mN}$$

Because these forces are distributed uniformly along each of the sides, we treat each force as if it were applied at the center of the side. The origin for the torque may be established anywhere since the sum of the forces is zero, and we choose the center of the loop. Thus,

$$\mathbf{T} = \mathbf{T}_1 + \mathbf{T}_2 + \mathbf{T}_3 + \mathbf{T}_4 = \mathbf{R}_1 \times \mathbf{F}_1 + \mathbf{R}_2 \times \mathbf{F}_2 + \mathbf{R}_3 \times \mathbf{F}_3 + \mathbf{R}_4 \times \mathbf{F}_4$$
$$= (-1\mathbf{a}_y) \times (-3.2\mathbf{a}_y - 2.4\mathbf{a}_z) + (0.5\mathbf{a}_x) \times (6.4\mathbf{a}_x)$$
$$+(1\mathbf{a}_y) \times (3.2\mathbf{a}_y + 2.4\mathbf{a}_z) + (-0.5\mathbf{a}_x) \times (-6.4\mathbf{a}_x)$$
$$= 2.4\mathbf{a}_x + 2.4\mathbf{a}_x = 4.8\mathbf{a}_x \text{ mN} \cdot \text{m}$$

Crossing the loop moment with the magnetic flux density is certainly easier.

D8.5. A conducting filamentary triangle joins points $A(3, 1, 1)$, $B(5, 4, 2)$, and $C(1, 2, 4)$. The segment AB carries a current of 0.2 A in the $\mathbf{a}_{AB}$ direction. There is present a magnetic field $\mathbf{B} = 0.2\mathbf{a}_x - 0.1\mathbf{a}_y + 0.3\mathbf{a}_z$ T. Find: (a) the force on segment BC; (b) the force on the triangular loop; (c) the torque on the loop about an origin at A; (d) the torque on the loop about an origin at C.

Ans. (a) $-0.08\mathbf{a}_x + 0.32\mathbf{a}_y + 0.16\mathbf{a}_z$ N; (b) 0; (c) $-0.16\mathbf{a}_x - 0.08\mathbf{a}_y + 0.08\mathbf{a}_z$ N · m; (d) $-0.16\mathbf{a}_x - 0.08\mathbf{a}_y + 0.08\mathbf{a}_z$ N · m

8.5 THE NATURE OF MAGNETIC MATERIALS

We are now in a position to combine our knowledge of the action of a magnetic field on a current loop with a simple model of an atom and obtain some appreciation of the difference in behavior of various types of materials in magnetic fields.

Although accurate quantitative results can only be predicted through the use of quantum theory, the simple atomic model, which assumes that there is a central positive nucleus surrounded by electrons in various circular orbits, yields reasonable quantitative results and provides a satisfactory qualitative theory. An electron in an orbit is analogous to a small current loop (in which the current is directed oppositely to the direction of electron travel) and, as such, experiences a torque in an external magnetic field, the torque tending to align the magnetic field produced by the orbiting electron with the external magnetic field. If there were no other magnetic moments to consider, we would then conclude that all the orbiting electrons in the material would shift in such a way as to add their magnetic fields to the applied field, and thus that the resultant magnetic field at any point in the material would be greater than it would be at that point if the material were not present.

A second moment, however, is attributed to *electron spin*. Although it is tempting to model this phenomenon by considering the electron to be spinning about its own axis and thus generating a magnetic dipole moment, satisfactory quantitative results are not obtained from such a theory. Instead, it is necessary to digest the mathematics of relativistic quantum theory to show that an electron may have a spin magnetic moment of about $\pm 9 \times 10^{-24}$ A · m^2; the plus and minus signs indicate that alignment aiding or opposing an external magnetic field is possible. In an atom with many electrons present, only the spins of those electrons in shells which are not completely filled will contribute to a magnetic moment for the atom.

A third contribution to the moment of an atom is caused by *nuclear spin*. Although this factor has a negligible effect on the overall magnetic properties of materials, it is the basis of the nuclear magnetic resonance imaging (MRI) procedure provided by many of the larger hospitals.

Thus each atom contains many different component moments, and their combination determines the magnetic characteristics of the material and provides its general magnetic classification. We describe briefly six different types of material: diamagnetic, paramagnetic, ferromagnetic, antiferromagnetic, ferrimagnetic, and superparamagnetic.

First we consider atoms in which the small magnetic fields produced by the motion of the electrons in their orbits and those produced by the electron spin combine to produce a net field of zero. Note that we are considering here the fields produced by the electron motion itself in the absence of any external magnetic field; we might also describe this material as one in which the permanent magnetic moment $\mathbf{m}_0$ of each atom is zero. Such a material is termed *diamagnetic*. It would seem, therefore, that an external magnetic field would produce no torque on the atom, no realignment of the dipole fields, and consequently an internal magnetic field that is the same as the applied field. With an error that only amounts to about one part in a hundred thousand, this is correct.

Consider an orbiting electron whose moment $\mathbf{m}$ is in the same direction as the applied field $\mathbf{B}_0$ (Figure 8.8). The magnetic field produces an outward force on the orbiting electron. Since the orbital radius is quantized and cannot change, the inward Coulomb force of attraction is also unchanged. The force imbalance created by the outward magnetic force must therefore be compensated for by a reduced orbital velocity. Hence, the orbital moment decreases, and a smaller internal field results.

If we had selected an atom for which $\mathbf{m}$ and $\mathbf{B}_0$ were opposed, the magnetic force would be inward, the velocity would increase, the orbital moment would increase, and greater cancellation of $\mathbf{B}_0$ would occur. Again a smaller internal field would result.

Metallic bismuth shows a greater diamagnetic effect than most other diamagnetic materials, among which are hydrogen, helium, the other "inert" gases, sodium chloride, copper, gold, silicon, germanium, graphite, and sulfur. We should also realize that the diamagnetic effect is present in all materials, because it arises from an interaction of the external magnetic field with every orbiting electron; however, it is overshadowed by other effects in the materials we shall consider next.

Now consider an atom in which the effects of the electron spin and orbital motion do not quite cancel. The atom as a whole has a small magnetic moment, but the random orientation of the atoms in a larger sample produces an *average* magnetic moment of zero. The material shows no magnetic effects in the absence of an

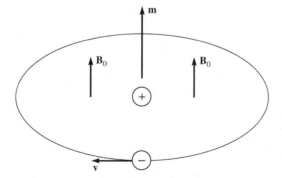

Figure 8.8 An orbiting electron is shown having a magnetic moment **m** in the same direction as an applied field **B**$_0$.

external field. When an external field is applied, however, there is a small torque on each atomic moment, and these moments tend to become aligned with the external field. This alignment acts to increase the value of **B** within the material over the external value. However, the diamagnetic effect is still operating on the orbiting electrons and may counteract the increase. If the net result is a decrease in **B**, the material is still called diamagnetic. However, if there is an increase in **B**, the material is termed *paramagnetic*. Potassium, oxygen, tungsten, and the rare earth elements and many of their salts, such as erbium chloride, neodymium oxide, and yttrium oxide, one of the materials used in masers, are examples of paramagnetic substances.

The remaining four classes of material, ferromagnetic, antiferromagnetic, ferrimagnetic, and superparamagnetic, all have strong atomic moments. Moreover, the interaction of adjacent atoms causes an alignment of the magnetic moments of the atoms in either an aiding or exactly opposing manner.

In *ferromagnetic* materials, each atom has a relatively large dipole moment, caused primarily by uncompensated electron spin moments. Interatomic forces cause these moments to line up in a parallel fashion over regions containing a large number of atoms. These regions are called *domains,* and they may have a variety of shapes and sizes ranging from one micrometer to several centimeters, depending on the size, shape, material, and magnetic history of the sample. Virgin ferromagnetic materials will have domains which each have a strong magnetic moment; the domain moments, however, vary in direction from domain to domain. The overall effect is therefore one of cancellation, and the material as a whole has no magnetic moment. Upon application of an external magnetic field, however, those domains which have moments in the direction of the applied field increase their size at the expense of their neighbors, and the internal magnetic field increases greatly over that of the external field alone. When the external field is removed, a completely random domain alignment is not usually attained, and a residual, or remnant, dipole field remains in the macroscopic structure. The fact that the magnetic moment of the material is different after the field has been removed, or that the magnetic state of the material is a function of its magnetic history, is called *hysteresis,* a subject which will be discussed again when magnetic circuits are studied in Section 8.8.

Ferromagnetic materials are not isotropic in single crystals, and we will therefore limit our discussion to polycrystalline materials, except for mentioning that one of the characteristics of anisotropic magnetic materials is magnetostriction, or the change in dimensions of the crystal when a magnetic field is impressed on it.

The only elements that are ferromagnetic at room temperature are iron, nickel, and cobalt, and they lose all their ferromagnetic characteristics above a temperature called the Curie temperature, which is 1043 K (770°C) for iron. Some alloys of these metals with each other and with other metals are also ferromagnetic, as for example alnico, an aluminum-nickel-cobalt alloy with a small amount of copper. At lower temperatures some of the rare earth elements, such as gadolinium and dysprosium, are ferromagnetic. It is also interesting that some alloys of nonferromagnetic metals are ferromagnetic, such as bismuth-manganese and copper-manganese-tin.

In *antiferromagnetic* materials, the forces between adjacent atoms cause the atomic moments to line up in an antiparallel fashion. The net magnetic moment is zero, and antiferromagnetic materials are affected only slightly by the presence of an external

Table 8.1 Characteristics of magnetic materials

Classification	Magnetic Moments	B Values	Comments
Diamagnetic	$\mathbf{m}_{orb} + \mathbf{m}_{spin} = 0$	$B_{int} < B_{appl}$	$B_{int} \doteq B_{appl}$
Paramagnetic	$\mathbf{m}_{orb} + \mathbf{m}_{spin}$ = small	$B_{int} > B_{appl}$	$B_{int} \doteq B_{appl}$
Ferromagnetic	$|\mathbf{m}_{spin}| \gg |\mathbf{m}_{orb}|$	$B_{int} \gg B_{appl}$	Domains
Antiferromagnetic	$|\mathbf{m}_{spin}| \gg |\mathbf{m}_{orb}|$	$B_{int} \doteq B_{appl}$	Adjacent moments oppose
Ferrimagnetic	$|\mathbf{m}_{spin}| \gg |\mathbf{m}_{orb}|$	$B_{int} > B_{appl}$	Unequal adjacent moments oppose; low σ
Superparamagnetic	$|\mathbf{m}_{spin}| \gg |\mathbf{m}_{orb}|$	$B_{int} > B_{appl}$	Nonmagnetic matrix; recording tapes

magnetic field. This effect was first discovered in manganese oxide, but several hundred antiferromagnetic materials have been identified since then. Many oxides, sulfides, and chlorides are included, such as nickel oxide (NiO), ferrous sulfide (FeS), and cobalt chloride ($CoCl_2$). Antiferromagnetism is only present at relatively low temperatures, often well below room temperature. The effect is not of engineering importance at present.

The *ferrimagnetic* substances also show an antiparallel alignment of adjacent atomic moments, but the moments are not equal. A large response to an external magnetic field therefore occurs, although not as large as that in ferromagnetic materials. The most important group of ferrimagnetic materials are the *ferrites*, in which the conductivity is low, several orders of magnitude less than that of semiconductors. The fact that these substances have greater resistance than the ferromagnetic materials results in much smaller induced currents in the material when alternating fields are applied, as for example in transformer cores that operate at the higher frequencies. The reduced currents (eddy currents) lead to lower ohmic losses in the transformer core. The iron oxide magnetite (Fe_3O_4), a nickel-zinc ferrite ($Ni_{1/2}Zn_{1/2}Fe_2O_4$), and a nickel ferrite ($NiFe_2O_4$) are examples of this class of materials. Ferrimagnetism also disappears above the Curie temperature.

Superparamagnetic materials are composed of an assembly of ferromagnetic particles in a nonferromagnetic matrix. Although domains exist within the individual particles, the domain walls cannot penetrate the intervening matrix material to the adjacent particle. An important example is the magnetic tape used in audiotape or videotape recorders.

Table 8.1 summarizes the characteristics of the six types of magnetic materials we have discussed.

8.6 MAGNETIZATION AND PERMEABILITY

To place our description of magnetic materials on a more quantitative basis, we will now devote a page or so to showing how the magnetic dipoles act as a distributed source for the magnetic field. Our result will be an equation that looks very much like Ampère's circuital law, $\oint \mathbf{H} \cdot d\mathbf{L} = I$. The current, however, will be the movement of *bound* charges (orbital electrons, electron spin, and nuclear spin), and the field, which has the dimensions of **H**, will be called the magnetization **M**. The current produced by the bound charges is called a *bound current* or *Amperian current*.

We begin by defining the magnetization **M** in terms of the magnetic dipole moment **m**. The bound current I_b circulates about a path enclosing a differential area $d\mathbf{S}$, establishing a dipole moment $(A \cdot m^2)$,

$$\mathbf{m} = I_b d\mathbf{S}$$

If there are n magnetic dipoles per unit volume and we consider a volume Δv, then the total magnetic dipole moment is found by the vector sum

$$\mathbf{m}_{\text{total}} = \sum_{i=1}^{n\Delta v} \mathbf{m}_i$$

Each of the $\mathbf{m}_i$ may be different. Next, we define the *magnetization* **M** as the *magnetic dipole moment per unit volume,*

$$\boxed{\mathbf{M} = \lim_{\Delta v \to 0} \frac{1}{\Delta v} \sum_{i=1}^{n\Delta v} \mathbf{m}_i = n\mathbf{m} \text{ (identical dipoles)}} \tag{19}$$

and see that its units must be the same as for **H**, amperes per meter. The second equality (not the first) in Eq. (19) applies to the special case in which all dipoles are identical.

Now consider the effect of some alignment of the magnetic dipoles as the result of the application of a magnetic field. We will investigate this alignment along a closed path, a short portion along the lower edge of which is shown in Figure 8.9. The figure shows several identical magnetic moments **m** that make an angle θ with the element of path $d\mathbf{L}$; each moment consists of a bound current I_b circulating about an area $d\mathbf{S}$. Therefore, over the path segment $d\mathbf{L}$, we are considering a small volume, $dS \cos\theta dL$, or $d\mathbf{S} \cdot d\mathbf{L}$, within which there are $nd\mathbf{S} \cdot d\mathbf{L}$ magnetic dipoles. In changing from a random orientation to this partial alignment, the bound current crossing the surface enclosed by the path (the plane of the page that lies above the path in Figure 8.9) has increased by I_b for each of the $nd\mathbf{S} \cdot d\mathbf{L}$ dipoles. Thus the differential change in the net bound current I_B over the segment $d\mathbf{L}$ will be

$$d I_B = n I_b d\mathbf{S} \cdot d\mathbf{L} = \mathbf{M} \cdot d\mathbf{L} \tag{20}$$

and within an entire closed contour,

$$I_B = \oint \mathbf{M} \cdot d\mathbf{L} \tag{21}$$

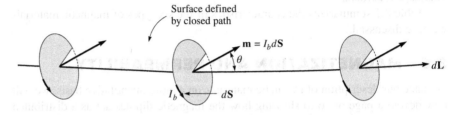

Figure 8.9 A section $d\mathbf{L}$ of a closed path along which magnetic dipoles have been partially aligned by some external magnetic field. The alignment has caused the bound current crossing the surface defined by the closed path to increase by $nI_b d\mathbf{S} \cdot d\mathbf{L}$ A.

Equation (21) merely says that if we go around a closed path and find dipole moments going our way more often than not, there will be a corresponding current composed of, for example, orbiting electrons crossing the interior surface.

This last expression has some resemblance to Ampère's circuital law, and we may now generalize the relationship between **B** and **H** so that it applies to media other than free space. Our present discussion is based on the forces and torques on differential current loops in a **B** field, and we therefore take **B** as our fundamental quantity and seek an improved definition of **H**. We thus write Ampère's circuital law in terms of the *total* current, bound plus free,

$$\oint \frac{\mathbf{B}}{\mu_0} \cdot d\mathbf{L} = I_T \tag{22}$$

where

$$I_T = I_B + I$$

and I is the total *free* current enclosed by the closed path. Note that the free current appears without subscript since it is the most important type of current and will be the only current appearing in Maxwell's equations.

Combining these last three equations, we obtain an expression for the free current enclosed,

$$I = I_T - I_B = \oint \left(\frac{\mathbf{B}}{\mu_0} - \mathbf{M} \right) \cdot d\mathbf{L} \tag{23}$$

We may now define **H** in terms of **B** and **M**,

$$\mathbf{H} = \frac{\mathbf{B}}{\mu_0} - \mathbf{M} \tag{24}$$

and we see that $\mathbf{B} = \mu_0\mathbf{H}$ in free space where the magnetization is zero. This relationship is usually written in a form that avoids fractions and minus signs:

$$\boxed{\mathbf{B} = \mu_0(\mathbf{H} + \mathbf{M})} \tag{25}$$

We may now use our newly defined **H** field in (23),

$$I = \oint \mathbf{H} \cdot d\mathbf{L} \tag{26}$$

obtaining Ampère's circuital law in terms of the free currents.

Using the several current densities, we have

$$I_B = \int_S \mathbf{J}_B \cdot d\mathbf{S}$$
$$I_T = \int_S \mathbf{J}_T \cdot d\mathbf{S}$$
$$I = \int_S \mathbf{J} \cdot d\mathbf{S}$$

With the help of Stokes' theorem, we may therefore transform (21), (26), and (22) into the equivalent curl relationships:

$$\nabla \times \mathbf{M} = \mathbf{J}_B$$

$$\nabla \times \frac{\mathbf{B}}{\mu_0} = \mathbf{J}_T$$

$$\boxed{\nabla \times \mathbf{H} = \mathbf{J}} \tag{27}$$

We will emphasize only (26) and (27), the two expressions involving the free charge, in the work that follows.

The relationship between **B**, **H**, and **M** expressed by (25) may be simplified for linear isotropic media where a magnetic susceptibility χ_m can be defined:

$$\boxed{\mathbf{M} = \chi_m \mathbf{H}} \tag{28}$$

Thus we have

$$\mathbf{B} = \mu_0(\mathbf{H} + \chi_m \mathbf{H})$$
$$= \mu_0 \mu_r \mathbf{H}$$

where

$$\mu_r = 1 + \chi_m \tag{29}$$

is defined as the *relative permeability* μ_r. We next define the *permeability* μ:

$$\mu = \mu_0 \mu_r \tag{30}$$

and this enables us to write the simple relationship between **B** and **H**,

$$\mathbf{B} = \mu \mathbf{H} \tag{31}$$

EXAMPLE 8.5

Given a ferrite material that we shall specify to be operating in a linear mode with $B = 0.05$ T, let us assume $\mu_r = 50$, and calculate values for χ_m, M, and H.

Solution. Because $\mu_r = 1 + \chi_m$, we have

$$\chi_m = \mu_r - 1 = 49$$

Also,

$$B = \mu_r \mu_0 H$$

and

$$H = \frac{0.05}{50 \times 4\pi \times 10^{-7}} = 796 \, \text{A/m}$$

The magnetization is $M = \chi_m H$, or 39,000 A/m. The alternate ways of relating B and H are, first,

$$B = \mu_0(H + M)$$

or

$$0.05 = 4\pi \times 10^{-7}(796 + 39{,}000)$$

showing that Amperian currents produce 49 times the magnetic field intensity that the free charges do; and second,

$$B = \mu_r \mu_0 H$$

or

$$0.05 = 50 \times 4\pi \times 10^{-7} \times 796$$

where we use a relative permeability of 50 and let this quantity account completely for the motion of the bound charges. We shall emphasize the latter interpretation in the chapters that follow.

The first two laws that we investigated for magnetic fields were the Biot-Savart law and Ampère's circuital law. Both were restricted to free space in their application. We may now extend their use to any homogeneous, linear, isotropic magnetic material that may be described in terms of a relative permeability μ_r.

Just as we found for anisotropic dielectric materials, the permeability of an anisotropic magnetic material must be given as a 3×3 matrix, and **B** and **H** are both 3×1 matrices. We have

$$B_x = \mu_{xx}H_x + \mu_{xy}H_y + \mu_{xz}H_z$$

$$B_y = \mu_{yx}H_x + \mu_{yy}H_y + \mu_{yz}H_z$$

$$B_z = \mu_{zx}H_x + \mu_{zy}H_y + \mu_{zz}H_z$$

For anisotropic materials, then, $\mathbf{B} = \mu\mathbf{H}$ is a matrix equation; however, $\mathbf{B} = \mu_0(\mathbf{H} + \mathbf{M})$ remains valid, although **B**, **H**, and **M** are no longer parallel in general. The most common anisotropic magnetic material is a single ferromagnetic crystal, although thin magnetic films also exhibit anisotropy. Most applications of ferromagnetic materials, however, involve polycrystalline arrays that are much easier to make.

Our definitions of susceptibility and permeability also depend on the assumption of linearity. Unfortunately, this is true only in the less interesting paramagnetic and diamagnetic materials for which the relative permeability rarely differs from unity by more than one part in a thousand. Some typical values of the susceptibility for diamagnetic materials are hydrogen, -2×10^{-5}; copper, -0.9×10^{-5}; germanium, -0.8×10^{-5}; silicon, -0.3×10^{-5}; and graphite, -12×10^{-5}. Several representative paramagnetic susceptibilities are oxygen, 2×10^{-6}; tungsten, 6.8×10^{-5}; ferric oxide (Fe_2O_3), 1.4×10^{-3}; and yttrium oxide (Y_2O_3), 0.53×10^{-6}. If we simply take the ratio of B to $\mu_0 H$ as the relative permeability of a ferromagnetic material, typical values of μ_r would range from 10 to 100,000. Diamagnetic, paramagnetic, and anti-ferromagnetic materials are commonly said to be nonmagnetic.

D8.6. Find the magnetization in a magnetic material where: (a) $\mu = 1.8 \times 10^{-5}$ H/m and $H = 120$ A/m; (b) $\mu_r = 22$, there are 8.3×10^{28} atoms/m^3, and each atom has a dipole moment of 4.5×10^{-27} A $\cdot$ m^2; (c) $B = 300$ μT and $\chi_m = 15$.

Ans. (a) 1599 A/m; (b) 374 A/m; (c) 224 A/m

D8.7. The magnetization in a magnetic material for which $\chi_m = 8$ is given in a certain region as $150z^2\mathbf{a}_x$ A/m. At $z = 4$ cm, find the magnitude of: (a) $\mathbf{J}_T$; (b) $\mathbf{J}$; (c) $\mathbf{J}_B$.

Ans. (a) 13.5 A/m^2; (b) 1.5 A/m^2; (c) 12 A/m^2

8.7 MAGNETIC BOUNDARY CONDITIONS

We should have no difficulty in arriving at the proper boundary conditions to apply to **B**, **H**, and **M** at the interface between two different magnetic materials, for we have solved similar problems for both conducting materials and dielectrics. We need no new techniques.

Figure 8.10 shows a boundary between two isotropic homogeneous linear materials with permeabilities μ_1 and μ_2. The boundary condition on the normal

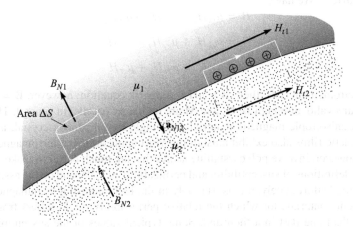

Figure 8.10 A gaussian surface and a closed path are constructed at the boundary between media 1 and 2, having permeabilities of μ_1 and μ_2, respectively. From this we determine the boundary conditions $B_{N1} = B_{N2}$ and $H_{t1} - H_{t2} = K$, the component of the surface current density directed into the page.

components is determined by allowing the surface to cut a small cylindrical gaussian surface. Applying Gauss's law for the magnetic field from Section 7.5,

$$\oint_S B \cdot d\mathbf{S} = 0$$

we find that

$$B_{N1}\,\Delta S - B_{N2}\,\Delta S = 0$$

or the normal component of **B** is continuous across the boundary ($B_{N2} = \mathrm{B}_{N1}$). A more formal statement of this is

$$\boxed{(\mathbf{B}_2 - \mathbf{B}_1) \cdot \mathbf{a}_{N12} = 0} \tag{32}$$

where $\mathbf{a}_{N12}$ is the unit normal at the boundary directed from region 1 to region 2.

The normal component of **B** is continuous, and therefore the normal component of **H** is discontinuous by the ratio μ_1/μ_2.

$$H_{N2} = \frac{\mu_1}{\mu_2} H_{N1} \tag{33}$$

The relationship between the normal components of **M**, of course, is fixed once the relationship between the normal components of **H** is known. For linear magnetic materials, the result is written simply as

$$M_{N2} = \chi_{m2}\frac{\mu_1}{\mu_2}H_{N1} = \frac{\chi_{m2}\mu_1}{\chi_{m1}\mu_2}M_{N1} \tag{34}$$

Next, Ampère's circuital law

$$\oint \mathbf{H} \cdot d\mathbf{L} = I$$

is applied about a small closed path in a plane normal to the boundary surface, as shown to the right in Figure 8.10. Taking a clockwise trip around the path, we find that

$$H_{t1}\,\Delta L - H_{t2}\,\Delta L = K\Delta L$$

where we assume that the boundary may carry a surface current **K** whose component normal to the plane of the closed path is K. Thus

$$H_{t1} - H_{t2} = K$$

The directions are specified more exactly by using the cross product to identify the tangential components,

$$\boxed{(\mathbf{H}_1 - \mathbf{H}_2) \times \mathbf{a}_{N12} = \mathbf{K}} \tag{35}$$

An equivalent formulation in terms of the vector tangential components may be more convenient for **H**:

$$\mathbf{H}_{t1} - \mathbf{H}_{t2} = \mathbf{a}_{N12} \times \mathbf{K}$$

For tangential **B**, we have

$$\frac{B_{t1}}{\mu_1} - \frac{B_{t2}}{\mu_2} = K \tag{36}$$

The boundary condition on the tangential component of the magnetization for linear materials is therefore

$$M_{t2} = \frac{\chi_{m2}}{\chi_{m1}} M_{t1} - \chi_{m2} K \tag{37}$$

The last three boundary conditions on the tangential components are much simpler, of course, if the surface current density is zero. This is a free current density, and it must be zero if neither material is a conductor.

EXAMPLE 8.6

To illustrate these relationships with an example, let us assume that $\mu = \mu_1 = 4\ \mu H/m$ in region 1 where $z > 0$, whereas $\mu_2 = 7\ \mu H/m$ in region 2 wherever $z < 0$. Moreover, let $\mathbf{K} = 80\mathbf{a}_x$ A/m on the surface $z = 0$. We establish a field, $\mathbf{B}_1 = 2\mathbf{a}_x - 3\mathbf{a}_y + \mathbf{a}_z$ mT, in region 1 and seek the value of $\mathbf{B}_2$.

Solution. The normal component of $\mathbf{B}_1$ is

$$\mathbf{B}_{N1} = (\mathbf{B}_1 \cdot \mathbf{a}_{N12})\,\mathbf{a}_{N12} = [(2\mathbf{a}_x - 3\mathbf{a}_y + \mathbf{a}_z) \cdot (-\mathbf{a}_z)](-\mathbf{a}_z) = \mathbf{a}_z\ \text{mT}$$

Thus,

$$\mathbf{B}_{N2} = \mathbf{B}_{N1} = \mathbf{a}_z\ \text{mT}$$

We next determine the tangential components:

$$\mathbf{B}_{t1} = \mathbf{B}_1 - \mathbf{B}_{N1} = 2\mathbf{a}_x - 3\mathbf{a}_y\ \text{mT}$$

and

$$\mathbf{H}_{t1} = \frac{\mathbf{B}_{t1}}{\mu_1} = \frac{(2\mathbf{a}_x - 3\mathbf{a}_y)\,10^{-3}}{4 \times 10^{-6}} = 500\mathbf{a}_x - 750\mathbf{a}_y\ \text{A/m}$$

Thus,

$$\mathbf{H}_{t2} = \mathbf{H}_{t1} - \mathbf{a}_{N12} \times \mathbf{K} = 500\mathbf{a}_x - 750\mathbf{a}_y - (-\mathbf{a}_z) \times 80\mathbf{a}_x$$
$$= 500\mathbf{a}_x - 750\mathbf{a}_y + 80\mathbf{a}_y = 500\mathbf{a}_x - 670\mathbf{a}_y\ \text{A/m}$$

and

$$\mathbf{B}_{t2} = \mu_2\mathbf{H}_{t2} = 7 \times 10^{-6}(500\mathbf{a}_x - 670\mathbf{a}_y) = 3.5\mathbf{a}_x - 4.69\mathbf{a}_y\ \text{mT}$$

Therefore,

$$\mathbf{B}_2 = \mathbf{B}_{N2} + \mathbf{B}_{t2} = 3.5\mathbf{a}_x - 4.69\mathbf{a}_y + \mathbf{a}_z\ \text{mT}$$

D8.8. Let the permittivity be 5 μH/m in region A where $x < 0$, and 20 μH/m in region B where $x > 0$. If there is a surface current density $\mathbf{K} = 150\mathbf{a}_y - 200\mathbf{a}_z$ A/m at $x = 0$, and if $H_A = 300\mathbf{a}_x - 400\mathbf{a}_y + 500\mathbf{a}_z$ A/m, find: (a) $|\mathbf{H}_{tA}|$; (b) $|\mathbf{H}_{NA}|$; (c) $|\mathbf{H}_{tB}|$; (d) $|\mathbf{H}_{NB}|$.

Ans. (a) 640 A/m; (b) 300 A/m; (c) 695 A/m; (d) 75 A/m

8.8 THE MAGNETIC CIRCUIT

In this section, we digress briefly to discuss the fundamental techniques involved in solving a class of magnetic problems known as magnetic circuits. As we will see shortly, the name arises from the great similarity to the dc-resistive-circuit analysis with which it is assumed we are all familiar. The only important difference lies in the non-linear nature of the ferromagnetic portions of the magnetic circuit; the methods which must be adopted are similar to those required in nonlinear electric circuits which contain diodes, thermistors, incandescent filaments, and other nonlinear elements.

As a convenient starting point, we identify those field equations on which resistive circuit analysis is based. At the same time we will point out or derive the analogous equations for the magnetic circuit. We begin with the electrostatic potential and its relationship to electric field intensity,

$$\mathbf{E} = -\nabla V \tag{38a}$$

The scalar magnetic potential has already been defined, and its analogous relation to the magnetic field intensity is

$$\boxed{\mathbf{H} = -\nabla V_m} \tag{38b}$$

In dealing with magnetic circuits, it is convenient to call V_m the *magnetomotive force,* or mmf, and we will acknowledge the analogy to the electromotive force, or emf, by doing so. The units of the mmf are, of course, amperes, but it is customary to recognize that coils with many turns are often employed by using the term "ampere-turns." Remember that no current may flow in any region in which V_m is defined.

The electric potential difference between points A and B may be written as

$$V_{AB} = \int_A^B \mathbf{E} \cdot d\mathbf{L} \tag{39a}$$

and the corresponding relationship between the mmf and the magnetic field intensity,

$$\boxed{V_{m\,AB} = \int_A^B \mathbf{H} \cdot d\mathbf{L}} \tag{39b}$$

was developed in Chapter 7, where we learned that the path selected must not cross the chosen barrier surface.

Ohm's law for the electric circuit has the point form

$$\mathbf{J} = \sigma \mathbf{E} \tag{40a}$$

and we see that the magnetic flux density will be the analog of the current density,

$$\boxed{\mathbf{B} = \mu \mathbf{H}} \tag{40b}$$

To find the total current, we must integrate:

$$I = \int_S \mathbf{J} \cdot d\mathbf{S} \tag{41a}$$

A corresponding operation is necessary to determine the total magnetic flux flowing through the cross section of a magnetic circuit:

$$\boxed{\Phi = \int_S \mathbf{B} \cdot d\mathbf{S}} \tag{41b}$$

We then defined resistance as the ratio of potential difference and current, or

$$V = IR \tag{42a}$$

and we shall now define *reluctance* as the ratio of the magnetomotive force to the total flux; thus

$$\boxed{V_m = \Phi \mathfrak{R}} \tag{42b}$$

where reluctance is measured in ampere-turns per weber (A · t/Wb). In resistors that are made of a linear isotropic homogeneous material of conductivity σ and have a uniform cross section of area S and length d, the total resistance is

$$R = \frac{d}{\sigma S} \tag{43a}$$

If we are fortunate enough to have such a linear isotropic homogeneous magnetic material of length d and uniform cross section S, then the total reluctance is

$$\boxed{\mathfrak{R} = \frac{d}{\mu S}} \tag{43b}$$

The only such material to which we shall commonly apply this relationship is air.

Finally, let us consider the analog of the source voltage in an electric circuit. We know that the closed line integral of $\mathbf{E}$ is zero,

$$\oint \mathbf{E} \cdot d\mathbf{L} = 0$$

In other words, Kirchhoff's voltage law states that the rise in potential through the source is exactly equal to the fall in potential through the load. The expression for magnetic phenomena takes on a slightly different form,

$$\oint \mathbf{H} \cdot d\mathbf{L} = I_{\text{total}}$$

for the closed line integral is not zero. Because the total current linked by the path is usually obtained by allowing a current I to flow through an N-turn coil, we may express this result as

$$\boxed{\oint \mathbf{H} \cdot d\mathbf{L} = NI} \tag{44}$$

In an electric circuit, the voltage source is a part of the closed path; in the magnetic circuit, the current-carrying coil will surround or link the magnetic circuit. In tracing a magnetic circuit, we will not be able to identify a pair of terminals at which the magnetomotive force is applied. The analogy is closer here to a pair of coupled circuits in which induced voltages exist (and in which we will see in Chapter 9 that the closed line integral of $\mathbf{E}$ is also not zero).

We will try out some of these ideas on a simple magnetic circuit. In order to avoid the complications of ferromagnetic materials at this time, we will assume that we have an air-core toroid with 500 turns, a cross-sectional area of 6 cm², a mean radius of 15 cm, and a coil current of 4 A. As we already know, the magnetic field is confined to the interior of the toroid, and if we consider the closed path of our magnetic circuit along the mean radius, we link 2000 A · t,

$$V_{m,\,\text{source}} = 2000 \text{ A} \cdot \text{t}$$

Although the field in the toroid is not quite uniform, we may assume that it is, for all practical purposes, and calculate the total reluctance of the circuit as

$$\mathfrak{R} = \frac{d}{\mu S} = \frac{2\pi(0.15)}{4\pi 10^{-7} \times 6 \times 10^{-4}} = 1.25 \times 10^9 \text{ A} \cdot \text{t/Wb}$$

Thus

$$\Phi = \frac{V_{m,S}}{\mathfrak{R}} = \frac{2000}{1.25 \times 10^9} = 1.6 \times 10^{-6} \text{ Wb}$$

This value of the total flux is in error by less than $\frac{1}{4}$ percent, in comparison with the value obtained when the exact distribution of flux over the cross section is used.

Hence

$$B = \frac{\Phi}{S} = \frac{1.6 \times 10^{-6}}{6 \times 10^{-4}} = 2.67 \times 10^{-3} \text{ T}$$

and finally,

$$H = \frac{B}{\mu} = \frac{2.67 \times 10^{-3}}{4\pi 10^{-7}} = 2120 \text{ A} \cdot \text{t/m}$$

As a check, we may apply Ampère's circuital law directly in this symmetrical problem,

$$H_\phi 2\pi r = NI$$

and obtain

$$H_\phi = \frac{NI}{2\pi r} = \frac{500 \times 4}{6.28 \times 0.15} = 2120\,\text{A/m}$$

at the mean radius.

Our magnetic circuit in this example does not give us any opportunity to find the mmf across different elements in the circuit, for there is only one type of material. The analogous electric circuit is, of course, a single source and a single resistor. We could make it look just as long as the preceding analysis, however, if we found the current density, the electric field intensity, the total current, the resistance, and the source voltage.

More interesting and more practical problems arise when ferromagnetic materials are present in the circuit. We begin by considering the relationship between B and H in such a material. We may assume that we are establishing a curve of B versus H for a sample of ferromagnetic material which is completely demagnetized; both B and H are zero. As we begin to apply an mmf, the flux density also rises, but not linearly, as the experimental data of Figure 8.11 show near the origin. After H reaches a value of about 100 A · t/m, the flux density rises more slowly and begins to saturate when H is several hundred A · t/m. Having reached partial saturation, we now turn to Figure 8.12, where we may continue our experiment at point x by reducing H. As we do so, the effects of hysteresis begin to show, and we do not retrace our original curve. Even after H is zero, $B = B_r$, the remnant flux density. As H is reversed, then brought back to zero, and the complete cycle traced several times, the hysteresis loop of Figure 8.12 is obtained. The mmf required to reduce the flux density to zero is identified as H_c, the coercive "force." For smaller maximum values of H, smaller

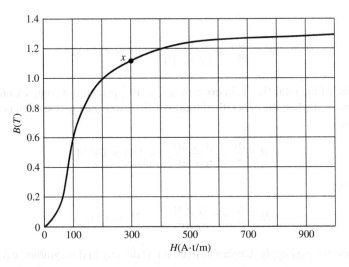

Figure 8.11 Magnetization curve of a sample of silicon sheet steel.

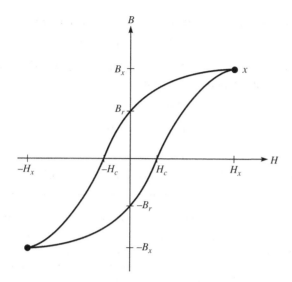

Figure 8.12   A hysteresis loop for silicon steel. The coercive force H_c and remnant flux density B_r are indicated.

hysteresis loops are obtained, and the locus of the tips is about the same as the virgin magnetization curve of Figure 8.11.

EXAMPLE 8.7

We may use the magnetization curve for silicon steel to solve a magnetic circuit problem that is slightly different from our previous example. We use a steel core in the toroid, except for an air gap of 2 mm. Magnetic circuits with air gaps occur because gaps are deliberately introduced in some devices, such as inductors, which must carry large direct currents, because they are unavoidable in other devices such as rotating machines, or because of unavoidable problems in assembly. There are still 500 turns about the toroid, and we ask what current is required to establish a flux density of 1 T everywhere in the core.

Solution. This magnetic circuit is analogous to an electric circuit containing a voltage source and two resistors, one of which is nonlinear. Because we are given the "current," it is easy to find the "voltage" across each series element, and hence the total "emf." In the air gap,

$$\mathfrak{R}_{air} = \frac{d_{air}}{\mu S} = \frac{2 \times 10^{-3}}{4\pi 10^{-7} \times 6 \times 10^{-4}} = 2.65 \times 10^6 \text{ A} \cdot \text{t/Wb}$$

Knowing the total flux,

$$\Phi = BS = 1(6 \times 10^{-4}) = 6 \times 10^{-4} \text{ Wb}$$

which is the same in both steel and air, we may find the mmf required for the gap,

$$V_{m,\text{air}} = (6 \times 10^{-4})(2.65 \times 10^6) = 1590 \; \text{A} \cdot \text{t}$$

Referring to Figure 8.11, a magnetic field strength of 200 A · t/m is required to produce a flux density of 1 T in the steel. Thus,

$$H_{\text{steel}} = 200 \; \text{A} \cdot \text{t}$$

$$V_{m,\text{steel}} = H_{\text{steel}} d_{\text{steel}} = 200 \times 0.30\pi$$

$$= 188 \; \text{A} \cdot \text{t}$$

The total mmf is therefore 1778 A · t, and a coil current of 3.56 A is required.

We have made several approximations in obtaining this answer. We have already mentioned the lack of a completely uniform cross section, or cylindrical symmetry; the path of every flux line is not of the same length. The choice of a "mean" path length can help compensate for this error in problems in which it may be more important than it is in our example. Fringing flux in the air gap is another source of error, and formulas are available by which we may calculate an effective length and cross-sectional area for the gap which will yield more accurate results. There is also a leakage flux between the turns of wire, and in devices containing coils concentrated on one section of the core, a few flux lines bridge the interior of the toroid. Fringing and leakage are problems that seldom arise in the electric circuit because the ratio of the conductivities of air and the conductive or resistive materials used is so high. In contrast, the magnetization curve for silicon steel shows that the ratio of H to B in the steel is about 200 up to the "knee" of the magnetization curve; this compares with a ratio in air of about 800,000. Thus, although flux prefers steel to air by the commanding ratio of 4000 to 1, this is not very close to the ratio of conductivities of, say, 10^{15} for a good conductor and a fair insulator.

EXAMPLE 8.8

As a last example, let us consider the reverse problem. Given a coil current of 4 A in the magnetic circuit of Example 8.7, what will the flux density be?

Solution. First let us try to linearize the magnetization curve by a straight line from the origin to $B = 1$, $H = 200$. We then have $B = H/200$ in steel and $B = \mu_0 H$ in air. The two reluctances are found to be 0.314×10^6 for the steel path and 2.65×10^6 for the air gap, or $2.96 \times 10^6 \text{A} \cdot \text{t/Wb}$ total. Since V_m is 2000 A · t, the flux is 6.76×10^{-4} Wb, and $B = 1.13$ T. A more accurate solution may be obtained by assuming several values of B and calculating the necessary mmf. Plotting the results enables us to determine the true value of B by interpolation. With this method we obtain $B = 1.10$ T. The good accuracy of the linear model results from the fact that the reluctance of the air gap in a magnetic circuit is often much greater than the reluctance of the ferromagnetic portion of the circuit. A relatively poor approximation for the iron or steel can thus be tolerated.

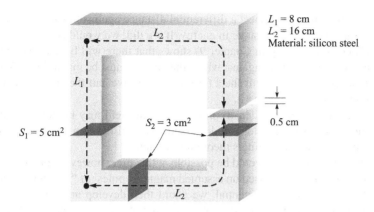

Figure 8.13 See Problem D8.9.

D8.9. Given the magnetic circuit of Figure 8.13, assume $B = 0.6$ T at the midpoint of the left leg and find: (a) $V_{m,\text{air}}$; (b) $V_{m,\text{steel}}$; (c) the current required in a 1300-turn coil linking the left leg.

Ans. (a) 3980 A · t; (b) 72 A · t; (c) 3.12 A

D8.10. The magnetization curve for material X under normal operating conditions may be approximated by the expression $B = (H/160)(0.25 + e^{-H/320})$, where H is in A/m and B is in T. If a magnetic circuit contains a 12 cm length of material X, as well as a 0.25-mm air gap, assume a uniform cross section of 2.5 cm^2 and find the total mmf required to produce a flux of (a) 10 μWb; (b) 100 μWb.

Ans. (a) 8.58 A · t; (b) 86.7 A · t

8.9 POTENTIAL ENERGY AND FORCES ON MAGNETIC MATERIALS

In the electrostatic field we first introduced the point charge and the experimental law of force between point charges. After defining electric field intensity, electric flux density, and electric potential, we were able to find an expression for the energy in an electrostatic field by establishing the work necessary to bring the prerequisite point charges from infinity to their final resting places. The general expression for energy is

$$W_E = \frac{1}{2} \int_{\text{vol}} \mathbf{D} \cdot \mathbf{E} \, dv \tag{45}$$

where a linear relationship between $\mathbf{D}$ and $\mathbf{E}$ is assumed.

This is not as easily done for the steady magnetic field. It would seem that we might assume two simple sources, perhaps two current sheets, find the force on one due to the other, move the sheet a differential distance against this force, and equate the necessary work to the change in energy. If we did, we would be wrong, because Faraday's law (coming up in Chapter 9) shows that there will be a voltage induced in the moving current sheet against which the current must be maintained. Whatever source is supplying the current sheet turns out to receive half the energy we are putting into the circuit by moving it.

In other words, energy density in the magnetic field may be determined more easily after time-varying fields are discussed. We will develop the appropriate expression in discussing Poynting's theorem in Chapter 11.

An alternate approach would be possible at this time, however, for we might define a magnetostatic field based on assumed magnetic poles (or "magnetic charges"). Using the scalar magnetic potential, we could then develop an energy expression by methods similar to those used in obtaining the electrostatic energy relationship. These new magnetostatic quantities we would have to introduce would be too great a price to pay for one simple result, and we will therefore merely present the result at this time and show that the same expression arises in the Poynting theorem later. The total energy stored in a steady magnetic field in which **B** is linearly related to **H** is

$$\boxed{W_H = \frac{1}{2}\int_{\text{vol}} \mathbf{B} \cdot \mathbf{H}\, dv} \qquad (46)$$

Letting $\mathbf{B} = \mu\mathbf{H}$, we have the equivalent formulations

$$W_H = \frac{1}{2}\int_{\text{vol}} \mu H^2\, dv \qquad (47)$$

or

$$W_H = \frac{1}{2}\int_{\text{vol}} \frac{B^2}{\mu}\, dv \qquad (48)$$

It is again convenient to think of this energy as being distributed throughout the volume with an energy density of $\frac{1}{2}\,\mathbf{B} \cdot \mathbf{H}\,\text{J/m}^3$, although we have no mathematical justification for such a statement.

In spite of the fact that these results are valid only for linear media, we may use them to calculate the forces on nonlinear magnetic materials if we focus our attention on the linear media (usually air) which may surround them. For example, suppose that we have a long solenoid with a silicon-steel core. A coil containing n turns/m with a current I surrounds it. The magnetic field intensity in the core is therefore nI A · t/m, and the magnetic flux density can be obtained from the magnetization curve for silicon steel. Let us call this value B_{st}. Suppose that the core is composed of two semi-infinite cylinders[2] that are just touching. We now apply a mechanical force to separate these two sections of the core while keeping the flux density constant.

---

[2] A semi-infinite cylinder is a cylinder of infinite length having one end located in finite space.

We apply a force F over a distance dL, thus doing work $F\,dL$. Faraday's law does not apply here, for the fields in the core have not changed, and we can therefore use the principle of virtual work to determine that the work we have done in moving one core appears as stored energy in the air gap we have created. By (48), this increase is

$$dW_H = F\,dL = \frac{1}{2}\frac{B_{st}^2}{\mu_0}S\,dL$$

where S is the core cross-sectional area. Thus

$$F = \frac{B_{st}^2 S}{2\mu_0}$$

If, for example, the magnetic field intensity is sufficient to produce saturation in the steel, approximately 1.4 T, the force is

$$F = 7.80 \times 10^5 S \ \text{N}$$

or about 113 lb_f/in^2.

> **D8.11.** (*a*) What force is being exerted on the pole faces of the circuit described in Problem D8.9 and Figure 8.13? (*b*) Is the force trying to open or close the air gap?
>
> **Ans.** (*a*) 1194 N; (*b*) As Wilhelm Eduard Weber would put it, "schliessen."

8.10 INDUCTANCE AND MUTUAL INDUCTANCE

Inductance is the last of the three familiar parameters from circuit theory that we are defining in more general terms. Resistance was defined in Chapter 5 as the ratio of the potential difference between two equipotential surfaces of a conducting material to the total current crossing either equipotential surface. The resistance is a function of conductor geometry and conductivity only. Capacitance was defined in the same chapter as the ratio of the total charge on either of two equipotential conducting surfaces to the potential difference between the surfaces. Capacitance is a function only of the geometry of the two conducting surfaces and the permittivity of the dielectric medium between or surrounding them.

8.10.1 Flux Linkage and Self-Inductance

As a prelude to defining inductance, we first need to introduce the concept of flux linkage. Consider a toroid of N turns in which a current I produces a total flux Φ. We assume first that this flux links or encircles each of the N turns, and we also see that each of the N turns links the total flux Φ. The *flux linkage* $N\Phi$ is defined as the product of the number of turns N and the flux Φ linking each of them.[3] For a coil having a single turn, the flux linkage is equal to the total flux.

Illustrations

---

[3] The symbol λ is commonly used for flux linkages. We will only occasionally use this concept, however, and we will continue to write it as $N\Phi$.

We now define *inductance* (or self-inductance) as the ratio of the total flux linkages to the current which they link,

$$L = \frac{N\Phi}{I} \qquad (49)$$

The current I flowing in the N-turn coil produces the total flux Φ and $N\Phi$ flux linkages, where we assume for the moment that the flux Φ links each turn. This definition is applicable only to magnetic media which are linear, so that the flux is proportional to the current. If ferromagnetic materials are present, there is no single definition of inductance which is useful in all cases, and we shall restrict our attention to linear materials.

The unit of inductance is the henry (H), equivalent to one weber-turn per ampere.

Let us apply (49) in a straightforward way to calculate the inductance per meter length of a coaxial cable of inner radius a and outer radius b. We may take the expression for total flux developed as Eq. (42) in Chapter 7,

$$\Phi = \frac{\mu_0 I d}{2\pi} \ln \frac{b}{a}$$

and obtain the inductance rapidly for a length d,

$$L = \frac{\mu_0 d}{2\pi} \ln \frac{b}{a} \text{ H}$$

or, on a per-meter basis,

$$L = \frac{\mu_0}{2\pi} \ln \frac{b}{a} \text{ H/m} \qquad (50)$$

In this case, $N = 1$ turn, and all the flux links all the current.

In the problem of a toroidal coil of N turns and a current I, as shown in Figure 7.12b, we have

$$B_\phi = \frac{\mu_0 N I}{2\pi \rho}$$

If the dimensions of the cross section are small compared with the mean radius of the toroid ρ_0, then the total flux is

$$\Phi = \frac{\mu_0 N I S}{2\pi \rho_0}$$

where S is the cross-sectional area. Multiplying the total flux by N, we have the flux linkages, and dividing by I, we have the inductance

$$L = \frac{\mu_0 N^2 S}{2\pi \rho_0} \qquad (51)$$

Once again we have assumed that all the flux links all the turns, and this is a good assumption for a toroidal coil of many turns packed closely together. Suppose, however, that our toroid has an appreciable spacing between turns, a short part of which might look like Figure 8.14. The flux linkages are no longer the product of the

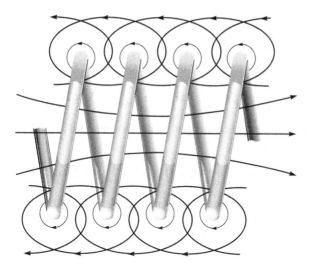

Figure 8.14 A portion of a coil showing partial flux linkages. The total flux linkages are obtained by adding the fluxes linking each turn.

flux at the mean radius times the total number of turns. In order to obtain the total flux linkages we must look at the coil on a turn-by-turn basis.

$$(N\Phi)_{\text{total}} = \Phi_1 + \Phi_2 + \cdots + \Phi_i + \cdots + \Phi_N$$
$$= \sum_{i=1}^{N} \Phi_i$$

where Φ_i is the flux linking the ith turn. Rather than doing this, we usually rely on experience and empirical quantities called winding factors and pitch factors to adjust the basic formula to apply to the real physical world.

8.10.2 Vector Potential and Inductance

An equivalent definition for inductance may be made using an energy point of view,

$$\boxed{L = \frac{2W_H}{I^2}} \tag{52}$$

where I is the total current flowing in the closed path and W_H is the energy in the magnetic field produced by the current. After using (52) to obtain several other general expressions for inductance, we will show that it is equivalent to (49). We first express the potential energy W_H in terms of the magnetic fields,

$$L = \frac{\int_{\text{vol}} \mathbf{B} \cdot \mathbf{H}\, dv}{I^2} \tag{53}$$

and then replace **B** with $\nabla \times \mathbf{A}$,

$$L = \frac{1}{I^2} \int_{\text{vol}} \mathbf{H} \cdot (\nabla \times \mathbf{A})\, dv$$

The vector identity

$$\nabla \cdot (\mathbf{A} \times \mathbf{H}) \equiv \mathbf{H} \cdot (\nabla \times \mathbf{A}) - \mathbf{A} \cdot (\nabla \times \mathbf{H}) \tag{54}$$

may be proved by expansion in rectangular coordinates. The inductance is then

$$L = \frac{1}{I^2}\left[\int_{vol}\nabla \cdot (\mathbf{A} \times \mathbf{H})\, dv + \int_{vol}\mathbf{A} \cdot (\nabla \times \mathbf{H})\, dv\right] \tag{55}$$

After applying the divergence theorem to the first integral and letting $\nabla \times \mathbf{H} = \mathbf{J}$ in the second integral, we have

$$L = \frac{1}{I^2}\left[\oint_S (\mathbf{A} \times \mathbf{H}) \cdot d\mathbf{S} + \int_{vol}\mathbf{A} \cdot \mathbf{J}\, dv\right]$$

The surface integral is zero, as the surface encloses the volume containing all the magnetic energy, and this requires that $\mathbf{A}$ and $\mathbf{H}$ be zero on the bounding surface. The inductance may therefore be written as

$$L = \frac{1}{I^2}\int_{vol}\mathbf{A} \cdot \mathbf{J}\, dv \tag{56}$$

Equation (56) expresses the inductance in terms of an integral of the values of $\mathbf{A}$ and $\mathbf{J}$ at every point. Because current density exists only within the conductor, the integrand is zero at all points *outside* the conductor, and the vector magnetic potential need not be determined there. The vector potential is that which arises from the current $\mathbf{J}$, and any other current source contributing a vector potential field in the region of the original current density is to be ignored for the present. Later we will see that this leads to a *mutual inductance*.

The vector magnetic potential $\mathbf{A}$ due to $\mathbf{J}$ is given by Eq. (51), Chapter 7,

$$\mathbf{A} = \int_{vol}\frac{\mu\mathbf{J}}{4\pi R}dv$$

and the inductance may therefore be expressed more basically as a rather formidable double volume integral,

$$L = \frac{1}{I^2}\int_{vol}\left(\int_{vol}\frac{\mu\mathbf{J}}{4\pi R}dv\right) \cdot \mathbf{J}\, dv \tag{57}$$

A slightly simpler integral expression is obtained by restricting our attention to current filaments of small cross section for which $\mathbf{J}\, dv$ may be replaced by $I\, d\mathbf{L}$ and the volume integral by a closed line integral along the axis of the filament,

$$\begin{aligned}L &= \frac{1}{I^2}\oint\left(\oint\frac{\mu I\, d\mathbf{L}}{4\pi R}\right) \cdot I\, d\mathbf{L} \\ &= \frac{\mu}{4\pi}\oint\left(\oint\frac{d\mathbf{L}}{R}\right) \cdot d\mathbf{L}\end{aligned} \tag{58}$$

Our only present interest in Eqs. (57) and (58) lies in their implication that the inductance is a function of the distribution of the current in space or the geometry of the conductor configuration.

To obtain our original definition of inductance (49), let us hypothesize a uniform current distribution in a filamentary conductor of small cross section so that $\mathbf{J}\, dv$

in (56) becomes $I\,d\mathbf{L}$,

$$L = \frac{1}{I}\oint \mathbf{A} \cdot d\mathbf{L} \tag{59}$$

For a small cross section, $d\mathbf{L}$ may be taken along the center of the filament. We now apply Stokes' theorem and obtain

$$L = \frac{1}{I}\int_S (\nabla \times \mathbf{A}) \cdot d\mathbf{S}$$

or

$$L = \frac{1}{I}\int_S \mathbf{B} \cdot d\mathbf{S}$$

or

$$L = \frac{\Phi}{I} \tag{60}$$

Retracing the steps by which (60) is obtained, we should see that the flux Φ is that portion of the total flux that passes through any and every open surface whose perimeter is the filamentary current path.

If we now let the filament make N identical turns about the total flux, an idealization that may be closely realized in some types of inductors, the closed line integral must consist of N laps about this common path, and (60) becomes

$$L = \frac{N\Phi}{I} \tag{61}$$

The flux Φ is now the flux crossing any surface whose perimeter is the path occupied by any *one* of the N turns. The inductance of an N-turn coil may still be obtained from (60), however, if we realize that the flux is that which crosses the complicated surface[4] whose perimeter consists of all N turns.

Use of any of the inductance expressions for a true filamentary conductor (having zero radius) leads to an infinite value of inductance, regardless of the configuration of the filament. Near the conductor, Ampère's circuital law shows that the magnetic field intensity varies inversely with the distance from the conductor, and a simple integration soon shows that an infinite amount of energy and an infinite amount of flux are contained within any finite cylinder about the filament. This difficulty is eliminated by specifying a small but finite filamentary radius.

The interior of any conductor also contains magnetic flux, and this flux links a variable fraction of the total current, depending on its location. These flux linkages lead to an *internal inductance*, which must be combined with the external inductance to obtain the total inductance. The internal inductance of a long, straight wire of circular cross section, radius a, and uniform current distribution is

$$\boxed{L_{a,\text{int}} = \frac{\mu}{8\pi} \quad \text{H/m}} \tag{62}$$

a result requested in Problem 8.43 at the end of this chapter.

---

[4] Somewhat like a spiral ramp.

In Chapter 11, we will see that the current distribution in a conductor at high frequencies tends to be concentrated near the surface. The internal flux is reduced, and it is usually sufficient to consider only the external inductance. At lower frequencies, however, internal inductance may become an appreciable part of the total inductance.

8.10.3 Mutual Inductance

We conclude by defining the *mutual inductance* between circuits 1 and 2, M_{12}, in terms of mutual flux linkages,

$$M_{12} = \frac{N_2 \Phi_{12}}{I_1} \qquad (63)$$

where Φ_{12} signifies the flux produced by I_1 which links the path of the filamentary current I_2, and N_2 is the number of turns in circuit 2. The mutual inductance, therefore, depends on the magnetic interaction between two currents. With either current alone, the total energy stored in the magnetic field can be found in terms of a single inductance, or self-inductance; with both currents having nonzero values, the total energy is a function of the two self-inductances and the mutual inductance. In terms of a mutual energy, it can be shown that (63) is equivalent to

$$M_{12} = \frac{1}{I_1 I_2} \int_{\text{vol}} (\mathbf{B}_1 \cdot \mathbf{H}_2) dv \qquad (64)$$

or

$$M_{12} = \frac{1}{I_1 I_2} \int_{\text{vol}} (\mu \mathbf{H}_1 \cdot \mathbf{H}_2) dv \qquad (65)$$

where $\mathbf{B}_1$ is the field resulting from I_1 (with $I_2 = 0$) and $\mathbf{H}_2$ is the field arising from I_2 (with $I_1 = 0$). Interchange of the subscripts does not change the right-hand side of (65), and therefore

$$M_{12} = M_{21} \qquad (66)$$

Mutual inductance is also measured in henrys, and we rely on the context to allow us to differentiate it from magnetization, also represented by M.

EXAMPLE 8.9

Calculate the self-inductances of and the mutual inductances between two coaxial solenoids of radius R_1 and R_2, $R_2 > R_1$, carrying currents I_1 and I_2 with n_1 and n_2 turns/m, respectively.

Solution. We first attack the mutual inductances. From Eq. (15), Chapter 7, we let $n_1 = N/d$, and obtain

$$\mathbf{H}_1 = n_1 I_1 \mathbf{a}_z \quad (0 < \rho < R_1)$$
$$= 0 \quad (\rho > R_1)$$

and

$$\mathbf{H}_2 = n_2 I_2 \mathbf{a}_z \quad (0 < \rho < R_2)$$
$$= 0 \quad (\rho > R_2)$$

Thus, for this uniform field

$$\Phi_{12} = \mu_0 n_1 I_1 \pi R_1^2$$

and

$$M_{12} = \mu_0 n_1 n_2 \pi R_1^2$$

Similarly,

$$\Phi_{21} = \mu_0 n_2 I_2 \pi R_1^2$$
$$M_{21} = \mu_0 n_1 n_2 \pi R_1^2 = M_{12}$$

If $n_1 = 50$ turns/cm, $n_2 = 80$ turns/cm, $R_1 = 2$ cm, and $R_2 = 3$ cm, then

$$M_{12} = M_{21} = 4\pi \times 10^{-7}(5000)(8000)\,\pi(0.02^2) = 63.2 \text{ mH/m}$$

The self-inductances are easily found. The flux produced in coil 1 by I_1 is

$$\Phi_{11} = \mu_0 n_1 I_1 \pi R_1^2$$

and thus

$$L_1 = \mu_0 n_1^2 S_1 d \quad \text{H}$$

The inductance per unit length is therefore

$$L_1 = \mu_0 n_1^2 S_1 \quad \text{H/m}$$

or

$$L_1 = 39.5 \quad \text{mH/m}$$

Similarly,

$$L_2 = \mu_0 n_2^2 S_2 = 22.7 \quad \text{mH/m}$$

We see, therefore, that there are many methods available for the calculation of self-inductance and mutual inductance. Unfortunately, even problems possessing a high degree of symmetry present very challenging integrals for evaluation, and only a few problems are available for us to try our skill on.

Inductance will be discussed in circuit terms in Chapter 10.

D8.12. Calculate the self-inductance of: (*a*) 3.5 m of coaxial cable with $a =$ 0.8 mm and $b = 4$ mm, filled with a material for which $\mu_r = 50$; (*b*) a toroidal coil of 500 turns, wound on a fiberglass form having a 2.5 × 2.5 cm square cross section and an inner radius of 2 cm; (*c*) a solenoid having 500 turns about a cylindrical core of 2 cm radius in which $\mu_r = 50$ for $0 < \rho < 0.5$ cm and $\mu_r = 1$ for $0.5 < \rho < 2$ cm; the length of the solenoid is 50 cm.

Ans. (*a*) 56.3 μH; (*b*) 1.01 mH; (*c*) 3.2 mH

D8.13. A solenoid is 50 cm long, 2 cm in diameter, and contains 1500 turns. The cylindrical core has a diameter of 2 cm and a relative permeability of 75. This coil is coaxial with a second solenoid, also 50 cm long, but with a 3 cm diameter and 1200 turns. Calculate: (a) L for the inner solenoid; (b) L for the outer solenoid; (c) M between the two solenoids.

Ans. (a) 133.2 mH; (b) 192 mH; (c) 106.6 mH

REFERENCES

1. Kraus, J. D., and D. A. Fleisch. (See References for Chapter 3.) Examples of the calculation of inductance are given on pp. 99–108.
2. Matsch, L. W. (See References for Chapter 6.) Chapter 3 is devoted to magnetic circuits and ferromagnetic materials.
3. Paul, C. R., K. W. Whites, and S. Y. Nasar. (See References for Chapter 7.) Magnetic circuits, including those with permanent magnets, are discussed on pp. 263–70.

CHAPTER 8   PROBLEMS

Quizzes

8.1 A point charge, $Q = -0.3$ μC and $m = 3 \times 10^{-16}$ kg, is moving through the field $\mathbf{E} = 30\mathbf{a}_z$ V/m. Use Eq. (1) and Newton's laws to develop the appropriate differential equations and solve them, subject to the initial conditions at $t = 0$, $\mathbf{v} = 3 \times 10^5 \mathbf{a}_x$ m/s at the origin. At $t = 3$ μs, find (a) the position $P(x, y, z)$ of the charge; (b) the velocity $\mathbf{v}$; (c) the kinetic energy of the charge.

8.2 Compare the magnitudes of the electric and magnetic forces on an electron that has attained a velocity of 10^7 m/s. Assume an electric field intensity of 10^5 V/m, and a magnetic flux density associated with that of the Earth's magnetic field in temperate latitudes, 0.5 gauss.

8.3 A point charge for which $Q = 2 \times 10^{-16}$ C and $m = 5 \times 10^{-26}$ kg is moving in the combined fields $\mathbf{E} = 100\mathbf{a}_x - 200\mathbf{a}_y + 300\mathbf{a}_z$ V/m and $\mathbf{B} = -3\mathbf{a}_x + 2\mathbf{a}_y - \mathbf{a}_z$ mT. If the charge velocity at $t = 0$ is $\mathbf{v}(0) = (2\mathbf{a}_x - 3\mathbf{a}_y - 4\mathbf{a}_z)10^5$ m/s (a) give the unit vector showing the direction in which the charge is accelerating at $t = 0$; (b) find the kinetic energy of the charge at $t = 0$.

8.4 Show that a charged particle in a uniform magnetic field describes a circular orbit with an orbital period that is independent of the radius. Find the relationship between the angular velocity and magnetic flux density for an electron (the *cyclotron frequency*).

8.5 A cylindrical shell of radius a is oriented along the z axis. The cylinder carries surface current of density $\mathbf{K} = K_0 \mathbf{a}_z$, where K_0 is a constant. Magnetic flux density $\mathbf{B} = -B_0 \mathbf{a}_\rho$ is applied at the surface over length d. Find the torque on the cylinder.

8.6 Show that the differential work in moving a current element $I\,d\mathbf{L}$ through a distance $d\mathbf{l}$ in a magnetic field $\mathbf{B}$ is the negative of that done in moving the element $I\,d\mathbf{l}$ through a distance $d\mathbf{L}$ in the same field.

8.7 A conducting strip of infinite length lies in the xy plane with its length oriented along the x axis, and where $-b/2 < y < b/2$ defines its width along y. Current I_1 flows down the strip in the positive x direction and is uniformly distributed over the width. Above the strip and parallel to it at $z = d$ is an infinitely long current filament that carries current I_2 in the positive x direction. Find the force of attraction between the two currents per unit length in x. Assume $d \ll b$.

8.8 Two conducting strips, having infinite length in the z direction, lie in the xz plane. One occupies the region $d/2 < x < b + d/2$ and carries surface current density $\mathbf{K} = K_0\mathbf{a}_z$; the other is situated at $-(b + d/2) < x < -d/2$ and carries surface current density $-K_0\mathbf{a}_z$. (a) Find the force per unit length in z that tends to separate the two strips. (b) Let b approach zero while maintaining constant current, $I = K_0 b$, and show that the force per unit length approaches $\mu_0 I^2/(2\pi d)$ N/m.

8.9 A current of $-100\mathbf{a}_z$ A/m flows on the conducting cylinder $\rho = 5$ mm, and $+500\mathbf{a}_z$ A/m is present on the conducting cylinder $\rho = 1$ mm. Find the magnitude of the total force per meter length that is acting to split the outer cylinder apart along its length.

8.10 A planar transmission line consists of two conducting planes of width b separated d m in air, carrying equal and opposite currents of I A. If $b \gg d$, find the force of repulsion per meter of length between the two conductors.

8.11 (a) Use Eq. (14), Section 8.3, to show that the force of attraction per unit length between two filamentary conductors in free space with currents $I_1\mathbf{a}_z$ at $x = 0$, $y = d/2$, and $I_2\mathbf{a}_z$ at $x = 0$, $y = -d/2$, is $\mu_0 I_1 I_2/(2\pi d)$. (b) Show how a simpler method can be used to check your result.

8.12 Two circular wire rings are parallel to each other, share the same axis, are of radius a, and are separated by distance d, where $d \ll a$. Each ring carries current I. Find the approximate force of attraction and indicate the relative orientations of the currents.

8.13 An infinitely long current filament is oriented along the x axis and carries current I_1 in the positive x direction. A second infinite filament is positioned at $z = z_0$ and carries current I_2 along its orientation in the $(\mathbf{a}_x + \mathbf{a}_y)/\sqrt{2}$ direction. (a) Find the differential magnetic force at any point on I_2 arising from I_1. (b) Find the net z-directed force acting on I_2.

8.14 A solenoid is 25 cm long, 3 cm in diameter, and carries 4 A dc in its 400 turns. Its axis is perpendicular to a uniform magnetic field of 0.8 Wb/m^2 in air. Using an origin at the center of the solenoid, calculate the torque acting on it.

8.15 A solid conducting filament extends from $x = -b$ to $x = b$ along the line $y = 2$, $z = 0$. This filament carries a current of 3 A in the $\mathbf{a}_x$ direction. An infinite filament on the z axis carries 5 A in the $\mathbf{a}_z$ direction. Obtain an expression for the torque exerted on the finite conductor about an origin located at $(0, 2, 0)$.

8.16 Assume that an electron is describing a circular orbit of radius a about a positively charged nucleus. (*a*) By selecting an appropriate current and area, show that the equivalent orbital dipole moment is $ea^2\omega/2$, where ω is the electron's angular velocity. (*b*) Show that the torque produced by a magnetic field parallel to the plane of the orbit is $ea^2\omega B/2$. (*c*) By equating the Coulomb and centrifugal forces, show that ω is $(4\pi\epsilon_0 m_e a^3/e^2)^{-1/2}$, where m_e is the electron mass. (*d*) Find values for the angular velocity, torque, and the orbital magnetic moment for a hydrogen atom, where a is about 6×10^{-11} m; let $B = 0.5$ T.

8.17 The hydrogen atom described in Problem 8.16 is now subjected to a magnetic field having the same direction as that of the atom. Show that the forces caused by B result in a decrease of the angular velocity by $eB/(2m_e)$ and a decrease in the orbital moment by $e^2 a^2 B/(4m_e)$. What are these decreases for the hydrogen atom in parts per million for an external magnetic flux density of 0.5 T?

8.18 Calculate the vector torque on the square loop shown in Figure 8.15 about an origin at A in the field B, given (*a*) $A(0, 0, 0)$ and $\mathbf{B} = 100\mathbf{a}_y$ mT; (*b*) $A(0, 0, 0)$ and $\mathbf{B} = 200\mathbf{a}_x + 100\mathbf{a}_y$ mT; (*c*) $A(1, 2, 3)$ and $\mathbf{B} = 200\mathbf{a}_x + 100\mathbf{a}_y - 300\mathbf{a}_z$ mT; (*d*) $A(1, 2, 3)$ and $\mathbf{B} = 200\mathbf{a}_x + 100\mathbf{a}_y - 300\mathbf{a}_z$ mT for $x \geq 2$ and $\mathbf{B} = 0$ elsewhere.

8.19 Given a material for which $\chi_m = 3.1$ and within which $\mathbf{B} = 0.4y\mathbf{a}_z$ T, find (*a*) **H**; (*b*) μ; (*c*) μ_r; (*d*) **M**; (*e*) **J**; (*f*) $\mathbf{J}_B$; (*g*) $\mathbf{J}_T$.

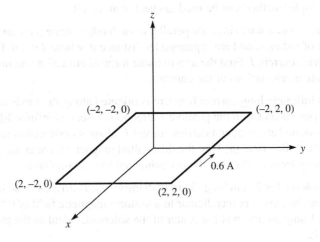

Figure 8.15   See Problem 8.18.

8.20 Find **H** in a material where (*a*) $\mu_r = 4.2$, there are 2.7×10^{29} atoms/m^3, and each atom has a dipole moment of $2.6 \times 10^{-30}\mathbf{a}_y$ A · m^2; (*b*) **M** = 270$\mathbf{a}_z$ A/m and $\mu = 2$ μH/m; (*c*) $\chi_m = 0.7$ and **B** = 2$\mathbf{a}_z$ T. (*d*) Find **M** in a material where bound surface current densities of 12$\mathbf{a}_z$ A/m and $-9\mathbf{a}_z$ A/m exist at $\rho = 0.3$ m and 0.4 m, respectively.

8.21 Find the magnitude of the magnetization in a material for which (*a*) the magnetic flux density is 0.02 Wb/m^2; (*b*) the magnetic field intensity is 1200 A/m and the relative permeability is 1.005; (*c*) there are 7.2×10^{28} atoms per cubic meter, each having a dipole moment of 4×10^{-30} A · m^2 in the same direction, and the magnetic susceptibility is 0.003.

8.22 Under some conditions, it is possible to approximate the effects of ferromagnetic materials by assuming linearity in the relationship of **B** and **H**. Let $\mu_r = 1000$ for a certain material of which a cylindrical wire of radius 1 mm is made. If $I = 1$ A and the current distribution is uniform, find (*a*) **B**, (*b*) **H**, (*c*) **M**, (*d*) **J**, and (*e*) $\mathbf{J}_B$ within the wire.

8.23 Calculate values for H_ϕ, B_ϕ, and M_ϕ at $\rho = c$ for a coaxial cable with $a = 2.5$ mm and $b = 6$ mm if it carries a current $I = 12$ A in the center conductor, and $\mu = 3$ μH/m for 2.5 mm $< \rho <$ 3.5 mm, $\mu = 5$ μH/m for 3.5 mm $< \rho <$ 4.5 mm, and $\mu = 10$ μH/m for 4.5 mm $< \rho <$ 6 mm. Use $c =$: (*a*) 3 mm; (*b*) 4 mm; (*c*) 5 mm.

8.24 Two current sheets, $K_0\mathbf{a}_y$ A/m at $z = 0$ and $-K_0\mathbf{a}_y$ A/m at $z = d$, are separated by an inhomogeneous material for which $\mu_r = az + 1$, where a is a constant. (*a*) Find expressions for **H** and **B** in the material. (*b*) Find the total flux that crosses a 1 m^2 area on the *yz* plane.

8.25 A conducting filament at $z = 0$ carries 12 A in the $\mathbf{a}_z$ direction. Let $\mu_r = 1$ for $\rho < 1$ cm, $\mu_r = 6$ for $1 < \rho < 2$ cm, and $\mu_r = 1$ for $\rho > 2$ cm. Find: (*a*) **H** everywhere; (*b*) **B** everywhere.

8.26 A long solenoid has a radius of 3 cm, 5000 turns/m, and carries current $I = 0.25$ A. The region $0 < \rho < a$ within the solenoid has $\mu_r = 5$, whereas $\mu_r = 1$ for $a < \rho < 3$ cm. Determine a so that (*a*) a total flux of 10 μWb is present; (*b*) the flux is equally divided between the regions $0 < \rho < a$ and $a < \rho < 3$ cm.

8.27 Let $\mu_{r1} = 2$ in region 1, defined by $2x + 3y - 4z > 1$, while $\mu_{r2} = 5$ in region 2 where $2x + 3y - 4z < 1$. In region 1, $\mathbf{H}_1 = 50\mathbf{a}_x - 30\mathbf{a}_y + 20\mathbf{a}_z$ A/m. Find (*a*) $\mathbf{H}_{N1}$; (*b*) $\mathbf{H}_{t1}$; (*c*) $\mathbf{H}_{t2}$; (*d*) $\mathbf{H}_{N2}$; (*e*) θ_1, the angle between $\mathbf{H}_1$ and $\mathbf{a}_{N21}$; (*f*) θ_2, the angle between $\mathbf{H}_2$ and $\mathbf{a}_{N21}$.

8.28 For values of B below the knee on the magnetization curve for silicon steel, approximate the curve by a straight line with $\mu = 5$ mH/m. The core shown in Figure 8.16 has areas of 1.6 cm^2 and lengths of 10 cm in each outer leg, and an area of 2.5 cm^2 and a length of 3 cm in the central leg. A coil of 1200 turns carrying 12 mA is placed around the central leg. Find B in the (*a*) center leg; (*b*) center leg if a 0.3 mm air gap is present in the center leg.

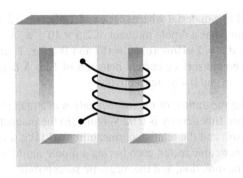

Figure 8.16   See Problem 8.28.

8.29 In Problem 8.28, the linear approximation suggested in the statement of the problem leads to flux density of 0.666 T in the central leg. Using this value of B and the magnetization curve for silicon steel, what current is required in the 1200-turn coil?

8.30 A rectangular core has fixed permeability $\mu r \gg 1$, a square cross section of dimensions $a \times a$, and has centerline dimensions around its perimeter of b and d. Coils 1 and 2, having turn numbers N_1 and N_2, are wound on the core. Consider a selected core cross-sectional plane as lying within the xy plane, such that the surface is defined by $0 < x < a$, $0 < y < a$. (a) With current I_1 in coil 1, use Ampere's circuital law to find the magnetic flux density as a function of position over the core cross section. (b) Integrate your result of part (a) to determine the total magnetic flux within the core. (c) Find the self-inductance of coil 1. (d) Find the mutual inductance between coils 1 and 2.

8.31 A toroid is constructed of a magnetic material having a cross-sectional area of 2.5 cm^2 and an effective length of 8 cm. There is also a short air gap of 0.25 mm length and an effective area of 2.8 cm^2. An mmf of 200 A · t is applied to the magnetic circuit. Calculate the total flux in the toroid if the magnetic material: (a) is assumed to have infinite permeability; (b) is assumed to be linear with $\mu_r = 1000$; (c) is silicon steel.

8.32 (a) Find an expression for the magnetic energy stored per unit length in a coaxial transmission line consisting of conducting sleeves of negligible thickness, having radii a and b. A medium of relative permeability μ_r fills the region between conductors. Assume current I flows in both conductors in opposite directions. (b) Obtain the inductance, L, per unit length of line by equating the energy to $(1/2)LI^2$.

8.33 A toroidal core has a square cross section, 2.5 cm $< \rho <$ 3.5 cm, -0.5 cm $< z < 0.5$ cm. The upper half of the toroid, $0 < z < 0.5$ cm, is constructed of a linear material for which $\mu_r = 10$, while the lower half, -0.5 cm $< z < 0$, has $\mu_r = 20$. An mmf of 150 A · t establishes a flux in the $\mathbf{a}_\phi$ direction. For $z > 0$, find: (a) $H_\phi(\rho)$; (b) $B_\phi(\rho)$; (c) $\Phi_{z>0}$. (d) Repeat for $z > 0$. (e) Find Φ_{total}.

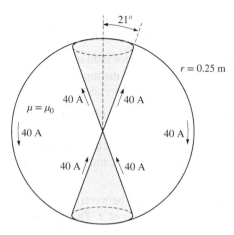

Figure 8.17   See Problem 8.35.

8.34 Determine the energy stored per unit length in the internal magnetic field of an infinitely long, straight wire of radius a, carrying uniform current I.

8.35 The cones $\theta = 21°$ and $\theta = 159°$ are conducting surfaces and carry total currents of 40 A, as shown in Figure 8.17. The currents return on a spherical conducting surface of 0.25 m radius. (a) Find **H** in the region $0 < r < 0.25$, $21° < \theta < 159°$, $0 < \phi < 2\pi$. (b) How much energy is stored in this region? (c) Find the inductance of the cone-sphere configuration. The inductance is that offered at the origin between the vertices of the cone.

8.36 The dimensions of the outer conductor of a coaxial cable are b and c, where $c > b$. Assuming $\mu = \mu_0$, find the magnetic energy stored per unit length in the region $b < \rho < c$ for a uniformly distributed total current I flowing in opposite directions in the inner and outer conductors.

8.37 A toroid has known reluctance $\mathfrak{R}$. Two windings having N_1 and N_2 turns are present. Find (a) the self-inductances of the two coils; (b) the mutual inductance between the coils.

8.38 A toroidal core has a rectangular cross section defined by the surfaces $\rho = 2$ cm, $\rho = 3$ cm, $z = 4$ cm, and $z = 4.5$ cm. The core material has a relative permeability of 80. The core is wound with two coils $N_1 = 1000$ and $N_2 = 2500$ turns of wire. Find (a) the self-inductances of the two coils; (b) the mutual inductance between coils.

8.39 Conducting planes in air at $z = 0$ and $z = d$ carry surface currents of $\pm K_0 \mathbf{a}_x$ A/m. (a) Find the energy stored in the magnetic field per unit length ($0 < x < 1$) in a width w ($0 < y < w$). (b) Calculate the inductance per unit length of this transmission line from $W_H = \frac{1}{2} L I^2$, where I is the total current in a width w in either conductor. (c) Calculate the total flux passing through the rectangle $0 < x < 1, 0 < z < d$, in the plane $y = 0$, and from this result again find the inductance per unit length.

8.40 A coaxial cable has conductor radii a and b, where $a < b$. Material of permeability $\mu_r \neq 1$ exists in the region $a < \rho < c$, whereas the region $c < \rho < b$ is air filled. Find an expression for the inductance per unit length.

8.41 A rectangular coil is composed of 150 turns of a filamentary conductor. Find the mutual inductance in free space between this coil and an infinite straight filament on the z axis if the four corners of the coil are located at: (a) (0, 1, 0), (0, 3, 0), (0, 3, 1), and (0, 1, 1); (b) (1, 1, 0), (1, 3, 0), (1, 3, 1), and (1, 1, 1).

8.42 Find the mutual inductance between two filaments forming circular rings of radii a and Δa, where $\Delta a \ll a$. The field should be determined by approximate methods. The rings are coplanar and concentric.

8.43 (a) Use energy relationships to show that the internal inductance of a nonmagnetic cylindrical wire of radius a carrying a uniformly distributed current I is $\mu_0/(8\pi)$ H/m. (b) Find the internal inductance if the portion of the conductor for which $\rho < c < a$ is removed.

8.44 Show that the external inductance per unit length of a two-wire transmission line carrying equal and opposite currents is approximately $(\mu/\pi) \ln(d/a)$ H/m, where a is the radius of each wire and d is the center-to-center wire spacing. On what basis is the approximation valid?

8.45 (a) Beginning with the definition of the scalar magnetic potential, $\mathbf{H} = -\nabla V_m$, and using Eq. (25), show that in a region having permanent magnetization $\mathbf{M}$ and zero current, a Poisson equation for magnetic potential can be developed:

$$\nabla^2 V_m = \frac{-\rho_m}{\mu_0}$$

where $\rho_m = -\mu_0 \nabla \cdot \mathbf{M}$ is an equivalent magnetic charge density. What happens when $\mathbf{M}$ is uniform? (b) Using the definition of the vector magnetic potential, $\mathbf{B} = \nabla \times \mathbf{A}$, and Eq. (25), show that when a permanent magnetization exists, and with zero current:

$$\nabla \times \nabla \times \mathbf{A} = \mu_0 \mathbf{J}_{eq}$$

where $\mathbf{J}_{eq} = \nabla \times \mathbf{M}$ in analogy to Eq. (57) in Chapter 7.

CHAPTER 9   Time-Varying Fields and Maxwell's Equations

术语

time-varying field　时变场，指场不仅是空间的函数还是时间的函数。

electromotive force（emf）　电动势。

motional emf　动生电动势。

displacement current　位移电流。位移电流密度为电位移D的时间变化率，故称为位移电流。

retarded potential　推迟位，指在时间上场点的位落后于源。

The basic relationships of the electrostatic field and the steady magnetic field were obtained in the previous eight chapters, and we are now ready to discuss time-varying fields. The discussion will be short, for vector analysis and vector calculus should now be more familiar tools; some of the relationships are unchanged, and most of the relationships are changed only slightly.

Two new concepts will be introduced: the electric field produced by a changing magnetic field and the magnetic field produced by a changing electric field. The first of these concepts resulted from experimental research by Michael Faraday and the second from the theoretical efforts of James Clerk Maxwell.

Maxwell actually was inspired by Faraday's experimental work and by the mental picture provided through the "lines of force" that Faraday introduced in developing his theory of electricity and magnetism. He was 40 years younger than Faraday, but they knew each other during the five years Maxwell spent in London as a young professor, a few years after Faraday had retired. Maxwell's theory was developed subsequent to his holding this university position while he was working alone at his home in Scotland. It occupied him for five years between the ages of 35 and 40.

The four basic equations of electromagnetic theory presented in this chapter bear his name. ■

9.1   FARADAY'S LAW

After Oersted[1] demonstrated in 1820 that an electric current affected a compass needle, Faraday professed his belief that if a current could produce a magnetic field, then a magnetic field should be able to produce a current. The concept of the "field"

Animations

---

[1] Hans Christian Oersted was professor of physics at the University of Copenhagen in Denmark.

was not available at that time, and Faraday's goal was to show that a current could be produced by "magnetism."

He worked on this problem intermittently over a period of 10 years, until he was finally successful in 1831.[2] He wound two separate windings on an iron toroid and placed a galvanometer in one circuit and a battery in the other. Upon closing the battery circuit, he noted a momentary deflection of the galvanometer; a similar deflection in the opposite direction occurred when the battery was disconnected. This, of course, was the first experiment he made involving a *changing* magnetic field, and he followed it with a demonstration that either a *moving* magnetic field or a moving coil could also produce a galvanometer deflection.

9.1.1 Faraday's Law in Point and Integral Forms

In terms of fields, we now say that a time-varying magnetic field produces an *electromotive force* (emf) that may establish a current in a suitable closed circuit. An electromotive force is merely a voltage that arises from conductors moving in a magnetic field or from changing magnetic fields, and we shall define it in this section. Faraday's law is customarily stated as

$$\text{emf} = -\frac{d\Phi}{dt} \text{ V} \qquad (1)$$

Equation (1) implies a closed path, although not necessarily a closed conducting path; the closed path, for example, might include a capacitor, or it might be a purely imaginary line in space. The magnetic flux is that flux which passes through any and every surface whose perimeter is the closed path, and $d\Phi/dt$ is the time rate of change of this flux.

A nonzero value of $d\Phi/dt$ may result from any of the following situations:

1. A time-changing flux linking a stationary closed path
2. Relative motion between a steady flux and a closed path
3. A combination of the two

The minus sign is an indication that the emf is in such a direction as to produce a current whose flux, if added to the original flux, would reduce the magnitude of the emf. This statement that the induced voltage acts to produce an opposing flux is known as *Lenz's law*.[3]

If the closed path is that taken by an *N*-turn filamentary conductor, it is often sufficiently accurate to consider the turns as coincident and let

$$\text{emf} = -N\frac{d\Phi}{dt} \qquad (2)$$

where Φ is now interpreted as the flux passing through any one of *N* coincident paths.

---

[2] Joseph Henry produced similar results at Albany Academy in New York at about the same time.

[3] Henri Frederic Emile Lenz was born in Germany but worked in Russia. He published his law in 1834.

We need to define emf as used in (1) or (2). The emf is obviously a scalar, and (perhaps not so obviously) a dimensional check shows that it is measured in volts. We define the emf as

$$\text{emf} = \oint \mathbf{E} \cdot d\mathbf{L} \tag{3}$$

and note that it is the voltage about a specific *closed path*. If any part of the path is changed, generally the emf changes. The departure from static results is clearly shown by (3), for an electric field intensity resulting from a static charge distribution must lead to zero potential difference about a closed path. In electrostatics, the line integral leads to a potential difference; with time-varying fields, the result is an emf or a voltage.

Replacing Φ in (1) with the surface integral of $\mathbf{B}$, we have

$$\text{emf} = \oint \mathbf{E} \cdot d\mathbf{L} = -\frac{d}{dt}\int_S \mathbf{B} \cdot d\mathbf{S} \tag{4}$$

where the fingers of our right hand indicate the direction of the closed path, and our thumb indicates the direction of $d\mathbf{S}$. A flux density $\mathbf{B}$ in the direction of $d\mathbf{S}$ and increasing with time thus produces an average value of $\mathbf{E}$ which is *opposite* to the positive direction about the closed path. The right-handed relationship between the surface integral and the closed line integral in (4) should always be kept in mind during flux integrations and emf determinations.

We will divide our investigation into two parts by first finding the contribution to the total emf made by a changing field within a stationary path (transformer emf), and then we will consider a moving path within a constant (motional, or generator, emf).

We first consider a stationary path. The magnetic flux is the only time-varying quantity on the right side of (4), and a partial derivative may be taken under the integral sign,

$$\text{emf} = \oint \mathbf{E} \cdot d\mathbf{L} = -\int_S \frac{\partial \mathbf{B}}{\partial t} \cdot d\mathbf{S} \tag{5}$$

Before we apply this simple result to an example, let us obtain the point form of this integral equation. Applying Stokes' theorem to the closed line integral, we have

$$\int_S (\nabla \times \mathbf{E}) \cdot d\mathbf{S} = -\int_S \frac{\partial \mathbf{B}}{\partial t} \cdot d\mathbf{S}$$

where the surface integrals may be taken over identical surfaces. The surfaces are perfectly general and may be chosen as differentials,

$$(\nabla \times \mathbf{E}) \cdot d\mathbf{S} = -\frac{\partial \mathbf{B}}{\partial t} \cdot d\mathbf{S}$$

and

$$\nabla \times \mathbf{E} = -\frac{\partial \mathbf{B}}{\partial t} \tag{6}$$

This is one of Maxwell's four equations as written in differential, or point, form, the form in which they are most generally used. Equation (5) is the integral form of this equation and is equivalent to Faraday's law as applied to a fixed path. If **B** is not a function of time, (5) and (6) evidently reduce to the electrostatic equations

$$\oint \mathbf{E} \cdot d\mathbf{L} = 0 \quad \text{(electrostatics)}$$

and

$$\nabla \times \mathbf{E} = 0 \quad \text{(electrostatics)}$$

9.1.2 EMF Arising from a Time-Varying Magnetic Field

As an example of the interpretation of (5) and (6), assume the existence of a simple magnetic field which increases exponentially with time within the cylindrical region $\rho < b$,

$$\mathbf{B} = B_0 e^{kt} \mathbf{a}_z \tag{7}$$

where $B_0 =$ constant. Choosing the circular path $\rho = a$, $a < b$, in the $z = 0$ plane, along which E_ϕ must be constant by symmetry, we then have from (5)

$$\text{emf} = 2\pi a E_\phi = -k B_0 e^{kt} \pi a^2$$

The emf around this closed path is $-k B_0 e^{kt} \pi a^2$. It is proportional to a^2 because the magnetic flux density is uniform and the flux passing through the surface at any instant is proportional to the area.

If we now replace a with ρ, $\rho < b$, the electric field intensity at any point is

$$\mathbf{E} = -\frac{1}{2} k B_0 e^{kt} \rho \mathbf{a}_\phi \tag{8}$$

We next try to obtain the same answer from (6), which becomes

$$(\nabla \times \mathbf{E})_z = -k B_0 e^{kt} = \frac{1}{\rho} \frac{\partial(\rho E_\phi)}{\partial \rho}$$

Multiplying by ρ and integrating from 0 to ρ (treating t as a constant, since the derivative is a partial derivative),

$$-\frac{1}{2} k B_0 e^{kt} \rho^2 = \rho E_\phi$$

or

$$\mathbf{E} = -\frac{1}{2} k B_0 e^{kt} \rho \mathbf{a}_\phi$$

once again.

If B_0 is considered positive, a filamentary conductor of resistance R would have a current flowing in the negative $\mathbf{a}_\phi$ direction, and this current would establish a flux within the circular loop in the negative $\mathbf{a}_z$ direction. Because E_ϕ increases exponentially with time, the current and flux do also, and thus they tend to reduce the time rate of increase of the applied flux and the resultant emf in accordance with Lenz's law.

Before leaving this example, it is well to point out that the given field **B** does not satisfy all of Maxwell's equations. Such fields are often assumed (*always* in ac-circuit problems) and cause no difficulty when they are interpreted properly. They

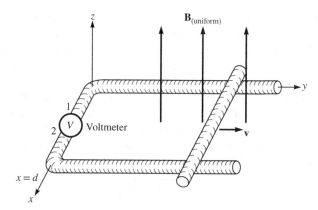

Figure 9.1   An example illustrating the application of Faraday's law to the case of a constant magnetic flux density **B** and a moving path. The shorting bar moves to the right with a velocity **v**, and the circuit is completed through the two rails and an extremely small high-resistance voltmeter. The voltmeter reading is $V_{12} = -Bvd$.

occasionally cause surprise, however. This particular field is discussed further in Problem 9.19 at the end of the chapter.

9.1.3 Motional EMF

Now consider the case of a time-constant flux and a moving closed path. Before we derive any special results from Faraday's law (1), we use the basic law to analyze the specific problem outlined in Figure 9.1. The closed circuit consists of two parallel conductors which are connected at one end by a high-resistance voltmeter of negligible dimensions and at the other end by a sliding bar moving at a velocity **v**. The magnetic flux density **B** is constant (in space and time) and is normal to the plane containing the closed path.

Let the position of the shorting bar be given by y; the flux passing through the surface within the closed path at any time t is then

$$\Phi = Byd$$

From (1), we obtain

$$\text{emf} = -\frac{d\Phi}{dt} = -B\frac{dy}{dt}d = -Bvd \qquad (9)$$

The emf is defined as $\oint \mathbf{E} \cdot d\mathbf{L}$ and we have a conducting path, so we may actually determine **E** at every point along the closed path. We found in electrostatics that the tangential component of **E** is zero at the surface of a conductor, and we shall show in Section 9.4 that the tangential component is zero at the surface of a *perfect* conductor ($\sigma = \infty$) for all time-varying conditions. This is equivalent to saying that a perfect conductor is a "short circuit." The entire closed path in Figure 9.1 may be considered a perfect conductor, with the exception of the voltmeter. The actual computation of $\oint \mathbf{E} \cdot d\mathbf{L}$ then must involve no contribution along the entire moving bar, both rails, and the voltmeter leads. Because we are integrating in a counterclockwise direction (keeping the interior of the positive side of the surface on our left as usual), the contribution $E \, \Delta L$ across the voltmeter must be $-Bvd$, showing that the electric field intensity in the instrument is directed from terminal 2 to terminal 1. For an up-scale reading, the positive terminal of the voltmeter should therefore be terminal 2.

The direction of the resultant small current flow may be confirmed by noting that the enclosed flux is reduced by a clockwise current in accordance with Lenz's law. The voltmeter terminal 2 is again seen to be the positive terminal.

We now consider this example using the concept of *motional emf*. The force on a charge Q moving at a velocity $\mathbf{v}$ in a magnetic field $\mathbf{B}$ is

$$\boxed{\mathbf{F} = Q\mathbf{v} \times \mathbf{B}}$$

or

$$\frac{\mathbf{F}}{Q} = \mathbf{v} \times \mathbf{B} \tag{10}$$

The sliding conducting bar is composed of positive and negative charges, and each experiences this force. The force per unit charge, as given by (10), is called the *motional* electric field intensity $\mathbf{E}_m$,

$$\boxed{\mathbf{E}_m = \mathbf{v} \times \mathbf{B}} \tag{11}$$

If the moving conductor were lifted off the rails, this electric field intensity would force electrons to one end of the bar (the far end) until the *static field* due to these charges just balanced the field induced by the motion of the bar. The resultant tangential electric field intensity would then be zero along the length of the bar.

The motional emf produced by the moving conductor is then

$$\text{emf} = \oint \mathbf{E}_m \cdot d\mathbf{L} = \oint (\mathbf{v} \times \mathbf{B}) \cdot d\mathbf{L} \tag{12}$$

where the last integral may have a nonzero value only along that portion of the path which is in motion, or along which $\mathbf{v}$ has some nonzero value. Evaluating the right side of (12), we obtain

$$\oint (\mathbf{v} \times \mathbf{B}) \cdot d\mathbf{L} = \int_d^0 vB \, dx = -Bvd$$

as before. This is the total emf, since $\mathbf{B}$ is not a function of time.

In the case of a conductor moving in a uniform constant magnetic field, we may therefore ascribe a motional electric field intensity $\mathbf{E}_m = \mathbf{v} \times \mathbf{B}$ to every portion of the moving conductor and evaluate the resultant emf by

$$\text{emf} = \oint \mathbf{E} \cdot d\mathbf{L} = \oint \mathbf{E}_m \cdot d\mathbf{L} = \oint (\mathbf{v} \times \mathbf{B}) \cdot d\mathbf{L} \tag{13}$$

If the magnetic flux density is also changing with time, then we must include both contributions, the transformer emf (5) and the motional emf (12),

$$\text{emf} = \oint \mathbf{E} \cdot d\mathbf{L} = -\int_S \frac{\partial \mathbf{B}}{\partial t} \cdot d\mathbf{S} + \oint (\mathbf{v} \times \mathbf{B}) \cdot d\mathbf{L} \tag{14}$$

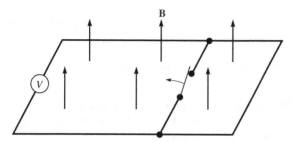

Figure 9.2 An apparent increase in flux linkages does not lead to an induced voltage when one part of a circuit is simply substituted for another by opening the switch. No indication will be observed on the voltmeter.

This expression is equivalent to the simple statement

$$\text{emf} = -\frac{d\Phi}{dt} \qquad (1)$$

and either can be used to determine these induced voltages.

Although (1) appears simple, there are a few contrived examples in which its proper application is quite difficult. These usually involve sliding contacts or switches; they always involve the substitution of one part of a circuit by a new part.[4] As an example, consider the simple circuit of Figure 9.2, which contains several perfectly conducting wires, an ideal voltmeter, a uniform constant field **B**, and a switch. When the switch is opened, there is obviously more flux enclosed in the voltmeter circuit; however, it continues to read zero. The change in flux has not been produced by either a time-changing **B** [first term of (14)] or a conductor moving through a magnetic field [second part of (14)]. Instead, a new circuit has been substituted for the old. Thus it is necessary to use care in evaluating the change in flux linkages.

The separation of the emf into the two parts indicated by (14), one due to the time rate of change of **B** and the other to the motion of the circuit, is somewhat arbitrary in that it depends on the relative velocity of the *observer* and the system. A field that is changing with both time and space may look constant to an observer moving with the field. This line of reasoning is developed more fully in applying the special theory of relativity to electromagnetic theory.[5]

D9.1. Within a certain region, $\epsilon = 10^{-11}$ F/m and $\mu = 10^{-5}$ H/m. If $B_x = 2 \times 10^{-4} \cos 10^5 t \sin 10^{-3} y$ T: (*a*) use $\nabla \times \mathbf{H} = \epsilon \dfrac{\partial \mathbf{E}}{\partial t}$ to find **E**; (*b*) find the total magnetic flux passing through the surface $x = 0$, $0 < y < 40$ m, $0 < z < 2$ m,

---

[4] See Bewley, in References at the end of the chapter, particularly pp. 12–19.

[5] This is discussed in several of the references listed in the References at the end of the chapter. See Panofsky and Phillips, pp. 142–51; Owen, pp. 231–45; and Harman in several places.

at $t = 1 \; \mu s$; (c) find the value of the closed line integral of **E** around the perimeter of the given surface.

Ans. (a) $-20\,000 \sin 10^5 t \cos 10^{-3} y \mathbf{a}_z$ V/m; (b) 0.318 mWb; (c) -3.19 V

D9.2. With reference to the sliding bar shown in Figure 9.1, let $d = 7$ cm, **B** $= 0.3 \mathbf{a}_z$ T, and $\mathbf{v} = 0.1 \mathbf{a}_y e^{20y}$ m/s. Let $y = 0$ at $t = 0$. Find: (a) $v(t = 0)$; (b) $y(t = 0.1)$; (c) $v(t = 0.1)$; (d) V_{12} at $t = 0.1$.

Ans. (a) 0.1 m/s; (b) 1.12 cm; (c) 0.125 m/s; (d) -2.63 mV

9.2 DISPLACEMENT CURRENT

Faraday's experimental law has been used to obtain one of Maxwell's equations in differential form,

$$\nabla \times \mathbf{E} = -\frac{\partial \mathbf{B}}{\partial t} \tag{15}$$

which shows us that a time-changing magnetic field produces an electric field. Remembering the definition of curl, we see that this electric field has the special property of circulation; its line integral about a general closed path is not zero. Now we turn our attention to the time-changing electric field.

9.2.1 Modifying Ampère's Law for Time-Varying Fields

We first look at the point form of Ampère's circuital law as it applies to steady magnetic fields,

$$\nabla \times \mathbf{H} = \mathbf{J} \tag{16}$$

and show its inadequacy for time-varying conditions by taking the divergence of each side,

$$\nabla \cdot \nabla \times \mathbf{H} \equiv 0 = \nabla \cdot \mathbf{J}$$

The divergence of the curl is identically zero, so $\nabla \cdot \mathbf{J}$ is also zero. However, the equation of continuity,

$$\nabla \cdot \mathbf{J} = -\frac{\partial \rho_v}{\partial t}$$

then shows us that (16) can be true only if $\partial \rho_v / \partial t = 0$. This is an unrealistic limitation, and (16) must be amended before we can accept it for time-varying fields. Suppose we add an unknown term **G** to (16),

$$\nabla \times \mathbf{H} = \mathbf{J} + \mathbf{G}$$

Again taking the divergence, we have

$$0 = \nabla \cdot \mathbf{J} + \nabla \cdot \mathbf{G}$$

Thus

$$\nabla \cdot \mathbf{G} = \frac{\partial \rho_v}{\partial t}$$

Replacing ρ_v with $\nabla \cdot \mathbf{D}$,

$$\nabla \cdot \mathbf{G} = \frac{\partial}{\partial t}(\nabla \cdot \mathbf{D}) = \nabla \cdot \frac{\partial \mathbf{D}}{\partial t}$$

from which we obtain the simplest solution for $\mathbf{G}$,

$$\mathbf{G} = \frac{\partial \mathbf{D}}{\partial t}$$

Ampère's circuital law in point form therefore becomes

$$\boxed{\nabla \times \mathbf{H} = \mathbf{J} + \frac{\partial \mathbf{D}}{\partial t}} \qquad (17)$$

Equation (17) has not been derived. It is merely a form we have obtained that does not disagree with the continuity equation. It is also consistent with all our other results, and we accept it as we did each experimental law and the equations derived from it. We are building a theory, and we have every right to our equations *until they are proved wrong*. This has not yet been done.

We now have a second one of Maxwell's equations and shall investigate its significance. The additional term $\partial \mathbf{D}/\partial t$ has the dimensions of current density, amperes per square meter. Because it results from a time-varying electric flux density (or displacement density), Maxwell termed it a *displacement current density*. We sometimes denote it by $\mathbf{J}_d$:

$$\nabla \times \mathbf{H} = \mathbf{J} + \mathbf{J}_d$$

$$\mathbf{J}_d = \frac{\partial \mathbf{D}}{\partial t}$$

This is the third type of current density we have met. Conduction current density,

$$\mathbf{J} = \sigma \mathbf{E}$$

is the motion of charge (usually electrons) in a region of zero net charge density, and convection current density,

$$\mathbf{J} = \rho_v \mathbf{v}$$

is the motion of volume charge density. Both are represented by $\mathbf{J}$ in (17). Bound current density is, of course, included in $\mathbf{H}$. In a nonconducting medium in which no volume charge density is present, $\mathbf{J} = 0$, and then

$$\nabla \times \mathbf{H} = \frac{\partial \mathbf{D}}{\partial t} \quad (\text{if } \mathbf{J} = 0) \qquad (18)$$

Notice the symmetry between (18) and (15):

$$\nabla \times \mathbf{E} = -\frac{\partial \mathbf{B}}{\partial t} \qquad (15)$$

Again, the analogy between the intensity vectors $\mathbf{E}$ and $\mathbf{H}$ and the flux density vectors $\mathbf{D}$ and $\mathbf{B}$ is apparent. We cannot place too much faith in this analogy,

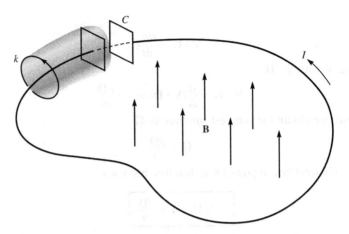

Figure 9.3 A filamentary conductor forms a loop connecting the two plates of a parallel-plate capacitor. A time-varying magnetic field inside the closed path produces an emf of $V_0 \cos \omega t$ around the closed path. The conduction current I is equal to the displacement current between the capacitor plates.

however, for it fails when we investigate forces on particles. The force on a charge is related to **E** and to **B**, and some good arguments may be presented showing an analogy between **E** and **B** and between **D** and **H**. We omit them, however, and merely say that the concept of displacement current was probably suggested to Maxwell by the symmetry first mentioned in this paragraph.[6]

The total displacement current crossing any given surface is expressed by the surface integral,

$$I_d = \int_S \mathbf{J}_d \cdot d\mathbf{S} = \int_S \frac{\partial \mathbf{D}}{\partial t} \cdot d\mathbf{S}$$

and we may obtain the time-varying version of Ampère's circuital law by integrating (17) over the surface $S,S,$

$$\int_S (\nabla \times \mathbf{H}) \cdot d\mathbf{S} = \int_S \mathbf{J} \cdot d\mathbf{S} + \int_S \frac{\partial \mathbf{D}}{\partial t} \cdot d\mathbf{S}$$

and applying Stokes' theorem,

$$\boxed{\oint \mathbf{H} \cdot d\mathbf{L} = I + I_d = I + \int_S \frac{\partial \mathbf{D}}{\partial t} \cdot d\mathbf{S}} \tag{19}$$

9.2.2 An Illustration of Displacement Current

What is the nature of displacement current density? Consider the simple circuit of Figure 9.3, which contains a filamentary loop and a parallel-plate capacitor. Within

---

[6] The analogy that relates **B** to **D** and **H** to **E** is strongly advocated by Fano, Chu, and Adler (see References for Chapter 5); the case for comparing **B** to **E** and **D** to **H** is presented in Halliday and Resnick (see References for this chapter).

the loop, a magnetic field varying sinusoidally with time is applied to produce an emf about the closed path (the filament plus the dashed portion between the capacitor plates), which we shall take as

$$\text{emf} = V_0 \cos \omega t$$

Using elementary circuit theory and assuming that the loop has negligible resistance and inductance, we may obtain the current in the loop as

$$I = -\omega C V_0 \sin \omega t$$

$$= -\omega \frac{\epsilon S}{d} V_0 \sin \omega t$$

where the quantities ϵ, S, and d pertain to the capacitor. We now apply Ampère's circuital law about the smaller closed circular path k and neglect displacement current for the moment:

$$\oint_k \mathbf{H} \cdot d\mathbf{L} = I_k$$

The path and the value of $\mathbf{H}$ along the path are both definite quantities (although difficult to determine), and $\oint_k \mathbf{H} \cdot d\mathbf{L}$ is a definite quantity. The current I_k is that current through every surface whose perimeter is the path k. If we choose a simple surface punctured by the filament, such as the plane circular surface defined by the circular path k, the current is evidently the conduction current. Suppose now we consider the closed path k as the mouth of a paper bag whose bottom passes between the capacitor plates. The bag is not pierced by the filament, and the conductor current is zero. Now we need to consider displacement current, for within the capacitor

$$D = \epsilon E = \epsilon \left(\frac{V_0}{d} \cos \omega t \right)$$

and therefore

$$I_d = \frac{\partial D}{\partial t} S = -\omega \frac{\epsilon S}{d} V_0 \sin \omega t$$

This is the same value as that of the conduction current in the filamentary loop. Therefore the application of Ampère's circuital law, including displacement current to the path k, leads to a definite value for the line integral of $\mathbf{H}$. This value must be equal to the total current crossing the chosen surface. For some surfaces the current is almost entirely conduction current, but for those surfaces passing between the capacitor plates, the conduction current is zero, and it is the displacement current which is now equal to the closed line integral of $\mathbf{H}$.

Physically, we should note that a capacitor stores charge and that the electric field between the capacitor plates is much greater than the small leakage fields outside. We therefore introduce little error when we neglect displacement current on all those surfaces which do not pass between the plates.

Displacement current is associated with time-varying electric fields and therefore exists in all imperfect conductors carrying a time-varying conduction current.

The last part of the following drill problem indicates the reason why this additional current was never discovered experimentally.

D9.3. Find the amplitude of the displacement current density: (*a*) adjacent to an automobile antenna where the magnetic field intensity of an FM signal is $H_x = 0.15 \cos[3.12(3 \times 10^8 t - y)]$ A/m; (*b*) in the airspace at a point within a large power distribution transformer where $\mathbf{B} = 0.8 \cos[1.257 \times 10^{-6}(3 \times 10^8 t - x)]\mathbf{a}_y$ T; (*c*) within a large, oil-filled power capacitor where $\epsilon_r = 5$ and $\mathbf{E} = 0.9 \cos[1.257 \times 10^{-6}(3 \times 10^8 t - z\sqrt{5})]\mathbf{a}_x$ MV/m; (*d*) in a metallic conductor at 60 Hz, if $\epsilon = \epsilon_0$, $\mu = \mu_0$, $\sigma = 5.8 \times 10^7$ S/m, and $\mathbf{J} = \sin(377t - 117.1z)\mathbf{a}_x$ MA/m^2.

Ans. (*a*) 0.468 A/m^2; (*b*) 0.800 A/m^2; (*c*) 0.0150 A/m^2; (*d*) 57.6 pA/m^2

9.3 MAXWELL'S EQUATIONS IN POINT FORM

We have already obtained two of Maxwell's equations for time-varying fields,

$$\nabla \times \mathbf{E} = -\frac{\partial \mathbf{B}}{\partial t} \tag{20}$$

and

$$\nabla \times \mathbf{H} = \mathbf{J} + \frac{\partial \mathbf{D}}{\partial t} \tag{21}$$

The remaining two equations are unchanged from their non-time-varying form:

$$\nabla \cdot \mathbf{D} = \rho_v \tag{22}$$

$$\nabla \cdot \mathbf{B} = 0 \tag{23}$$

Equation (22) essentially states that charge density is a source (or sink) of electric flux lines. Note that we can no longer say that *all* electric flux begins and terminates on charge, because the point form of Faraday's law (20) shows that **E**, and hence **D**, may have circulation if a changing magnetic field is present. Thus the lines of electric flux may form closed loops. However, the converse is still true, and every coulomb of charge must have one coulomb of electric flux diverging from it.

Equation (23) again acknowledges the fact that "magnetic charges," or poles, are not known to exist. Magnetic flux is always found in closed loops and never diverges from a point source.

These four equations form the basis of all electromagnetic theory. They are partial differential equations and relate the electric and magnetic fields to each other and to their sources, charge, and current density. The auxiliary equations relating **D** and **E**,

$$\boxed{\mathbf{D} = \epsilon \mathbf{E}} \tag{24}$$

relating **B** and **H**,

$$\boxed{\mathbf{B} = \mu \mathbf{H}} \tag{25}$$

defining conduction current density,

$$\boxed{\mathbf{J} = \sigma \mathbf{E}} \tag{26}$$

and defining convection current density in terms of the volume charge density ρv,

$$\boxed{\mathbf{J} = \rho_v \mathbf{v}} \tag{27}$$

are also required to define and relate the quantities appearing in Maxwell's equations.

The potentials V and **A** have not been included because they are not strictly necessary, although they are extremely useful. They will be discussed at the end of this chapter.

If we do not have "nice" materials to work with, then we should replace (24) and (25) with the relationships involving the polarization and magnetization fields,

$$\boxed{\mathbf{D} = \epsilon_0 \mathbf{E} + \mathbf{P}} \tag{28}$$

$$\boxed{\mathbf{B} = \mu_0 (\mathbf{H} + \mathbf{M})} \tag{29}$$

For linear materials we may relate **P** to **E**

$$\mathbf{P} = \chi_e \epsilon_0 \mathbf{E} \tag{30}$$

and **M** to **H**

$$\mathbf{M} = \chi_m \mathbf{H} \tag{31}$$

Finally, because of its fundamental importance we should include the Lorentz force equation, written in point form as the force per unit volume,

$$\boxed{\mathbf{f} = \rho_v (\mathbf{E} + \mathbf{v} \times \mathbf{B})} \tag{32}$$

The chapters that follow are devoted to the application of Maxwell's equations to several simple problems.

9.4 MAXWELL'S EQUATIONS IN INTEGRAL FORM

The integral forms of Maxwell's equations are usually easier to recognize in terms of the experimental laws from which they have been obtained by a generalization process. Experiments must treat physical macroscopic quantities, and their results therefore are expressed in terms of integral relationships. A differential equation always represents a theory. We now collect the integral forms of Maxwell's equations from Section 9.3.

Integrating (20) over a surface and applying Stokes' theorem, we obtain Faraday's law,

$$\oint \mathbf{E} \cdot d\mathbf{L} = -\int_S \frac{\partial \mathbf{B}}{\partial t} \cdot d\mathbf{S} \tag{33}$$

and the same process applied to (21) yields Ampère's circuital law,

$$\oint \mathbf{H} \cdot d\mathbf{L} = I + \int_S \frac{\partial \mathbf{D}}{\partial t} \cdot d\mathbf{S} \tag{34}$$

Gauss's laws for the electric and magnetic fields are obtained by integrating (22) and (23) throughout a volume and using the divergence theorem:

$$\oint_S \mathbf{D} \cdot d\mathbf{S} = \int_{vol} \rho_v dv \tag{35}$$

$$\oint_S \mathbf{B} \cdot d\mathbf{S} = 0 \tag{36}$$

These four integral equations enable us to find the boundary conditions on $\mathbf{B}$, $\mathbf{D}$, $\mathbf{H}$, and $\mathbf{E}$, which are necessary to evaluate the constants obtained in solving Maxwell's equations in partial differential form. These boundary conditions are in general unchanged from their forms for static or steady fields, and the same methods may be used to obtain them. Between any two real physical media (where $\mathbf{K}$ must be zero on the boundary surface), (33) enables us to relate the tangential $\mathbf{E}$-field components,

$$E_{t1} = E_{t2} \tag{37}$$

and from (34),

$$H_{t1} = H_{t2} \tag{38}$$

The surface integrals produce the boundary conditions on the normal components,

$$D_{N1} - D_{N2} = \rho_S \tag{39}$$

and

$$B_{N1} = B_{N2} \tag{40}$$

It is often desirable to idealize a physical problem by assuming a perfect conductor for which σ is infinite but $\mathbf{J}$ is finite. From Ohm's law, then, in a perfect conductor,

$$\mathbf{E} = 0$$

and it follows from the point form of Faraday's law that

$$\mathbf{H} = 0$$

for time-varying fields. The point form of Ampère's circuital law then shows that the finite value of $\mathbf{J}$ is

$$\mathbf{J} = 0$$

and current must be carried on the conductor surface as a surface current $\mathbf{K}$. Thus, if region 2 is a perfect conductor, (37) to (40) become, respectively,

$$E_{t1} = 0 \tag{41}$$
$$H_{t1} = K \quad (\mathbf{H}_{t1} = \mathbf{K} \times \mathbf{a}_N) \tag{42}$$
$$D_{N1} = \rho_s \tag{43}$$
$$B_{N1} = 0 \tag{44}$$

where $\mathbf{a}_N$ is an outward normal at the conductor surface.

Note that surface charge density is considered a physical possibility for either dielectrics, perfect conductors, or imperfect conductors, but that surface *current* density is assumed only in conjunction with perfect conductors.

The preceding boundary conditions are a very necessary part of Maxwell's equations. All real physical problems have boundaries and require the solution of Maxwell's equations in two or more regions and the matching of these solutions at the boundaries. In the case of perfect conductors, the solution of the equations within the conductor is trivial (all time-varying fields are zero), but the application of the boundary conditions (41) to (44) may be very difficult.

Certain fundamental properties of wave propagation are evident when Maxwell's equations are solved for an *unbounded* region. This problem is treated in Chapter 11. It represents the simplest application of Maxwell's equations because it is the only problem which does not require the application of any boundary conditions.

D9.5. The unit vector $0.64\mathbf{a}_x + 0.6\mathbf{a}_y - 0.48\mathbf{a}_z$ is directed from region 2 ($\epsilon_r =$ 2, $\mu_r = 3$, $\sigma_2 = 0$) toward region 1 ($\epsilon_{r1} = 4$, $\mu_{r1} = 2$, $\sigma_1 = 0$). If $\mathbf{B}_1 = (\mathbf{a}_x - 2\mathbf{a}_y + 3\mathbf{a}_z)$ sin $300t$ T at point P in region 1 adjacent to the boundary, find the amplitude at P of: (a) $\mathbf{B}_{N1}$; (b) $\mathbf{B}_{t1}$; (c) $\mathbf{B}_{N2}$; (d) $\mathbf{B}_2$.

Ans. (a) 2.00 T; (b) 3.16 T; (c) 2.00 T; (d) 5.15 T

D9.6. The surface $y = 0$ is a perfectly conducting plane, whereas the region $y > 0$ has $\epsilon_r = 5$, $\mu_r = 3$, and $\sigma = 0$. Let $\mathbf{E} = 20 \cos(2 \times 10^8 t - 2.58z)\mathbf{a}_y$ V/m for $y > 0$, and find at $t = 6$ ns; (a) ρ_S at $P(2, 0, 0.3)$; (b) $\mathbf{H}$ at P; (c) $\mathbf{K}$ at P.

Ans. (a) 0.81 nC/m^2; (b) $-62.3\mathbf{a}_x$ mA/m; (c) $-62.3\mathbf{a}_z$ mA/m

9.5 THE RETARDED POTENTIALS

Animations

The time-varying potentials, usually called *retarded* potentials for a reason that we will see shortly, find their greatest application in radiation problems (to be addressed in Chapter 14) in which the distribution of the source is known approximately. We should remember that the scalar electric potential V may be expressed in terms of a static charge distribution,

$$V = \int_{\text{vol}} \frac{\rho_v \, dv}{4\pi\epsilon R} \quad \text{(static)} \tag{45}$$

and the vector magnetic potential may be found from a current distribution which is constant with time,

$$\mathbf{A} = \int_{\text{vol}} \frac{\mu \mathbf{J} \, dv}{4\pi R} \quad \text{(dc)} \tag{46}$$

The differential equations satisfied by V,

$$\nabla^2 V = -\frac{\rho_v}{\epsilon} \quad \text{(static)} \tag{47}$$

and $\mathbf{A}$,

$$\nabla^2 \mathbf{A} = -\mu\mathbf{J} \quad \text{(dc)} \tag{48}$$

may be regarded as the point forms of the integral equations (45) and (46), respectively.

Having found V and $\mathbf{A}$, the fundamental fields are then simply obtained by using the gradient,

$$\mathbf{E} = -\nabla V \quad \text{(static)} \tag{49}$$

or the curl,

$$\mathbf{B} = \nabla \times \mathbf{A} \quad \text{(dc)} \tag{50}$$

We now wish to define suitable time-varying potentials which are consistent with the preceding expressions when only static charges and direct currents are involved.

Equation (50) apparently is still consistent with Maxwell's equations. These equations state that $\nabla \cdot \mathbf{B} = 0$, and the divergence of (50) leads to the divergence of the curl that is identically zero. We will therefore tentatively accept (50) as satisfactory for time-varying fields and turn our attention to (49).

The inadequacy of (49) is obvious because application of the curl operation to each side and recognition of the curl of the gradient as being identically zero confront us with $\nabla \times \mathbf{E} = 0$. However, the point form of Faraday's law states that $\nabla \times \mathbf{E}$ is not generally zero, so we may effect an improvement by adding an unknown term to (49),

$$\mathbf{E} = -\nabla V + \mathbf{N}$$

taking the curl,

$$\nabla \times \mathbf{E} = 0 + \nabla \times \mathbf{N}$$

using the point form of Faraday's law,

$$\nabla \times \mathbf{N} = -\frac{\partial \mathbf{B}}{\partial t}$$

and using (50), giving us

$$\nabla \times \mathbf{N} = -\frac{\partial}{\partial t}(\nabla \times \mathbf{A})$$

or

$$\nabla \times \mathbf{N} = -\nabla \times \frac{\partial \mathbf{A}}{\partial t}$$

The simplest solution of this equation is

$$\mathbf{N} = -\frac{\partial \mathbf{A}}{\partial t}$$

and this leads to

$$\boxed{\mathbf{E} = -\nabla V - \frac{\partial \mathbf{A}}{\partial t}} \tag{51}$$

We still must check (50) and (51) by substituting them into the remaining two of Maxwell's equations:

$$\nabla \times \mathbf{H} = \mathbf{J} + \frac{\partial \mathbf{D}}{\partial t}$$
$$\nabla \cdot \mathbf{D} = \rho_v$$

Doing this, we obtain the more complicated expressions

$$\frac{1}{\mu} \nabla \times \nabla \times \mathbf{A} = \mathbf{J} + \epsilon\left(-\nabla \frac{\partial V}{\partial t} - \frac{\partial^2 \mathbf{A}}{\partial t^2}\right)$$

and

$$\epsilon\left(-\nabla \cdot \nabla V - \frac{\partial}{\partial t} \nabla \cdot \mathbf{A}\right) = \rho_v$$

or

$$\nabla(\nabla \cdot \mathbf{A}) - \nabla^2 \mathbf{A} = \mu \mathbf{J} - \mu\epsilon\left(\nabla\frac{\partial V}{\partial t} + \frac{\partial^2 \mathbf{A}}{\partial t^2}\right) \tag{52}$$

and

$$\nabla^2 V + \frac{\partial}{\partial t}(\nabla \cdot \mathbf{A}) = -\frac{\rho_v}{\epsilon} \tag{53}$$

There is no apparent inconsistency in (52) and (53). Under static or dc conditions $\nabla \cdot \mathbf{A} = 0$, and (52) and (53) reduce to (48) and (47), respectively. We will therefore assume that the time-varying potentials may be defined in such a way that $\mathbf{B}$ and $\mathbf{E}$ may be obtained from them through (50) and (51). These latter two equations do not serve, however, to define $\mathbf{A}$ and V *completely*. They represent necessary, but not sufficient, conditions. Our initial assumption was merely that $\mathbf{B} = \nabla \times \mathbf{A}$, and a vector cannot be defined by giving its curl alone. Suppose, for example, that we have a very simple vector potential field in which A_y and A_z are zero. Expansion of (50) leads to

$$B_x = 0$$
$$B_y = \frac{\partial A_x}{\partial z}$$
$$B_z = -\frac{\partial A_x}{\partial y}$$

and we see that no information is available about the manner in which A_x varies with x. This information could be found if we also knew the value of the divergence of $\mathbf{A}$, for in our example

$$\nabla \cdot \mathbf{A} = \frac{\partial A_x}{\partial x}$$

Finally, we should note that our information about $\mathbf{A}$ is given only as partial derivatives and that a space-constant term might be added. In all physical problems in which the region of the solution extends to infinity, this constant term must be zero, for there can be no fields at infinity.

Generalizing from this simple example, we may say that a vector field is defined completely when both its curl and divergence are given and when its value is known at any one point (including infinity). We are therefore at liberty to specify the divergence of $\mathbf{A}$, and we do so with an eye on (52) and (53), seeking the simplest expressions. We define

$$\nabla \cdot \mathbf{A} = -\mu\epsilon \frac{\partial V}{\partial t} \tag{54}$$

and (52) and (53) become

$$\nabla^2 \mathbf{A} = -\mu\mathbf{J} + \mu\epsilon \frac{\partial^2 \mathbf{A}}{\partial t^2} \tag{55}$$

and

$$\nabla^2 V = -\frac{\rho_v}{\epsilon} + \mu\epsilon \frac{\partial^2 V}{\partial t^2} \tag{56}$$

These equations are related to the wave equation, which will be discussed in Chapters 10 and 11. They show considerable symmetry, and we should be highly pleased with our definitions of V and $\mathbf{A}$,

$$\mathbf{B} = \nabla \times \mathbf{A} \tag{50}$$

$$\nabla \cdot \mathbf{A} = -\mu\epsilon\frac{\partial V}{\partial t} \tag{54}$$

$$\mathbf{E} = -\nabla V - \frac{\partial \mathbf{A}}{\partial t} \tag{51}$$

The integral equivalents of (45) and (46) for the time-varying potentials follow from the definitions (50), (51), and (54), but we shall merely present the final results and indicate their general nature. In Chapter 11, we will find that any electromagnetic disturbance will travel at a velocity

$$v = \frac{1}{\sqrt{\mu\epsilon}}$$

through any homogeneous medium described by μ and ϵ. In the case of free space, this velocity turns out to be the velocity of light, approximately 3×10^8 m/s. It is logical, then, to suspect that the potential at any point is due not to the value of the charge density at some distant point at the same instant, but to its value at some previous time, because the effect propagates at a finite velocity. Thus (45) becomes

$$\boxed{V = \int_{\text{vol}} \frac{[\rho_v]}{4\pi\epsilon R} dv} \tag{57}$$

where $[\rho_v]$ indicates that every t appearing in the expression for ρ_v has been replaced by a *retarded* time,

$$t' = t - \frac{R}{v}$$

Thus, if the charge density throughout space were given by

$$\rho_v = e^{-r}\cos\omega t$$

then

$$[\rho_v] = e^{-r}\cos\left[\omega\left(t - \frac{R}{v}\right)\right]$$

where R is the distance between the differential element of charge being considered and the point at which the potential is to be determined.

The retarded vector magnetic potential is given by

$$\boxed{\mathbf{A} = \int_{\text{vol}} \frac{\mu[\mathbf{J}]}{4\pi R} dv} \tag{58}$$

The use of a retarded time has resulted in the time-varying potentials being given the name of retarded potentials. In Chapter 14 we will apply (58) to the simple situation of a differential current element in which I is a sinusoidal function of time. Other simple applications of (58) are considered in several problems at the end of this chapter.

We may summarize the use of the potentials by stating that a knowledge of the distribution of ρ_v and $\mathbf{J}$ throughout space theoretically enables us to determine V and $\mathbf{A}$ from (57) and (58). The electric and magnetic fields are then obtained by applying (50) and (51). If the charge and current distributions are unknown, or reasonable approximations cannot be made for them, these potentials usually offer no easier path toward the solution than does the direct application of Maxwell's equations.

D9.7. A point charge of $4 \cos 10^8 \pi t\ \mu C$ is located at $P_+(0, 0, 1.5)$, whereas $-4 \cos 10^8 \pi t\ \mu C$ is at $P_-(0, 0, -1.5)$, both in free space. Find V at $P(r = 450, \theta, \phi = 0)$ at $t = 15$ ns for $\theta =:$ (a) $0°$; (b) $90°$; (c) $45°$.

Ans. (a) 159.8 V; (b) 0; (c) 143 V

REFERENCES

1. Bewley, L. V. *Flux Linkages and Electromagnetic Induction.* New York: Macmillan, 1952. This little book discusses many of the paradoxical examples involving induced (?) voltages.

2. Faraday, M. *Experimental Researches in Electricity.* London: B. Quaritch, 1839, 1855. Very interesting reading of early scientific research. A more recent and available source is *Great Books of the Western World,* vol. 45, Encyclopaedia Britannica, Inc., Chicago, 1952.

3. Halliday, D., R. Resnick, and J. Walker. *Fundamentals of Physics.* 5th ed. New York: John Wiley & Sons, 1997. This text is widely used in the first university-level course in physics.

4. Harman, W. W. *Fundamentals of Electronic Motion.* New York: McGraw-Hill, 1953. Relativistic effects are discussed in a clear and interesting manner.

5. Nussbaum, A. *Electromagnetic Theory for Engineers and Scientists.* Englewood Cliffs, N.J.: Prentice-Hall, 1965. See the rocket-generator example beginning on p. 211.

6. Owen, G. E. *Electromagnetic Theory.* Boston: Allyn and Bacon, 1963. Faraday's law is discussed in terms of the frame of reference in Chapter 8.

7. Panofsky, W. K. H., and M. Phillips. *Classical Electricity and Magnetism.* 2d ed. Reading, Mass.: Addison-Wesley, 1962. Relativity is treated at a moderately advanced level in Chapter 15.

CHAPTER 9   PROBLEMS

9.1 In Figure 9.4, let $B = 0.2 \cos 120\pi t$ T, and assume that the conductor joining the two ends of the resistor is perfect. It may be assumed that the magnetic field produced by $I(t)$ is negligible. Find (a) $V_{ab}(t)$; (b) $I(t)$.

9.2 In the example described by Figure 9.1, replace the constant magnetic flux density with the time-varying quantity $\mathbf{B} = B_0 \sin \omega t\ \mathbf{a}_z$. Assume that $\mathbf{v}$ is constant and that the displacement y of the bar is zero at $t = 0$. Find the emf at any time, t.

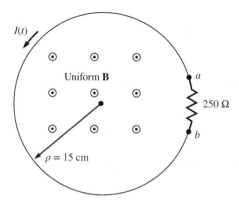

Figure 9.4 See Problem 9.1.

9.3 Given $\mathbf{H} = 300\mathbf{a}_z \cos(3 \times 10^8 t - y)$ A/m in free space, find the emf developed in the general $\mathbf{a}_\phi$ direction about the closed path having corners at (a) $(0, 0, 0)$, $(1, 0, 0)$, $(1, 1, 0)$, and $(0, 1, 0)$; (b) $(0, 0, 0)$ $(2\pi, 0, 0)$, $(2\pi, 2\pi, 0)$, and $(0, 2\pi, 0)$.

9.4 A rectangular loop of wire containing a high-resistance voltmeter has corners initially at $(a/2, b/2, 0)$, $(-a/2, b/2, 0)$, $(-a/2, -b/2, 0)$, and $(a/2, -b/2, 0)$. The loop begins to rotate about the x axis at constant angular velocity ω, with the first-named corner moving in the $\mathbf{a}_z$ direction at $t = 0$. Assume a uniform magnetic flux density $\mathbf{B} = B_0\mathbf{a}_z$. Determine the induced emf in the rotating loop and specify the direction of the current.

9.5 The location of the sliding bar in Figure 9.5 is given by $x = 5t + 2t^3$, and the separation of the two rails is 20 cm. Let $\mathbf{B} = 0.8x^2\mathbf{a}_z$ T. Find the voltmeter reading at (a) $t = 0.4$ s; (b) $x = 0.6$ m.

9.6 Let the wire loop of Problem 9.4 be stationary in its $t = 0$ position and find the induced emf that results from a magnetic flux density given by $\mathbf{B}(y, t) = B_0 \cos(\omega t - \beta y)\mathbf{a}_z$, where ω and β are constants.

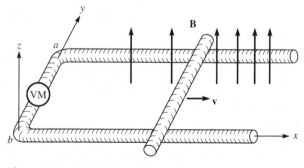

Figure 9.5 See Problem 9.5.

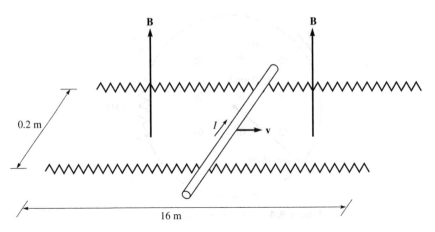

Figure 9.6   See Problem 9.7.

9.7   The rails in Figure 9.6 each have a resistance of 2.2 Ω/m. The bar moves to the right at a constant speed of 9 m/s in a uniform magnetic field of 0.8 T. Find $I(t)$, $0 < t < 1$ s, if the bar is at $x = 2$ m at $t = 0$ and (a) a 0.3 Ω resistor is present across the left end with the right end open-circuited; (b) a 0.3 Ω resistor is present across each end.

9.8   A perfectly conducting filament is formed into a circular ring of radius a. At one point, a resistance R is inserted into the circuit, and at another a battery of voltage V_0 is inserted. Assume that the loop current itself produces negligible magnetic field. (a) Apply Faraday's law, Eq. (4), evaluating each side of the equation carefully and independently to show the equality; (b) repeat part a, assuming the battery is removed, the ring is closed again, and a linearly increasing **B** field is applied in a direction normal to the loop surface.

9.9   A square filamentary loop of wire is 25 cm on a side and has a resistance of 125 Ω per meter length. The loop lies in the $z = 0$ plane with its corners at $(0, 0, 0)$, $(0.25, 0, 0)$, $(0.25, 0.25, 0)$, and $(0, 0.25, 0)$ at $t = 0$. The loop is moving with a velocity $v_y = 50$ m/s in the field $B_z = 8 \cos(1.5 \times 10^8 t - 0.5x)$ μT. Develop a function of time that expresses the ohmic power being delivered to the loop.

9.10   (a) Show that the ratio of the amplitudes of the conduction current density and the displacement current density is $\sigma/\omega\epsilon$ for the applied field $E = E_m \cos \omega t$. Assume $\mu = \mu_0$. (b) What is the amplitude ratio if the applied field is $E = E_m e^{-t/\tau}$, where τ is real?

9.11   Let the internal dimensions of a coaxial capacitor be $a = 1.2$ cm, $b = 4$ cm, and $l = 40$ cm. The homogeneous material inside the capacitor has the parameters $\epsilon = 10^{-11}$ F/m, $\mu = 10^{-5}$ H/m, and $\sigma = 10^{-5}$ S/m. If the electric field intensity is $\mathbf{E} = (10^6/\rho) \cos 10^5 t \mathbf{a}_\rho$ V/m, find (a) **J**; (b) the total conduction current I_c through the capacitor; (c) the total displacement current I_d through the capacitor; (d) the ratio of the amplitude of I_d to that of I_c, the quality factor of the capacitor.

9.12 The magnetic flux density $\mathbf{B} = B_0 \cos(\omega t) \cos(k_0 z) \, \mathbf{a}_y$ Wb/m^2 exists in free space. B_0 and k_0 are constants. Find (a) the displacement current density; (b) the electric field intensity; (c) k_0.

9.13 In free space it is known that $\mathbf{E} = E_0/r \sin \theta \cos(\omega t - k_0 r)\mathbf{a}_\theta = E_\theta \, \mathbf{a}_\theta$. Show that a component of $\mathbf{H}$ in the $\mathbf{a}_\varphi$ direction arises, where $H_\varphi = E_\theta \, (\epsilon_0/\mu_0)^{1/2}$. Do this by applying Eqs. (20) and (21) and requiring consistency.

9.14 A voltage source $V_0 \sin \omega t$ is connected between two concentric conducting spheres, $r = a$ and $r = b$, $b > a$, where the region between them is a material for which $\epsilon = \epsilon_r\epsilon_0$, $\mu = \mu_0$, and $\sigma = 0$. Find the total displacement current through the dielectric and compare it with the source current as determined from the capacitance (Section 6.3) and circuit-analysis methods.

9.15 Use each of Maxwell's equations in point form to obtain as much information as possible about (a) $\mathbf{H}$, if $\mathbf{E} = 0$; (b) $\mathbf{E}$, if $\mathbf{H} = 0$.

9.16 Derive the continuity equation from Maxwell's equations.

9.17 The electric field intensity in the region $0 < x < 5$, $0 < y < \pi/12$, $0 < z < 0.06$ m in free space is given by $\mathbf{E} = C \sin 12y \sin az \cos 2 \times 10^{10}t\mathbf{a}_x$ V/m. Beginning with the $\nabla \times \mathbf{E}$ relationship, use Maxwell's equations to find a numerical value for a, if it is known that a is greater than zero.

9.18 The parallel-plate transmission line shown in Figure 9.7 has dimensions $b = 4$ cm and $d = 8$ mm, while the medium between the plates is characterized by $\mu_r = 1$, $\epsilon_r = 20$, and $\sigma = 0$. Neglect fields outside the dielectric. Given the field $\mathbf{H} = 5 \cos(10^9 t - \beta z)\mathbf{a}_y$ A/m, use Maxwell's equations to help find (a) β, if $\beta > 0$; (b) the displacement current density at $z = 0$; (c) the total displacement current crossing the surface $x = 0.5d$, $0 < y < b$, $0 < z < 0.1$ m in the $\mathbf{a}_x$ direction.

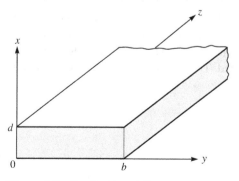

Figure 9.7 See Problem 9.18.

9.19 In Section 9.1, Faraday's law was used to show that the field $\mathbf{E} = -\frac{1}{2}kB_0 e^{kt}$ $\rho\mathbf{a}_\phi$ results from the changing magnetic field $\mathbf{B} = B_0 e^{kt}\mathbf{a}_z$. (a) Show that these fields do not satisfy Maxwell's other curl equation. (b) If we let $B_0 = 1$ T and $k = 10^6$ s^{-1}, we are establishing a fairly large magnetic flux density in 1 μs. Use the $\nabla \times \mathbf{H}$ equation to show that the rate at which B_z should (but does not) change with ρ is only about 5×10^{-6} T per meter in free space at $t = 0$.

9.20 Given Maxwell's equations in point form, assume that all fields vary as e^{st} and write the equations without explicitly involving time.

9.21 (a) Show that under static field conditions, Eq. (55) reduces to Ampère's circuital law. (b) Verify that Eq. (51) becomes Faraday's law when we take the curl.

9.22 In a sourceless medium in which $\mathbf{J} = 0$ and $\rho_v = 0$, assume a rectangular coordinate system in which $\mathbf{E}$ and $\mathbf{H}$ are functions only of z and t. The medium has permittivity ϵ and permeability μ. (a) If $\mathbf{E} = E_x\mathbf{a}_x$ and $\mathbf{H} = H_y\mathbf{a}_y$, begin with Maxwell's equations and determine the second-order partial differential equation that E_x must satisfy. (b) Show that $E_x = E_0 \cos(\omega t - \beta z)$ is a solution of that equation for a particular value of β. (c) Find β as a function of given parameters.

9.23 Referring to the parallel plate transmission line of Figure 9.7, the electric field between plates is given as $\mathbf{E}(z,t) = -E_0 \cos(\omega t - \beta z)\,\mathbf{a}_y$. Find the vector potential $\mathbf{A}(y,z,t)$ if it is given that $A(0,z,t) = 0$. Assume $b \gg d$ so that x variation is negligible.

9.24 A vector potential is given as $\mathbf{A} = A_0 \cos(\omega t - kz)\,\mathbf{a}_y$. (a) Assuming as many components as possible are zero, find $\mathbf{H}$, $\mathbf{E}$, and V. (b) Specify k in terms of A_0, ω, and the constants of the lossless medium, ϵ and μ.

9.25 In a region where $\mu_r = \epsilon_r = 1$ and $\sigma = 0$, the retarded potentials are given by $V = x(z - ct)$ V and $\mathbf{A} = x\left(\frac{z}{c} - t\right)\mathbf{a}_z$ Wb/m, where $c = 1\sqrt{\mu_0\epsilon_0}$. (a) Show that $\nabla \cdot \mathbf{A} = -\mu\epsilon\dfrac{\partial V}{\partial t}$. (b) Find $\mathbf{B}$, $\mathbf{H}$, $\mathbf{E}$, and $\mathbf{D}$. (c) Show that these results satisfy Maxwell's equations if $\mathbf{J}$ and ρ_v are zero.

9.26 Write Maxwell's equations in point form in terms of $\mathbf{E}$ and $\mathbf{H}$ as they apply to a sourceless medium, where $\mathbf{J}$ and ρ_v are both zero. Replace ϵ with μ, μ with ϵ, $\mathbf{E}$ with $\mathbf{H}$, and $\mathbf{H}$ with $-\mathbf{E}$, and show that the equations are unchanged. This is a more general expression of the *duality principle* in circuit theory.

CHAPTER 10 Transmission Lines

术语

transmission line　传输线，电压与电流沿线传播，并伴有衰减以及反射等现象的物理问题。这个传播过程可以称为传输线上的波过程。

lumped element　集总（参数）元件，是指电路原理中的电阻、电感、电容元件；区别于本章传输线的分布参数。

distributed element　分布（参数）元件。

propagation　传播。

telegraphist's equation　电报方程。

characteristic impedance　特性阻抗。

sinusoidal voltage　正弦电压。

frequency domain　频域。

phase constant　相位常数，表征波在单位距离的相位变化；单位是弧度 / 米（rad / m）。

complex analysis　复数分析。

phasor form　相量形式。

shunt admittance　并联导纳。

propagation constant　传播常数。

attenuation coefficient　衰减系数，表征波传播单位距离的幅值衰减；单位是奈培 / 米（Nepers / m，Np / m）

skin effect　趋肤或集肤效应。

decibel　分贝，即贝尔（Bel）单位的十分之一（deci-Bel，dB）。

instantaneous power　瞬时功率。

complex conjugate　复（数）共轭。

traveling wave　行波。

standing wave　驻波。

transient analysis　瞬态分析，即在时间域内分析，区别于正弦稳态的相量形式（phasor form）分析。

steady state　稳态。

Transmission lines are used to transmit electric energy and signals from one point to another, specifically from a source to a load. Examples include the connection between a transmitter and an antenna, connections between computers in a network, or connections between a hydroelectric generating plant and a substation several hundred miles away. Other familiar examples include the interconnects between components of a stereo system and the connection between a cable service provider and your television set. Examples that are less familiar include the connections between devices on a circuit board that are designed to operate at high frequencies.

What all of these examples have in common is that the devices to be connected are separated by distances on the order of a wavelength or much larger, whereas in basic circuit analysis methods, connections between elements are assumed to have negligible length. The latter condition enabled us, for example, to take for granted that the voltage across a resistor on one side of a circuit was exactly in phase with the voltage source on the other side, or, more generally, that the time measured at the source location is precisely the same time as measured at all other points in the circuit. When distances are sufficiently large between source and receiver, time delay effects become appreciable, leading to delay-induced phase differences. In short, we deal with *wave phenomena* on transmission lines in the same manner that we deal

导读

1. 传输线可以起到电能或信号的定向传输作用。若线的长度接近或大于信号的波长，则需要考虑信号沿线传播需要的时间。传输是一个信号从源到负载的波的转播过程，为此应将其视为本章介绍的传输线进行分布参数电路的分析。其中的一个重要特性是像空间电磁波在媒质交界面上会发生反射与透射一样，信号在参数不同的传输线的接合处或线的端部也会有反射现象。这是集总参数电路所不能表征的特性，其只适用于分析系统尺寸远小于信号波长的问题。

2. 时域与频域分析法的关系：对于时变激励求响应的问题可直接以时间为变量列写方程进行求解，称为时域分析法。也可将激励信号转换为许多不同频率的信号之和，将每个频率的激励视为正弦稳态问题，采用相量法进行求解，然后再将每个频率下的解转换到时域中进行叠加，得到原问题的时域解，这个过程称为频域分析法。注意：频域分析法只适用于线性系统，若系统中有非线性元件或参数，频域分析法就不成立了。

with point-to-point energy propagation in free space or in dielectrics.

The basic elements in a circuit, such as resistors, capacitors, inductors, and the connections between them, are considered *lumped* elements if the time delay in traversing the elements is negligible. On the other hand, if the elements or interconnections are large enough, it may be necessary to consider them as *distributed* elements. This means that their resistive, capacitive, and inductive characteristics must be evaluated on a per-unit-distance basis. Transmission lines have this property in general, and thus they become circuit elements in themselves, possessing impedances that contribute to the circuit problem. The basic rule is that one must consider elements as distributed if the propagation delay across the element dimension is on the order of the shortest time interval of interest. In the time-harmonic case, this condition would lead to a measurable phase difference between each end of the device in question.

In this chapter, we investigate wave phenomena in transmission lines. Our objectives include:

(1) to understand how to treat transmission lines as circuit elements possessing complex impedances that are functions of line length and frequency,

(2) to understand wave propagation on lines, including cases in which losses may occur,

(3) to learn methods of combining different transmission lines to accomplish a desired objective, and

(4) to understand transient phenomena on lines. ∎

10.1 PHYSICAL DESCRIPTION OF TRANSMISSION LINE PROPAGATION

To obtain a feel for the manner in which waves propagate on transmission lines, the following demonstration may be helpful. Consider a *lossless* line, as shown in Figure 10.1. By lossless, we mean that all power that is launched into the line at the input end eventually arrives at the output end. A battery having voltage V_0 is connected to the input by closing switch S_1 at time $t = 0$. When the switch is closed, the effect is to launch voltage, $V^+ = V_0$. This voltage does not instantaneously appear everywhere on the line, but rather begins to travel from the battery toward the load resistor, R, at a certain velocity. The *wavefront*, represented by the vertical dashed line in Figure 10.1, represents the instantaneous boundary between the section of the line that has been charged to V_0 and the remaining section that is yet to be charged. It also represents the boundary between the section of the line that carries the charging current, I^+, and the remaining section that carries no current. Both current and voltage are discontinuous across the wavefront.

As the line charges, the wavefront moves from left to right at velocity v, which

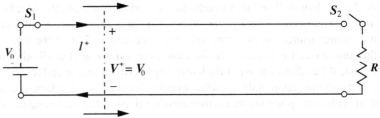

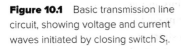

Figure 10.1 Basic transmission line circuit, showing voltage and current waves initiated by closing switch S_1.

is to be determined. On reaching the far end, all or a fraction of the wave voltage and current will reflect, depending on what the line is attached to. For example, if the resistor at the far end is left disconnected (switch S_2 is open), then all of the wavefront voltage will be reflected. If the resistor is connected, then some fraction of the incident voltage will reflect. The details of this will be treated in Section 10.9. Of interest at the moment are the factors that determine the wave velocity. The key to understanding and quantifying this is to note that the conducting transmission line will possess capacitance and inductance that are expressed on a per-unit-length basis. We have already derived expressions for these and evaluated them in Chapters 6 and 8 for certain transmission line geometries. Knowing these line characteristics, we can construct a model for the transmission line using lumped capacitors and inductors, as shown in Figure 10.2. The ladder network thus formed is referred to as a *pulse-forming network,* for reasons that will soon become clear.[1]

Consider now what happens when connecting the same switched voltage source to the network. Referring to Figure 10.2, on closing the switch at the battery location, current begins to increase in L_1, allowing C_1 to charge. As C_1 approaches full charge, current in L_2 begins to increase, allowing C_2 to charge next. This progressive charging process continues down the network, until all three capacitors are fully charged. In the network, a "wavefront" location can be identified as the point between two adjacent capacitors that exhibit the most difference between their charge levels. As the charging process continues, the wavefront moves from left to right. Its speed depends on how fast each inductor can reach its full-current state and, simultaneously, by how fast each capacitor can charge to full voltage. The wave is faster if the values of L_i and C_i are lower. We therefore expect the wave velocity to be inversely proportional to a function involving the product of inductance and capacitance. In the lossless transmission line, it turns out (as will be shown) that the wave velocity is given by $v = 1/\sqrt{LC}$, where L and C are specified per unit length.

导读

3. 行波（traveling wave）的概念：传输线上的入射波称为正向行波，反射波称为反向行波；行波的称谓来源于波的"行进"现象，即在不同时刻观察波的空间形态时，随着时间的推移，空间波形看似波在向一个方向移动。

4. 驻波（standing wave）的概念：对于正弦稳态下传输线上的波，入射波与反射波的合成可能形成传输线上的驻波，其呈现为不同时刻观察波的空间形态，看不到波的"行进"现象，而是"驻足"停留的现象，表现为不同时刻虽然各点的波的幅值在变化，但任何时刻的最小值与最大值都处在相同位置。

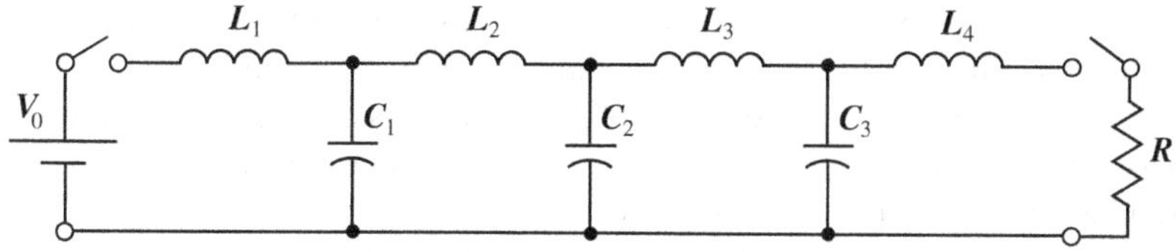

Figure 10.2 Lumped-element model of a transmission line. All inductance values are equal, as are all capacitance values.

Similar behavior is seen in the line and network when either is *initially charged.* In this case, the battery remains connected, and a resistor can be connected (by a switch) across the output end, as shown in Figure 10.2. In the case of the ladder network, the capacitor nearest the shunted end (C_3) will discharge through the resistor first, followed by the next-nearest capacitor, and so on. When the network is completely discharged, a voltage pulse has been formed across the resistor, and so we see why this ladder configuration is called a pulse-forming network. Essentially identical behavior is seen in a charged transmission line when connecting a resistor between conductors at the output end. The switched voltage exercises, as used in these discussions, are examples of transient problems on transmission lines. Transients will be treated in detail in Section 10.14. In the beginning, line responses to sinusoidal signals are emphasized.

---

[1] Designs and applications of pulse-forming networks are discussed in Reference1.

Finally, we surmise that the existence of voltage and current across and within the transmission line conductors implies the existence of electric and magnetic fields in the space around the conductors. Consequently, we have two possible approaches to the analysis of transmission lines: (1) We can solve Maxwell's equations subject to the line configuration to obtain the fields, and with these find general expressions for the wave power, velocity, and other parameters of interest. (2) Or we can (for now) avoid the fields and solve for the voltage and current using an appropriate circuit model. It is the latter approach that we use in this chapter; the contribution of field theory is solely in the prior (and assumed) evaluation of the inductance and capacitance parameters. We will find, however, that circuit models become inconvenient or useless when losses in transmission lines are to be fully characterized, or when analyzing more complicated wave behavior (i.e., moding) which may occur as frequencies get high. The loss issues will be taken up in Section 10.5. Moding phenomena will be considered in Chapter 13.

10.2 THE TRANSMISSION LINE EQUATIONS

Our first goal is to obtain the differential equations, known as the *wave equations,* which the voltage or current must satisfy on a uniform transmission line. To do this, we construct a circuit model for an incremental length of line, write two circuit equations, and use these to obtain the wave equations.

Our circuit model contains the *primary constants* of the transmission line. These include the inductance, L, and capacitance, C, as well as the shunt conductance, G, and series resistance, R—all of which have values that are specified *per unit length*. The shunt conductance is used to model leakage current through the dielectric that may occur throughout the line length; the assumption is that the dielectric may possess conductivity, σ_d, in addition to a dielectric constant, ϵ_r, where the latter affects the capacitance. The series resistance is associated with any finite conductivity, σ_c, in the conductors. Either one of the latter parameters, R and G, will be responsible for power loss in transmission. In general, both are functions of frequency. Knowing the frequency and the dimensions, we can determine the values of R, G, L, and C by using formulas developed in earlier chapters.

We assume propagation in the $\mathbf{a}_z$ direction. Our model consists of a line section of length Δz containing resistance $R\Delta z$, inductance $L\Delta z$, conductance $G\Delta z$, and capacitance $C\Delta z$, as indicated in Figure 10.3. Because the section of the line looks the same from either end, we divide the series elements in half to produce a symmetrical network. We could equally well have placed half the conductance and half the capacitance at each end.

Our objective is to determine the manner and extent to which the output voltage and current are changed from their input values in the limit as the length approaches a very small value. We will consequently obtain a pair of differential equations that describe the rates of change of voltage and current with respect to z. In Figure 10.3, the input and output voltages and currents differ respectively by quantities ΔV and ΔI, which are to be determined. The two equations are obtained by successive applications of Kirchoff's voltage law (KVL) and Kirchoff's current law (KCL).

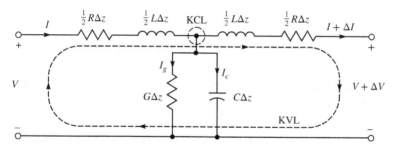

Figure 10.3 Lumped-element model of a short transmission line section with losses. The length of the section is Δz. Analysis involves applying Kirchoff's voltage and current laws (KVL and KCL) to the indicated loop and node, respectively.

First, KVL is applied to the loop that encompasses the entire section length, as shown in Figure 10.3:

$$V = \frac{1}{2}RI\Delta z + \frac{1}{2}L\frac{\partial I}{\partial t}\Delta z + \frac{1}{2}L\left(\frac{\partial I}{\partial t} + \frac{\partial \Delta I}{\partial t}\right)\Delta z$$
$$+ \frac{1}{2}R(I + \Delta I)\Delta z + (V + \Delta V) \tag{1}$$

We can solve Eq. (1) for the ratio, $\Delta V/\Delta z$, obtaining:

$$\frac{\Delta V}{\Delta z} = -\left(RI + L\frac{\partial I}{\partial t} + \frac{1}{2}L\frac{\partial \Delta I}{\partial t} + \frac{1}{2}R\Delta I\right) \tag{2}$$

Next, we write:

$$\Delta I = \frac{\partial I}{\partial z}\Delta z \quad \text{and} \quad \Delta V = \frac{\partial V}{\partial z}\Delta z \tag{3}$$

which are then substituted into (2) to result in

$$\frac{\partial V}{\partial z} = -\left(1 + \frac{\Delta z}{2}\frac{\partial}{\partial z}\right)\left(RI + L\frac{\partial I}{\partial t}\right) \tag{4}$$

Now, in the limit as Δz approaches zero (or a value small enough to be negligible), (4) simplifies to the final form:

$$\boxed{\frac{\partial V}{\partial z} = -\left(RI + L\frac{\partial I}{\partial t}\right)} \tag{5}$$

Equation (5) is the first of the two equations that we are looking for. To find the second equation, we apply KCL to the upper central node in the circuit of Figure 10.3, noting from the symmetry that the voltage at the node will be $V + \Delta V/2$:

$$I = I_g + I_c + (I + \Delta I) = G\Delta z\left(V + \frac{\Delta V}{2}\right)$$
$$+ C\Delta z\frac{\partial}{\partial t}\left(V + \frac{\Delta V}{2}\right) + (I + \Delta I) \tag{6}$$

Then, using (3) and simplifying, we obtain

$$\frac{\partial I}{\partial z} = -\left(1 + \frac{\Delta z}{2}\frac{\partial}{\partial z}\right)\left(GV + C\frac{\partial V}{\partial t}\right) \tag{7}$$

Again, we obtain the final form by allowing Δz to be reduced to a negligible magnitude. The result is

$$\boxed{\frac{\partial I}{\partial z} = -\left(GV + C\frac{\partial V}{\partial t}\right)} \tag{8}$$

The coupled differential equations, (5) and (8), describe the evolution of current and voltage in any transmission line. Historically, they have been referred to as the *telegraphist's equations*. Their solution leads to the wave equation for the transmission line, which we now undertake. We begin by differentiating Eq. (5) with respect to z and Eq. (8) with respect to t, obtaining:

$$\frac{\partial^2 V}{\partial z^2} = -R\frac{\partial I}{\partial z} - L\frac{\partial^2 I}{\partial t \partial z} \tag{9}$$

and

$$\frac{\partial^2 I}{\partial z \partial t} = -G\frac{\partial V}{\partial t} - C\frac{\partial^2 V}{\partial t^2} \tag{10}$$

Next, Eqs. (8) and (10) are substituted into (9). After rearranging terms, the result is:

$$\boxed{\frac{\partial^2 V}{\partial z^2} = LC\frac{\partial^2 V}{\partial t^2} + (LG + RC)\frac{\partial V}{\partial t} + RGV} \tag{11}$$

An analogous procedure involves differentiating Eq. (5) with respect to t and Eq. (8) with respect to z. Then, Eq. (5) and its derivative are substituted into the derivative of (8) to obtain an equation for the current that is in identical form to that of (11):

$$\boxed{\frac{\partial^2 I}{\partial z^2} = LC\frac{\partial^2 I}{\partial t^2} + (LG + RC)\frac{\partial I}{\partial t} + RGI} \tag{12}$$

Equations (11) and (12) are the *general wave equations* for the transmission line. Their solutions under various conditions form a major part of our study.

10.3 LOSSLESS PROPAGATION

Lossless propagation means that power is not dissipated or otherwise deviated as the wave travels down the transmission line; all power at the input end eventually reaches the output end. More realistically, any mechanisms that would cause losses to occur have negligible effect. In our model, lossless propagation occurs when $R = G = 0$.

Under this condition, only the first term on the right-hand side of either Eq. (11) or Eq. (12) survives. Eq. (11), for example, becomes

$$\frac{\partial^2 V}{\partial z^2} = LC\frac{\partial^2 V}{\partial t^2} \qquad (13)$$

In considering the voltage function that will satisfy (13), it is most expedient to simply state the solution, and then show that it is correct. The solution of (13) is of the form:

$$V(z,t) = f_1\left(t - \frac{z}{v}\right) + f_2\left(t + \frac{z}{v}\right) = V^+ + V^- \qquad (14)$$

where v, the *wave velocity,* is a constant. The expressions $(t \pm z/v)$ are the arguments of functions f_1 and f_2. The identities of the functions themselves are not critical to the solution of (13). Therefore, f_1 and f_2 can be *any* function.

The arguments of f_1 and f_2 indicate, respectively, travel of the functions in the forward and backward z directions. We assign the symbols V^+ and V^- to identify the forward and backward voltage wave components. To understand the behavior, consider for example the value of f_1 (whatever this might be) at the zero value of its argument, occurring when $z = t = 0$. Now, as time increases to positive values (as it must), and if we are to keep track of $f_1(0)$, then the value of z must also increase to keep the argument $(t - z/v)$ equal to zero. The function f_1 therefore moves (or propagates) in the positive z direction. Using similar reasoning, the function f_2 will propagate in the *negative z* direction, as z in the argument $(t + z/v)$ must *decrease* to offset the increase in t. Therefore we associate the argument $(t - z/v)$ with *forward z* propagation, and the argument $(t + z/v)$ with *backward z* travel. This behavior occurs irrespective of what f_1 and f_2 are. As is evident in the argument forms, the propagation velocity is v in both cases.

Animations

We next verify that functions having the argument forms expressed in (14) are solutions to (13). First, we take partial derivatives of f_1, for example with respect to z and t. Using the chain rule, the z partial derivative is

$$\frac{\partial f_1}{\partial z} = \frac{\partial f_1}{\partial(t - z/v)}\frac{\partial(t - z/v)}{\partial z} = -\frac{1}{v}f_1' \qquad (15)$$

where it is apparent that the primed function, f_1', denotes the derivative of f_1 with respect to its argument. The partial derivative with respect to time is

$$\frac{\partial f_1}{\partial t} = \frac{\partial f_1}{\partial(t - z/v)}\frac{\partial(t - z/v)}{\partial t} = f_1' \qquad (16)$$

Next, the second partial derivatives with respect to z and t can be taken using similar reasoning:

$$\frac{\partial^2 f_1}{\partial z^2} = \frac{1}{v^2}f_1'' \quad \text{and} \quad \frac{\partial^2 f_1}{\partial t^2} = f_1'' \qquad (17)$$

where f_1'' is the second derivative of f_1 with respect to its argument. The results in (17) can now be substituted into (13), obtaining

$$\frac{1}{v^2}f_1'' = LCf_1''$$

(18)

We now identify the wave velocity for lossless propagation, which is the condition for equality in (18):

$$\boxed{v = \frac{1}{\sqrt{LC}}}$$

(19)

Performing the same procedure using f_2 (and its argument) leads to the same expression for v.

The form of v as expressed in Eq. (19) confirms our original expectation that the wave velocity would be in some inverse proportion to L and C. The same result will be true for current, as Eq. (12) under lossless conditions would lead to a solution of the form identical to that of (14), with velocity given by (19). What is not known yet, however, is the relation *between* voltage and current.

We have already found that voltage and current are related through the telegraphist's equations, (5) and (8). These, under lossless conditions ($R = G = 0$), become

$$\boxed{\frac{\partial V}{\partial z} = -L\frac{\partial I}{\partial t}}$$

(20)

$$\boxed{\frac{\partial I}{\partial z} = -C\frac{\partial V}{\partial t}}$$

(21)

Using the voltage function, we can substitute (14) into (20) and use the methods demonstrated in (15) to write

$$\frac{\partial I}{\partial t} = -\frac{1}{L}\frac{\partial V}{\partial z} = \frac{1}{Lv}(f_1' - f_2')$$

(22)

We next integrate (22) over time, obtaining the current in terms of its forward and backward propagating components:

$$\boxed{I(z,t) = \frac{1}{Lv}\left[f_1\left(t - \frac{z}{v}\right) - f_2\left(t + \frac{z}{v}\right)\right] = I^+ + I^-}$$

(23)

In performing this integration, all integration constants are set to zero. The reason for this, as demonstrated by (20) and (21), is that a time-varying voltage must lead to a time-varying current, with the reverse also true. The factor $1/Lv$ appearing in (23) multiplies voltage to obtain current, and so we identify the product Lv as the *characteristic impedance*, Z_0, of the lossless line. Z_0 is defined as the ratio of the voltage

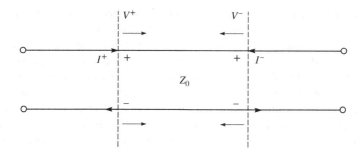

Figure 10.4 Current directions in waves having positive voltage polarity.

to the current in a single propagating wave. Using (19), we write the characteristic impedance as

$$Z_0 = Lv = \sqrt{\frac{L}{C}} \tag{24}$$

By inspecting (14) and (23), we now note that

$$V^+ = Z_0 I^+ \tag{25a}$$

and

$$V^- = -Z_0 I^- \tag{25b}$$

The significance of the preceding relations can be seen in Figure 10.4. The figure shows forward- and backward-propagating voltage waves, V^+ and V^-, both of which have positive polarity. The currents that are associated with these voltages will flow in opposite directions. We define *positive current* as having a *clockwise* flow in the line, and *negative current* as having a *counterclockwise* flow. The minus sign in (25b) thus assures that negative current will be associated with a backward-propagating wave that has positive polarity. This is a general convention, applying to lines with losses also. Propagation with losses is studied by solving (11) under the assumption that either R or G (or both) are not zero. We will do this in Section 10.7 under the special case of sinusoidal voltages and currents. Sinusoids in lossless transmission lines are considered in Section 10.4.

10.4 LOSSLESS PROPAGATION OF SINUSOIDAL VOLTAGES

An understanding of sinusoidal waves on transmission lines is important because any signal that is transmitted in practice can be decomposed into a discrete or continuous summation of sinusoids. This is the basis of *frequency domain* analysis of signals on lines. In such studies, the effect of the transmission line on any signal can be determined by noting the effects on the frequency components. This means that one can effectively propagate the spectrum of a given signal, using frequency-dependent line parameters, and then reassemble the frequency components into the resultant signal in time domain. Our objective in this section is to obtain an understanding of sinusoidal propagation and the implications on signal behavior for the lossless line case.

We begin by assigning sinusoidal functions to the voltage functions in Eq. (14). Specifically, we consider a specific frequency, $f = \omega/2\pi$, and write $f_1 = f_2 = V_0 \cos(\omega t + \phi)$. By convention, the cosine function is chosen; the sine is obtainable, as we know, by setting $\phi = -\pi/2$. We next replace t with $(t \pm z/v_p)$, obtaining

$$\mathcal{V}(z,t) = |V_0|\cos[\omega(t \pm z/v_p) + \phi] = |V_0|\cos[\omega t \pm \beta z + \phi] \tag{26}$$

where we have assigned a new notation to the velocity, which is now called the *phase velocity*, v_p. This is applicable to a pure sinusoid (having a single frequency) and will be found to depend on frequency in some cases. Choosing, for the moment, $\phi = 0$, we obtain the two possibilities of forward or backward z travel by choosing the minus or plus sign in (26). The two cases are:

$$\boxed{\mathcal{V}_f(z, t) = |V_0| \cos(\omega t - \beta z)} \quad \text{(forward } z \text{ propagation)} \tag{27a}$$

and

$$\boxed{\mathcal{V}_b(z, t) = |V_0| \cos(\omega t + \beta z)} \quad \text{(backward } z \text{ propagation)} \tag{27b}$$

where the magnitude factor, $|V_0|$, is the value of $\mathcal{V}$ at $z = 0$, $t = 0$. We define the *phase constant* β, obtained from (26), as

$$\boxed{\beta \equiv \frac{\omega}{v_p}} \tag{28}$$

We refer to the solutions expressed in (27a) and (27b) as the *real instantaneous* forms of the transmission-line voltage. They are the mathematical representations of what one would experimentally measure. The terms ωt and βz, appearing in these equations, have units of angle and are usually expressed in radians. We know that ω is the radian time frequency, measuring phase shift *per unit time,* and it has units of rad/s. In a similar way, we see that β will be interpreted as a *spatial* frequency, which in the present case measures the phase shift *per unit distance* along the z direction. Its units are rad/m. If we were to fix the time at $t = 0$, Eqs. (27a) and (27b) would become

$$\mathcal{V}_f(z, 0) = \mathcal{V}_b(z, 0) = |V_0|\cos(\beta z) \tag{29}$$

which we identify as a simple periodic function that repeats every incremental distance λ, known as the *wavelength*. The requirement is that $\beta\lambda = 2\pi$, and so

$$\boxed{\lambda = \frac{2\pi}{\beta} = \frac{v_p}{f}} \tag{30}$$

We next consider a point (such as a wave crest) on the cosine function of Eq. (27*a*), the occurrence of which requires the argument of the cosine to be an integer multiple of 2π. Considering the *m*th crest of the wave, the condition at $t = 0$ becomes

$$\beta z = 2m\pi$$

To keep track of this point on the wave, we require that the entire cosine argument be the same multiple of 2π for all time. From (27*a*) the condition becomes

$$\omega t - \beta z = \omega(t - z/v_p) = 2m\pi \qquad (31)$$

Again, with increasing time, the position z must also increase in order to satisfy (31). Consequently the wave crest (and the entire wave) travels in the positive z direction at velocity v_p. Eq. (27*b*), having cosine argument ($\omega t + \beta z$), describes a wave that travels in the *negative z* direction, since as time increases, z must now *decrease* to keep the argument constant. Similar behavior is found for the wave current, but complications arise from line-dependent phase differences that occur between current and voltage. These issues are best addressed once we are familiar with complex analysis of sinusoidal signals.

10.5 COMPLEX ANALYSIS OF SINUSOIDAL WAVES

Expressing sinusoidal waves as complex functions is useful (and essentially indispensable) because it greatly eases the evaluation and visualization of phase that will be found to accumulate by way of many mechanisms. In addition, we will find many cases in which two or more sinusoidal waves must be combined to form a resultant wave—a task made much easier if complex analysis is used.

Expressing sinusoidal functions in complex form is based on the Euler identity:

$$e^{\pm jx} = \cos(x) \pm j \sin(x) \qquad (32)$$

from which we may write the cosine and sine, respectively, as the real and imaginary parts of the complex exponent:

$$\cos(x) = \text{Re}[e^{\pm jx}] = \frac{1}{2}(e^{jx} + e^{-jx}) = \frac{1}{2}e^{jx} + c.c. \qquad (33a)$$

$$\sin(x) = \pm\text{Im}[e^{\pm jx}] = \frac{1}{2j}(e^{jx} - e^{-jx}) = \frac{1}{2j}e^{jx} + c.c. \qquad (33b)$$

where $j \equiv \sqrt{-1}$, and where *c.c.* denotes the complex conjugate of the preceding term. The conjugate is formed by changing the sign of j wherever it appears in the complex expression.

We may next apply (33*a*) to our voltage wave function, Eq. (26):

$$\mathcal{V}(z,t) = |V_0| \cos[\omega t \pm \beta z + \phi] = \frac{1}{2}\underbrace{(|V_0|e^{j\phi})}_{V_0} e^{\pm j\beta z} e^{j\omega t} + c.c. \qquad (34)$$

Note that we have arranged the phases in (34) such that we identify the *complex amplitude* of the wave as $V_0 = (|V_0|e^{j\phi})$. In future usage, a single symbol (V_0 in the present example) will usually be used for the voltage or current amplitudes, with the understanding that these will generally be complex (having magnitude and phase).

Two additional definitions follow from Eq. (34). First, we define the *complex instantaneous* voltage as:

$$V_c(z,t) = V_0 e^{\pm j\beta z} e^{j\omega t} \qquad (35)$$

The *phasor* voltage is then formed by dropping the $e^{j\omega t}$ factor from the complex instantaneous form:

$$V_s(z) = V_0 e^{\pm j\beta z} \qquad (36)$$

The phasor voltage can be defined provided we have *sinusoidal steady-state* conditions— meaning that V_0 is independent of time. This has in fact been our assumption all along, because a time-varying amplitude would imply the existence of other frequency components in our signal. Again, we are treating only a single-frequency wave. The significance of the phasor voltage is that we are effectively letting time stand still and observing the stationary wave in space at $t = 0$. The processes of evaluating relative phases between various line positions and of combining multiple waves is made much simpler in phasor form. Again, this works only if all waves under consideration have the same frequency. With the definitions in (35) and (36), the real instantaneous voltage can be constructed using (34):

$$\mathcal{V}(z,t) = |V_0|\cos[\omega t \pm \beta z + \phi] = \text{Re}[V_c(z,t)] = \tfrac{1}{2}V_c + c.c. \qquad (37a)$$

Or, in terms of the phasor voltage:

$$\mathcal{V}(z,t) = |V_0|\cos[\omega t \pm \beta z + \phi] = \text{Re}[V_s(z)e^{j\omega t}] = \tfrac{1}{2}V_s(z)e^{j\omega t} + c.c. \qquad (37b)$$

In words, we may obtain our real sinusoidal voltage wave by multiplying the phasor voltage by $e^{j\omega t}$ (reincorporating the time dependence) and then taking the real part of the resulting expression. It is imperative that one becomes familiar with these relations and their meaning before proceeding further.

EXAMPLE 10.1

Two voltage waves having equal frequencies and amplitudes propagate in opposite directions on a lossless transmission line. Determine the total voltage as a function of time and position.

Solution. Because the waves have the same frequency, we can write their combination using their phasor forms. Assuming phase constant, β, and real amplitude, V_0, the two wave voltages combine in this way:

$$V_{sT}(z) = V_0 e^{-j\beta z} + V_0 e^{+j\beta z} = 2V_0\cos(\beta z)$$

In real instantaneous form, this becomes

$$\mathcal{V}(z, t) = \text{Re}[2\,V_0\cos(\beta z)e^{j\omega t}] = 2\,V_0\cos(\beta z)\cos(\omega t)$$

We recognize this as a *standing wave,* in which the amplitude varies, as $\cos(\beta z)$, and oscillates in time, as $\cos(\omega t)$. Zeros in the amplitude (nulls) occur at fixed locations, $z_n = (m\pi)/(2\beta)$ where m is an odd integer. We extend the concept in Section 10.10, where we explore the *voltage standing wave ratio* as a measurement technique.

10.6 TRANSMISSION LINE EQUATIONS AND THEIR SOLUTIONS IN PHASOR FORM

We now apply our results of the previous section to the transmission line equations, beginning with the general wave equation, (11). This is rewritten as follows, for the real instantaneous voltage, $\mathcal{V}(z, t)$:

$$\frac{\partial^2 \mathcal{V}}{\partial z^2} = LC\frac{\partial^2 \mathcal{V}}{\partial t^2} + (LG + RC)\frac{\partial \mathcal{V}}{\partial t} + RG\mathcal{V} \tag{38}$$

We next substitute $\mathcal{V}(z, t)$ as given by the far right-hand side of (37b), noting that the complex conjugate term (*c.c.*) will form a separate redundant equation. We also use the fact that the operator $\partial/\partial t$, when applied to the complex form, is equivalent to multiplying by a factor of $j\omega$. After substitution, and after all time derivatives are taken, the factor $e^{j\omega t}$ divides out. We are left with the wave equation in terms of the phasor voltage:

$$\frac{d^2 V_s}{dz^2} = -\omega^2 LC V_s + j\omega(LG + RC)V_s + RG V_s \tag{39}$$

Rearranging terms leads to the simplified form:

$$\frac{d^2 V_s}{dz^2} = \underbrace{(R + j\omega L)}_{Z}\underbrace{(G + j\omega C)}_{Y} V_s = \gamma^2 V_s \tag{40}$$

where Z and Y, as indicated, are respectively the *net series impedance* and the *net shunt admittance* in the transmission line—both as per-unit-distance measures. The *propagation constant* in the line is defined as

$$\gamma = \sqrt{(R + j\omega L)(G + j\omega C)} = \sqrt{ZY} = \alpha + j\beta \tag{41}$$

The significance of the term will be explained in Section 10.7. For our immediate purposes, the solution of (40) will be

$$V_s(z) = V_0^+ e^{-\gamma z} + V_0^- e^{+\gamma z} \tag{42a}$$

The wave equation for current will be identical in form to (40). We therefore expect the phasor current to be in the form:

$$I_s(z) = I_0^+ e^{-\gamma z} + I_0^- e^{\gamma z} \tag{42b}$$

The relation between the current and voltage waves is now found, as before, through the telegraphist's equations, (5) and (8). In a manner consistent with Eq. (37b), we write the sinusoidal current as

$$\mathcal{I}(z, t) = |I_0| \cos(\omega t \pm \beta z + \xi) = \frac{1}{2} \underbrace{(|I_0|e^{j\xi})}_{I_0} e^{\pm j\beta z} e^{j\omega t} + \text{c.c.} = \frac{1}{2} I_s(z) e^{j\omega t} + \text{c.c.} \tag{43}$$

Substituting the far right-hand sides of (37b) and (43) into (5) and (8) transforms the latter equations as follows:

$$\frac{\partial \mathcal{V}}{\partial z} = -\left(R\mathcal{I} + L\frac{\partial \mathcal{I}}{\partial t}\right) \quad \Rightarrow \quad \boxed{\frac{dV_s}{dz} = -(R + j\omega L)I_s = -ZI_s} \tag{44a}$$

and

$$\frac{\partial \mathcal{I}}{\partial z} = -\left(G\mathcal{V} + C\frac{\partial \mathcal{V}}{\partial t}\right) \quad \Rightarrow \quad \boxed{\frac{dI_s}{dz} = -(G + j\omega C)V_s = -YV_s} \tag{44b}$$

We can now substitute (42a) and (42b) into either (44a) or (44b) [we will use (44a)] to find:

$$-\gamma V_0^+ e^{-\gamma z} + \gamma V_0^- e^{\gamma z} = -Z(I_0^+ e^{-\gamma z} + I_0^- e^{\gamma z}) \tag{45}$$

Next, equating coefficients of $e^{-\gamma z}$ and $e^{\gamma z}$, we find the general expression for the line characteristic impedance:

$$Z_0 = \frac{V_0^+}{I_0^+} = -\frac{V_0^-}{I_0^-} = \frac{Z}{\gamma} = \frac{Z}{\sqrt{ZY}} = \sqrt{\frac{Z}{Y}} \tag{46}$$

Incorporating the expressions for Z and Y, we find the characteristic impedance in terms of our known line parameters:

$$\boxed{Z_0 = \sqrt{\frac{R + j\omega L}{G + j\omega C}} = |Z_0|e^{j\theta}} \tag{47}$$

Note that with the voltage and current as given in (37b) and (43), we would identify the phase of the characteristic impedance, $\theta = \phi - \xi$.

EXAMPLE 10.2

A lossless transmission line is 80 cm long and operates at a frequency of 600 MHz. The line parameters are $L = 0.25~\mu H/m$ and $C = 100~pF/m$. Find the characteristic impedance, the phase constant, and the phase velocity.

Solution. Because the line is lossless, both R and G are zero. The characteristic impedance is

$$Z_0 = \sqrt{\frac{L}{C}} = \sqrt{\frac{0.25 \times 10^{-6}}{100 \times 10^{-12}}} = 50 \ \Omega$$

Because $\gamma = \alpha + j\beta = \sqrt{(R + j\omega L)(G + j\omega C)} = j\omega\sqrt{LC}$, we see that

$$\beta = \omega\sqrt{LC} = 2\pi(600 \times 10^6)\sqrt{(0.25 \times 10^{-6})(100 \times 10^{-12})} = 18.85 \text{ rad/m}$$

Also,

$$v_p = \frac{\omega}{\beta} = \frac{2\pi(600 \times 10^6)}{18.85} = 2 \times 10^8 \text{ m/s}$$

10.7 LOW-LOSS PROPAGATION

Having obtained the phasor forms of voltage and current in a general transmission line [Eqs. (42a) and (42b)], we can now look more closely at the significance of these results. First we incorporate (41) into (42a) to obtain

$$\boxed{V_s(z) = V_0^+ e^{-\alpha z} e^{-j\beta z} + V_0^- e^{\alpha z} e^{j\beta z}} \tag{48}$$

Next, multiplying (48) by $e^{j\omega t}$ and taking the real part gives the real instantaneous voltage:

$$\boxed{\mathcal{V}(z, t) = V_0^+ e^{-\alpha z} \cos(\omega t - \beta z) + V_0^- e^{\alpha z} \cos(\omega t + \beta z)} \tag{49}$$

In this exercise, we have assigned V_0^+ and V_0^- to be real. Eq. (49) is recognized as describing forward- and backward-propagating waves that diminish in amplitude with distance according to $e^{-\alpha z}$ for the forward wave, and $e^{\alpha z}$ for the backward wave. Both waves are said to *attenuate* with propagation distance at a rate determined by the *attenuation coefficient,* α, expressed in units of nepers/m [Np/m].[2]

The phase constant, β, found by taking the imaginary part of (41), is likely to be a somewhat complicated function, and will in general depend on R and G. Nevertheless, β is still defined as the ratio ω/v_p, and the wavelength is still defined as the distance that provides a phase shift of 2π rad, so that $\lambda = 2\pi/\beta$. By inspecting (41), we observe that losses in propagation are avoided (or $\alpha = 0$) only when $R = G = 0$. In that case, (41) gives $\gamma = j\beta = j\omega\sqrt{LC}$, and so $v_p = 1/\sqrt{LC}$, as we found before.

Expressions for α and β when losses are small can be readily obtained from (41). In the *low-loss approximation,* we require $R \ll \omega L$ and $G \ll \omega C$, a condition that is

---

[2] The term *neper* was selected (by some poor speller) to honor John Napier, a Scottish mathematician who first proposed the use of logarithms.

often true in practice. Before we apply these conditions, Eq. (41) can be written in the form:

$$\gamma = \alpha + j\beta = [(R + j\omega L)(G + j\omega C)]^{1/2}$$

$$= j\omega\sqrt{LC}\left[\left(1 + \frac{R}{j\omega L}\right)^{1/2}\left(1 + \frac{G}{j\omega C}\right)^{1/2}\right] \tag{50}$$

The low-loss approximation then allows us to use the first three terms in the binomial series:

$$\sqrt{1+x} \doteq 1 + \frac{x}{2} - \frac{x^2}{8} \quad (x \ll 1) \tag{51}$$

We use (51) to expand the terms in large parentheses in (50), obtaining:

$$\gamma \doteq j\omega\sqrt{LC}\left[\left(1 + \frac{R}{j2\omega L} + \frac{R^2}{8\omega^2 L^2}\right)\left(1 + \frac{G}{j2\omega C} + \frac{G^2}{8\omega^2 C^2}\right)\right] \tag{52}$$

All products in (52) are then carried out, neglecting the terms involving RG^2, R^2G, and R^2G^2, as these will be negligible compared to all others. The result is

$$\gamma = \alpha + j\beta \doteq j\omega\sqrt{LC}\left[1 + \frac{1}{j2\omega}\left(\frac{R}{L} + \frac{G}{C}\right) + \frac{1}{8\omega^2}\left(\frac{R^2}{L^2} - \frac{2RG}{LC} + \frac{G^2}{C^2}\right)\right] \tag{53}$$

Now, separating real and imaginary parts of (53) yields α and β:

$$\boxed{\alpha \doteq \frac{1}{2}\left(R\sqrt{\frac{C}{L}} + G\sqrt{\frac{L}{C}}\right)} \tag{54a}$$

and

$$\boxed{\beta \doteq \omega\sqrt{LC}\left[1 + \frac{1}{8}\left(\frac{G}{\omega C} - \frac{R}{\omega L}\right)^2\right]} \tag{54b}$$

We note that α scales in direct proportion to R and G, as would be expected. We also note that the terms in (54b) that involve R and G lead to a phase velocity, $v_p = \omega/\beta$, that is frequency-dependent. Moreover, the *group velocity*, $v_g = d\omega/d\beta$, will also depend on frequency, and will lead to signal distortion, as we will explore in Chapter 12. Note that with nonzero R and G, phase and group velocities that are constant with frequency can be obtained when $R/L = G/C$, known as *Heaviside's condition*. In this case, (54b) becomes $\beta \doteq \omega\sqrt{LC}$, and the line is said to be *distortionless*. Further complications occur when accounting for possible frequency dependencies within R, G, L, and C. Consequently, conditions of low-loss or distortion-free propagation will usually occur over limited frequency ranges. As a rule, loss increases with increasing frequency, mostly because of the increase in R with frequency. The nature of this latter effect, known as *skin effect* loss, requires field theory to understand and quantify. We will study this in Chapter 11, and we will apply it to transmission line structures in Chapter 13.

Finally, we can apply the low-loss approximation to the characteristic impedance, Eq. (47). Using (51), we find

$$Z_0 = \sqrt{\frac{R + j\omega L}{G + j\omega C}} = \sqrt{\frac{j\omega L\left(1 + \frac{R}{j\omega L}\right)}{j\omega C\left(1 + \frac{G}{j\omega C}\right)}} \doteq \sqrt{\frac{L}{C}\left[\frac{\left(1 + \frac{R}{j2\omega L} + \frac{R^2}{8\omega^2 L^2}\right)}{\left(1 + \frac{G}{j2\omega C} + \frac{G^2}{8\omega^2 C^2}\right)}\right]} \qquad (55)$$

Next, we multiply (55) by a factor of 1, in the form of the complex conjugate of the denominator of (55) divided by itself. The resulting expression is simplified by neglecting all terms on the order of $R^2 G$, $G^2 R$, and higher. Additionally, the approximation, $1/(1 + x) \doteq 1 - x$, where $x \ll 1$ is used. The result is

$$Z_0 \doteq \sqrt{\frac{L}{C}}\left\{1 + \frac{1}{2\omega^2}\left[\frac{1}{4}\left(\frac{R}{L} + \frac{G}{C}\right)^2 - \frac{G^2}{C^2}\right] + \frac{j}{2\omega}\left(\frac{G}{C} - \frac{R}{L}\right)\right\} \qquad (56)$$

Note that when Heaviside's condition (again, $R/L = G/C$) holds, Z_0 simplifies to just $\sqrt{L/C}$, as is true when both R and G are zero.

EXAMPLE 10.3

Suppose in a certain transmission line $G = 0$, but R is finite valued and satisfies the low-loss requirement, $R \ll \omega L$. Use Eq. (56) to write the approximate magnitude and phase of Z_0.

Solution. With $G = 0$, the imaginary part of (56) is much greater than the second term in the real part [proportional to $(R/\omega L)^2$]. Therefore, the characteristic impedance becomes

$$Z_0(G = 0) \doteq \sqrt{\frac{L}{C}}\left(1 - j\frac{R}{2\omega L}\right) = |Z_0|e^{j\theta}$$

where $|Z_0| \doteq \sqrt{L/C}$, and $\theta = \tan^{-1}(-R/2\omega L)$.

D10.1. At an operating radian frequency of 500 Mrad/s, typical circuit values for a certain transmission line are: $R = 0.2\ \Omega/m$, $L = 0.25\ \mu H/m$, $G = 10\ \mu S/m$, and $C = 100$ pF/m. Find: (a) α; (b) β; (c) λ; (d) v_p; (e) Z_0.

Ans. (a) 2.25 mNp/m; (b) 2.50 rad/m; (c) 2.51 m; (d) 2×10^8 m/sec; (e) $50.0 - j0.0350\ \Omega$

10.8 POWER TRANSMISSION AND THE USE OF DECIBELS IN LOSS CHARACTERIZATION

Having found the sinusoidal voltage and current in a lossy transmission line, we next evaluate the power transmitted over a specified distance as a function of voltage and current amplitudes. We start with the *instantaneous* power, given simply as the product of the real voltage and current. Consider the forward-propagating term in (49), where again, the amplitude, $V_0^+ = |V_0|$, is taken to be real. The current waveform will be similar, but will generally be shifted in phase. Both current and voltage attenuate according to the factor $e^{-\alpha z}$. The instantaneous power therefore becomes:

$$P(z,t) = V(z,t)I(z,t) = |V_0||I_0|e^{-2\alpha z}\cos(\omega t - \beta z)\cos(\omega t - \beta z + \theta) \tag{57}$$

Usually, the *time-averaged* power, $\langle P \rangle$, is of interest. We find this through:

$$\langle P \rangle = \frac{1}{T}\int_0^T |V_0||I_0|e^{-2\alpha z}\cos(\omega t - \beta z)\cos(\omega t - \beta z + \theta)\, dt \tag{58}$$

where $T = 2\pi/\omega$ is the time period for one oscillation cycle. Using a trigonometric identity, the product of cosines in the integrand can be written as the sum of individual cosines at the sum and difference frequencies:

$$\langle P \rangle = \frac{1}{T}\int_0^T \frac{1}{2}|V_0||I_0|e^{-2\alpha z}[\cos(2\omega t - 2\beta z + \theta) + \cos(\theta)]\, dt \tag{59}$$

The first cosine term integrates to zero, leaving the $\cos\theta$ term. The remaining integral easily evaluates as

$$\langle P \rangle = \frac{1}{2}|V_0||I_0|e^{-2\alpha z}\cos\theta = \frac{1}{2}\frac{|V_0|^2}{|Z_0|}e^{-2\alpha z}\cos\theta \ [\text{W}] \tag{60}$$

The same result can be obtained directly from the phasor voltage and current. We begin with these, expressed as

$$V_s(z) = V_0 e^{-\alpha z}e^{-j\beta z} \tag{61}$$

and

$$I_s(z) = I_0 e^{-\alpha z}e^{-j\beta z} = \frac{V_0}{Z_0}e^{-\alpha z}e^{-j\beta z} \tag{62}$$

where $Z_0 = |Z_0|e^{j\theta}$. We now note that the time-averaged power as expressed in (60) can be obtained from the phasor forms through:

$$\boxed{\langle P \rangle = \frac{1}{2}\text{Re}\{V_s I_s^*\}} \tag{63}$$

where again, the asterisk (*) denotes the complex conjugate (applied in this case to the current phasor only). Using (61) and (62) in (63), it is found that

$$\langle P \rangle = \frac{1}{2}\mathcal{R}e\left\{V_0 e^{-\alpha z}e^{-j\beta z}\frac{V_0^*}{|Z_0|e^{-j\theta}}e^{-\alpha z}e^{+j\beta z}\right\}$$

$$= \frac{1}{2}\mathcal{R}e\left\{\frac{V_0 V_0^*}{|Z_0|}e^{-2\alpha z}e^{j\theta}\right\} = \frac{1}{2}\frac{|V_0|^2}{|Z_0|}e^{-2\alpha z}\cos\theta \tag{64}$$

which we note is identical to the time-integrated result in (60). Eq. (63) applies to any single-frequency wave.

An important result of the preceding exercise is that power attenuates as $e^{-2\alpha z}$, or

$$\boxed{\langle \mathcal{P}(z) \rangle = \langle \mathcal{P}(0) \rangle e^{-2\alpha z}} \tag{65}$$

Power drops at twice the exponential rate with distance as either voltage or current.

A convenient measure of power loss is in *decibel* units. This is based on expressing the power decrease as a power of 10. Specifically, we write

$$\frac{\langle \mathcal{P}(z) \rangle}{\langle \mathcal{P}(0) \rangle} = e^{-2\alpha z} = 10^{-\kappa \alpha z} \tag{66}$$

where the constant, κ, is to be determined. Setting $\alpha z = 1$, we find

$$e^{-2} = 10^{-\kappa} \quad \Rightarrow \quad \kappa = \log_{10}(e^2) = 0.869 \tag{67}$$

Now, by definition, the power loss in decibels (dB) is

$$\boxed{\text{Power loss (dB)} = 10 \log_{10}\left[\frac{\langle \mathcal{P}(0) \rangle}{\langle \mathcal{P}(z) \rangle}\right] = 8.69 \alpha z} \tag{68}$$

where we note that inverting the power ratio in the argument of the log function [as compared to the ratio in (66)] yields a positive number for the dB loss. Also, noting that $\langle \mathcal{P} \rangle \propto |V_0|^2$, we may write, equivalently:

$$\boxed{\text{Power loss (dB)} = 10 \log_{10}\left[\frac{\langle \mathcal{P}(0) \rangle}{\langle \mathcal{P}(z) \rangle}\right] = 20 \log_{10}\left[\frac{|V_0(0)|}{|V_0(z)|}\right]} \tag{69}$$

where $|V_0(z)| = |V_0(0)| e^{-\alpha z}$.

EXAMPLE 10.4

A 20-m length of transmission line is known to produce a 2.0-dB drop in power from end to end.

(*a*) What fraction of the input power reaches the output?
(*b*) What fraction of the input power reaches the midpoint of the line?
(*c*) What exponential attenuation coefficient, α, does this represent?

Solution. (*a*) The power fraction will be

$$\frac{\langle \mathcal{P}(20) \rangle}{\langle \mathcal{P}(0) \rangle} = 10^{-0.2} = 0.63$$

(*b*) 2 dB in 20 m implies a loss rating of 0.1 dB/m. So, over a 10-m span, the loss is 1.0 dB. This represents the power fraction, $10^{-0.1} = 0.79$.

(*c*) The exponential attenuation coefficient is found through

$$\alpha = \frac{2.0 \text{ dB}}{(8.69 \text{ dB/Np})(20 \text{ m})} = 0.012 \text{ [Np/m]}$$

A final point addresses the question: Why use decibels? The most compelling reason is that when evaluating the accumulated loss for several lines and devices that are all end-to-end connected, the net loss in dB for the entire span is just the sum of the dB losses of the individual elements.

> **D10.2.** Two transmission lines are to be joined end to end. Line 1 is 30 m long and is rated at 0.1 dB/m. Line 2 is 45 m long and is rated at 0.15 dB/m. The joint is not done well and imparts a 3-dB loss. What percentage of the input power reaches the output of the combination?
>
> **Ans.** 5.3%

10.9 WAVE REFLECTION AT DISCONTINUITIES

The concept of wave reflection was introduced in Section 10.1. As implied there, the need for a reflected wave originates from the necessity to satisfy all voltage and current boundary conditions at the ends of transmission lines and at locations at which two dissimilar lines are connected to each other. The consequences of reflected waves are usually less than desirable, in that some of the power that was intended to be transmitted to a load, for example, reflects and propagates back to the source. Conditions for achieving *no* reflected waves are therefore important to understand.

The basic reflection problem is illustrated in Figure 10.5. In it, a transmission line of characteristic impedance Z_0 is terminated by a load having complex impedance, $Z_L = R_L + jX_L$. If the line is lossy, then we know that Z_0 will also be complex. For convenience, we assign coordinates such that the load is at location $z = 0$. Therefore, the line occupies the region $z < 0$. A voltage wave is presumed to be incident on the load and is expressed in phasor form for all z:

$$V_i(z) = V_{0i}e^{-\alpha z}e^{-j\beta z} \tag{70a}$$

When the wave reaches the load, a reflected wave is generated that back-propagates:

$$V_r(z) = V_{0r}e^{+\alpha z}e^{+j\beta z} \tag{70b}$$

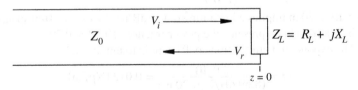

Figure 10.5 Voltage wave reflection from a complex load impedance.

The phasor voltage at the load is now the sum of the incident and reflected voltage phasors, evaluated at $z = 0$:

$$V_L = V_{0i} + V_{0r} \tag{71}$$

Additionally, the current through the load is the sum of the incident and reflected currents, also at $z = 0$:

$$I_L = I_{0i} + I_{0r} = \frac{1}{Z_0}[V_{0i} - V_{0r}] = \frac{V_L}{Z_L} = \frac{1}{Z_L}[V_{0i} + V_{0r}] \tag{72}$$

We can now solve for the ratio of the reflected voltage amplitude to the incident voltage amplitude, defined as the *reflection coefficient,* Γ:

$$\boxed{\Gamma \equiv \frac{V_{0r}}{V_{0i}} = \frac{Z_L - Z_0}{Z_L + Z_0} = |\Gamma|e^{j\phi_r}} \tag{73}$$

where we emphasize the complex nature of Γ—meaning that, in general, a reflected wave will experience a reduction in amplitude and a phase shift, relative to the incident wave.

Now, using (71) with (73), we may write

$$V_L = V_{0i} + \Gamma V_{0i} \tag{74}$$

from which we find the *transmission coefficient,* defined as the ratio of the load voltage amplitude to the incident voltage amplitude:

$$\boxed{\tau \equiv \frac{V_L}{V_{0i}} = 1 + \Gamma = \frac{2Z_L}{Z_0 + Z_L} = |\tau|e^{j\phi_t}} \tag{75}$$

A point that may at first cause some alarm is that if Γ is a positive real number, then $\tau > 1$; the voltage amplitude at the load is thus greater than the incident voltage. Although this would seem counterintuitive, it is not a problem because the load current will be lower than that in the incident wave. We will find that this always results in an average *power* at the load that is less than or equal to that in the incident wave. An additional point concerns the possibility of loss in the line. The incident wave amplitude that is used in (73) and (75) is always the amplitude that occurs *at the load*—after loss has occurred in propagating from the input.

Usually, the main objective in transmitting power to a load is to configure the line/load combination such that there is no reflection. The load therefore receives all the transmitted power. The condition for this is $\Gamma = 0$, which means that the load impedance must be equal to the line impedance. In such cases the load is said to be *matched* to the line (or vice versa). Various impedance-matching methods exist, many of which will be explored later in this chapter.

Finally, the fractions of the incident wave *power* that are reflected and dissipated by the load need to be determined. The incident power is found from (64), where this time we position the load at $z = L$, with the line input at $z = 0$.

$$\langle P_i \rangle = \frac{1}{2}\text{Re}\left\{ \frac{V_0 V_0^*}{|Z_0|}e^{-2\alpha L}e^{j\theta} \right\} = \frac{1}{2}\frac{|V_0|^2}{|Z_0|}e^{-2\alpha L}\cos\theta \tag{76a}$$

The reflected power is then found by substituting the reflected wave voltage into (76a), where the latter is obtained by multiplying the incident voltage by Γ:

$$\langle P_r \rangle = \frac{1}{2}\mathrm{Re}\left\{\frac{(\Gamma V_0)(\Gamma^* V_0^*)}{|Z_0|}e^{-2\alpha L}e^{j\theta}\right\} = \frac{1}{2}\frac{|\Gamma|^2|V_0|^2}{|Z_0|}e^{-2\alpha L}\cos\theta \qquad (76b)$$

The reflected power fraction at the load is now determined by the ratio of (76b) to (76a):

$$\boxed{\frac{\langle P_r \rangle}{\langle P_i \rangle} = \Gamma\Gamma^* = |\Gamma|^2} \qquad (77a)$$

The fraction of the incident power that is transmitted into the load (or dissipated by it) is therefore

$$\boxed{\frac{\langle P_t \rangle}{\langle P_i \rangle} = 1 - |\Gamma|^2} \qquad (77b)$$

Illustrations

The reader should be aware that the transmitted power fraction is *not* $|\tau|^2$, as one might be tempted to conclude.

In situations involving the connection of two semi-infinite transmission lines having different characteristic impedances, reflections will occur at the junction, with the second line being treated as the load. For a wave incident from line 1 (Z_{01}) to line 2 (Z_{02}), we find

$$\boxed{\Gamma = \frac{Z_{02} - Z_{01}}{Z_{02} + Z_{01}}} \qquad (78)$$

The fraction of the power that propagates into the second line is then $1 - |\Gamma|^2$.

EXAMPLE 10.5

A 50-Ω lossless transmission line is terminated by a load impedance, $Z_L = 50 - j75\ \Omega$. If the incident power is 100 mW, find the power dissipated by the load.

Solution. The reflection coefficient is

$$\Gamma = \frac{Z_L - Z_0}{Z_L + Z_0} = \frac{50 - j75 - 50}{50 - j75 + 50} = 0.36 - j0.48 = 0.60\,e^{-j.93}$$

Then

$$\langle P_t \rangle = (1 - |\Gamma|^2)\langle P_i \rangle = [1 - (0.60)^2](100) = 64\ \mathrm{mW}$$

EXAMPLE 10.6

Two lossy lines are to be joined end to end. The first line is 10 m long and has a loss rating of 0.20 dB/m. The second line is 15 m long and has a loss rating of 0.10 dB/m. The reflection coefficient at the junction (line 1 to line 2) is $\Gamma = 0.30$. The input power (to line 1) is 100 mW. (*a*) Determine the total loss of the combination in dB. (*b*) Determine the power transmitted to the output end of line 2.

Solution. (*a*) The dB loss of the joint is

$$L_j(\text{dB}) = 10 \log_{10}\left(\frac{1}{1 - |\Gamma|^2}\right) = 10 \log_{10}\left(\frac{1}{1 - 0.09}\right) = 0.41 \, \text{dB}$$

The total loss of the link in dB is now

$$L_t(\text{dB}) = (0.20)(10) + 0.41 + (0.10)(15) = 3.91 \, \text{dB}$$

(*b*) The output power will be $P_{\text{out}} = 100 \times 10^{-0.391} = 41$ mW.

10.10 VOLTAGE STANDING WAVE RATIO

In many instances, characteristics of transmission line performance are amenable to measurement. Included in these are measurements of unknown load impedances, or input impedances of lines that are terminated by known or unknown load impedances. Such techniques rely on the ability to measure voltage amplitudes that occur as functions of position within a line, usually designed for this purpose. A typical apparatus consists of a *slotted line,* which is a lossless coaxial transmission line having a longitudinal gap in the outer conductor along its entire length. The line is positioned between the sinusoidal voltage source and the impedance that is to be measured. Through the gap in the slotted line, a voltage probe may be inserted to measure the voltage amplitude between the inner and outer conductors. As the probe is moved along the length of the line, the maximum and minimum voltage amplitudes are noted, and their ratio, known as the *voltage standing wave ratio,* or VSWR, is determined. The significance of this measurement and its utility form the subject of this section.

To understand the meaning of the voltage measurements, we consider a few special cases. First, if the slotted line is terminated by a matched impedance, then no reflected wave occurs; the probe will indicate the same voltage amplitude at every point. Of course, the instantaneous voltages that the probe samples will differ in phase by $\beta(z_2 - z_1)$ rad as the probe is moved from $z = z_1$ to $z = z_2$, but the system is insensitive to the phase of the field. The equal-amplitude voltages are characteristic of an unattenuated traveling wave.

Second, if the slotted line is terminated by an open or short circuit (or in general a purely imaginary load impedance), the total voltage in the line is a standing wave and, as was shown in Example 10.1, the voltage probe provides no output when it is located at the nodes; these occur periodically with half-wavelength spacing. As the probe position is changed, its output varies as $|\cos(\beta z + \phi)|$, where z is the distance from the load, and where the phase, ϕ, depends on the load impedance.

For example, if the load is a short circuit, the requirement of zero voltage at the short leads to a null occurring there, and so the voltage in the line will vary as $|\sin(\beta z)|$ (where $\phi = \pm\pi/2$).

A more complicated situation arises when the reflected voltage is neither 0 nor 100 percent of the incident voltage. Some energy is absorbed by the load and some is reflected. The slotted line, therefore, supports a voltage that is composed of both a traveling wave and a standing wave. It is customary to describe this voltage as a standing wave, even though a traveling wave is also present. We will see that the voltage does not have zero amplitude at any point for all time, and the degree to which the voltage is divided between a traveling wave and a true standing wave is expressed by the ratio of the maximum amplitude found by the probe to the minimum amplitude (VSWR). This information, along with the positions of the voltage minima or maxima with respect to that of the load, enable one to determine the load impedance. The VSWR also provides a measure of the quality of the termination. Specifically, a perfectly matched load yields a VSWR of exactly 1. A totally reflecting load produces an infinite VSWR.

To derive the specific form of the total voltage, we begin with the forward and backward-propagating waves that occur within the slotted line. The load is positioned at $z = 0$, and so all positions within the slotted line occur at negative values of z. Taking the input wave amplitude as V_0, the total phasor voltage is

$$V_{sT}(z) = V_0 e^{-j\beta z} + \Gamma V_0 e^{j\beta z} \tag{79}$$

The line, being lossless, has real characteristic impedance, Z_0. The load impedance, Z_L, is in general complex, which leads to a complex reflection coefficient:

$$\boxed{\Gamma = \frac{Z_L - Z_0}{Z_L + Z_0} = |\Gamma|e^{j\phi}} \tag{80}$$

If the load is a short circuit ($Z_L = 0$), ϕ is equal to π; if Z_L is real and less than Z_0, ϕ is also equal to π; and if Z_L is real and greater than Z_0, ϕ is zero. Using (80), we may rewrite (79) in the form:

$$V_{sT}(z) = V_0\left(e^{-j\beta z} + |\Gamma|e^{j(\beta z+\phi)}\right) = V_0 e^{j\phi/2}\left(e^{-j\beta z}e^{-j\phi/2} + |\Gamma|e^{j\beta z}e^{j\phi/2}\right) \tag{81}$$

To express (81) in a more useful form, we can apply the algebraic trick of adding and subtracting the term $V_0(1 - |\Gamma|)e^{-j\beta z}$:

$$V_{sT}(z) = V_0(1 - |\Gamma|)e^{-j\beta z} + V_0|\Gamma|e^{j\phi/2}\left(e^{-j\beta z}e^{-j\phi/2} + e^{j\beta z}e^{j\phi/2}\right) \tag{82}$$

The last term in parentheses in (82) becomes a cosine, and we write

$$\boxed{V_{sT}(z) = V_0(1 - |\Gamma|)e^{-j\beta z} + 2V_0|\Gamma|e^{j\phi/2}\cos(\beta z + \phi/2)} \tag{83}$$

The important characteristics of this result are most easily seen by converting it to real instantaneous form:

$$\mathcal{V}(z, t) = \text{Re}[V_{sT}(z)e^{j\omega t}] = \underbrace{V_0(1 - |\Gamma|)\cos(\omega t - \beta z)}_{\text{traveling wave}}$$

$$+ \underbrace{2|\Gamma|V_0\cos(\beta z + \phi/2)\cos(\omega t + \phi/2)}_{\text{standing wave}} \qquad (84)$$

Equation (84) is recognized as the sum of a traveling wave of amplitude $(1 - |\Gamma|)\,V_0$ and a standing wave having amplitude $2|\Gamma|V_0$. We can visualize events as follows: The portion of the incident wave that reflects and back-propagates in the slotted line interferes with an equivalent portion of the incident wave to form a standing wave. The rest of the incident wave (which does not interfere) is the traveling wave part of (84). The maximum amplitude observed in the line is found where the amplitudes of the two terms in (84) add directly to give $(1 + |\Gamma|)\,V_0$. The minimum amplitude is found where the standing wave achieves a null, leaving only the traveling wave amplitude of $(1 - |\Gamma|)V_0$. The fact that the two terms in (84) combine in this way with the proper phasing is not immediately apparent, but the following arguments will show that this does occur.

To obtain the minimum and maximum voltage amplitudes, we may revisit the first part of Eq. (81):

$$V_{sT}(z) = V_0\left(e^{-j\beta z} + |\Gamma|e^{j(\beta z + \phi)}\right) \qquad (85)$$

First, the minimum voltage amplitude is obtained when the two terms in (85) subtract directly (having a phase difference of π). This occurs at locations

$$z_{\min} = -\frac{1}{2\beta}(\phi + (2m + 1)\pi) \quad (m = 0, 1, 2, \ldots) \qquad (86)$$

Note again that all positions within the slotted line occur at negative values of z. Substituting (86) into (85) leads to the minimum amplitude:

$$V_{sT}(z_{\min}) = V_0(1 - |\Gamma|) \qquad (87)$$

The same result is obtained by substituting (86) into the real voltage, (84). This produces a null in the standing wave part, and we obtain

$$\mathcal{V}(z_{\min}, t) = \pm V_0(1 - |\Gamma|)\sin(\omega t + \phi/2) \qquad (88)$$

The voltage oscillates (through zero) in time, with amplitude $V_0(1 - |\Gamma|)$. The plus and minus signs in (88) apply to even and odd values of m in (86), respectively.

Next, the maximum voltage amplitude is obtained when the two terms in (85) add in-phase. This will occur at locations given by

$$z_{\max} = -\frac{1}{2\beta}(\phi + 2m\pi) \quad (m = 0, 1, 2, \ldots) \qquad (89)$$

On substituting (89) into (85), we obtain

$$V_{sT}(z_{\max}) = V_0(1 + |\Gamma|) \tag{90}$$

As before, we may substitute (89) into the real instantaneous voltage (84). The effect is to produce a maximum in the standing wave part, which then adds in-phase to the running wave. The result is

$$V(z_{\max}, t) = \pm V_0(1 + |\Gamma|)\cos(\omega t + \phi/2) \tag{91}$$

where the plus and minus signs apply to positive and negative values of m in (89), respectively. Again, the voltage oscillates through zero in time, with amplitude $V_0(1 + |\Gamma|)$.

Note that a voltage maximum is located at the load ($z = 0$) if $\phi = 0$; moreover, $\phi = 0$ when Γ is real and positive. This occurs for real Z_L when $Z_L > Z_0$. Thus there is a voltage maximum at the load when the load impedance is greater than Z_0 and both impedances are real. With $\phi = 0$, maxima also occur at $z_{\max} = -m\pi/\beta = -m\lambda/2$. For a zero-load impedance, $\phi = \pi$, and the maxima are found at $z_{\max} = -\pi/(2\beta)$, $-3\pi/(2\beta)$, or $z_{\max} = -\lambda/4$, $-3\lambda/4$, and so forth.

The minima are separated by multiples of one half-wavelength (as are the maxima), and for a zero load impedance, the first minimum occurs when $-\beta z = 0$, or at the load. In general, a voltage minimum is found at $z = 0$ whenever $\phi = \pi$; this occurs if $Z_L < Z_0$ where Z_L is real. The general results are illustrated in Figure 10.6.

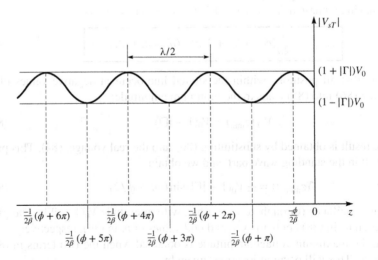

Figure 10.6 Plot of the magnitude of V_{sT} as found from Eq. (85) as a function of position, z, in front of the load (at $z = 0$). The reflection coefficient phase is ϕ, which leads to the indicated locations of maximum and minimum voltage amplitude, as found from Eqs. (86) and (89).

Finally, we introduce the symbol s to denote the VSWR. Voltage standing wave ratio is defined as:

$$s \equiv \frac{V_{sT}(z_{max})}{V_{sT}(z_{min})} = \frac{1 + |\Gamma|}{1 - |\Gamma|} \tag{92}$$

Since the absolute voltage amplitudes have divided out, our measured VSWR permits the immediate evaluation of $|\Gamma|$. The phase of Γ is then found by measuring the location of the first maximum or minimum with respect to the load, and then using (86) or (89) as appropriate. Once Γ is known, the load impedance can be found, assuming Z_0 is known.

Illustrations

D10.3. What voltage standing wave ratio results when $\Gamma = \pm 1/2$?

Ans. 3

EXAMPLE 10.7

Slotted line measurements yield a VSWR of 5, a 15-cm spacing between successive voltage maxima, and the first maximum at a distance of 7.5 cm in front of the load. Determine the load impedance, assuming a 50-Ω impedance for the slotted line.

Solution. The 15-cm spacing between maxima is $\lambda/2$, implying a wavelength of 30 cm. Because the slotted line is air-filled, the frequency is $f = c/\lambda = 1$ GHz. The first maximum at 7.5 cm is thus at a distance of $\lambda/4$ from the load, which means that a voltage minimum occurs at the load. Thus Γ will be real and negative. We use (92) to write

$$|\Gamma| = \frac{s - 1}{s + 1} = \frac{5 - 1}{5 + 1} = \frac{2}{3}$$

So

$$\Gamma = -\frac{2}{3} = \frac{Z_L - Z_0}{Z_L + Z_0}$$

which we solve for Z_L to obtain

$$Z_L = \frac{1}{5} Z_0 = \frac{50}{5} = 10\ \Omega$$

10.11 TRANSMISSION LINES OF FINITE LENGTH

A new type of problem emerges when considering the propagation of sinusoidal voltages on finite-length lines that have loads that are not impedance matched. In such cases, numerous reflections occur at the load and at the generator, setting up a multiwave bidirectional voltage distribution in the line. As always, the objective is to determine the net power transferred to the load in steady state, but we must now include the effect of the numerous forward- and backward-reflected waves.

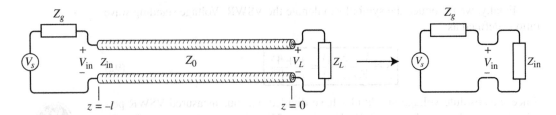

Figure 10.7   Finite-length transmission line configuration and its equivalent circuit.

Figure 10.7 shows the basic problem. The line, assumed to be lossless, has characteristic impedance Z_0 and is of length l. The sinusoidal voltage source at frequency ω provides phasor voltage V_s. Associated with the souce is a complex internal impedance, Z_g, as shown. The load impedance, Z_L, is also assumed to be complex and is located at $z = 0$. The line thus exists along the negative z axis. The easiest method of approaching the problem is not to attempt to analyze every reflection individually, but rather to recognize that in steady state, there will exist one net forward wave and one net backward wave, representing the superposition of all waves that are incident on the load and all waves that are reflected from it. We may thus write the total voltage in the line as

$$V_{sT}(z) = V_0^+ e^{-j\beta z} + V_0^- e^{j\beta z} \tag{93}$$

in which V_0^+ and V_0^- are complex amplitudes, composed respectively of the sum of all individual forward and backward wave amplitudes and phases. In a similar way, we may write the total current in the line:

$$I_{sT}(z) = I_0^+ e^{-j\beta z} + I_0^- e^{j\beta z} \tag{94}$$

We now define the *wave impedance, $Z_w(z)$*, as the ratio of the total phasor voltage to the total phasor current. Using (93) and (94), this becomes:

$$Z_w(z) \equiv \frac{V_{sT}(z)}{I_{sT}(z)} = \frac{V_0^+ e^{-j\beta z} + V_0^- e^{j\beta z}}{I_0^+ e^{-j\beta z} + I_0^- e^{j\beta z}} \tag{95}$$

We next use the relations $V_0^- = \Gamma V_0^+, I_0^+ = V_0^+/Z_0$, and $I_0^- = -V_0^-/Z_0$. Eq. (95) simplifies to

$$Z_w(z) = Z_0 \left[\frac{e^{-j\beta z} + \Gamma e^{j\beta z}}{e^{-j\beta z} - \Gamma e^{j\beta z}} \right] \tag{96}$$

Now, using the Euler identity, (32), and substituting $\Gamma = (Z_L - Z_0)/(Z_L + Z_0)$, Eq. (96) becomes

$$Z_w(z) = Z_0 \left[\frac{Z_L \cos(\beta z) - j Z_0 \sin(\beta z)}{Z_0 \cos(\beta z) - j Z_L \sin(\beta z)} \right] \tag{97}$$

The wave impedance at the line input is now found by evaluating (97) at $z = -l$, obtaining

$$Z_{in} = Z_0 \left[\frac{Z_L \cos(\beta l) + j Z_0 \sin(\beta l)}{Z_0 \cos(\beta l) + j Z_L \sin(\beta l)} \right] \qquad (98)$$

This is the quantity that we need in order to create the equivalent circuit in Figure 10.7.

One special case is that in which the line length is a half-wavelength, or an integer multiple thereof. In that case,

$$\beta l = \frac{2\pi}{\lambda} \frac{m\lambda}{2} = m\pi \quad (m = 0, 1, 2, \ldots)$$

Using this result in (98), we find

$$Z_{in}(l = m\lambda/2) = Z_L \qquad (99)$$

For a half-wave line, the equivalent circuit can be constructed simply by removing the line completely and placing the load impedance at the input. This simplification works, of course, provided the line length is indeed an integer multiple of a half-wavelength. Once the frequency begins to vary, the condition is no longer satisfied, and (98) must be used in its general form to find Z_{in}.

Another important special case is that in which the line length is an odd multiple of a quarter wavelength:

$$\beta l = \frac{2\pi}{\lambda}(2m + 1)\frac{\lambda}{4} = (2m + 1)\frac{\pi}{2} \quad (m = 0, 1, 2, \ldots)$$

Using this result in (98) leads to

$$Z_{in}(l = \lambda/4) = \frac{Z_0^2}{Z_L} \qquad (100)$$

An immediate application of (100) is to the problem of joining two lines having different characteristic impedances. Suppose the impedances are (from left to right) Z_{01} and Z_{03}. At the joint, we may insert an additional line whose characteristic impedance is Z_{02} and whose length is $\lambda/4$. We thus have a sequence of joined lines whose impedances progress as Z_{01}, Z_{02}, and Z_{03}, in that order. A voltage wave is now incident from line 1 onto the joint between Z_{01} and Z_{02}. Now the effective load at the far end of line 2 is Z_{03}. The input impedance to line 2 at any frequency is now

$$Z_{in} = Z_{02} \frac{Z_{03} \cos \beta_2 l + j Z_{02} \sin \beta_2 l}{Z_{02} \cos \beta_2 l + j Z_{03} \sin \beta_2 l} \qquad (101)$$

Then, since the length of line 2 is $\lambda/4$,

$$Z_{in}(\text{line } 2) = \frac{Z_{02}^2}{Z_{03}} \qquad (102)$$

Reflections at the Z_{01}–Z_{02} interface will not occur if $Z_{in} = Z_{01}$. Therefore, we can match the junction (allowing complete transmission through the three-line sequence) if Z_{02} is chosen so that

$$Z_{02} = \sqrt{Z_{01}Z_{03}}$$

(103)

This technique is called *quarter-wave matching* and again is limited to the frequency (or narrow band of frequencies) such that $l \doteq (2m + 1)\lambda/4$. We will encounter more examples of these techniques when we explore electromagnetic wave reflection in Chapter 12. Meanwhile, further examples that involve the use of the input impedance and the VSWR are presented in Section 10.12.

10.12 SOME TRANSMISSION LINE EXAMPLES

In this section, we apply many of the results that we obtained in the previous sections to several typical transmission line problems. We simplify our work by restricting our attention to the lossless line.

We begin by assuming a two-wire 300 Ω line ($Z_0 = 300$ Ω), such as the lead-in wire from the antenna to a television or FM receiver. The circuit is shown in Figure 10.8. The line is 2 m long, and the values of L and C are such that the velocity on the line is 2.5×10^8 m/s. We will terminate the line with a receiver having an input resistance of 300 Ω and represent the antenna by its Thevenin equivalent $Z = 300$ Ω in series with $V_s = 60$ V at 100 MHz. This antenna voltage is larger by a factor of about 10^5 than it would be in a practical case, but it also provides simpler values to work with; in order to think practical thoughts, divide currents or voltages by 10^5, divide powers by 10^{10}, and leave impedances alone.

Because the load impedance is equal to the characteristic impedance, the line is matched; the reflection coefficient is zero, and the standing wave ratio is unity. For the given velocity and frequency, the wavelength on the line is $v/f = 2.5$ m, and the phase constant is $2\pi/\lambda = 0.8\pi$ rad/m; the attenuation constant is zero. The electrical length of the line is $\beta l = (0.8\pi)2$, or 1.6π rad. This length may also be expressed as 288°, or 0.8 wavelength.

The input impedance offered to the voltage source is 300 Ω, and since the internal impedance of the source is 300 Ω, the voltage at the input to the line is half of 60 V, or 30 V. The source is matched to the line and delivers the maximum available

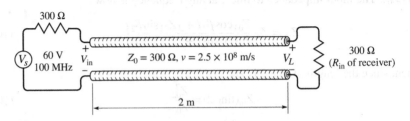

Figure 10.8 A transmission line that is matched at both ends produces no reflections and thus delivers maximum power to the load.

power to the line. Because there is no reflection and no attenuation, the voltage at the load is 30 V, but it is delayed in phase by 1.6π rad. Thus,

$$V_{\text{in}} = 30 \cos(2\pi 10^8 t)\,\text{V}$$

whereas

$$V_L = 30 \cos(2\pi 10^8 t - 1.6\pi)\,\text{V}$$

The input current is

$$I_{\text{in}} = \frac{V_{\text{in}}}{300} = 0.1 \cos(2\pi 10^8 t)\,\text{A}$$

while the load current is

$$I_L = 0.1 \cos(2\pi 10^8 t - 1.6\pi)\,\text{A}$$

The average power delivered to the input of the line by the source must all be delivered to the load by the line,

$$P_{\text{in}} = P_L = \frac{1}{2} \times 30 \times 0.1 = 1.5\ \text{W}$$

Now let us connect a second receiver, also having an input resistance of 300 Ω, across the line in parallel with the first receiver. The load impedance is now 150 Ω, the reflection coefficient is

$$\Gamma = \frac{150 - 300}{150 + 300} = -\frac{1}{3}$$

and the standing wave ratio on the line is

$$s = \frac{1 + \frac{1}{3}}{1 - \frac{1}{3}} = 2$$

The input impedance is no longer 300 Ω but is now

$$Z_{\text{in}} = Z_0 \frac{Z_L \cos\beta l + jZ_0 \sin\beta l}{Z_0 \cos\beta l + jZ_L \sin\beta l} = 300 \frac{150 \cos 288° + j300 \sin 288°}{300 \cos 288° + j150 \sin 288°}$$
$$= 510\angle{-23.8°} = 466 - j206\ \Omega$$

which is a capacitive impedance. Physically, this means that this length of line stores more energy in its electric field than in its magnetic field. The input current phasor is thus

$$I_{s,in} = \frac{60}{300 + 466 - j206} = 0.0756\angle 15.0°\ \text{A}$$

and the power supplied to the line by the source is

$$P_{\text{in}} = \frac{1}{2} \times (0.0756)^2 \times 466 = 1.333\ \text{W}$$

Since there are no losses in the line, 1.333 W must also be delivered to the load. Note that this is less than the 1.50 W which we were able to deliver to a matched load; moreover, this power must divide equally between two receivers, and thus each receiver now receives only 0.667 W. Because the input impedance of each receiver is 300 Ω, the voltage across the receiver is easily found as

$$0.667 = \frac{1}{2}\frac{|V_{s,L}|^2}{300}$$

$$|V_{s,L}| = 20\,\text{V}$$

in comparison with the 30 V obtained across the single load.

Before we leave this example, let us ask ourselves several questions about the voltages on the transmission line. Where is the voltage a maximum and a minimum, and what are these values? Does the phase of the load voltage still differ from the input voltage by 288°? Presumably, if we can answer these questions for the voltage, we could do the same for the current.

Equation (89) serves to locate the voltage maxima at

$$z_{max} = -\frac{1}{2\beta}(\phi + 2m\pi) \quad (m = 0, 1, 2, \ldots)$$

where $\Gamma = |\Gamma|e^{j\phi}$. Thus, with $\beta = 0.8\pi$ and $\phi = \pi$, we find

$$z_{max} = -0.625 \quad \text{and} \quad -1.875\,\text{m}$$

while the minima are $\lambda/4$ distant from the maxima;

$$z_{min} = 0 \quad \text{and} \quad -1.25\,\text{m}$$

and we find that the load voltage (at $z = 0$) is a voltage minimum. This, of course, verifies the general conclusion we reached earlier: a voltage minimum occurs at the load if $Z_L < Z_0$, and a voltage maximum occurs if $Z_L > Z_0$, where both impedances are pure resistances.

The minimum voltage on the line is thus the load voltage, 20 V; the maximum voltage must be 40 V, since the standing wave ratio is 2. The voltage at the input end of the line is

$$V_{s,in} = I_{s,in}\,Z_{in} = (0.0756\angle15.0°)(510\angle-23.8°) = 38.5\angle-8.8°$$

The input voltage is almost as large as the maximum voltage anywhere on the line because the line is about three-quarters of a wavelength long, a length which would place the voltage maximum at the input when $Z_L < Z_0$.

Finally, it is of interest to determine the load voltage in magnitude *and phase*. We begin with the total voltage in the line, using (93).

$$V_{sT} = \left(e^{-j\beta z} + \Gamma e^{j\beta z}\right)V_0^+ \tag{104}$$

We may use this expression to determine the voltage at any point on the line in terms of the voltage at any other point. Because we know the voltage at the input to the line, we let $z = -l$,

$$V_{s,in} = \left(e^{j\beta l} + \Gamma e^{-j\beta l}\right)V_0^+ \tag{105}$$

and solve for V_0^+,

$$V_0^+ = \frac{V_{s,in}}{e^{j\beta l} + \Gamma e^{-j\beta l}} = \frac{38.5\angle-8.8°}{e^{j1.6\pi} - \frac{1}{3}e^{-j1.6\pi}} = 30.0\angle72.0°\,\text{V}$$

We may now let $z = 0$ in (104) to find the load voltage,

$$V_{s,L} = (1 + \Gamma)V_0^+ = 20\angle 72° = 20\angle -288°$$

The amplitude agrees with our previous value. The presence of the reflected wave causes $V_{s,\text{in}}$ and $V_{s,L}$ to differ in phase by about $-279°$ instead of $-288°$.

EXAMPLE 10.8

In order to provide a slightly more complicated example, let us now place a purely capacitive impedance of $-j300$ Ω in parallel with the two 300 Ω receivers. We are to find the input impedance and the power delivered to each receiver.

Solution. The load impedance is now 150 Ω in parallel with $-j300$ Ω, or

$$Z_L = \frac{150(-j300)}{150 - j300} = \frac{-j300}{1 - j2} = 120 - j60 \ \Omega$$

We first calculate the reflection coefficient and the VSWR:

$$\Gamma = \frac{120 - j60 - 300}{120 - j60 + 300} = \frac{-180 - j60}{420 - j60} = 0.447\angle -153.4°$$

$$s = \frac{1 + 0.447}{1 - 0.447} = 2.62$$

Thus, the VSWR is higher and the mismatch is therefore worse. Let us next calculate the input impedance. The electrical length of the line is still $288°$, so that

$$Z_{\text{in}} = 300 \frac{(120 - j60)\ \cos 288° \ + j300\ \sin 288°}{300\ \cos 288° + j(120 - j60)\ \sin 288°} = 755 - j138.5 \ \Omega$$

This leads to a source current of

$$I_{s,\text{in}} = \frac{V_{Th}}{Z_{Th} + Z_{\text{in}}} = \frac{60}{300 + 755 - j138.5} = 0.0564\angle 7.47° \ \text{A}$$

Therefore, the average power delivered to the input of the line is $P_{\text{in}} = \frac{1}{2}(0.0564)^2(755) = 1.200$ W. Since the line is lossless, it follows that $P_L = 1.200$ W, and each receiver gets only 0.6 W.

EXAMPLE 10.9

As a final example, our line is now terminated with a purely capacitive impedance, $Z_L = -j300$ Ω. We seek the reflection coefficient, the VSWR, and the power delivered to the load.

Solution. Obviously, we cannot deliver any average power to the load since it is a pure reactance. As a consequence, the reflection coefficient is

$$\Gamma = \frac{-j300 - 300}{-j300 + 300} = -j1 = 1\angle -90°$$

and the reflected wave is equal in amplitude to the incident wave. Hence, it should not surprise us to see that the VSWR is

$$s = \frac{1 + |-j1|}{1 - |-j1|} = \infty$$

and the input impedance is a pure reactance,

$$Z_{\text{in}} = 300 \frac{-j300 \cos 288° + j300 \sin 288°}{300 \cos 288° + j(-j300) \sin 288°} = j589$$

Thus, no average power can be delivered to the input impedance by the source, and therefore no average power can be delivered to the load.

Although we could continue to find many other facts and figures for these examples, much of the work may be done more easily for problems of this type by using graphical techniques. We encounter these in Section 10.13.

> **D10.4.** A 50-ohm lossless line has a length of 0.4λ. The operating frequency is 300 MHz. A load $Z_L = 40 + j30\ \Omega$ is connected at $z = 0$, and the Thevenin-equivalent source at $z = -l$ is $12\angle0°$ V in series with $Z_{Th} = 50 + j0\ \Omega$. Find: (a) Γ; (b) s; (c) Z_{in}.
>
> **Ans.** (a) $0.333\angle90°$; (b) 2.00; (c) $25.5 + j\,5.90\ \Omega$

> **D10.5.** For the transmission line of Problem D10.4, also find: (a) the phasor voltage at $z = -l$; (b) the phasor voltage at $z = 0$; (c) the average power delivered to Z_L.
>
> **Ans.** (a) $4.14\angle8.58°$ V; (b) $6.32\angle-125.6°$ V; (c) 0.320 W

10.13 GRAPHICAL METHODS: THE SMITH CHART

Transmission line problems often involve manipulations with complex numbers, making the time and effort required for a solution several times greater than are needed for a similar sequence of operations on real numbers. One means of reducing the labor without seriously affecting the accuracy is by using transmission-line charts. Probably the most widely used one is the Smith chart.[3]

Basically, this diagram shows curves of constant resistance and constant reactance; these may represent either an input impedance or a load impedance. The latter, of course, is the input impedance of a zero-length line. An indication of location along the line is also provided, usually in terms of the fraction of a wavelength from a voltage maximum or minimum. Although they are not specifically shown on the chart, the standing-wave ratio and the magnitude and angle of the reflection

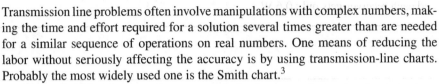

---

[3] P. H. Smith, "Transmission Line Calculator," *Electronics*, vol. 12, pp. 29–31, January 1939.

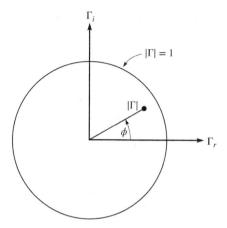

Figure 10.9 The polar coordinates of the Smith chart are the magnitude and phase angle of the reflection coefficient; the rectangular coordinates are the real and imaginary parts of the reflection coefficient. The entire chart lies within the circle $|\Gamma| = 1$.

coefficient are very quickly determined. As a matter of fact, the diagram is constructed within a circle of unit radius, using polar coordinates, with radius variable $|\Gamma|$ and counterclockwise angle variable ϕ, where $\Gamma = |\Gamma| e^{j\phi}$. Figure 10.9 shows this circle. Since $|\Gamma| < 1$, all our information must lie on or within the unit circle. Peculiarly enough, the reflection coefficient itself will not be plotted on the final chart, for these additional contours would make the chart very difficult to read.

The basic relationship upon which the chart is constructed is

$$\Gamma = \frac{Z_L - Z_0}{Z_L + Z_0} \tag{106}$$

The impedances that we plot on the chart will be *normalized* with respect to the characteristic impedance. The normalized load impedance, z_L, is

$$z_L = r + jx = \frac{Z_L}{Z_0} = \frac{R_L + jX_L}{Z_0}$$

and thus

$$\Gamma = \frac{z_L - 1}{z_L + 1}$$

or

$$z_L = \frac{1 + \Gamma}{1 - \Gamma} \tag{107}$$

In polar form, we have used $|\Gamma|$ and ϕ as the magnitude and angle of Γ. With Γ_r and Γ_i as the real and imaginary parts of Γ, we write

$$\Gamma = \Gamma_r + j\Gamma_i \tag{108}$$

Thus

$$r + jx = \frac{1 + \Gamma_r + j\Gamma_i}{1 - \Gamma_r - j\Gamma_i} \tag{109}$$

The real and imaginary parts of this equation are

$$r = \frac{1 - \Gamma_r^2 - \Gamma_i^2}{(1 - \Gamma_r)^2 + \Gamma_i^2} \tag{110}$$

$$x = \frac{2\Gamma_i}{(1 - \Gamma_r)^2 + \Gamma_i^2} \tag{111}$$

After several lines of elementary algebra, we may write (110) and (111) in forms which readily display the nature of the curves on Γ_r, Γ_i axes,

$$\left(\Gamma_r - \frac{r}{1+r}\right)^2 + \Gamma_i^2 = \left(\frac{1}{1+r}\right)^2 \tag{112}$$

$$(\Gamma_r - 1)^2 + \left(\Gamma_i - \frac{1}{x}\right)^2 = \left(\frac{1}{x}\right)^2 \tag{113}$$

The first equation describes a family of circles, where each circle is associated with a specific value of resistance r. For example, if $r = 0$, the radius of this zero-resistance circle is seen to be unity, and it is centered at the origin ($\Gamma_r = 0$, $\Gamma_i = 0$). This checks, for a pure reactance termination leads to a reflection coefficient of unity magnitude. On the other hand, if $r = \infty$, then $z_L = \infty$ and we have $\Gamma = 1 + j0$. The circle described by (112) is centered at $\Gamma_r = 1$, $\Gamma_i = 0$ and has zero radius. It is therefore the point $\Gamma = 1 + j0$, as we decided it should be. As another example, the circle for $r = 1$ is centered at $\Gamma_r = 0.5$, $\Gamma_i = 0$ and has a radius of 0.5. This circle is shown in Figure 10.10, along with circles for $r = 0.5$ and $r = 2$. All circles are centered on the Γ_r axis and pass through the point $\Gamma = 1 + j0$.

Equation (113) also represents a family of circles, but each of these circles is defined by a particular value of x, rather than r. If $x = \infty$, then $z_L = \infty$, and $\Gamma = 1 + j0$ again. The circle described by (113) is centered at $\Gamma = 1 + j0$ and has zero radius; it is therefore the point $\Gamma = 1 + j0$. If $x = +1$, then the circle is centered at $\Gamma = 1 + j1$ and has unit radius. Only one-quarter of this circle lies within the boundary curve $|\Gamma| = 1$, as shown in Figure 10.11. A similar quarter-circle appears below the Γ_r axis for $x = -1$. The portions of other circles for $x = 0.5$, -0.5, 2, and -2 are also shown. The "circle" representing $x = 0$ is the Γ_r axis; this is also labeled in Figure 10.11.

The two families of circles both appear on the Smith chart, as shown in Figure 10.12. It is now evident that if we are given Z_L, we may divide by Z_0 to

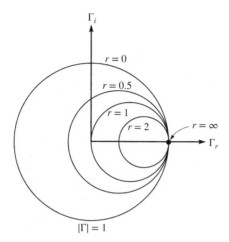

Figure 10.10 Constant-r circles are shown on the Γ_r, Γ_i plane. The radius of any circle is $1/(1 + r)$.

obtain z_L, locate the appropriate r and x circles (interpolating as necessary), and determine Γ by the intersection of the two circles. Because the chart does not have concentric circles showing the values of $|\Gamma|$, it is necessary to measure the radial distance from the origin to the intersection with dividers or a compass and use an auxiliary scale to find $|\Gamma|$. The graduated line segment below the chart in Figure 10.12 serves this purpose. The angle of Γ is ϕ, and it is the counterclockwise angle from the Γ_r axis. Again, radial lines showing the angle would clutter up the chart badly, so the

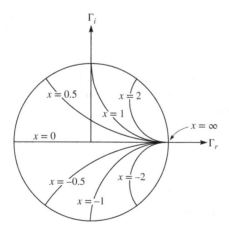

Figure 10.11 The portions of the circles of constant x lying within $|\Gamma| = 1$ are shown on the Γ_r, Γ_i axes. The radius of a given circle is $1/|x|$.

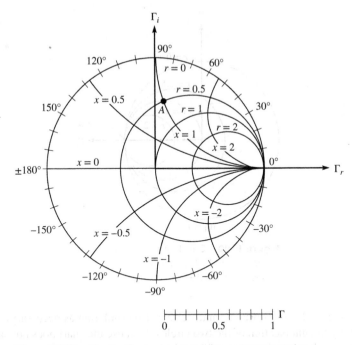

Figure 10.12 The Smith chart contains the constant-r circles and constant-x circles, an auxiliary radial scale to determine $|\Gamma|$, and an angular scale on the circumference for measuring ϕ.

angle is indicated on the circumference of the circle. A straight line from the origin through the intersection may be extended to the perimeter of the chart. As an example, if $Z_L = 25 + j50\ \Omega$ on a 50-Ω line, $z_L = 0.5 + j1$, and point A on Figure 10.12 shows the intersection of the $r = 0.5$ and $x = 1$ circles. The reflection coefficient is approximately 0.62 at an angle ϕ of 83°.

The Smith chart is completed by adding a second scale on the circumference by which distance along the line may be computed. This scale is in wavelength units, but the values placed on it are not obvious. To obtain them, we first divide the voltage at any point along the line,

$$V_s = V_0^+ (e^{-j\beta z} + \Gamma e^{j\beta z})$$

by the current

$$I_s = \frac{V_0^+}{Z_0}(e^{-j\beta z} - \Gamma e^{j\beta z})$$

obtaining the normalized input impedance

$$z_{in} = \frac{V_s}{Z_0 I_s} = \frac{e^{-j\beta z} + \Gamma e^{j\beta z}}{e^{-j\beta z} - \Gamma e^{j\beta z}}$$

Replacing z with $-l$ and dividing numerator and denominator by $e^{j\beta l}$, we have the general equation relating normalized input impedance, reflection coefficient, and line length,

$$z_{in} = \frac{1 + \Gamma e^{-j2\beta l}}{1 - \Gamma e^{-j2\beta l}} = \frac{1 + |\Gamma| e^{j(\phi - 2\beta l)}}{1 - |\Gamma| e^{j(\phi - 2\beta l)}} \qquad (114)$$

Note that when $l = 0$, we are located at the load, and $z_{in} = (1 + \Gamma)/(l - \Gamma) = z_L$, as shown by (107).

Equation (114) shows that the input impedance at any point $z = -l$ can be obtained by replacing Γ, the reflection coefficient of the load, with $\Gamma e^{-j2\beta l}$. That is, we decrease the angle of Γ by $2\beta l$ radians as we move from the load to the line input. Only the angle of Γ is changed; the magnitude remains constant.

Thus, as we proceed from the load z_L to the input impedance z_{in}, we move *toward* the generator a distance l on the transmission line, but we move through a *clockwise* angle of $2\beta l$ on the Smith chart. Since the magnitude of Γ stays constant, the movement toward the source is made along a constant-radius circle. One lap around the chart is accomplished whenever βl changes by π rad, or when l changes by one-half wavelength. This agrees with our earlier discovery that the input impedance of a half-wavelength lossless line is equal to the load impedance.

The Smith chart is thus completed by the addition of a scale showing a change of 0.5λ for one circumnavigation of the unit circle. For convenience, two scales are usually given, one showing an increase in distance for clockwise movement and the other an increase for counterclockwise travel. These two scales are shown in Figure 10.13. Note that the one marked "wavelengths toward generator" (wtg) shows increasing values of l/λ for clockwise travel, as described previously. The zero point of the wtg scale is rather arbitrarily located to the left. This corresponds to input impedances having phase angles of $0°$ and $R_L < Z_0$. We have also seen that voltage minima are always located here.

EXAMPLE 10.10

The use of the transmission line chart is best shown by example. Again consider a load impedance, $Z_L = 25 + j50 \ \Omega$, terminating a 50-Ω line. The line length is 60 cm and the operating frequency is such that the wavelength on the line is 2 m. We desire the input impedance.

Solution. We have $z_L = 0.5 + j1$, which is marked as A on Figure 10.14, and we read $\Gamma = 0.62\angle 82°$. By drawing a straight line from the origin through A to the circumference, we note a reading of 0.135 on the wtg scale. We have $l/\lambda = 0.6/2 = 0.3$, and it is, therefore, 0.3λ from the load to the input. We therefore find z_{in} on the $|\Gamma| = 0.62$ circle opposite a wtg reading of $0.135 + 0.300 = 0.435$. This construction is shown in Figure 10.14, and the point locating the input impedance is marked B. The normalized input impedance is read as $0.28 - j0.40$, and thus $Z_{in} = 14 - j20$. A more accurate analytical calculation gives $Z_{in} = 13.7 - j20.2$.

Information concerning the location of the voltage maxima and minima is also readily obtained on the Smith chart. We already know that a maximum or minimum

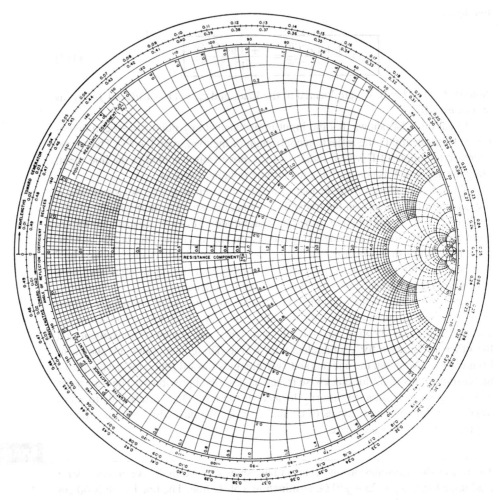

Figure 10.13 A photographic reduction of one version of a useful Smith chart (*Source: Emeloid Company, Hillside, NJ.*) For accurate work, larger charts are available wherever fine technical books are sold.

Source: The Emeloid Company, Hillside, NJ.

must occur at the load when Z_L is a pure resistance; if $R_L > Z_0$ there is a maximum at the load, and if $R_L < Z_0$ there is a minimum. We may extend this result now by noting that we could cut off the load end of a transmission line at a point where the input impedance is a pure resistance and replace that section with a resistance R_{in}; there would be no changes on the generator portion of the line. It follows, then, that the location of voltage maxima and minima must be at those points where Z_{in} is a pure resistance. Purely resistive input impedances must occur on the $x = 0$ line (the Γ_r axis) of the Smith chart. Voltage maxima or current minima occur when $r > 1$, or at wtg $= 0.25$, and voltage minima or current maxima occur when $r < 1$, or at wtg $= 0$.

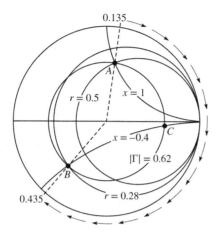

Figure 10.14 Normalized input impedance produced by a normalized load impedance $z_L = 0.5 + j1$ on a line 0.3λ long is $z_{in} = 0.28 - j0.40$.

In Example 10.10, then, the maximum at wtg $= 0.250$ must occur $0.250 - 0.135 = 0.115$ wavelengths toward the generator from the load. This is a distance of 0.115×200, or 23 cm from the load.

We should also note that because the standing wave ratio produced by a resistive load R_L is either R_L/R_0 or R_0/R_L, whichever is greater than unity, the value of s may be read directly as the value of r at the intersection of the $|\Gamma|$ circle and the r axis, $r > 1$. In our example, this intersection is marked point C, and $r = 4.2$; thus, $s = 4.2$.

Transmission line charts may also be used for *normalized admittances*, although there are several slight differences in such use. We let $y_L = Y_L/Y_0 = g + jb$ and use the r circles as g circles and the x circles as b circles. The two differences are, first, the line segment where $g > 1$ and $b = 0$ corresponds to a voltage minimum; and second, $180°$ must be added to the angle of Γ as read from the perimeter of the chart. We will use the Smith chart in this way in Section 10.14.

Special charts are also available for non-normalized lines, particularly 50 Ω charts and 20 mS charts.

D10.6. A load $Z_L = 80 - j100$ Ω is located at $z = 0$ on a lossless 50-Ω line. The operating frequency is 200 MHz, and the wavelength on the line is 2 m. (*a*) If the line is 0.8 m in length, use the Smith chart to find the input impedance. (*b*) What is s? (*c*) What is the distance from the load to the nearest voltage maximum? (*d*) What is the distance from the input to the nearest point at which the remainder of the line could be replaced by a pure resistance?

Ans. (*a*) $79 + j99$ Ω: (*b*) 4.50; (*c*) 0.0397 m; (*d*) 0.760 m

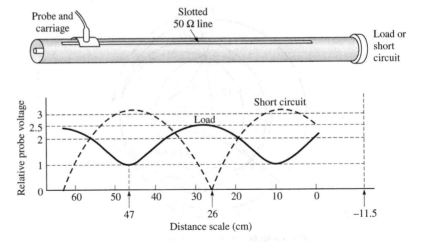

Figure 10.15 A sketch of a coaxial slotted line. The distance scale is on the slotted line. With the load in place, $s = 2.5$, and the minimum occurs at a scale reading of 47 cm. For a short circuit, the minimum is located at a scale reading of 26 cm. The wavelength is 75 cm.

We next consider two examples of practical transmission line problems. The first is the determination of load impedance from experimental data, and the second is the design of a single-stub matching network.

It is assumed that we have made experimental measurements on a 50-Ω slotted line that show there is a voltage standing wave ratio of 2.5. This has been determined by moving a sliding carriage back and forth along the line to determine maximum and minimum voltage readings. A scale provided on the track along which the carriage moves indicates that a *minimum* occurs at a scale reading of 47.0 cm, as shown in Figure 10.15. The zero point of the scale is arbitrary and does not correspond to the location of the load. The location of the minimum is usually specified instead of the maximum because it can be determined more accurately than that of the maximum; think of the sharper minima on a rectified sine wave. The frequency of operation is 400 MHz, so the wavelength is 75 cm. In order to pinpoint the location of the load, we remove it and replace it with a short circuit; the position of the minimum is then determined as 26.0 cm.

We know that the short circuit must be located an integral number of half-wavelengths from the minimum; let us arbitrarily locate it one half-wavelength away at $26.0 - 37.5 = -11.5$ cm on the scale. Since the short circuit has replaced the load, the load is also located at -11.5 cm. Our data thus show that the minimum is $47.0 - (-11.5) = 58.5$ cm from the load, or subtracting one-half wavelength, a minimum is 21.0 cm from the load. The voltage *maximum* is thus $21.0 - (37.5/2) = 2.25$ cm from the load, or $2.25/75 = 0.030$ wavelength from the load.

With this information, we can now turn to the Smith chart. At a voltage maximum, the input impedance is a pure resistance equal to sR_0; on a normalized basis,

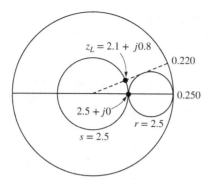

Figure 10.16 If $z_{in} = 2.5 + j0$ on a line 0.3 wavelengths long, then $z_L = 2.1 + j0.8$.

$z_{in} = 2.5$. We therefore enter the chart at $z_{in} = 2.5$ and read 0.250 on the wtg scale. Subtracting 0.030 wavelength to reach the load, we find that the intersection of the $s = 2.5$ (or $|\Gamma| = 0.429$) circle and the radial line to 0.220 wavelength is at $z_L = 2.1 + j0.8$. The construction is sketched on the Smith chart of Figure 10.16. Thus $Z_L = 105 + j40\ \Omega$, a value that assumes its location at a scale reading of -11.5 cm, or an integral number of half-wavelengths from that position. Of course, we may select the "location" of our load at will by placing the short circuit at the point that we wish to consider the load location. Since load locations are not well defined, it is important to specify the point (or plane) at which the load impedance is determined.

As a final example, let us try to match this load to the 50-Ω line by placing a short-circuited stub of length d_{11} a distance d from the load (see Figure 10.17). The stub line has the same characteristic impedance as the main line. The lengths d and d_{11} are to be determined.

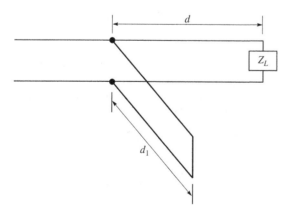

Figure 10.17 A short-circuited stub of length d_1, located at a distance d from a load Z_L, is used to provide a matched load to the left of the stub.

The input impedance to the stub is a pure reactance; when combined in parallel with the input impedance of the length d containing the load, the resultant input impedance must be $1 + j0$. Because it is much easier to combine admittances in parallel than impedances, we can rephrase our goal in admittance language: the input admittance of the length d containing the load must be $1 + jb_{in}$ for the addition of the input admittance of the stub jb_{stub} to produce a total admittance of $1 + j0$. Hence the stub admittance is $-jb_{in}$. We will therefore use the Smith chart as an admittance chart instead of an impedance chart.

The impedance of the load is $2.1 + j0.8$, and its location is at -11.5 cm. The admittance of the load is therefore $1/(2.1 + j0.8)$, and this value may be determined by adding one-quarter wavelength on the Smith chart, as Z_{in} for a quarter-wavelength line is R_0^2/Z_L, or $z_{in} = 1/z_L$, or $y_{in} = z_L$. Entering the chart (Figure 10.18) at $z_L = 2.1 + j0.8$, we read 0.220 on the wtg scale; we add (or subtract) 0.250 and find the admittance $0.41 - j0.16$ corresponding to this impedance. This point is still located on the $s = 2.5$ circle. Now, at what point or points on this circle is the real part of the admittance equal to unity? There are two answers, $1 + j0.95$ at wtg $= 0.16$, and $1 - j0.95$ at wtg $= 0.34$, as shown in Figure 10.18. We select the former value since this leads to the shorter stub. Hence $y_{stub} = -j0.95$, and the stub location corresponds to wtg $= 0.16$. Because the load admittance was found at wtg $= 0.470$, then we must move $(0.5 - 0.47) + 0.16 = 0.19$ wavelength to get to the stub location.

Finally, we may use the chart to determine the necessary length of the short-circuited stub. The input conductance is zero for any length of short-circuited stub, so we are restricted to the perimeter of the chart. At the short circuit, $y = \infty$ and wtg $= 0.250$. We find that $b_{in} = -0.95$ is achieved at wtg $= 0.379$, as shown in Figure 10.18. The stub is therefore $0.379 - 0.250 = 0.129$ wavelength, or 9.67 cm long.

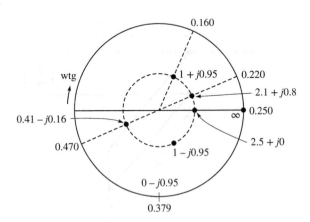

Figure 10.18 A normalized load, $z_L = 2.1 + j0.8$, is matched by placing a 0.129-wavelength short-circuited stub 0.19 wavelengths from the load.

D10.7. Standing wave measurements on an air-filled lossless 75-Ω line show maxima of 18 V and minima of 5 V. The first voltage minimum is located at a scale reading of 17 cm; the second minimum occurs at 37 cm. Find: (a) s; (b) λ; (c) f; (d) Γ_L; (e) Z_L.

Ans. (a) 3.60; (b) 0.400 m; (c) 750 MHz; (d) 0.57∠130; (e) 24.2 + j32.6 Ω

D10.8. A normalized load, $z_L = 2 - j1$, is located at $z = 0$ on a lossless 50-Ω line. Let the wavelength be 100 cm. (a) A short-circuited stub is to be located at $z = -d$. What is the shortest suitable value for d? (b) What is the shortest possible length of the stub? Find s: (c) on the main line for $z < -d$; (d) on the main line for $-d < z < 0$; (e) on the stub.

Ans. (a) 12.5 cm; (b) 12.5 cm; (c) 1.00; (d) 2.62; (e) ∞

10.14 TRANSIENT ANALYSIS

Throughout most of this chapter, we have considered the operation of transmission lines under steady-state conditions, in which voltage and current were sinusoidal and at a single frequency. In this section we move away from the simple time-harmonic case and consider transmission line responses to voltage step functions and pulses, grouped under the general heading of *transients*. These situations were briefly considered in Section 10.2 with regard to switched voltages and currents. Line operation in transient mode is important to study because it allows us to understand how lines can be used to store and release energy (in pulse-forming applications, for example). Pulse propagation is important in general since digital signals, composed of sequences of pulses, are widely used.

Animations

10.14.1 Formulation of the Transient Problem

We will confine our discussion to the propagation of transients in lines that are lossless and have no dispersion, so that the basic behavior and analysis methods may be learned. We must remember, however, that transient signals are necessarily composed of many frequencies, as Fourier analysis will show. Consequently, the question of dispersion in the line arises, since, as we have found, line propagation constants and reflection coefficients at complex loads will be frequency-dependent. So, in general, pulses are likely to broaden with propagation distance, and pulse shapes may change when reflecting from a complex load. These issues will not be considered in detail here, but they are readily addressed when the precise frequency dependences of β and Γ are known. In particular, $\beta(\omega)$ can be found by evaluating the imaginary part of γ, as given in Eq. (41), which would in general include the frequency dependences of R, C, G, and L arising from various mechanisms. For example, the skin effect (which affects both the conductor resistance and the internal inductance) will result in frequency-dependent R and L. Once $\beta(\omega)$ is known, pulse broadening can be evaluated using the methods to be presented in Chapter 12.

We begin our basic discussion of transients by considering a lossless transmission line of length l terminated by a matched load, $R_L = Z_0$, as shown in Figure 10.19a.

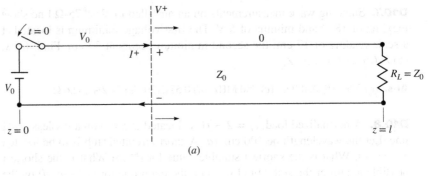

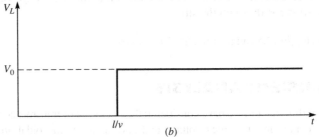

Figure 10.19 (a) Closing the switch at time $t = 0$ initiates voltage and current waves V^+ and I^+. The leading edge of both waves is indicated by the dashed line, which propagates in the lossless line toward the load at velocity v. In this case, $V^+ = V_0$; the line voltage is V^+ everywhere to the left of the leading edge, where current is $I^+ = V^+/Z_0$. To the right of the leading edge, voltage and current are both zero. Clockwise current, indicated here, is treated as positive and will occur when V^+ is positive. (b) Voltage across the load resistor as a function of time, showing the one-way transit time delay, l/v.

At the front end of the line is a battery of voltage V_0, which is connected to the line by closing a switch. At time $t = 0$, the switch is closed, and the line voltage at $z = 0$ becomes equal to the battery voltage. This voltage, however, does not appear across the load until adequate time has elapsed for the propagation delay. Specifically, at $t = 0$, a voltage wave is initiated in the line at the battery end, which then propagates toward the load. The leading edge of the wave, labeled V^+ in Figure 10.19, is of value $V^+ = V_0$. It can be thought of as a propagating step function, because at all points to the left of V^+, the line voltage is V_0; at all points to the right (not yet reached by the leading edge), the line voltage is zero. The wave propagates at velocity v, which in general is the group velocity in the line.[4] The wave reaches the load at time $t = l/v$ and then

---

[4] Because we have a step function (composed of many frequencies) as opposed to a sinusoid at a single frequency, the wave will propagate at the group velocity. In a lossless line with no dispersion as considered in this section, $\beta = \omega\sqrt{LC}$, where L and C are constant with frequency. In this case, we would find that the group and phase velocities are equal; that is, $d\omega/d\beta = \omega/\beta = v = 1/\sqrt{LC}$. We will thus write the velocity as v, knowing it to be both v_p and v_g.

does not reflect, as the load is matched. The transient phase is thus over, and the load voltage is equal to the battery voltage. A plot of load voltage as a function of time is shown in Figure 10.19b, indicating the propagation delay of $t = l/v$.

Associated with the voltage wave V^+ is a current wave whose leading edge is of value I^+. This wave is a propagating step function as well, whose value at all points to the left of V^+ is $I^+ = V^+/Z_0$; at all points to the right, current is zero. A plot of current through the load as a function of time will thus be identical in form to the voltage plot of Figure 10.19b, except that the load current at $t = l/v$ will be $I_L = V^+/Z_0 = V_0/R_L$.

We next consider a more general case, in which the load of Figure 10.19a is again a resistor but is *not matched* to the line ($R_L \neq Z_0$). Reflections will thus occur at the load, complicating the problem. At $t = 0$, the switch is closed as before and a voltage wave, $V_1^+ = V_0$, propagates to the right. Upon reaching the load, however, the wave will now reflect, producing a back-propagating wave, V_1^-. The relation between V_1^- and V_1^+ is through the reflection coefficient at the load:

$$\frac{V_1^-}{V_1^+} = \Gamma_L = \frac{R_L - Z_0}{R_L + Z_0} \tag{115}$$

As V_1^- propagates back toward the battery, it leaves behind its leading edge a total voltage of $V_1^+ + V_1^-$. Voltage V_1^+ exists everywhere ahead of the V_1^- wave until it reaches the battery, whereupon the entire line now is charged to voltage $V_1^+ + V_1^-$. At the battery, the V_1^- wave reflects to produce a new forward wave, V_2^+. The ratio of V_2^+ and V_1^- is found through the reflection coefficient at the battery:

$$\frac{V_2^+}{V_1^-} = \Gamma_g = \frac{Z_g - Z_0}{Z_g + Z_0} = \frac{0 - Z_0}{0 + Z_0} = -1 \tag{116}$$

where the impedance at the generator end, Z_g, is that of the battery, or zero.

V_2^+ (equal to $-V_1^-$) now propagates to the load, where it reflects to produce backward wave $V_2^- = \Gamma_L V_2^+$. This wave then returns to the battery, where it reflects with $\Gamma_g = -1$, and the process repeats. Note that with each round trip the wave voltage is reduced in magnitude because $|\Gamma_L| < 1$. Because of this the propagating wave voltages will eventually approach zero, and steady state is reached.

The voltage across the load resistor can be found at any given time by summing the voltage waves that have reached the load and have reflected from it up to that time. After many round trips, the load voltage will be, in general,

$$V_L = V_1^+ + V_1^- + V_2^+ + V_2^- + V_3^+ + V_3^- + \cdots$$
$$= V_1^+ \left(1 + \Gamma_L + \Gamma_g \Gamma_L + \Gamma_g \Gamma_L^2 + \Gamma_g^2 \Gamma_L^2 + \Gamma_g^2 \Gamma_L^3 + \cdots \right)$$

With a simple factoring operation, the preceding equation becomes

$$V_L = V_1^+ (1 + \Gamma_L)\left(1 + \Gamma_g \Gamma_L + \Gamma_g^2 \Gamma_L^2 + \cdots \right) \tag{117}$$

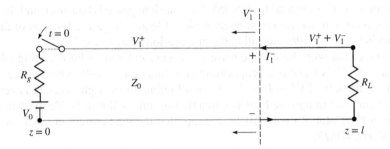

Figure 10.20 With series resistance at the battery location, voltage division occurs when the switch is closed, such that $V_0 = V_{rg} + V_1^+$. Shown is the first reflected wave, which leaves voltage $V_1^+ + V_1^-$ behind its leading edge. Associated with the wave is current I_1^-, which is $-V_1^-/Z_0$. Counterclockwise current is treated as negative and will occur when V_1^- is positive.

Allowing time to approach infinity, the second term in parentheses in (117) becomes the power series expansion for the expression $1/(1 - \Gamma_g\Gamma_L)$. Thus, in steady state we obtain

$$V_L = V_1^+\left(\frac{1 + \Gamma_L}{1 - \Gamma_g\Gamma_L}\right) \tag{118}$$

In our present example, $V_1^+ = V_0$ and $\Gamma_g = -1$. Substituting these into (118), we find the expected result in steady state: $V_L = V_0$.

A more general situation would involve a nonzero impedance at the battery location, as shown in Figure 10.20. In this case, a resistor of value R_g is positioned in series with the battery. When the switch is closed, the battery voltage appears across the series combination of R_g and the line characteristic impedance, Z_0. The value of the initial voltage wave, V_1^+, is thus found through simple voltage division, or

$$\boxed{V_1^+ = \frac{V_0 Z_0}{R_g + Z_0}} \tag{119}$$

With this initial value, the sequence of reflections and the development of the voltage across the load occurs in the same manner as determined by (117), with the steady-state value determined by (118). The value of the reflection coefficient at the generator end, determined by (116), is $\Gamma_g = (R_g - Z_0)/(R_g + Z_0)$.

10.14.2 Voltage Reflection Diagram

A useful way of keeping track of the voltage at any point in the line is through a *voltage reflection diagram*. Such a diagram for the line of Figure 10.20 is shown in Figure 10.21a. It is a two-dimensional plot in which position on the line, z, is shown on the horizontal axis. Time is plotted on the vertical axis and is conveniently expressed as it relates to position and velocity through $t = z/v$. A vertical line, located at $z = l$, is drawn, which, together with the ordinate, defines the z axis boundaries of the transmission line. With the switch located at the battery position, the initial voltage

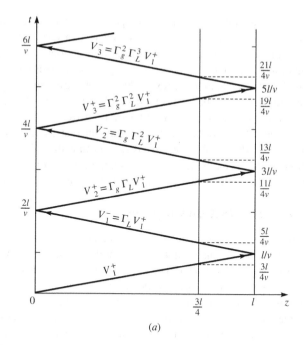

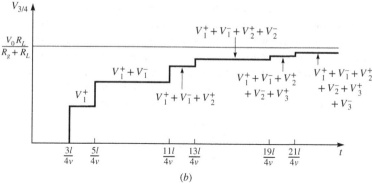

Figure 10.21 (a) Voltage reflection diagram for the line of Figure 10.20. A reference line, drawn at $z = 3l/4$, is used to evaluate the voltage at that position as a function of time. (b) The line voltage at $z = 3l/4$ as determined from the reflection diagram of (a). Note that the voltage approaches the expected $V_0 R_L/(R_g + R_L)$ as time approaches infinity.

wave, V_1^+, starts at the origin, or lower-left corner of the diagram ($z = t = 0$). The location of the leading edge of V_1^+ as a function of time is shown as the diagonal line that joins the origin to the point along the right-hand vertical line that corresponds to time $t = l/v$ (the one-way transit time). From there (the load location), the position of the leading edge of the reflected wave, V_1^-, is shown as a "reflected" line that joins the $t = l/v$ point on the right boundary to the $t = 2l/v$ point on the ordinate. From there (at the battery location), the wave reflects again, forming V_2^+, shown as a line parallel to that for V_1^+. Subsequent reflected waves are shown, and their values are labeled.

The voltage as a function of time at a given position in the line can now be determined by adding the voltages in the waves as they intersect a vertical line drawn at the desired location. This addition is performed starting at the bottom of the diagram ($t = 0$) and progressing upward (in time). Whenever a voltage wave crosses the vertical line, its value is added to the total at that time. For example, the voltage at a location three-fourths the distance from the battery to the load is plotted in Figure 10.21b. To obtain this plot, the line $z = (3/4)l$ is drawn on the diagram. Whenever a wave crosses this line, the voltage in the wave is added to the voltage that has accumulated at $z = (3/4)l$ over all earlier times. This general procedure enables one to easily determine the voltage at any specific time and location. In doing so, the terms in (117) that have occurred up to the chosen time are being added, but with information on the time at which each term appears.

10.14.3   Current Reflection Diagram

Line current can be found in a similar way through a *current reflection diagram*. It is easiest to construct the current diagram directly from the voltage diagram by determining a value for current that is associated with each voltage wave. In dealing with current, it is important to keep track of the *sign* of the current because it relates to the voltage waves and their polarities. Referring to Figures 10.19a and 10.20, we use the convention in which current associated with a *forward-z* traveling voltage wave of positive polarity is positive. This would result in current that flows in the clockwise direction, as shown in Figure 10.19a. Current associated with a *backward-z* traveling voltage wave of positive polarity (thus flowing counterclockwise) is negative. Such a case is illustrated in Figure 10.20. In our two-dimensional transmission-line drawings, we assign positive polarity to voltage waves propagating in *either* direction if the upper conductor carries a positive charge and the lower conductor a negative charge. In Figures 10.19a and 10.20, both voltage waves are of positive polarity, so their associated currents will be net positive for the forward wave and net negative for the backward wave. In general, we write

$$I^+ = \frac{V^+}{Z_0}$$
(120)

and

$$I^- = -\frac{V^-}{Z_0}$$
(121)

Finding the current associated with a backward-propagating voltage wave immediately requires a minus sign, as (121) indicates.

Figure 10.22a shows the current reflection diagram that is derived from the voltage diagram of Figure 10.21a. Note that the current values are labeled in terms of the voltage values, with the appropriate sign added as per (120) and (121). Once the current diagram is constructed, current at a given location and time can be found in exactly the same manner as voltage is found using the voltage diagram. Figure 10.22b shows the current as a function of time at the $z = (3/4)l$ position, determined by summing the current wave values as they cross the vertical line drawn at that location.

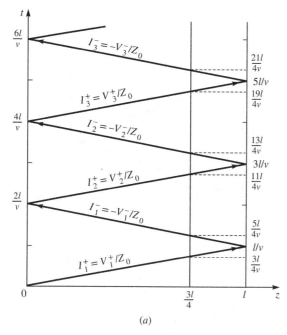

(a)

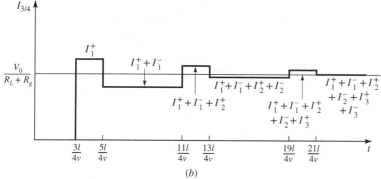

(b)

Figure 10.22 (a) Current reflection diagram for the line of Figure 10.20 as obtained from the voltage diagram of Figure 10.21a. (b) Current at the z = 3l/4 position as determined from the current reflection diagram, showing the expected steady-state value of $V_0/(R_L + R_g)$.

EXAMPLE 10.11

In Figure 10.20, $R_g = Z_0 = 50\ \Omega$, $R_L = 25\ \Omega$, and the battery voltage is $V_0 = 10$ V. The switch is closed at time $t = 0$. Determine the voltage at the load resistor and the current in the battery as functions of time.

Solution. Voltage and current reflection diagrams are shown in Figure 10.23a and b. At the moment the switch is closed, half the battery voltage appears across the

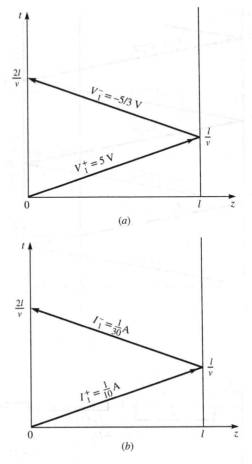

Figure 10.23 Voltage (a) and current (b) reflection diagrams for Example 10.11.

50-Ω resistor, with the other half comprising the initial voltage wave. Thus $V_1^+ = (1/2)V_0 = 5$ V. The wave reaches the 25-Ω load, where it reflects with reflection coefficient

$$\Gamma_L = \frac{25 - 50}{25 + 50} = -\frac{1}{3}$$

So $V_1^- = -(1/3)V_1^+ = -5/3$ V. This wave returns to the battery, where it encounters reflection coefficient $\Gamma_g = 0$. Thus, no further waves appear; steady state is reached.

Once the voltage wave values are known, the current reflection diagram can be constructed. The values for the two current waves are

$$I_1^+ = \frac{V_1^+}{Z_0} = \frac{5}{50} = \frac{1}{10} \text{A}$$

and

$$I_1^- = -\frac{V_1^-}{Z_0} = -\left(-\frac{5}{3}\right)\left(\frac{1}{50}\right) = \frac{1}{30}\,\text{A}$$

Note that no attempt is made here to derive I_1^- from I_1^+. They are both obtained independently from their respective voltages.

The voltage at the load as a function of time is now found by summing the voltages along the vertical line at the load position. The resulting plot is shown in Figure 10.24*a*. Current in the battery is found by summing the currents along the vertical axis, with the resulting plot shown as Figure 10.24*b*. Note that in steady state, we treat the circuit as lumped, with the battery in series with the 50- and 25-Ω resistors. Therefore, we expect to see a steady-state current through the battery (and everywhere else) of

$$I_B(\text{steady state}) = \frac{10}{50 + 25} = \frac{1}{7.5}\,\text{A}$$

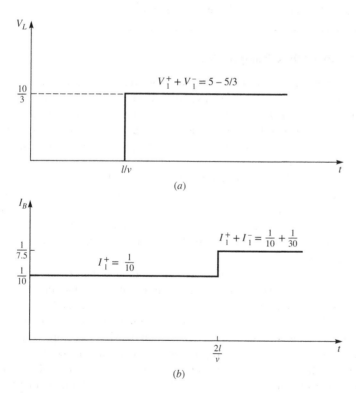

(a)

(b)

Figure 10.24 Voltage across the load (*a*) and current in the battery (*b*) as determined from the reflection diagrams of Figure 10.23 (Example 10.11).

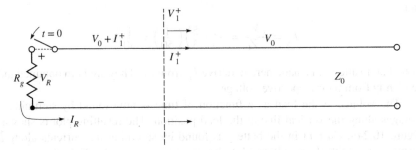

Figure 10.25 In an initially charged line, closing the switch as shown initiates a voltage wave of opposite polarity to that of the initial voltage. The wave thus depletes the line voltage and will fully discharge the line in one round trip if $R_g = Z_0$.

This value is also found from the current reflection diagram for $t > 2l/v$. Similarly, the steady-state load voltage should be

$$V_L(\text{steady state}) = V_0 \frac{R_L}{R_g + R_L} = \frac{(10)(25)}{50 + 25} = \frac{10}{3} \text{ V}$$

which is found also from the voltage reflection diagram for $t > l/v$.

10.14.4   Initially Charged Lines

Another type of transient problem involves lines that are *initially charged*. In these cases, the manner in which the line discharges through a load is of interest. Consider the situation shown in Figure 10.25, in which a charged line of characteristic impedance Z_0 is discharged through a resistor of value R_g when a switch at the resistor location is closed.[5] We consider the resistor at the $z = 0$ location; the other end of the line is open (as would be necessary) and is located at $z = l$.

When the switch is closed, current I_R begins to flow through the resistor, and the line discharge process begins. This current does not immediately flow everywhere in the transmission line but begins at the resistor and establishes its presence at more distant parts of the line as time progresses. By analogy, consider a long line of automobiles at a red light. When the light turns green, the cars at the front move through the intersection first, followed successively by those further toward the rear. The point that divides cars in motion and those standing still is, in fact, a wave that propagates toward the back of the line. In the transmission line, the flow of charge progresses in a similar way. A voltage wave, V_1^+, is initiated and propagates to the right. To the left of its leading edge, charge is in motion; to the right of the leading edge, charge is stationary and carries its original density. Accompanying the charge in motion to the left of V_1^+ is a drop in the charge density as the discharge process occurs, and so the line voltage to the left of V_1^+ is partially reduced. This voltage will be given by the sum of the initial voltage, V_0, and V_1^+, which means that V_1^+ must in

---

[5] Even though this is a load resistor, we will call it R_g because it is located at the front (generator) end of the line.

fact be negative (or of opposite sign to V_0). The line discharge process is analyzed by keeping track of V_1^+ as it propagates and undergoes multiple reflections at the two ends. Voltage and current reflection diagrams are used for this purpose in much the same way as before.

Referring to Figure 10.25, we see that for positive V_0 the current flowing through the resistor will be counterclockwise and hence negative. We also know that continuity requires that the resistor current be equal to the current associated with the voltage wave, or

$$I_R = I_1^+ = \frac{V_1^+}{Z_0}$$

Now the resistor voltage will be

$$V_R = V_0 + V_1^+ = -I_R R_g = -I_1^+ R_g = -\frac{V_1^+}{Z_0} R_g$$

where the minus signs arise from the fact that V_R (having positive polarity) is produced by the negative current, I_R. We solve for V_1^+ to obtain

$$\boxed{V_1^+ = \frac{-V_0 Z_0}{Z_0 + R_g}} \tag{122}$$

Having found V_1^+, we can set up the voltage and current reflection diagrams. The diagram for voltage is shown in Figure 10.26. Note that the initial condition of voltage V_0 everywhere on the line is accounted for by assigning voltage V_0 to the horizontal axis of the voltage diagram. The diagram is otherwise drawn as before, but with $\Gamma_L = 1$ (at the open-circuited load end). Variations in how the line discharges thus depend on the resistor value at the switch end, R_g, which determines the reflection coefficient, Γ_g, at that location. The current reflection diagram is derived from the voltage diagram in the usual way. There is no initial current to consider.

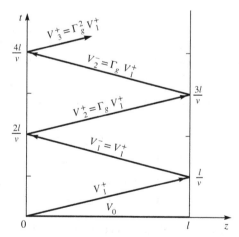

Figure 10.26 Voltage reflection diagram for the charged line of Figure 10.25, showing the initial condition of V_0 everywhere on the line at $t = 0$.

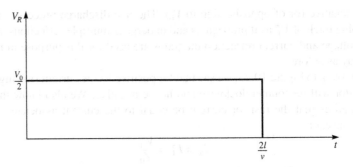

Figure 10.27   Voltage across the resistor as a function of time, as determined from the reflection diagram of Figure 10.26, in which $R_g = Z_0$ ($\Gamma = 0$).

A special case of practical importance is that in which the resistor is matched to the line, or $R_g = Z_0$. In this case, Eq. (122) gives $V_1^+ = -V_0/2$. The line fully discharges in one round trip of V_1^+ and produces a voltage across the resistor of value $V_R = V_0/2$, which persists for time $T = 2l/v$. The resistor voltage as a function of time is shown in Figure 10.27. The transmission line in this application is known as a *pulse-forming line*; pulses that are generated in this way are well formed and of low noise, provided the switch is sufficiently fast. Commercial units are available that are capable of generating high-voltage pulses of widths on the order of a few nanoseconds, using thyratron-based switches.

When the resistor is not matched to the line, full discharge still occurs, but does so over several reflections, leading to a complicated pulse shape.

EXAMPLE 10.12

In the charged line of Figure 10.25, the characteristic impedance is $Z_0 = 100\ \Omega$, and $R_g = 100/3\ \Omega$. The line is charged to an initial voltage, $V_0 = 160$ V, and the switch is closed at time $t = 0$. Determine and plot the voltage and current through the resistor for time $0 < t < 8l/v$ (four round trips).

Solution. With the given values of R_g and Z_0, Eq. (47) gives $\Gamma_g = -1/2$. Then, with $\Gamma_L = 1$, and using (122), we find

$$V_1^+ = V_1^- = -3/4\,V_0 = -120\text{ V}$$
$$V_2^+ = V_2^- = \Gamma_g V_1^- = +60\text{ V}$$
$$V_3^+ = V_3^- = \Gamma_g V_2^- = -30\text{ V}$$
$$V_4^+ = V_4^- = \Gamma_g V_3^- = +15\text{ V}$$

Using these values on the voltage reflection diagram, we evaluate the voltage in time at the resistor location by moving up the left-hand vertical axis, adding voltages as we progress, and beginning with $V_0 + V_1^+$ at $t = 0$. Note that when we add voltages along the vertical axis, we are encountering the intersection points between incident and reflected waves, which occur (in time) at each integer multiple of $2l/v$. So, when moving up the axis, we add the voltages of *both* waves to our total at each occurrence. The voltage within each time interval is thus:

$$
\begin{aligned}
V_R &= V_0 + V_1^+ = 40 \text{ V} & (0 < t < 2l/v) \\
&= V_0 + V_1^+ + V_1^- + V_2^+ = -20 \text{ V} & (2l/v < t < 4l/v) \\
&= V_0 + V_1^+ + V_1^- + V_2^+ + V_2^- + V_3^+ = 10 \text{ V} & (4l/v < t < 6l/v) \\
&= V_0 + V_1^+ + V_1^- + V_2^+ + V_2^- + V_3^+ + V_3^- + V_4^+ = -5 \text{ V} & (6l/v < t < 8l/v)
\end{aligned}
$$

The resulting voltage plot over the desired time range is shown in Figure 10.28*a*.

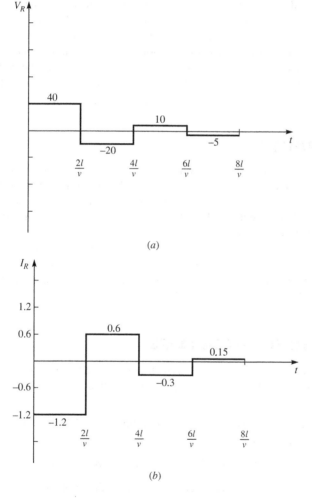

(a)

(b)

Figure 10.28 Resistor voltage (*a*) and current (*b*) as functions of time for the line of Figure 10.25, with values as specified in Example 10.12.

The current through the resistor is most easily obtained by dividing the voltages in Figure 10.28a by $-R_g$. As a demonstration, we can also use the current diagram of Figure 10.22a to obtain this result. Using (120) and (121), we evaluate the current waves as follows:

$$I_1^+ = V_1^+/Z_0 = -1.2 \text{ A}$$
$$I_1^- = -V_1^-/Z_0 = +1.2 \text{ A}$$
$$I_2^+ = -I_2^- = V_2^+/Z_0 = +0.6 \text{ A}$$
$$I_3^+ = -I_3^- = V_3^+/Z_0 = -0.30 \text{ A}$$
$$I_4^+ = -I_4^- = V_4^+/Z_0 = +0.15 \text{ A}$$

Using these values on the current reflection diagram, Figure 10.22a, we add up currents in the resistor in time by moving up the left-hand axis, as we did with the voltage diagram. The result is shown in Figure 10.28b. As a further check to the correctness of our diagram construction, we note that current at the open end of the line ($Z = l$) must always be zero. Therefore, summing currents up the right-hand axis must give a zero result for all time. The reader is encouraged to verify this.

REFERENCES

1. White, H. J., P. R. Gillette, and J. V. Lebacqz. "The Pulse-Forming Network." Chapter 6 in *Pulse Generators,* edited by G. N, Glasoe and J. V. Lebacqz. New York: Dover, 1965.

2. Brown, R. G., R. A. Sharpe, W. L. Hughes, and R. E. Post. *Lines, Waves, and Antennas.* 2d ed. New York: The Ronald Press Company, 1973. Transmission lines are covered in the first six chapters, with numerous examples.

3. Cheng, D. K. *Field and Wave Electromagnetics.* 2d ed. Reading, Mass.: Addison-Wesley, 1989. Provides numerous examples of Smith chart problems and transients.

4. Seshadri, S. R. *Fundamentals of Transmission Lines and Electromagnetic Fields.* Reading, Mass.: Addison-Wesley, 1971.

CHAPTER 10   PROBLEMS

10.1 The parameters of a certain transmission line operating at $\omega = 6 \times 10^8$ rad/s are $L = 0.350 \ \mu$H/m, $C = 40$ pF/m, $G = 0$, and $R = 15.0 \ \Omega$/m. Find α, β, λ, and Z_0.

10.2 A sinusoidal wave on a transmission line is specified by voltage and current in phasor form:

$$V_s(z) = V_0 e^{\alpha z} e^{j\beta z} \quad \text{and} \quad I_s(z) = I_0 e^{\alpha z} e^{j\beta z} e^{j\phi}$$

where V_0 and I_0 are both real. (a) In which direction does this wave propagate and why? (b) It is found that $\alpha = 0$, $Z_0 = 50 \ \Omega$, and the wave velocity is $v_p = 2.5 \times 10^8$ m/s, with $\omega = 10^8$ s^{-1}. Evaluate R, G, L, C, λ, and ϕ.

10.3 A voltage pulse propagates within a lossless transmission line of characteristic impedance $Z_0 = 50\ \Omega$. The pulse is gaussian in shape, having a voltage envelope given by $V(t) = V_0 e^{-t^2/2T^2}$ where $V_0 = 10$ V and $T = 20$ ns. The pulse is incident on a 100-Ω load at the far end of the line. Determine the energy in joules that is dissipated by the load.

10.4 A sinusoidal voltage wave of amplitude V_0, frequency ω, and phase constant β propagates in the forward z direction toward the open load end in a lossless transmission line of characteristic impedance Z_0. At the end, the wave totally reflects with zero phase shift, and the reflected wave now interferes with the incident wave to yield a standing wave pattern over the line length (as per Example 10.1). Determine the standing wave pattern for the *current* in the line. Express the result in real instantaneous form and simplify.

10.5 Two voltage waves of equal amplitude V_0 and radian frequency ω propagate in the forward z direction in a lossy transmission line having attenuation coefficient α, and characteristic impedance $Z_0 = |Z_0| e^{j\delta}$. One wave is shifted from the other by ϕ radians. (*a*) Find an expression for the net voltage wave formed by the superposition of the two voltages. Your result should be a single wave function in real instantaneous form. (*b*) Find an expression for the net current in the line, again in the form of a single wave function. (*c*) Find an expression for the average power in the line.

10.6 A 50-Ω load is attached to a 50-m section of the transmission line of Problem 10.1, and a 100-W signal is fed to the input end of the line. (*a*) Evaluate the distributed line loss in dB/m. (*b*) Evaluate the reflection coefficient at the load. (*c*) Evaluate the power that is dissipated by the load resistor. (*d*) What power drop in dB does the dissipated power in the load represent when compared to the original input power? (*e*) On partial reflection from the load, how much power returns to the input and what dB drop does this represent when compared to the original 100-W input power?

10.7 A transmitter and receiver are connected using a cascaded pair of transmission lines. At the operating frequency, line 1 has a measured loss of 0.1 dB/m, and line 2 is rated at 0.2 dB/m. The link is composed of 40 m of line 1 joined to 25 m of line 2. At the joint, a splice loss of 2 dB is measured. If the transmitted power is 100 mW, what is the received power?

10.8 An absolute measure of power is the dBm scale, in which power is specified in decibels relative to one milliwatt. Specifically, $P(\text{dBm}) = 10 \log_{10}[P(\text{mW})/1\ \text{mW}]$. Suppose that a receiver is rated as having a *sensitivity* of -20 dBm, indicating the *mimimum* power that it must receive in order to adequately interpret the transmitted electronic data. Suppose this receiver is at the load end of a 50-Ω transmission line having 100-m length and loss rating of 0.09 dB/m. The receiver impedance is 75 Ω and so is not matched to the line. What is the minimum required input power to the line in (*a*) dBm, (*b*) mW?

10.9 A 100-m transmission line is used to propagate a signal from a transmitter to a receiver whose input impedance is 50 Ω. The transmitter is capable of launching 12 dBm average power at the input end of the line (see Problem 10.8 for the definition of dBm). The line is lossy, having characteristic impedance $Z_0 = 75 + j10$ Ω, and power loss coefficient $A = 0.05$ dB/m. Find the power that enters the receiver in both dBm and in mW.

10.10 Two lossless transmission lines having different characteristic impedances are to be joined end to end. The impedances are $Z_{01} = 100$ Ω and $Z_{03} = 25$ Ω. The operating frequency is 1 GHz. (a) Find the required characteristic impedance, Z_{02}, of a quarter-wave section to be inserted between the two, which will impedance-match the joint, thus allowing total power transmission through the three lines. (b) The capacitance per unit length of the intermediate line is found to be 100 pF/m. Find the shortest length in meters of this line that is needed to satisfy the impedance-matching condition. (c) With the three-segment setup as found in parts (a) and (b), the frequency is now doubled to 2 GHz. Find the input impedance at the line-1-to-line-2 junction, seen by waves incident from line 1. (d) Under the conditions of part (c), and with power incident from line 1, evaluate the standing wave ratio that will be measured in line 1, and the fraction of the incident power from line 1 that is reflected and propagates back to the line 1 input.

10.11 Two voltage waves of equal amplitude V_0, which have different frequencies ω and 3ω (with corresponding phase constants β and 3β), propagate in the forward z direction in a lossless transmission line. Find an expression for the net voltage wave formed by the superposition of the given voltages. To do this, express both waves in complex instantaneous form (retaining the time-dependent terms). Combine the two algebraically, and convert the result to real instantaneous form to yield a single-term wave function. Plot your result as a function of βz at $t = 0$.

10.12 In a circuit in which a sinusoidal voltage source drives its internal impedance in series with a load impedance, maximum power transfer to the load occurs when the source and load impedances form a complex conjugate pair. Suppose the source (with its internal impedance) now drives a complex load of impedance $Z_L = R_L + jX_L$ that has been moved to the end of a lossless transmission line of length ℓ having characteristic impedance Z_0. If the source impedance is $Z_g = R_g + jX_g$, write an equation that can be solved for the required line length, ℓ, such that the displaced load will receive the maximum power.

10.13 The skin effect mechanism in transmission lines is responsible for the increase with frequency of the line resistance per meter, R. Specifically, resistance scales as the square root of frequency f according to $R = A_0 f^{1/2}$ where A_0 is a constant. Consider a low-loss line in which the attenuation coefficient is approximated by Eq. (54a), the phase constant by the first term in (54b), and the characteristic impedance by Eq. (24). The conductance

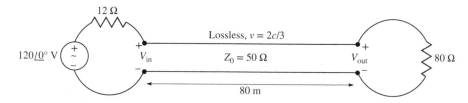

Figure 10.29 See Problem 10.15.

per unit length, G, is zero. The 50-Ω line has a measured power loss of 10.0 dB over a 100-m length at $f = 100$ MHz. (*a*) Find the value of A_0 and the line resistance per meter at 100 MHz. (*b*) At the line input, a 10-W power transmitter is attached. At the far end of the line, a 100-Ω load impedance is attached. How much power is dissipated by the load at frequency 400 MHz?

10.14 A lossless transmission line having characteristic impedance $Z_0 = 50\ \Omega$ is driven by a source at the input end that consists of the series combination of a 10-V sinusoidal generator and a 50-Ω resistor. The line is one-quarter wavelength long. At the other end of the line, a load impedance $Z_L = 50 - j50\ \Omega$ is attached. (*a*) Evaluate the input impedance to the line seen by the voltage source-resistor combination. (*b*) Evaluate the power that is dissipated by the load. (*c*) Evaluate the voltage amplitude that appears across the load.

10.15 For the transmission line represented in Figure 10.29, find $V_{s,\text{out}}$ if $f =$ (*a*) 60 Hz; (*b*) 500 kHz.

10.16 A 100-Ω lossless transmission line is connected to a second line of 40-Ω impedance, whose length is $\lambda/4$. The other end of the short line is terminated by a 25-Ω resistor. A sinusoidal wave (of frequency f) having 50 W average power is incident from the 100-Ω line. (*a*) Evaluate the input impedance to the quarter-wave line. (*b*) Determine the steady-state power that is dissipated by the resistor. (*c*) Now suppose that the operating frequency is lowered to one-half its original value. Determine the new input impedance, Z'_{in}, for this case. (*d*) For the new frequency, calculate the power in watts that returns to the input end of the line after reflection.

10.17 Determine the average power absorbed by each resistor in Figure 10.30.

10.18 The line shown in Figure 10.31 is lossless. Find s on both sections 1 and 2.

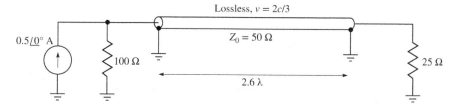

Figure 10.30 See Problem 10.17.

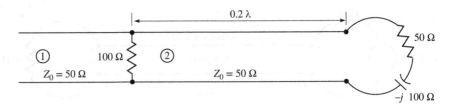

Figure 10.31   See Problem 10.18.

10.19 A lossless transmission line is 50 cm in length and operates at a frequency of 100 MHz. The line parameters are $L = 0.2 \ \mu H/m$ and $C = 80$ pF/m. The line is terminated in a short circuit at $z = 0$, and there is a load $Z_L = 50 + j20 \ \Omega$ across the line at location $z = -20$ cm. What average power is delivered to Z_L if the input voltage is $100\angle 0°$ V?

10.20 (a) Determine s on the transmission line of Figure 10.32. Note that the dielectric is air. (b) Find the input impedance. (c) If $\omega L = 10 \ \Omega$, find I_s. (d) What value of L will produce a maximum value for $|I_s|$ at $\omega = 1$ Grad/s? For this value of L, calculate the average power (e) supplied by the source; (f) delivered to $Z_L = 40 + j30 \ \Omega$.

10.21 A lossless line having an air dielectric has a characteristic impedance of 400 Ω. The line is operating at 200 MHz and $Z_{in} = 200 - j200 \ \Omega$. Use analytic methods or the Smith chart (or both) to find (a) s; (b) Z_L, if the line is 1 m long; (c) the distance from the load to the nearest voltage maximum.

10.22 A lossless 50-Ω line is terminated by an unknown load impedance. A VSWR of 5.0 is measured, and the first voltage minimum occurs at 0.10 wavelengths in front of the load. Using the Smith chart, find (a) the load impedance; (b) the magnitude and phase of the reflection coefficient; (c) the shortest length of line necessary to achieve an entirely resistive input impedance.

10.23 The normalized load on a lossless transmission line is $2 + j1$. Let $\lambda = 20$ m and make use of the Smith chart to find (a) the shortest distance from the load to a point at which $z_{in} = r_{in} + j0$, where $r_{in} > 0$; (b) z_{in} at this point. (c) The line is cut at this point and the portion containing z_L is thrown away. A resistor $r = r_{in}$ of part (a) is connected across the line. What is s on the remainder of the line? (d) What is the shortest distance from this resistor to a point at which $z_{in} = 2 + j1$?

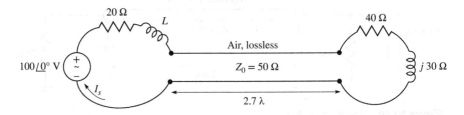

Figure 10.32   See Problem 10.20.

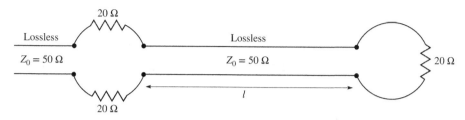

Figure 10.33   See Problem 10.24.

10.24 With the aid of the Smith chart, plot a curve of $|Z_{\text{in}}|$ versus l for the transmission line shown in Figure 10.33. Cover the range $0 < l/\lambda < 0.25$.

10.25 A 300-Ω transmission line is short-circuited at $z = 0$. A voltage maximum, $|V|_{\text{max}} = 10$ V, is found at $z = -25$ cm, and the minimum voltage, $|V|_{\text{min}} = 0$, is at $z = -50$ cm. Use the Smith chart to find Z_L (with the short circuit replaced by the load) if the voltage readings are (*a*) $|V|_{\text{max}} = 12$ V at $z = -5$ cm, and $|V|_{\text{min}} = 5$ V; (*b*) $|V|_{\text{max}} = 17$ V at $z = -20$ cm, and $|V|_{\text{min}} = 0$.

10.26 A 75-Ω lossless line is of length 1.2 λ. It is terminated by an unknown load impedance. The input end of the 75-Ω line is attached to the load end of a lossless 50-Ω line. A VSWR of 4 is measured on the 75-Ω line, on which the first voltage minimum occurs at a distance of 0.15 λ in front of the junction between the two lines. Use the Smith chart to find the unknown load impedance.

10.27 The characteristic admittance ($Y_0 = 1/Z_0$) of a lossless transmission line is 20 mS. The line is terminated in a load $Y_L = 40 - j20$ mS. Use the Smith chart to find (*a*) s; (*b*) Y_{in} if $l = 0.15\lambda$; (*c*) the distance in wavelengths from Y_L to the nearest voltage maximum.

10.28 The wavelength on a certain lossless line is 10 cm. If the normalized input impedance is $z_{\text{in}} = 1 + j2$, use the Smith chart to determine (*a*) s; (*b*) z_L, if the length of the line is 12 cm; (*c*) x_L, if $z_L = 2 + jx_L$ where $x_L > 0$.

10.29 A standing wave ratio of 2.5 exists on a lossless 60-Ω line. Probe measurements locate a voltage minimum on the line whose location is marked by a small scratch on the line. When the load is replaced by a short circuit, the minima are 25 cm apart, and one minimum is located at a point 7 cm toward the source from the scratch. Find Z_L.

10.30 A two-wire line constructed of lossless wire of circular cross section is gradually flared into a coupling loop that looks like an eggbeater. At the point X, indicated by the arrow in Figure 10.34, a short circuit is placed across the line. A probe is moved along the line and indicates that the first voltage minimum to the left of X is 16 cm from X. With the short circuit removed, a voltage minimum is found 5 cm to the left of X, and a voltage maximum is located that is 3 times the voltage of the minimum. Use the Smith chart to determine (*a*) f; (*b*) s; (*c*) the normalized input impedance of the eggbeater as seen looking to the right at point X.

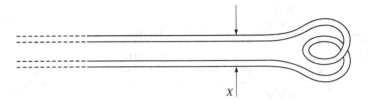

Figure 10.34   See Problem 10.30.

10.31 In order to compare the relative sharpness of the maxima and minima of a standing wave, assume a load $z_L = 4 + j0$ is located at $z = 0$. Let $|V|_{min} = 1$ and $\lambda = 1$ m. Determine the width of the (*a*) minimum where $|V| < 1.1$; (*b*) maximum where $|V| > 4/1.1$.

10.32 In Figure 10.17, let $Z_L = 250\ \Omega$ and $Z_0 = 50\ \Omega$. Find the shortest attachment distance d and the shortest length d_1 of a short-circuited stub line that will provide a perfect match on the main line to the left of the stub. Express all answers in wavelengths.

10.33 In Figure 10.17, let $Z_L = 100 + j150\ \Omega$ and $Z_0 = 100\ \Omega$. Find the shortest length d_1 of a short-circuited stub and the shortest distance d that it may be located from the load to provide a perfect match on the main line to the left of the stub. (*b*) Repeat for an open-circuited stub. Express all answers in wavelengths.

10.34 The lossless line shown in Figure 10.35 is operating with $\lambda = 100$ cm. If $d_1 = 10$ cm, $d = 25$ cm, and the line is matched to the left of the stub, what is Z_L?

10.35 A load, $Z_L = 25 + j75\ \Omega$, is located at $z = 0$ on a lossless two-wire line for which $Z_0 = 50\ \Omega$ and $v = c$. (*a*) If $f = 300$ MHz, find the shortest distance $d\ (z = -d)$ at which the input admittance has a real part equal to $1/Z_0$ and a negative imaginary part. (*b*) What value of capacitance C should be connected across the line at that point to provide unity standing wave ratio on the remaining portion of the line?

10.36 The two-wire lines shown in Figure 10.36 are all lossless and have $Z_0 = 200\ \Omega$. Find d and the shortest possible value for d_1 to provide a matched load if $\lambda = 100$ cm.

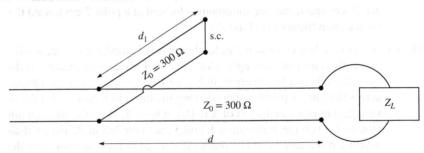

Figure 10.35   See Problem 10.34.

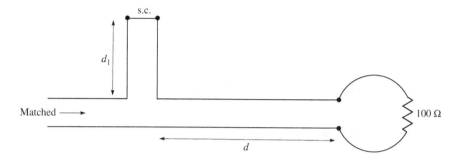

Figure 10.36 See Problem 10.36.

10.37 In the transmission line of Figure 10.20, $R_g = Z_0 = 50\ \Omega$, and $R_L = 25\ \Omega$. Determine and plot the voltage at the load resistor and the current in the battery as functions of time by constructing appropriate voltage and current reflection diagrams.

10.38 Repeat Problem 10.37, with $Z_0 = 50\ \Omega$, and $R_L = R_g = 25\ \Omega$. Carry out the analysis for the time period $0 < t < 8l/v$.

10.39 In the transmission line of Figure 10.20, $Z_0 = 50\ \Omega$, and $R_L = R_g = 25\ \Omega$. The switch is closed at $t = 0$ and *is opened again* at time $t = l/4v$, thus creating a rectangular voltage *pulse* in the line. Construct an appropriate voltage reflection diagram for this case, and use it to make a plot of the voltage at the load resistor as a function of time for $0 < t < 8l/v$ (note that the effect of opening the switch is to initiate a second voltage wave, whose value is such that it leaves a net current of zero in its wake).

10.40 In the charged line of Figure 10.25, the characteristic impedance is $Z_0 = 100\ \Omega$, and $R_g = 300\ \Omega$. The line is charged to initial voltage, $V_0 = 160$ V, and the switch is closed at $t = 0$. Determine and plot the voltage and current through the resistor for time $0 < t < 8l/v$ (four round-trips). This problem accompanies Example 10.12 as the other special case of the basic charged-line problem, in which now $R_g > Z_0$.

10.41 In the transmission line of Figure 10.37, the switch is located *midway* down the line and is closed at $t = 0$. Construct a voltage reflection diagram for this case, where $R_L = Z_0$. Plot the load resistor voltage as a function of time.

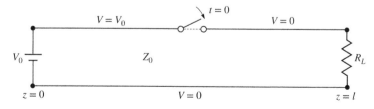

Figure 10.37 See Problem 10.41.

368   ENGINEERING ELECTROMAGNETICS

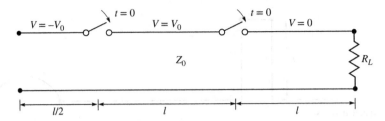

Figure 10.38   See Problem 10.42.

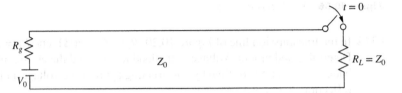

Figure 10.39   See Problem 10.43.

10.42 A simple *frozen wave generator* is shown in Figure 10.38. Both switches are closed simultaneously at $t = 0$. Construct an appropriate voltage reflection diagram for the case in which $R_L = Z_0$. Determine and plot the load resistor voltage as a function of time.

10.43 In Figure 10.39, $R_L = Z_0$ and $R_g = Z_0/3$. The switch is closed at $t = 0$. Determine and plot as functions of time (*a*) the voltage across R_L; (*b*) the voltage across R_g; (*c*) the current through the battery.

CHAPTER 11 The Uniform Plane Wave

术语

uniform plane wave　均匀平面波。

transverse plane　横平面，其法向是波传播的方向。

transverse electromagnetic wave（TEM wave）　横电磁波，电场与磁场矢量都在横平面内。

penetration depth　透入深度。

poynting theorem　坡印廷定理。

wave polarization　波的极化，用来描述空间一点上波的特性，定义为该点电场强度 E 矢量随着时间的变化或旋转特性。

wavenumber　波数，在无损媒质中为相位常数，表征波在单位距离的相位变化；由相位变化可以得到单位距离内有多少个波，单位是弧度 / 米（rad / m）。

This chapter is concerned with the application of Maxwell's equations to the problem of electromagnetic wave propagation. The uniform plane wave represents the simplest case, and while it is appropriate for an introduction, it is of great practical importance. Waves encountered in practice can often be assumed to be of this form. In this study, we will explore the basic principles of electromagnetic wave propagation, and we will come to understand the physical processes that determine the speed of propagation and the extent to which attenuation may occur. We will derive and use the Poynting theorem to find the power carried by a wave. Finally, we will learn how to describe wave polarization. ■

11.1 WAVE PROPAGATION IN FREE SPACE

We begin with a quick study of Maxwell's equations, in which we look for clues of wave phenomena. In Chapter 10, we saw how voltages and currents propagate as waves in transmission lines, and we know that the existence of voltages and currents implies the existence of electric and magnetic fields. So we can identify a transmission line as a structure that confines the fields while enabling them to travel along its length as waves. It can be argued that it is the fields that generate the voltage and current in a transmission line wave, and—if there is no structure on which the voltage and current can exist—the fields will exist nevertheless, and will propagate. In free space, the fields are not bounded by any confining structure, and so they may assume *any* magnitude and direction, as initially determined by the device (such as an antenna) that generates them.

11.1.1 Wave Equation for the Uniform Plane Wave

When considering electromagnetic waves in free space, we note that the medium is *sourceless*, meaning that $\rho_v = \mathbf{J} = 0$. Under these conditions, Maxwell's equations may be written in terms of **E** and **H** only as

$$\nabla \times \mathbf{H} = \epsilon_0 \frac{\partial \mathbf{E}}{\partial t} \tag{1}$$

$$\nabla \times \mathbf{E} = -\mu_0 \frac{\partial \mathbf{H}}{\partial t} \tag{2}$$

$$\nabla \cdot \mathbf{E} = 0 \tag{3}$$

$$\nabla \cdot \mathbf{H} = 0 \tag{4}$$

It is possible to infer wave motion from these four equations without actually solving them. Equation (1) states that if electric field **E** is changing with time at some point, then magnetic field **H** has curl at that point; therefore **H** varies spatially in a direction normal to its orientation direction. Also, if **E** is changing with time, then **H** will in general also change with time, although not necessarily in the same way. Next, we see from Eq. (2) that a time-varying **H** generates **E**, which, having curl, varies spatially in the direction normal to its orientation. We now have once more a changing electric field, our original hypothesis, but this field is present a small distance away from the point of the original disturbance. We might guess (correctly) that the velocity with which the effect moves away from the original point is the velocity of light, but this must be checked by a more detailed examination of Maxwell's equations.

We postulate the existence of a *uniform plane wave,* in which both fields, **E** and **H**, lie in the *transverse plane*—that is, the plane whose normal is the direction of propagation. Furthermore, and by definition, both fields are of constant magnitude in the transverse plane. For this reason, such a wave is sometimes called a *transverse electromagnetic* (TEM) wave. The required spatial variation of both fields in the direction normal to their orientations will therefore occur only in the direction of travel—or normal to the transverse plane. Assume, for example, that $\mathbf{E} = E_x \mathbf{a}_x$, or that the electric field is *polarized* in the x direction. If we further assume that wave travel is in the z direction, we allow spatial variation of **E** only with z. Using Eq. (2), we note that with these restrictions, the curl of **E** reduces to a single term:

$$\nabla \times \mathbf{E} = \frac{\partial E_x}{\partial z} \mathbf{a}_y = -\mu_0 \frac{\partial \mathbf{H}}{\partial t} = -\mu_0 \frac{\partial H_y}{\partial t} \mathbf{a}_y \tag{5}$$

The direction of the curl of **E** in (5) determines the direction of **H**, which we observe to be along the y direction. Therefore, in a uniform plane wave, the directions of **E** and **H** and the direction of travel are mutually orthogonal. Using the y-directed magnetic field, and the fact that it varies only in z, simplifies Eq. (1) to read

$$\nabla \times \mathbf{H} = -\frac{\partial H_y}{\partial z} \mathbf{a}_x = \epsilon_0 \frac{\partial \mathbf{E}}{\partial t} = \epsilon_0 \frac{\partial E_x}{\partial t} \mathbf{a}_x \tag{6}$$

Equations (5) and (6) can be more succinctly written:

$$\frac{\partial E_x}{\partial z} = -\mu_0 \frac{\partial H_y}{\partial t} \tag{7}$$

$$\frac{\partial H_y}{\partial z} = -\epsilon_0 \frac{\partial E_x}{\partial t} \tag{8}$$

These equations compare directly with the telegraphist's equations for the lossless transmission line [Eqs. (20) and (21) in Chapter 10]. Further manipulations of (7) and (8) proceed in the same manner as was done with the telegraphist's equations. Specifically, we differentiate (7) with respect to z, obtaining:

$$\frac{\partial^2 E_x}{\partial z^2} = -\mu_0 \frac{\partial^2 H_y}{\partial t \partial z} \tag{9}$$

Then, (8) is differentiated with respect to t:

$$\frac{\partial^2 H_y}{\partial z \partial t} = -\epsilon_0 \frac{\partial^2 E_x}{\partial t^2} \tag{10}$$

Substituting (10) into (9) results in

$$\frac{\partial^2 E_x}{\partial z^2} = \mu_0 \epsilon_0 \frac{\partial^2 E_x}{\partial t^2} \tag{11}$$

This equation, in direct analogy to Eq. (13) in Chapter 10, we identify as the wave equation for our x-polarized TEM electric field in free space. From Eq. (11), we further identify the propagation velocity:

$$v = \frac{1}{\sqrt{\mu_0 \epsilon_0}} = 3 \times 10^8 \text{ m/s} = c \tag{12}$$

where c denotes the velocity of light in free space. A similar procedure, involving differentiating (7) with t and (8) with z, yields the wave equation for the magnetic field; it is identical in form to (11):

$$\frac{\partial^2 H_y}{\partial z^2} = \mu_0 \epsilon_0 \frac{\partial^2 H_y}{\partial t^2} \tag{13}$$

11.1.2 Solutions of the Wave Equation

As was discussed in Chapter 10, the solution to equations of the form of (11) and (13) will be forward- and backward-propagating waves having the general form [in this case for Eq. (11)]:

$$E_x(z, t) = f_1(t - z/v) + f_2(t + z/v) \tag{14}$$

where again f_1 and f_2 can be any function whose argument is of the form $t \pm z/v$.

From here, we immediately specialize to sinusoidal functions of a specified frequency and write the solution to (11) in the form of forward- and backward-propagating cosines. Because the waves are sinusoidal, we denote their velocity as the *phase velocity*, v_p. The waves are written as:

$$
\begin{aligned}
E_x(z,t) &= \mathscr{E}_x(z,t) + \mathscr{E}_x'(z,t) \\
&= |E_{x0}|\cos\left[\omega\left(t - z/v_p\right) + \phi_1\right] + |E_{x0}'|\cos\left[\omega\left(t + z/v_p\right) + \phi_2\right] \\
&= \underbrace{|E_{x0}|\cos\left[\omega t - k_0 z + \phi_1\right]}_{\text{forward } z \text{ travel}} + \underbrace{|E_{x0}'|\cos\left[\omega t + k_0 z + \phi_2\right]}_{\text{backward } z \text{ travel}}
\end{aligned} \tag{15}
$$

In writing the second line of (15), we have used the fact that the waves are traveling in free space, in which case the phase velocity $v_p = c$. Additionally, the *wavenumber* in free space is defined as

$$
\boxed{k_0 \equiv \frac{\omega}{c} \text{ rad/m}} \tag{16}
$$

In a manner consistant with our transmission line studies, we refer to the solutions expressed in (15) as the *real instantaneous* forms of the electric field. They are the mathematical representations of what one would experimentally measure. The terms ωt and $k_0 z$, appearing in (15), have units of angle and are usually expressed in radians. We know that ω is the radian time frequency, measuring phase shift *per unit time;* it has units of *rad/s*. In a similar way, we see that k_0 will be interpreted as a *spatial* frequency, which in the present case measures the phase shift *per unit distance* along the z direction in rad/m. We note that k_0 is the phase constant for lossless propagation of uniform plane waves in free space. The *wavelength* in free space is the distance over which the spatial phase shifts by 2π radians, assuming fixed time, or

$$
k_0 z = k_0 \lambda = 2\pi \quad \rightarrow \quad \boxed{\lambda = \frac{2\pi}{k_0}} \quad \text{(free space)} \tag{17}
$$

The manner in which the waves propagate is the same as we encountered in transmission lines. Specifically, suppose we consider some point (such as a wave crest) on the forward-propagating cosine function of Eq. (15). For a crest to occur, the argument of the cosine must be an integer multiple of 2π. Considering the *m*th crest of the wave, the condition becomes

$$
k_0 z = 2m\pi
$$

So let us now consider the point on the cosine that we have chosen, and see what happens as time is allowed to increase. Our requirement is that the entire cosine argument be the same multiple of 2π for all time, in order to keep track of the chosen point. Our condition becomes

$$
\omega t - k_0 z = \omega(t - z/c) = 2m\pi \tag{18}
$$

As time increases, the position z must also increase in order to satisfy (18). The wave crest (and the entire wave) moves in the positive z direction at phase velocity c (in free space). Using similar reasoning, the wave in Eq. (15) having cosine argument

$(\omega t + k_0 z)$ describes a wave that moves in the negative z direction, since as time increases, z must now decrease to keep the argument constant. For simplicity, we will restrict our attention in this chapter to only the positive z traveling wave.

As was done for transmission line waves, we express the real instantaneous fields of Eq. (15) in terms of their phasor forms. Using the forward-propagating field in (15), we write:

$$\mathcal{E}_x(z,t) = \frac{1}{2}\underbrace{\left|E_{x0}\right|e^{j\phi_1}}_{E_{x0}}e^{-jk_0 z}e^{j\omega t} + c.c. = \frac{1}{2}E_{xs}e^{j\omega t} + c.c. = \mathcal{R}e[E_{xs}e^{j\omega t}] \qquad (19)$$

where $c.c.$ denotes the complex conjugate, and where we identify the *phasor electric field* as $E_{xs} = E_{x0}e^{-jk_0 z}$. As indicated in (19), E_{x0} is the *complex* amplitude (which includes the phase, ϕ_1).

EXAMPLE 11.1

Let us express $\mathcal{E}_y(z,t) = 100 \cos(10^8 t - 0.5z + 30°)$V/m as a phasor.

Solution. We first go to exponential notation,

$$\mathcal{E}_y(z,t) = \mathcal{R}e[100e^{j(10^8 t - 0.5z + 30°)}]$$

and then drop Re and suppress $e^{j10^8 t}$, obtaining the phasor

$$E_{ys}(z) = 100e^{-j0.5z + j30°}$$

Note that a mixed nomenclature is used for the angle in this case; that is, $0.5z$ is in radians, while $30°$ is in degrees. Given a scalar component or a vector expressed as a phasor, we may easily recover the time-domain expression.

EXAMPLE 11.2

Given the complex amplitude of the electric field of a uniform plane wave, $\mathbf{E}_0 = 100\mathbf{a}_x + 20\angle 30°\mathbf{a}_y$V/m, construct the phasor and real instantaneous fields if the wave is known to propagate in the forward z direction in free space and has frequency of 10 MHz.

Solution. We begin by constructing the general phasor expression:

$$\mathbf{E}_s(z) = [100\mathbf{a}_x + 20e^{j30°}\mathbf{a}_y]e^{-jk_0 z}$$

where $k_0 = \omega/c = 2\pi \times 10^7/3 \times 10^8 = 0.21$ rad/m. The real instantaneous form is then found through the rule expressed in Eq. (19):

$$\mathcal{E}(z,t) = \mathcal{R}e[100e^{-j0.21z}e^{j2\pi\times 10^7 t}\mathbf{a}_x + 20e^{j30°}e^{-j0.21z}e^{j2\pi\times 10^7 t}\mathbf{a}_y]$$

$$= \mathcal{R}e[100e^{j(2\pi\times 10^7 t - 0.21z)}\mathbf{a}_x + 20e^{j(2\pi\times 10^7 t - 0.21z + 30°)}\mathbf{a}_y]$$

$$= 100 \cos(2\pi \times 10^7 t - 0.21z)\mathbf{a}_x + 20 \cos(2\pi \times 10^7 t - 0.21z + 30°)\mathbf{a}_y$$

11.1.3 Vector Helmholtz Equation in Free Space

It is evident that taking the partial derivative of any field quantity with respect to time is equivalent to multiplying the corresponding phasor by $j\omega$. As an example, we can express Eq. (8) (using sinusoidal fields) as

$$\frac{\partial \mathcal{H}_y}{\partial z} = -\epsilon_0 \frac{\partial \mathcal{E}_x}{\partial t} \tag{20}$$

where, in a manner consistent with (19):

$$\mathcal{E}_x(z,t) = \frac{1}{2} E_{xs}(z) e^{j\omega t} + c.c. \qquad \text{and} \qquad \mathcal{H}_y(z,t) = \frac{1}{2} H_{ys}(z) e^{j\omega t} + c.c. \tag{21}$$

On substituting the fields in (21) into (20), the latter equation simplifies to

$$\boxed{\frac{dH_{ys}(z)}{dz} = -j\omega\epsilon_0 E_{xs}(z)} \tag{22}$$

In obtaining this equation, we note first that the complex conjugate terms in (21) produce their own separate equation, redundant with (22); second, the $e^{j\omega t}$ factors, common to both sides, have divided out; third, the partial derivative with z becomes the total derivative, since the phasor, H_{ys}, depends only on z.

We next apply this result to Maxwell's equations, to obtain them in phasor form. Substituting the field as expressed in (21) into Eqs. (1) through (4) results in

$$\boxed{\begin{aligned}
\nabla \times \mathbf{H}_s &= j\omega\epsilon_0 \mathbf{E}_s &(23)\\
\nabla \times \mathbf{E}_s &= -j\omega\mu_0 \mathbf{H}_s &(24)\\
\nabla \cdot \mathbf{E}_s &= 0 &(25)\\
\nabla \cdot \mathbf{H}_s &= 0 &(26)
\end{aligned}}$$

It should be noted that (25) and (26) are no longer independent relationships, for they can be obtained by taking the divergence of (23) and (24), respectively.

Eqs. (23) through (26) may be used to obtain the sinusoidal steady-state vector form of the wave equation in free space. We begin by taking the curl of both sides of (24):

$$\nabla \times \nabla \times \mathbf{E}_s = -j\omega\mu_0 \nabla \times \mathbf{H}_s = \nabla(\nabla \cdot \mathbf{E}_s) - \nabla^2 \mathbf{E}_s \tag{27}$$

where the last equality is an identity, which defines the *vector Laplacian* of $\mathbf{E}_s$:

$$\boxed{\nabla^2 \mathbf{E}_s = \nabla(\nabla \cdot \mathbf{E}_s) - \nabla \times \nabla \times \mathbf{E}_s}$$

From (25), we note that $\nabla \cdot \mathbf{E}_s = 0$. Using this, and substituting (23) in (27), we obtain

$$\boxed{\nabla^2 \mathbf{E}_s = -k_0^2 \mathbf{E}_s} \tag{28}$$

where again, $k_0 = \omega/c = \omega\sqrt{\mu_0\epsilon_0}$. Equation (28) is known as the vector Helmholtz equation in free space.[1] It is fairly formidable when expanded, even in rectangular coordinates, for three scalar phasor equations result (one for each vector component), and each equation has four terms. The x component of (28) becomes, still using the del-operator notation,

$$\boxed{\nabla^2 E_{xs} = -k_0^2 E_{xs}} \tag{29}$$

and the expansion of the operator leads to the second-order partial differential equation

$$\frac{\partial^2 E_{xs}}{\partial x^2} + \frac{\partial^2 E_{xs}}{\partial y^2} + \frac{\partial^2 E_{xs}}{\partial z^2} = -k_0^2 E_{xs}$$

Again, assuming a uniform plane wave in which E_{xs} does not vary with x or y, the two corresponding derivatives are zero, and we obtain

$$\boxed{\frac{d^2 E_{xs}}{dz^2} = -k_0^2 E_{xs}} \tag{30}$$

the solution of which we already know:

$$\boxed{E_{xs}(z) = E_{x0}e^{-jk_0z} + E'_{x0}e^{jk_0z}} \tag{31}$$

11.1.4 Relation Between E and H: Intrinsic Impedance

We now return to Maxwell's equations, (23) through (26), and determine the form of the **H** field. Given $\mathbf{E}_s$, $\mathbf{H}_s$ is most easily obtained from (24):

$$\nabla \times \mathbf{E}_s = -j\omega\mu_0\mathbf{H}_s \tag{24}$$

which is greatly simplified for a single E_{xs} component varying only with z,

$$\frac{dE_{xs}}{dz} = -j\omega\mu_0 H_{ys}$$

Using (31) for E_{xs}, we have

$$\begin{aligned}
H_{ys} &= -\frac{1}{j\omega\mu_0}\left[(-jk_0)E_{x0}e^{-jk_0z} + (jk_0)E'_{x0}e^{jk_0z}\right] \\
&= E_{x0}\sqrt{\frac{\epsilon_0}{\mu_0}}e^{-jk_0z} - E'_{x0}\sqrt{\frac{\epsilon_0}{\mu_0}}e^{jk_0z} = H_{y0}e^{-jk_0z} + H'_{y0}e^{jk_0z}
\end{aligned} \tag{32}$$

In real instantaneous form, this becomes

$$\boxed{H_y(z,t) = E_{x0}\sqrt{\frac{\epsilon_0}{\mu_0}}\cos(\omega t - k_0 z) - E'_{x0}\sqrt{\frac{\epsilon_0}{\mu_0}}\cos(\omega t + k_0 z)} \tag{33}$$

where E_{x0} and E'_{x0} are assumed real.

---

[1] Hermann Ludwig Ferdinand von Helmholtz (1821–1894) was a professor at the University of Berlin working in the fields of physiology, electrodynamics, and optics. Hertz was one of his students.

In general, we find from (32) that the electric and magnetic field amplitudes of the forward-propagating wave in free space are related through

$$E_{x0} = \sqrt{\frac{\mu_0}{\epsilon_0}} H_{y0} = \eta_0 H_{y0} \qquad (34a)$$

We also find the backward-propagating wave amplitudes are related through

$$E'_{x0} = -\sqrt{\frac{\mu_0}{\epsilon_0}} H'_{y0} = -\eta_0 H'_{y0} \qquad (34b)$$

where the *intrinsic impedance* of free space is defined as

$$\eta_0 = \sqrt{\frac{\mu_0}{\epsilon_0}} = 377 \doteq 120\pi \ \Omega \qquad (35)$$

The dimension of η_0 in ohms is immediately evident from its definition as the ratio of E (in units of V/m) to H (in units of A/m). It is in direct analogy to the characteristic impedance, Z_0, of a transmission line, where we defined the latter as the ratio of voltage to current in a traveling wave. We note that the difference between (34a) and (34b) is a minus sign. This is consistent with the transmission line analogy that led to Eqs. (25a) and (25b) in Chapter 10. Those equations accounted for the definitions of positive and negative current associated with forward and backward voltage waves. In a similar way, Eq. (34a) specifies that in a forward-z propagating uniform plane wave whose electric field vector lies in the positive x direction at a given point in time and space, the magnetic field vector lies in the *positive y* direction at the same space and time coordinates. In the case of a backward-z propagating wave having a positive x-directed electric field, the magnetic field vector lies in the *negative y* direction. The physical significance of this has to do with the definition of power flow in the wave, as specified through the Poynting vector, $\mathbf{S} = \mathbf{E} \times \mathbf{H}$ (in watts/m²). The cross product of $\mathbf{E}$ with $\mathbf{H}$ must give the correct wave propagation direction, and so the need for the minus sign in (34b) is apparent. Issues relating to power transmission will be addressed in Section 11.3.

Some feeling for the way in which the fields vary in space may be obtained from Figures 11.1a and 11.1b. The electric field intensity in Figure 11.1a is shown at $t = 0$, and the instantaneous value of the field is depicted along three lines, the z axis and arbitrary lines parallel to the z axis in the $x = 0$ and $y = 0$ planes. Since the field is uniform in planes perpendicular to the z axis, the variation along all three of the lines is the same. One complete cycle of the variation occurs in a wavelength, λ. The values of H_y at the same time and positions are shown in Figure 11.1b.

A uniform plane wave cannot exist physically, for it extends to infinity in two dimensions at least and represents an infinite amount of energy. The distant field of a transmitting antenna, however, is essentially a uniform plane wave in some limited region; for example, a radar signal impinging on a distant target is closely a uniform plane wave.

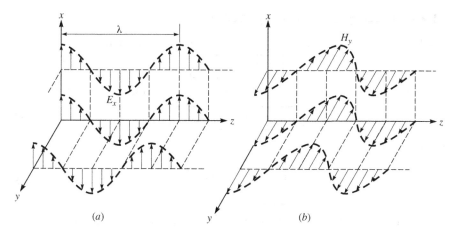

Figure 11.1 (*a*) Arrows represent the instantaneous values of $E_{x0}\cos[\omega(t-z/c)]$ at $t=0$ along the z axis, along an arbitrary line in the $x=0$ plane parallel to the z axis, and along an arbitrary line in the $y=0$ plane parallel to the z axis. (*b*) Corresponding values of H_y are indicated. Note that E_x and H_y are in phase at any point in time.

Although we have considered only a wave varying sinusoidally in time and space, a suitable combination of solutions to the wave equation may be made to achieve a wave of any desired form, but which satisfies (14). The summation of an infinite number of harmonics through the use of a Fourier series can produce a periodic wave of square or triangular shape in both space and time. Nonperiodic waves may be obtained from our basic solution by Fourier integral methods. These topics are among those considered in the more advanced books on electromagnetic theory.

D11.1. The electric field amplitude of a uniform plane wave propagating in the $\mathbf{a}_z$ direction is 250 V/m. If $\mathbf{E}=E_x\mathbf{a}_x$ and $\omega=1.00$ Mrad/s, find: (*a*) the frequency; (*b*) the wavelength; (*c*) the period; (*d*) the amplitude of $\mathbf{H}$.

Ans. (*a*) 159 kHz; (*b*) 1.88 km; (*c*) 6.28 μs; (*d*) 0.663 A/m

D11.2. Let $H_s=(2\angle-40°\mathbf{a}_x-3\angle20°\mathbf{a}_y)e^{-j0.07z}$A/m for a uniform plane wave traveling in free space. Find: (*a*) ω; (*b*) H_x at $P(1,2,3)$ at $t=31$ ns; (*c*) |$\mathbf{H}$| at $t=0$ at the origin.

Ans. (*a*) 21.0 Mrad/s; (*b*) 1.934 A/m; (*c*) 3.22 A/m

11.2 WAVE PROPAGATION IN DIELECTRICS

We now extend our analytical treatment of the uniform plane wave to propagation in a dielectric of permittivity ϵ and permeability μ. The medium is assumed to be homogeneous (having constant μ and ϵ with position) and isotropic (in which μ and ϵ are invariant with field orientation).

11.2.1 Propagation in Lossy Media

The Helmholtz equation in a homogeneous and isotropic medium is

$$\nabla^2 \mathbf{E}_s = -k^2 \mathbf{E}_s \tag{36}$$

where the wavenumber is a function of the material properties, as described by μ and ϵ:

$$k = \omega \sqrt{\mu \epsilon} = k_0 \sqrt{\mu_r \epsilon_r} \tag{37}$$

For E_{xs} we have

$$\frac{d^2 E_{xs}}{dz^2} = -k^2 E_{xs} \tag{38}$$

An important feature of wave propagation in a dielectric is that k can be complex-valued, and as such it is referred to as the complex *propagation constant. k* becomes complex when loss or gain mechanisms are present in the medium, as will be explained. A general solution of (38), in fact, allows the possibility of a complex k, and it is customary to write it in terms of its real and imaginary parts in the following way:

$$jk = \alpha + j\beta \tag{39}$$

A solution to (38) will be:

$$E_{xs} = E_{x0} e^{-jkz} = E_{x0} e^{-\alpha z} e^{-j\beta z} \tag{40}$$

Multiplying (40) by $e^{j\omega t}$ and taking the real part yields a form of the field that can be more easily visualized:

$$E_x = E_{x0} e^{-\alpha z} \cos(\omega t - \beta z) \tag{41}$$

We recognize this as a uniform plane wave that propagates in the forward z direction with phase constant β, but which (for positive α) loses amplitude with increasing z according to the factor $e^{-\alpha z}$. Thus the general effect of a complex-valued k is to yield a traveling wave that changes its amplitude with distance. If α is positive, it is called the *attenuation coefficient*. If α is negative, the wave grows in amplitude with distance, and α is called the *gain coefficient*. The latter effect would occur, for example, in laser amplifiers. In the present and future discussions in this book, we will consider only passive media, in which losses are present, thus producing a positive α.

The attenuation coefficient is measured in nepers per meter (Np/m) so that the exponent of e can be measured in the dimensionless units of nepers. Thus, if $\alpha = 0.01$ Np/m, the crest amplitude of the wave at $z = 50$ m will be $e^{-0.5}/e^{-0} = 0.607$ of its value at $z = 0$. In traveling a distance $1/\alpha$ in the $+z$ direction, the amplitude of the wave is reduced by the familiar factor of e^{-1}, or 0.368.

The ways in which physical processes in a material can affect the wave electric field are described through a *complex permittivity* of the form

$$\epsilon = \epsilon' - j\epsilon'' = \epsilon_0(\epsilon_r' - j\epsilon_r'') \tag{42}$$

Two important mechanisms that give rise to a complex permittivity (and thus result in wave losses) are bound electron or ion oscillations and dipole relaxation, both of which are discussed in Appendix E. An additional mechanism is the conduction of free electrons or holes, which we will explore at length in the next section.

Losses arising from the response of the medium to the magnetic field can occur as well, and these are modeled through a *complex permeability*, $\mu = \mu' - j\mu'' = \mu_0 (\mu_r' - j\mu_r'')$. Examples of such media include *ferrimagnetic* materials, or *ferrites*. The magnetic response is usually very weak compared to the dielectric response in most materials of interest for wave propagation; in such materials $\mu \approx \mu_0$. Consequently, our discussion of loss mechanisms will be confined to those described through the complex permittivity, and we will assume that μ is entirely real in our treatment.

We can substitute (42) into (37), which results in

$$k = \omega\sqrt{\mu(\epsilon' - j\epsilon'')} = \omega\sqrt{\mu\epsilon'}\sqrt{1 - j\frac{\epsilon''}{\epsilon'}} \qquad (43)$$

Note the presence of the second radical factor in (43), which becomes unity (and real) as ϵ'' vanishes. With nonzero ϵ'', k is complex, and so losses occur which are quantified through the attenuation coefficient, α, in (39). The phase constant, β (and consequently the wavelength and phase velocity), will also be affected by ϵ''. α and β are found by taking the real and imaginary parts of jk from (43). We obtain:

$$\alpha = \mathscr{R}e\{jk\} = \omega\sqrt{\frac{\mu\epsilon'}{2}}\left(\sqrt{1 + \left(\frac{\epsilon''}{\epsilon'}\right)^2} - 1\right)^{1/2} \qquad (44)$$

$$\beta = \mathscr{I}m\{jk\} = \omega\sqrt{\frac{\mu\epsilon'}{2}}\left(\sqrt{1 + \left(\frac{\epsilon''}{\epsilon'}\right)^2} + 1\right)^{1/2} \qquad (45)$$

We see that a nonzero α (and hence loss) results if the imaginary part of the permittivity, ϵ'', is present. We also observe in (44) and (45) the presence of the ratio ϵ''/ϵ', which is called the *loss tangent*. The meaning of the term will be demonstrated when we investigate the specific case of conductive media. The practical importance of the ratio lies in its magnitude compared to unity, which enables simplifications to be made in (44) and (45).

Whether or not losses occur, we see from (41) that the wave phase velocity is given by

$$v_p = \frac{\omega}{\beta} \qquad (46)$$

The wavelength is the distance required to effect a phase change of 2π radians

$$\beta\lambda = 2\pi$$

which leads to the fundamental definition of wavelength,

$$\lambda = \frac{2\pi}{\beta} \tag{47}$$

Because we have a uniform plane wave, the magnetic field is found through

$$H_{ys} = \frac{E_{x0}}{\eta} e^{-\alpha z} e^{-j\beta z}$$

where the intrinsic impedance is now a complex quantity,

$$\eta = \sqrt{\frac{\mu}{\epsilon' - j\epsilon''}} = \sqrt{\frac{\mu}{\epsilon'}} \frac{1}{\sqrt{1 - j(\epsilon''/\epsilon')}} \tag{48}$$

The electric and magnetic fields are no longer in phase.

A special case is that of a lossless medium, or *perfect dielectric*, in which $\epsilon'' = 0$, and so $\epsilon = \epsilon'$. From (44), this leads to $\alpha = 0$, and from (45),

$$\beta = \omega\sqrt{\mu\epsilon'} \qquad \text{(lossless medium)} \tag{49}$$

With $\alpha = 0$, the real field assumes the form

$$E_x = E_{x0}\cos(\omega t - \beta z) \tag{50}$$

We may interpret this as a wave traveling in the $+z$ direction at a phase velocity v_p, where

$$v_p = \frac{\omega}{\beta} = \frac{1}{\sqrt{\mu\epsilon'}} = \frac{c}{\sqrt{\mu_r\epsilon'_r}}$$

The wavelength is

$$\lambda = \frac{2\pi}{\beta} = \frac{2\pi}{\omega\sqrt{\mu\epsilon'}} = \frac{1}{f\sqrt{\mu\epsilon'}} = \frac{c}{f\sqrt{\mu_r\epsilon'_r}} = \frac{\lambda_0}{\sqrt{\mu_r\epsilon'_r}} \qquad \text{(lossless medium)} \tag{51}$$

where λ_0 is the free-space wavelength. Note that $\mu_r\epsilon'_r > 1$, and therefore the wavelength is shorter and the velocity is lower in all real media than they are in free space.

Associated with E_x is the magnetic field intensity

$$H_y = \frac{E_{x0}}{\eta}\cos(\omega t - \beta z)$$

where the intrinsic impedance is

$$\eta = \sqrt{\frac{\mu}{\epsilon}} \tag{52}$$

The two fields are once again perpendicular to each other, perpendicular to the direction of propagation, and in phase with each other everywhere. Note that when **E** is crossed into **H**, the resultant vector is in the direction of propagation. We shall see the reason for this when we discuss the Poynting vector.

EXAMPLE 11.3

Let us apply these results to a 1-MHz plane wave propagating in fresh water. At this frequency, losses in water are negligible, which means that we can assume that $\epsilon'' \doteq 0$. In water, $\mu_r = 1$ and at 1 MHz, $\epsilon_r' = 81$.

Solution. We begin by calculating the phase constant. Using (45) with $\epsilon'' = 0$, we have

$$\beta = \omega\sqrt{\mu\epsilon'} = \omega\sqrt{\mu_0\epsilon_0}\,\sqrt{\epsilon_r'} = \frac{\omega\sqrt{\epsilon_r'}}{c} = \frac{2\pi \times 10^6\sqrt{81}}{3.0 \times 10^8} = 0.19 \text{ rad/m}$$

Using this result, we can determine the wavelength and phase velocity:

$$\lambda = \frac{2\pi}{\beta} = \frac{2\pi}{.19} = 33 \text{ m}$$

$$v_p = \frac{\omega}{\beta} = \frac{2\pi \times 10^6}{.19} = 3.3 \times 10^7 \text{ m/s}$$

The wavelength in air would have been 300 m. Continuing our calculations, we find the intrinsic impedance using (48) with $\epsilon'' = 0$:

$$\eta = \sqrt{\frac{\mu}{\epsilon'}} = \frac{\eta_0}{\sqrt{\epsilon_r'}} = \frac{377}{9} = 42 \; \Omega$$

If we let the electric field intensity have a maximum amplitude of 0.1 V/m, then

$$E_x = 0.1\cos(2\pi10^6 t - .19z) \text{ V/m}$$

$$H_y = \frac{E_x}{\eta} = (2.4 \times 10^{-3})\cos(2\pi10^6 t - .19z) \text{ A/m}$$

D11.3. A 9.375-GHz uniform plane wave is propagating in polyethylene (see Appendix C). If the amplitude of the electric field intensity is 500 V/m and the material is assumed to be lossless, find: (a) the phase constant; (b) the wavelength in the polyethylene; (c) the velocity of propagation; (d) the intrinsic impedance; (e) the amplitude of the magnetic field intensity.

Ans. (a) 295 rad/m; (b) 2.13 cm; (c) 1.99×10^8 m/s; (d) 251 Ω; (e) 1.99 A/m

EXAMPLE 11.4

We again consider plane wave propagation in water, but at the much higher microwave frequency of 2.5 GHz. At frequencies in this range and higher, dipole relaxation and resonance phenomena in the water molecules become important.[2] Real and imaginary parts of the permittivity are present, and both vary with frequency. At frequencies below that of visible light, the two mechanisms together produce a value of ϵ'' that increases with increasing frequency, reaching a maximum in the vicinity of 10^{13} Hz. ϵ' decreases with increasing frequency, reaching a minimum also in the vicinity of 10^{13} Hz. Reference 3 provides specific details. At 2.5 GHz, dipole relaxation effects dominate. The permittivity values are $\epsilon_r' = 78$ and $\epsilon_r'' = 7$. From (44), we have

$$\alpha = \frac{(2\pi \times 2.5 \times 10^9)\sqrt{78}}{(3.0 \times 10^8)\sqrt{2}} \left(\sqrt{1 + \left(\frac{7}{78}\right)^2} - 1\right)^{1/2} = 21 \text{ Np/m}$$

This first calculation demonstrates the operating principle of the *microwave oven*. Almost all foods contain water, and so they can be cooked when incident microwave radiation is absorbed and converted into heat. Note that the field will attenuate to a value of e^{-1} times its initial value at a distance of $1/\alpha = 4.8$ cm. This distance is called the *penetration depth* of the material, and of course it is frequency-dependent. The 4.8-cm depth is reasonable for cooking food, since it would lead to a temperature rise that is fairly uniform throughout the depth of the material. At much higher frequencies, where ϵ'' is larger, the penetration depth decreases, and too much power is absorbed at the surface; at lower frequencies, the penetration depth increases, and not enough overall absorption occurs. Commercial microwave ovens operate at frequencies in the vicinity of 2.5 GHz.

Using (45), in a calculation very similar to that for α, we find $\beta = 464$ rad/m. The wavelength is $\lambda = 2\pi/\beta = 1.4$ cm, whereas in free space this would have been $\lambda_0 = c/f = 12$ cm.

Using (48), the intrinsic impedance is found to be

$$\eta = \frac{377}{\sqrt{78}} \frac{1}{\sqrt{1 - j(7/78)}} = 43 + j1.9 = 43\angle2.6° \ \Omega$$

and E_x leads H_y in time by 2.6° at every point.

11.2.2 Propagation in Conducting Media

We next consider the case of conductive materials, in which currents are formed by the motion of free electrons or holes under the influence of an electric field. The governing relation is $\mathbf{J} = \sigma \mathbf{E}$, where σ is the material conductivity. With finite conductivity, the wave loses power through resistive heating of the material. We look for an interpretation of the complex permittivity as it relates to the conductivity.

---

[2] These mechanisms and how they produce a complex permittivity are described in Appendix D. Additionally, the reader is referred to pp. 73–84 in Reference 1 and pp. 678–82 in Reference 2 for general treatments of relaxation and resonance effects on wave propagation. Discussions and data that are specific to water are presented in Reference 3, pp. 314–16.

Consider the Maxwell curl equation (23) which, using (42), becomes:

$$\nabla \times \mathbf{H}_s = j\omega(\epsilon' - j\epsilon'')\,\mathbf{E}_s = \omega\epsilon''\mathbf{E}_s + j\omega\epsilon'\mathbf{E}_s \qquad (53)$$

This equation can be expressed in a more familiar way, in which conduction current is included:

$$\nabla \times \mathbf{H}_s = \mathbf{J}_s + j\omega\epsilon\mathbf{E}_s \qquad (54)$$

We next use $\mathbf{J}_s = \sigma\,\mathbf{E}_s$, and interpret ϵ in (54) as ϵ'. The latter equation becomes:

$$\boxed{\nabla \times \mathbf{H}_s = (\sigma + j\omega\epsilon')\,\mathbf{E}_s = \mathbf{J}_{\sigma s} + \mathbf{J}_{ds}} \qquad (55)$$

which we have expressed in terms of conduction current density, $\mathbf{J}_{\sigma s} = \sigma\,\mathbf{E}_s$, and displacement current density, $\mathbf{J}_{ds} = j\omega\epsilon'\mathbf{E}_s$. Comparing Eqs. (53) and (55), we find that in a conductive medium:

$$\boxed{\epsilon'' = \frac{\sigma}{\omega}} \qquad (56)$$

We next turn our attention to the case of a dielectric material in which the loss is very small. The criterion by which we should judge whether or not the loss is small is the magnitude of the loss tangent, ϵ''/ϵ'. This parameter will have a direct influence on the attenuation coefficient, α, as seen from Eq. (44). In the case of conducting media, to which (56) applies, the loss tangent becomes $\sigma/\omega\epsilon'$. By inspecting (55), we see that the ratio of conduction current density to displacement current density magnitudes is

$$\boxed{\frac{J_{\sigma s}}{J_{ds}} = \frac{\epsilon''}{j\epsilon'} = \frac{\sigma}{j\omega\epsilon'}} \qquad (57)$$

That is, these two vectors point in the same direction in space, but they are 90° out of phase in time. Displacement current density leads conduction current density by 90°, just as the current through a capacitor leads the current through a resistor in parallel with it by 90° in an ordinary electric circuit. This phase relationship is shown in Figure 11.2. The angle θ (not to be confused with the polar angle in spherical coordinates) may therefore be identified as the angle by which the displacement current density leads the total current density, and

$$\boxed{\tan\theta = \frac{\epsilon''}{\epsilon'} = \frac{\sigma}{\omega\epsilon'}} \qquad (58)$$

The reasoning behind the term *loss tangent* is thus evident. Problem 11.16 at the end of the chapter indicates that the Q of a capacitor (its quality factor, not its charge) that incorporates a lossy dielectric is the reciprocal of the loss tangent.

11.2.3 Good Dielectric Approximation

If the loss tangent is small, then we may obtain useful approximations for the attenuation and phase constants, and the intrinsic impedance. The criterion for a small

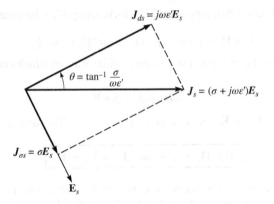

Figure 11.2 The time-phase relationship between $\mathbf{J}_{ds}$, $\mathbf{J}_{\sigma s}$, $\mathbf{J}_s$, and $\mathbf{E}_s$. The tangent of θ is equal to $\sigma/\omega\epsilon'$, and $90° - \theta$ is the common power-factor angle, or the angle by which $\mathbf{J}_s$ leads $\mathbf{E}_s$.

loss tangent is $\epsilon''/\epsilon' \ll 1$, which we say identifies the medium as a *good dielectric*. Considering a conductive material, for which $\epsilon'' = \sigma/\omega$, (43) becomes

$$jk = j\omega\sqrt{\mu\epsilon'}\sqrt{1 - j\frac{\sigma}{\omega\epsilon'}} \tag{59}$$

We may expand the second radical using the binomial theorem

$$(1 + x)^n = 1 + nx + \frac{n(n-1)}{2!}x^2 + \frac{n(n-1)(n-2)}{3!}x^3 + \cdots$$

where $|x| \ll 1$. We identify x as $-j\sigma/\omega\epsilon'$ and n as 1/2, and thus

$$jk = j\omega\sqrt{\mu\epsilon'}\left[1 - j\frac{\sigma}{2\omega\epsilon'} + \frac{1}{8}\left(\frac{\sigma}{\omega\epsilon'}\right)^2 + \cdots\right] = \alpha + j\beta$$

Now, for a good dielectric,

$$\alpha = \mathscr{R}e(jk) \doteq j\omega\sqrt{\mu\epsilon'}\left(-j\frac{\sigma}{2\omega\epsilon'}\right) = \frac{\sigma}{2}\sqrt{\frac{\mu}{\epsilon'}} \tag{60a}$$

and

$$\beta = \mathscr{I}m(jk) \doteq \omega\sqrt{\mu\epsilon'}\left[1 + \frac{1}{8}\left(\frac{\sigma}{\omega\epsilon'}\right)^2\right] \tag{60b}$$

Equations (60a) and (60b) can be compared directly with the transmission line α and β under low-loss conditions, as expressed in Eqs. (54a) and (54b) in Chapter 10. In this comparison, we associate σ with G, μ with L, and ϵ with C. Note that in plane wave propagation in media with no boundaries, there can be no quantity that is analogous to the transmission line conductor resistance parameter, R. In many cases, the second term in (60b) is small enough, so

$$\boxed{\beta \doteq \omega\sqrt{\mu\epsilon'}}$$
(61)

Applying the binomial expansion to (48), we obtain, for a good dielectric,

$$\eta \doteq \sqrt{\frac{\mu}{\epsilon'}}\left[1 - \frac{3}{8}\left(\frac{\sigma}{\omega\epsilon'}\right)^2 + j\frac{\sigma}{2\omega\epsilon'}\right]$$
(62a)

or

$$\boxed{\eta \doteq \sqrt{\frac{\mu}{\epsilon'}}\left(1 + j\frac{\sigma}{2\omega\epsilon'}\right)}$$
(62b)

The conditions under which these approximations can be used depend on the desired accuracy, measured by how much the results deviate from those given by the exact formulas, (44) and (45). Deviations of no more than a few percent occur if $\sigma/\omega\epsilon' < 0.1$.

EXAMPLE 11.5

As a comparison, we repeat the computations of Example 11.4, using the approximation formulas (60a), (61), and (62b).

Solution. First, the loss tangent in this case is $\epsilon''/\epsilon' = 7/78 = 0.09$. Using (60), with $\epsilon'' = \sigma/\omega$, we have

$$\alpha \doteq \frac{\omega\epsilon''}{2}\sqrt{\frac{\mu}{\epsilon'}} = \frac{1}{2}(7 \times 8.85 \times 10^{12})(2\pi \times 2.5 \times 10^9)\frac{377}{\sqrt{78}} = 21\,\text{cm}^{-1}$$

We then have, using (61),

$$\beta \doteq (2\pi \times 2.5 \times 10^9)\sqrt{78}/(3 \times 10^8) = 464\,\text{rad/m}$$

Finally, with (62b),

$$\eta \doteq \frac{377}{\sqrt{78}}\left(1 + j\frac{7}{2 \times 78}\right) = 43 + j1.9$$

These results are identical (within the accuracy limitations as determined by the given numbers) to those of Example 11.4. Small deviations will be found, as the reader can verify by repeating the calculations of both examples and expressing the results to four or five significant figures. As we know, this latter practice would not be meaningful because the given parameters were not specified with such accuracy. Such is often the case, since measured values are not always known with high precision. Depending on how precise these values are, one can sometimes use a more relaxed judgment on when the approximation formulas can be used by allowing loss tangent values that can be larger than 0.1 (but still less than 1).

D11.4. Given a nonmagnetic material having $\epsilon_r' = 3.2$ and $\sigma = 1.5 \times 10^{-4}$ S/m, find numerical values at 3 MHz for the (a) loss tangent; (b) attenuation constant; (c) phase constant; (d) intrinsic impedance.

Ans. (a) 0.28; (b) 0.016 Np/m; (c) 0.11 rad/m; (d) 207 $\angle 7.8°$ Ω

D11.5. Consider a material for which $\mu_r = 1$, $\epsilon_r' = 2.5$, and the loss tangent is 0.12. If these three values are constant with frequency in the range 0.5 MHz $\leq f \leq$ 100 MHz, calculate: (a) σ at 1 and 75 MHz; (b) λ at 1 and 75 MHz; (c) v_p at 1 and 75 MHz.

Ans. (a) 1.67×10^{-5} and 1.25×10^{-3} S/m; (b) 190 and 2.53 m; (c) 1.90×108 m/s twice

11.3 POYNTING'S THEOREM AND WAVE POWER

In order to find the power flow associated with an electromagnetic wave, it is necessary to develop a power theorem for the electromagnetic field known as the Poynting theorem. It was originally postulated in 1884 by an English physicist, John H. Poynting.

The development begins with one of Maxwell's curl equations, in which we assume that the medium may be conductive:

$$\nabla \times \mathbf{H} = \mathbf{J} + \frac{\partial \mathbf{D}}{\partial t} \tag{63}$$

Next, we take the scalar product of both sides of (63) with $\mathbf{E}$,

$$\mathbf{E} \cdot \nabla \times \mathbf{H} = \mathbf{E} \cdot \mathbf{J} + \mathbf{E} \cdot \frac{\partial \mathbf{D}}{\partial t} \tag{64}$$

We then introduce the following vector identity, which may be proved by expansion in rectangular coordinates:

$$\nabla \cdot (\mathbf{E} \times \mathbf{H}) = -\mathbf{E} \cdot \nabla \times \mathbf{H} + \mathbf{H} \cdot \nabla \times \mathbf{E} \tag{65}$$

Using (65) in the left side of (64) results in

$$\mathbf{H} \cdot \nabla \times \mathbf{E} - \nabla \cdot (\mathbf{E} \times \mathbf{H}) = \mathbf{J} \cdot \mathbf{E} + \mathbf{E} \cdot \frac{\partial \mathbf{D}}{\partial t} \tag{66}$$

where the curl of the electric field is given by the other Maxwell curl equation:

$$\nabla \times \mathbf{E} = -\frac{\partial \mathbf{B}}{\partial t}$$

Therefore

$$-\mathbf{H} \cdot \frac{\partial \mathbf{B}}{\partial t} - \nabla \cdot (\mathbf{E} \times \mathbf{H}) = \mathbf{J} \cdot \mathbf{E} + \mathbf{E} \cdot \frac{\partial \mathbf{D}}{\partial t}$$

or

$$-\nabla \cdot (\mathbf{E} \times \mathbf{H}) = \mathbf{J} \cdot \mathbf{E} + \epsilon \mathbf{E} \cdot \frac{\partial \mathbf{E}}{\partial t} + \mu \mathbf{H} \cdot \frac{\partial \mathbf{H}}{\partial t} \tag{67}$$

The two time derivatives in (67) can be rearranged as follows:

$$\epsilon \mathbf{E} \cdot \frac{\partial \mathbf{E}}{\partial t} = \frac{\partial}{\partial t}\left(\frac{1}{2}\mathbf{D} \cdot \mathbf{E}\right) \tag{68a}$$

and

$$\mu \mathbf{H} \cdot \frac{\partial \mathbf{H}}{\partial t} = \frac{\partial}{\partial t}\left(\frac{1}{2}\mathbf{B} \cdot \mathbf{H}\right) \tag{68b}$$

With these, Eq. (67) becomes

$$\boxed{-\nabla \cdot (\mathbf{E} \times \mathbf{H}) = \mathbf{J} \cdot \mathbf{E} + \frac{\partial}{\partial t}\left(\frac{1}{2}\mathbf{D} \cdot \mathbf{E}\right) + \frac{\partial}{\partial t}\left(\frac{1}{2}\mathbf{B} \cdot \mathbf{H}\right)} \tag{69}$$

Finally, we integrate (69) throughout a volume:

$$-\int_{\text{vol}} \nabla \cdot (\mathbf{E} \times \mathbf{H})\, dv = \int_{\text{vol}} \mathbf{J} \cdot \mathbf{E}\, dv + \int_{\text{vol}} \frac{\partial}{\partial t}\left(\frac{1}{2}\mathbf{D} \cdot \mathbf{E}\right) dv + \int_{\text{vol}} \frac{\partial}{\partial t}\left(\frac{1}{2}\mathbf{B} \cdot \mathbf{H}\right) dv$$

The divergence theorem is then applied to the left-hand side, thus converting the volume integral there into an integral over the surface that encloses the volume. On the right-hand side, the operations of spatial integration and time differentiation are interchanged. The final result is

$$\boxed{-\oint_{\text{area}} (\mathbf{E} \times \mathbf{H}) \cdot d\mathbf{S} = \int_{\text{vol}} \mathbf{J} \cdot \mathbf{E}\, dv + \frac{d}{dt}\int_{\text{vol}} \frac{1}{2}\mathbf{D} \cdot \mathbf{E}\, dv + \frac{d}{dt}\int_{\text{vol}} \frac{1}{2}\mathbf{B} \cdot \mathbf{H}\, dv} \tag{70}$$

Equation (70) is known as Poynting's theorem. On the right-hand side, the first integral is the total (but instantaneous) ohmic power dissipated within the volume. The second integral is the total energy stored in the electric field, and the third integral is the stored energy in the magnetic field.[3] Since time derivatives are taken of the second and third integrals, those results give the time rates of increase of energy stored within the volume, or the instantaneous power going to increase the stored energy. The sum of the expressions on the right must therefore be the total power flowing *into* this volume, and so the total power flowing *out* of the volume is

$$\boxed{\oint_{\text{area}} (\mathbf{E} \times \mathbf{H}) \cdot d\mathbf{S} \quad \text{W}} \tag{71}$$

where the integral is over the closed surface surrounding the volume. The cross product $\mathbf{E} \times \mathbf{H}$ is known as the Poynting vector, $\mathbf{S}$,

$$\boxed{\mathbf{S} = \mathbf{E} \times \mathbf{H} \quad \text{W/m}^2} \tag{72}$$

which is interpreted as an instantaneous power density, measured in watts per square meter (W/m²). The direction of the vector $\mathbf{S}$ indicates the direction of the

---

[3] This is the expression for magnetic field energy that we have been anticipating since Chapter 8.

instantaneous power flow at a point, and many of us think of the Poynting vector as a "pointing" vector. This homonym, while accidental, is correct.[4]

Because **S** is given by the cross product of **E** and **H**, the direction of power flow at any point is normal to both the **E** and **H** vectors. This certainly agrees with our experience with the uniform plane wave, for propagation in the $+z$ direction was associated with an E_x and H_y component,

$$E_x \mathbf{a}_x \times H_y \mathbf{a}_y = S_z \mathbf{a}_z$$

In a perfect dielectric, the **E** and **H** field amplitudes are given by

$$E_x = E_{x0}\cos(\omega t - \beta z)$$
$$H_y = \frac{E_{x0}}{\eta}\cos(\omega t - \beta z)$$

where η is real. The power density amplitude is therefore

$$S_z = \frac{E_{x0}^2}{\eta}\cos^2(\omega t - \beta z) \tag{73}$$

In the case of a lossy dielectric, E_x and H_y are not in time phase. We have

$$E_x = E_{x0}e^{-\alpha z}\cos(\omega t - \beta z)$$

If we let

$$\eta = |\eta|\angle\theta_\eta$$

then we may write the magnetic field intensity as

$$H_y = \frac{E_{x0}}{|\eta|}e^{-\alpha z}\cos(\omega t - \beta z - \theta_\eta)$$

Thus,

$$S_z = E_x H_y = \frac{E_{x0}^2}{|\eta|}e^{-2\alpha z}\cos(\omega t - \beta z)\cos(\omega t - \beta z - \theta_\eta) \tag{74}$$

Because we are dealing with a sinusoidal signal, the time-average power density, $\langle S_z \rangle$, is the quantity that will ultimately be measured. To find this, we integrate (74) over one cycle and divide by the period $T = 1/f$. Additionally, the identity $\cos A \cos B = 1/2 \cos(A + B) + 1/2 \cos(A - B)$ is applied to the integrand, and we obtain:

$$\langle S_z \rangle = \frac{1}{T}\int_0^T \frac{1}{2}\frac{E_{x0}^2}{|\eta|}e^{-2\alpha z}[\cos(2\omega t - 2\beta z - 2\theta_\eta) + \cos\theta_\eta]dt \tag{75}$$

---

[4] Note that the vector symbol **S** is used for the Poynting vector, and is not to be confused with the differential area vector, $d\mathbf{S}$. The latter, as we know, is the product of the outward normal to the surface and the differential area.

The second-harmonic component of the integrand in (75) integrates to zero, leaving only the contribution from the dc component. The result is

$$\langle S_z \rangle = \frac{1}{2} \frac{E_{x0}^2}{|\eta|} e^{-2\alpha z} \cos\theta_\eta \tag{76}$$

Note that the power density attenuates as $e^{-2\alpha z}$, whereas E_x and H_y fall off as $e^{-\alpha z}$.

We may finally observe that the preceding expression can be obtained very easily by using the phasor forms of the electric and magnetic fields. In vector form, this is

$$\langle \mathbf{S} \rangle = \frac{1}{2} \mathscr{R}e(\mathbf{E}_s \times \mathbf{H}_s^*) \quad \text{W/m}^2 \tag{77}$$

In the present case

$$\mathbf{E}_s = E_{x0} e^{-j\beta z} \mathbf{a}_x$$

and

$$\mathbf{H}_s^* = \frac{E_{x0}}{\eta^*} e^{+j\beta z} \mathbf{a}_y = \frac{E_{x0}}{|\eta|} e^{j\theta} e^{+j\beta z} \mathbf{a}_y$$

where E_{x0} has been assumed real. Eq. (77) applies to any sinusoidal electromagnetic wave and gives both the magnitude and direction of the time-average power density.

D11.6. At frequencies of 1, 100, and 3000 MHz, the dielectric constant of ice made from pure water has values of 4.15, 3.45, and 3.20, respectively, while the loss tangent is 0.12, 0.035, and 0.0009, also respectively. If a uniform plane wave with an amplitude of 100 V/m at $z = 0$ is propagating through such ice, find the time-average power density at $z = 0$ and $z = 10$ m for each frequency.

Ans. 27.1 and 25.7 W/m^2; 24.7 and 6.31 W/m^2; 23.7 and 8.63 W/m^2

11.4 PROPAGATION IN GOOD CONDUCTORS

As an additional study of propagation with loss, we will investigate the behavior of a *good conductor* when a uniform plane wave is established in it. Such a material satisfies the general high-loss criterion, in which the loss tangent, $\epsilon''/\epsilon' \gg 1$. Applying this to a good conductor leads to the more specific criterion, $\sigma/(\omega\epsilon') \gg 1$. As before, we have an interest in losses that occur on wave transmission *into* a good conductor, and we will find new approximations for the phase constant, attenuation coefficient, and intrinsic impedance. New to us, however, is a modification of the basic problem, appropriate for good conductors. This concerns waves associated with electromagnetic fields existing in an external dielectric that adjoins the conductor surface; in this case, the waves propagate *along* the surface. That portion of the overall field that exists within the conductor will suffer dissipative loss arising from the conduction currents it generates. The overall field therefore attenuates with increasing distance of travel along the surface. This is the mechanism for the resistive transmission line loss that we studied in Chapter 10, and which is embodied in the line resistance parameter, R.

11.4.1 Good Conductor Approximations

As implied, a good conductor has a high conductivity and large conduction currents. The energy represented by the wave traveling through the material therefore decreases as the wave propagates because ohmic losses are continuously present. When we discussed the loss tangent, we saw that the ratio of conduction current density to the displacement current density in a conducting material is given by $\sigma/\omega\epsilon'$. Choosing a poor metallic conductor and a very high frequency as a conservative example, this ratio[5] for nichrome ($\sigma \doteq 10^6$) at 100 MHz is about 2×10^8. We therefore have a situation where $\sigma/\omega\epsilon' \gg 1$, and we should be able to make several very good approximations to find α, β, and η for a good conductor.

The general expression for the propagation constant is, from (59),

$$jk = j\omega \sqrt{\mu\epsilon'} \sqrt{1 - j\frac{\sigma}{\omega\epsilon'}}$$

which we immediately simplify to obtain

$$jk = j\omega \sqrt{\mu\epsilon'} \sqrt{-j\frac{\sigma}{\omega\epsilon'}}$$

or

$$jk = j\sqrt{-j\omega\mu\sigma}$$

But

$$-j = 1\angle-90°$$

and

$$\sqrt{1\angle-90°} = 1\angle-45° = \frac{1}{\sqrt{2}}(1 - j)$$

Therefore

$$jk = j(1 - j)\sqrt{\frac{\omega\mu\sigma}{2}} = (1 + j)\sqrt{\pi f\mu\sigma} = \alpha + j\beta \qquad (78)$$

Hence

$$\boxed{\alpha = \beta = \sqrt{\pi f\mu\sigma}} \qquad (79)$$

Regardless of the parameters μ and σ of the conductor or of the frequency of the applied field, α and β are equal. If we again assume only an E_x component traveling in the $+z$ direction, then

$$E_x = E_{x0}e^{-z\sqrt{\pi f\mu\sigma}}\cos(\omega t - z\sqrt{\pi f\mu\sigma}) \qquad (80)$$

---

[5] It is customary to take $\epsilon' = \epsilon_0$ for metallic conductors.

We may tie this field in the conductor to an external field at the conductor surface. We let the region $z > 0$ be the good conductor and the region $z < 0$ be a perfect dielectric. At the boundary surface $z = 0$, (80) becomes

$$E_x = E_{x0}\cos \omega t \qquad (z = 0)$$

We consider this to be the source field that establishes the fields within the conductor. Since displacement current is negligible,

$$\mathbf{J} = \sigma \mathbf{E}$$

Thus, the conduction current density at any point within the conductor is directly related to $\mathbf{E}$:

$$J_x = \sigma E_x = \sigma E_{x0}e^{-z\sqrt{\pi f\mu\sigma}}\cos(\omega t - z\sqrt{\pi f\mu\sigma}) \qquad (81)$$

11.4.2 Skin Effect

Equations (80) and (81) contain a wealth of information. Considering first the negative exponential term, we find an exponential decrease in the conduction current density and electric field intensity with penetration into the conductor (away from the source). The exponential factor is unity at $z = 0$ and decreases to $e^{-1} = 0.368$ when

$$z = \frac{1}{\sqrt{\pi f\mu\sigma}}$$

This distance is denoted by δ and is termed the *depth of penetration*, or the *skin depth*,

$$\boxed{\delta = \frac{1}{\sqrt{\pi f\mu\sigma}} = \frac{1}{\alpha} = \frac{1}{\beta}} \qquad (82)$$

It is an important parameter in describing conductor behavior in electromagnetic fields. To get some idea of the magnitude of the skin depth, let us consider copper, $\sigma = 5.8 \times 10^7$ S/m, at several different frequencies. We have

$$\delta_{\text{Cu}} = \frac{0.066}{\sqrt{f}}$$

At a power frequency of 60 Hz, $\delta_{\text{Cu}} = 8.53$ mm. Remembering that the power density carries an exponential term $e^{-2\alpha z}$, we see that the power density is multiplied by a factor of $0.368^2 = 0.135$ for every 8.53 mm of distance into the copper.

At a microwave frequency of 10,000 MHz, δ is 6.61×10^{-4} mm. Stated more generally, all fields in a good conductor such as copper are essentially zero at distances greater than a few skin depths from the surface. Any current density or electric field intensity established at the surface of a good conductor decays rapidly as we progress into the conductor. Electromagnetic energy is not transmitted in the interior of a conductor; it travels in the region surrounding the conductor, while the conductor merely guides the waves. We will consider guided propagation in more detail in Chapter 13.

Suppose we have a copper bus bar in the substation of an electric utility company which we wish to have carry large currents, and we therefore select dimensions of 2 by 4 inches. Then much of the copper is wasted, for the fields are greatly reduced in one skin depth, about 8.5 mm.[6] A hollow conductor with a wall thickness of about 12 mm would be a much better design. Although we are applying the results of an analysis for an infinite planar conductor to one of finite dimensions, the fields are attenuated in the finite-size conductor in a similar (but not identical) fashion.

The extremely short skin depth at microwave frequencies shows that only the surface coating of the guiding conductor is important. A piece of glass with an evaporated silver surface 3 μm thick is an excellent conductor at these frequencies.

Expressions for the velocity and wavelength within a good conductor can be found. From (82), we already have

$$\alpha = \beta = \frac{1}{\delta} = \sqrt{\pi f \mu \sigma}$$

Then, as

$$\beta = \frac{2\pi}{\lambda}$$

we find the wavelength to be

$$\lambda = 2\pi\delta \qquad (83)$$

Also, recalling that

$$v_p = \frac{\omega}{\beta}$$

we have

$$\boxed{v_p = \omega\delta} \qquad (84)$$

For copper at 60 Hz, $\lambda = 5.36$ cm and $v_p = 3.22$ m/s, or about 7.2 mi/h! A lot of us can run faster than that. In free space, of course, a 60 Hz wave has a wavelength of 3100 mi and travels at the velocity of light.

EXAMPLE 11.6

We again consider wave propagation in water, but this time, seawater. The primary difference between seawater and fresh water is of course the salt content. Sodium chloride dissociates in water to form Na^+ and Cl^- ions, which, being charged, will move when forced by an electric field. Seawater is in this manner conductive, and so it will attenuate electromagnetic waves by this mechanism. At frequencies in the vicinity of 10^7 Hz and below, the bound charge effects in water discussed earlier are negligible, and losses in seawater arise principally from the salt-associated conductivity. We consider an incident wave of frequency 1 MHz. We wish to find the skin depth, wavelength, and phase velocity. In seawater, $\sigma = 4$ S/m, and $\epsilon_r' = 81$.

---

[6] This utility company operates at 60 Hz.

Solution. We first evaluate the loss tangent, using the given data:

$$\frac{\sigma}{\omega\epsilon'} = \frac{4}{(2\pi \times 10^6)(81)(8.85 \times 10^{-12})} = 8.9 \times 10^2 \gg 1$$

Seawater is therefore a good conductor at 1 MHz (and at frequencies lower than this). The skin depth is

$$\delta = \frac{1}{\sqrt{\pi f \mu \sigma}} = \frac{1}{\sqrt{(\pi \times 10^6)(4\pi \times 10^{-7})(4)}} = 0.25 \text{ m} = 25 \text{ cm}$$

Now

$$\lambda = 2\pi\delta = 1.6 \text{ m}$$

and

$$v_p = \omega\delta = (2\pi \times 10^6)(0.25) = 1.6 \times 10^6 \text{ m/s}$$

In free space, these values would have been $\lambda = 300$ m and of course $v = c$.

With a 25-cm skin depth, it is obvious that radio frequency communication in seawater is quite impractical. Notice, however, that δ varies as $1/\sqrt{f}$, so that things will improve at lower frequencies. For example, if we use a frequency of 10 Hz (in the ELF, or extremely low frequency range), the skin depth is increased over that at 1 MHz by a factor of $\sqrt{10^6/10}$, so that

$$\delta(10 \text{ Hz}) \doteq 80 \text{ m}$$

The corresponding wavelength is $\lambda = 2\pi\delta \doteq 500$ m. Frequencies in the ELF range were used for many years in submarine communications. Signals were transmitted from gigantic ground-based antennas (required because the free-space wavelength associated with 10 Hz is 3×10^7 m). The signals were then received by submarines, from which a suspended wire antenna of length shorter than 500 m is sufficient. The drawback is that signal data rates at ELF are slow enough that a single word can take several minutes to transmit. Typically, ELF signals would be used to tell the submarine to initiate emergency procedures, or to come near the surface in order to receive a more detailed message via satellite.

11.4.3 Intrinsic Impedance and Power Density in Good Conductors

We next turn our attention to finding the magnetic field, H_y, associated with E_x. To do so, we need an expression for the intrinsic impedance of a good conductor. We begin with Eq. (48), Section 11.2, with $\epsilon'' = \sigma/\omega$,

$$\boxed{\eta = \sqrt{\frac{j\omega\mu}{\sigma + j\omega\epsilon'}}}$$

Since $\sigma \gg \omega\epsilon'$, we have

$$\eta = \sqrt{\frac{j\omega\mu}{\sigma}}$$

which may be written as

$$\eta = \frac{\sqrt{2}\angle 45^\circ}{\sigma\delta} = \frac{(1+j)}{\sigma\delta} \tag{85}$$

Thus, if we write (80) in terms of the skin depth,

$$E_x = E_{x0}e^{-z/\delta}\cos\left(\omega t - \frac{z}{\delta}\right) \tag{86}$$

then

$$H_y = \frac{\sigma\delta E_{x0}}{\sqrt{2}}e^{-z/\delta}\cos\left(\omega t - \frac{z}{\delta} - \frac{\pi}{4}\right) \tag{87}$$

and we see that the maximum amplitude of the magnetic field intensity occurs one-eighth of a cycle later than the maximum amplitude of the electric field intensity at every point.

From (86) and (87) we may obtain the time-average Poynting vector by applying (77),

$$\langle S_z \rangle = \frac{1}{2}\frac{\sigma\delta E_{x0}^2}{\sqrt{2}}e^{-2z/\delta}\cos\left(\frac{\pi}{4}\right)$$

or

$$\langle S_z \rangle = \frac{1}{4}\sigma\delta E_{x0}^2 e^{-2z/\delta}$$

We again note that in a distance of one skin depth the power density is only $e^{-2} = 0.135$ of its value at the surface.

11.4.4 Skin Effect Resistance in Conductors

We are now prepared to address the problem of frequency-dependent resistance in conductors, which is an important factor in the operation of transmission lines and waveguides. Consider a section of conductor having length L and width b, as shown in Figure 11.3, and which forms the entrance cross-section for a plane wave incident

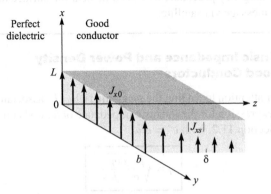

Figure 11.3   The current density $J_x = J_{x0}e^{-z/\delta}e^{-jz/\delta}$ decreases in magnitude as the wave propagates into the conductor. The average power loss in the region $0 < x < L$, $0 < y < b$, $z > 0$, is $\delta b L J_{x0}^2/4\sigma$ watts.

from $z < 0$. The incident electric field is x-polarized and thus generates current density $\mathbf{J}$ in that direction, which diminishes with increasing z as the wave attenuates. The total average power loss in a width $0 < y < b$ and length $0 < x < L$ in the direction of the current as shown and at depth z is obtained by finding the power crossing the conductor surface within this area,

$$P_L = \int_{\text{area}} \langle S_z \rangle da = \int_0^b \int_0^L \frac{1}{4}\sigma\delta E_{x0}^2 e^{-2z/\delta}\Big|_{z=0} dx\,dy = \frac{1}{4}\sigma\delta bL\,E_{x0}^2$$

In terms of the current density J_{x0} at the surface,

$$J_{x0} = \sigma E_{x0}$$

we have

$$\boxed{P_L = \frac{1}{4\sigma}\delta bL\,J_{x0}^2} \tag{88}$$

Now let us see what power loss would result if the *total* current in a width b were distributed *uniformly* in one skin depth. To find the total current, we integrate the current density over the infinite depth of the conductor,

$$I = \int_0^\infty \int_0^b J_x\,dy\,dz$$

where

$$J_x = J_{x0}e^{-z/\delta}\cos\left(\omega t - \frac{z}{\delta}\right)$$

or in complex exponential notation to simplify the integration,

$$J_{xs} = J_{x0}e^{-z/\delta}e^{-jz/\delta}$$
$$= J_{x0}e^{-(1+j)z/\delta}$$

Therefore,

$$I_s = \int_0^\infty \int_0^b J_{x0}e^{-(1+j)z/\delta}\,dy\,dz$$
$$= J_{x0}be^{-(1+j)z/\delta}\frac{-\delta}{1+j}\Big|_0^\infty$$
$$= \frac{J_{x0}b\delta}{1+j}$$

and

$$I = \frac{J_{x0}b\delta}{\sqrt{2}}\cos\left(\omega t - \frac{\pi}{4}\right)$$

If this current is distributed with a uniform density J' throughout the cross section $0 < y < b, 0 < z < \delta$, then

$$J' = \frac{J_{x0}}{\sqrt{2}}\cos\left(\omega t - \frac{\pi}{4}\right)$$

The ohmic power loss per unit volume is $\mathbf{J}\cdot\mathbf{E}$, and thus the total instantaneous power dissipated in the volume under consideration is

$$P_{Li}(t) = \frac{1}{\sigma}(J')^2 bL\delta = \frac{J_{x0}^2}{2\sigma}bL\delta\cos^2\left(\omega t - \frac{\pi}{4}\right)$$

The time-average power loss is easily obtained, since the average value of the cosine-squared factor is one-half,

$$P_L = \frac{1}{4\sigma} J_{x0}^2 bL\delta \qquad (89)$$

Comparing (88) and (89), we see that they are identical. Thus the average power loss in a conductor with skin effect present may be calculated by assuming that the total current is distributed uniformly in one skin depth. In terms of resistance, we may say that the resistance of a width b and length L of an infinitely thick slab with skin effect is the same as the resistance of a rectangular slab of width b, length L, and thickness δ without skin effect, or with uniform current distribution.

We may apply this to a conductor of circular cross section with little error, provided that the radius a is much greater than the skin depth. The resistance at a high frequency where there is a well-developed skin effect is therefore found by considering a slab of width equal to the circumference $2\pi a$ and thickness δ. Hence

$$R = \frac{L}{\sigma S} = \frac{L}{2\pi a\sigma\delta} \qquad (90)$$

A round copper wire of 1 mm radius and 1 km length has a resistance at direct current of

$$R_{dc} = \frac{10^3}{\pi 10^{-6}(5.8 \times 10^7)} = 5.48 \ \Omega$$

At 1 MHz, the skin depth is 0.066 mm. Thus $\delta \ll a$, and the resistance at 1 MHz is found by (90),

$$R = \frac{10^3}{2\pi 10^{-3}(5.8 \times 10^7)(0.066 \times 10^{-3})} = 41.5 \ \Omega$$

D11.7. A steel pipe is constructed of a material for which $\mu_r = 180$ and $\sigma = 4 \times 10^6$ S/m. The two radii are 5 and 7 mm, and the length is 75 m. If the total current $I(t)$ carried by the pipe is 8 cos ωt A, where $\omega = 1200\pi$ rad/s, find: (a) the skin depth; (b) the effective resistance; (c) the dc resistance; (d) the time-average power loss.

Ans. (a) 0.766 mm; (b) 0.557 Ω; (c) 0.249 Ω; (d) 17.82 W

11.5 WAVE POLARIZATION

In the previous sections, we have treated uniform plane waves in which the electric and magnetic field vectors were assumed to lie in fixed directions. Specifically, with the wave propagating along the z axis, **E** was taken to lie along x, which then required **H** to lie along y. This orthogonal relationship between **E**, **H**, and **S** is always true for a uniform plane wave. The directions of **E** and **H** within the plane perpendicular to $\mathbf{a}_z$ may change, however, as functions of time and position, depending on how the wave was generated or on what type of medium it is propagating through. Thus a complete description of an electro-magnetic wave would not only include parameters such as its wavelength, phase velocity, and power, but also a statement of the instantaneous orientation of its field vectors. We define the *wave polarization* as the time-dependent electric field vector orientation at a fixed point in space. A more complete characterization of a wave's polarization would in fact include specifying the field orientation at *all* points because some waves demonstrate spatial variations in their polarization. Specifying only the electric field direction is sufficient, since magnetic field is readily found from **E** using Maxwell's equations.

11.5.1 Linear Polarization

In the waves we have previously studied, **E** was in a fixed straight orientation for all times and positions. Such a wave is said to be *linearly polarized*. We have taken **E** to lie along the x axis, but the field could be oriented in any fixed direction in the xy plane and be linearly polarized. For positive z propagation, the wave would in general have its electric field phasor expressed as

$$\mathbf{E}_s = (E_{x0}\mathbf{a}_x + E_{y0}\mathbf{a}_y)e^{-\alpha z}e^{-j\beta z} \tag{91}$$

where E_{x0} and E_{y0} are constant amplitudes along x and y. The magnetic field is readily found by determining its x and y components directly from those of $\mathbf{E}_s$. Specifically, $\mathbf{H}_s$ for the wave of Eq. (91) is

$$\mathbf{H}_s = [H_{x0}\mathbf{a}_x + H_{y0}\mathbf{a}_y]e^{-\alpha z}e^{-j\beta z} = \left[-\frac{E_{y0}}{\eta}\mathbf{a}_x + \frac{E_{x0}}{\eta}\mathbf{a}_y\right]e^{-\alpha z}e^{-j\beta z} \tag{92}$$

The two fields are sketched in Figure 11.4. The figure demonstrates the reason for the minus sign in the term involving E_{y0} in Eq. (92). The direction of power flow, given by $\mathbf{E} \times \mathbf{H}$, is in the positive z direction in this case. A component of **E** in the

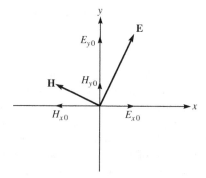

Figure 11.4 Electric and magnetic field configuration for a general linearly polarized plane wave propagating in the forward z direction (out of the page). Field components correspond to those in Eqs. (91) and (92).

positive y direction would require a component of **H** in the negative x direction—thus the minus sign. Using (91) and (92), the power density in the wave is found using (77):

$$\langle \mathbf{S}_z \rangle = \frac{1}{2} = \mathcal{R}e\{\mathbf{E}_s \times \mathbf{H}_s^*\} = \frac{1}{2}\mathcal{R}e\{E_{x0}H_{y0}^*(\mathbf{a}_x \times \mathbf{a}_y) + E_{y0}H_{x0}^*(\mathbf{a}_y \times \mathbf{a}_x)\}e^{-2\alpha z}$$

$$= \frac{1}{2}\mathcal{R}e\left\{\frac{E_{x0}E_{x0}^*}{\eta^*} + \frac{E_{y0}E_{y0}^*}{\eta^*}\right\}e^{-2\alpha z}\mathbf{a}_z$$

$$= \frac{1}{2}\mathcal{R}e\left\{\frac{1}{\eta^*}\right\}(|E_{x0}|^2 + |E_{y0}|^2)e^{-2\alpha z}\mathbf{a}_z \ \text{W/m}^2$$

This result demonstrates the idea that our linearly polarized plane wave can be considered as two distinct plane waves having x and y polarizations, whose electric fields are combining *in phase* to produce the total **E**. The same is true for the magnetic field components. This is a critical point in understanding wave polarization, in that *any polarization state can be described in terms of mutually perpendicular components of the electric field and their relative phasing.*

11.5.2 Phase-Displaced Field Components: Elliptical Polarization

We next consider the effect of a phase difference, ϕ, between E_{x0} and E_{y0}, where $\phi < \pi/2$. For simplicity, we will consider propagation in a lossless medium. The total field in phasor form is

$$\boxed{\mathbf{E}_s = (E_{x0}\mathbf{a}_x + E_{y0}e^{j\phi}\mathbf{a}_y)e^{-j\beta z}} \tag{93}$$

Again, to aid in visualization, we convert this wave to real instantaneous form by multiplying by $e^{j\omega t}$ and taking the real part:

$$\boxed{\mathbf{E}(z,t) = E_{x0}\cos(\omega t - \beta z)\mathbf{a}_x + E_{y0}\cos(\omega t - \beta z + \phi)\mathbf{a}_y} \tag{94}$$

where we have assumed that E_{x0} and E_{y0} are real. Suppose we set $t = 0$, in which case (94) becomes [using $\cos(-x) = \cos(x)$]

$$\mathbf{E}(z,0) = E_{x0}\cos(\beta z)\mathbf{a}_x + E_{y0}\cos(\beta z - \phi)\mathbf{a}_y \tag{95}$$

The component magnitudes of $\mathbf{E}(z, 0)$ are plotted as functions of z in Figure 11.5. Since time is fixed at zero, the wave is frozen in position. An observer can move along the z axis, measuring the component magnitudes and thus the orientation of the total electric field at each point. Let's consider a crest of E_x, indicated as point a in Figure 11.5. If ϕ were zero, E_y would have a crest at the same location. Since ϕ is not zero (and positive), the crest of E_y that would otherwise occur at point a is now displaced to point b farther down z. The two points are separated by distance ϕ/β. E_y thus *lags behind* E_x when we consider the *spatial* dimension.

Now suppose the observer stops at some location on the z axis, and time is allowed to move forward. Both fields now move in the positive z direction, as (94) indicates. But point b reaches the observer first, followed by point a. So we see that E_y leads E_x when we consider the *time* dimension. In either case (fixed t and varying z, or vice versa) the observer notes that the net field rotates about the z axis while

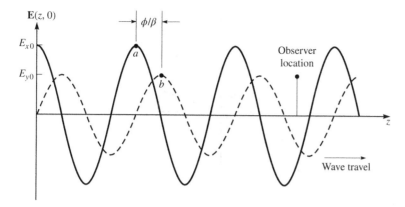

Figure 11.5 Plots of the electric field component magnitudes in Eq. (95) as functions of z. Note that the y component lags behind the x component in z. As time increases from zero, both waves travel to the right, as per Eq. (94). Thus, to an observer at a fixed location, the y component leads in time.

its magnitude changes. Considering a starting point in z and t, at which the field has a given orientation and magnitude, the wave will return to the same orientation and magnitude at a distance of one wavelength in z (for fixed t) or at a time $t = 2\pi/\omega$ later (at a fixed z).

For illustration purposes, if we take the length of the field vector as a measure of its magnitude, we find that at a fixed position, the tip of the vector traces out the shape of an ellipse over time $t = 2\pi/\omega$. The wave is said to be *elliptically polarized*. Elliptical polarization is in fact the most general polarization state of a wave, since it encompasses any magnitude and phase difference between E_x and E_y. Linear polarization is a special case of elliptical polarization in which the phase difference is zero.

11.5.3 Circular Polarization

Another special case of elliptical polarization occurs when $E_{x0} = E_{y0} = E_0$ and when $\phi = \pm\pi/2$. The wave in this case exhibits *circular polarization*. To see this, we incorporate these restrictions into Eq. (94) to obtain

$$\mathbf{E}(z,t) = E_0[\cos(\omega t - \beta z)\mathbf{a}_x + \cos(\omega t - \beta z \pm \pi/2)\mathbf{a}_y]$$
$$= E_0[\cos(\omega t - \beta z)\mathbf{a}_x \mp \sin(\omega t - \beta z)\mathbf{a}_y] \qquad (96)$$

If we consider a fixed position along z (such as $z = 0$) and allow time to vary, (96), with $\phi = +\pi/2$, becomes

$$\boxed{\mathbf{E}(0,t) = E_0[\cos(\omega t)\mathbf{a}_x - \sin(\omega t)\mathbf{a}_y]} \qquad (97)$$

If we choose $-\pi/2$ in (96), we obtain

$$\boxed{\mathbf{E}(0,t) = E_0[\cos(\omega t)\mathbf{a}_x + \sin(\omega t)\mathbf{a}_y]} \qquad (98)$$

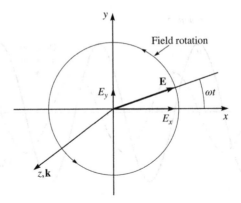

Figure 11.6 Electric field in the *xy* plane of a right circularly polarized plane wave, as described by Eq. (98). As the wave propagates in the forward *z* direction, the field vector rotates counterclockwise in the *xy* plane.

The field vector of Eq. (98) rotates in the counterclockwise direction in the *xy* plane, while maintaining constant amplitude E_0, and so the tip of the vector traces out a circle. Figure 11.6 shows this behavior.

Choosing $+\pi/2$ leads to (97), whose field vector rotates in the clockwise direction. The *handedness* of the circular polarization is associated with the rotation and propagation directions in the following manner: The wave exhibits *left circular polarization* (l.c.p.) if, when orienting the left hand with the thumb in the direction of propagation, the fingers curl in the rotation direction of the field with time. The wave exhibits *right circular polarization* (r.c.p.) if, with the right-hand thumb in the propagation direction, the fingers curl in the field rotation direction.[7] Thus, with forward *z* propagation, (97) describes a left circularly polarized wave, and (98) describes a right circularly polarized wave. The same convention is applied to elliptical polarization, in which the descriptions *left elliptical polarization* and *right elliptical polarization* are used.

Using (96), the instantaneous angle of the field from the *x* direction can be found for any position along *z* through

$$\theta(z,t) = \tan^{-1}\left(\frac{E_y}{E_x}\right) = \tan^{-1}\left(\frac{\mp\sin(\omega t - \beta z)}{\cos(\omega t - \beta z)}\right) = \mp(\omega t - \beta z) \tag{99}$$

where again the minus sign (yielding l.c.p. for positive *z* travel) applies for the choice of $\phi = +\pi/2$ in (96); the plus sign (yielding r.c.p. for positive *z* travel) is used if

---

[7] This convention is reversed by some workers (most notably in optics) who emphasize the importance of the *spatial* field configuration. Note that r.c.p. by our definition is formed by propagating a spatial field that is in the shape of a *left-handed* screw, and for that reason it is sometimes called left circular polarization (see Figure 11.7). Left circular polarization as we define it results from propagating a spatial field in the shape of a right-handed screw, and it is called right circular polarization by the spatial enthusiasts. Caution is obviously necessary in interpreting what is meant when polarization handedness is stated in an unfamiliar text.

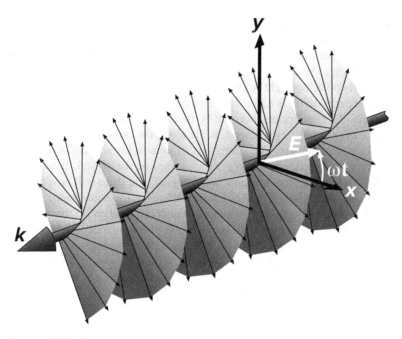

Figure 11.7 Representation of a right circularly polarized wave.
The electric field vector (in white) will rotate toward the y axis as the
entire wave moves through the xy plane in the direction of **k**. This
counterclockwise rotation (when looking toward the wave source) satisfies
the temporal right-handed rotation convention as described in the text.
The wave, however, appears as a left-handed screw, and for this reason it
is called left circular polarization in the other convention.

$\phi = -\pi/2$. If we choose $z = 0$, the angle becomes simply ωt, which reaches 2π (one complete rotation) at time $t = 2\pi/\omega$. If we choose $t = 0$ and allow z to vary, we form a corkscrew-like field pattern. One way to visualize this is to consider a spiral staircase–shaped pattern, in which the field lines (stairsteps) are perpendicular to the z (or staircase) axis. The relationship between this spatial field pattern and the resulting time behavior at fixed z as the wave propagates is shown in an artist's conception in Figure 11.7.

The handedness of the polarization is changed by reversing the pitch of the corkscrew pattern. The spiral staircase model is only a visualization aid. It must be remembered that the wave is still a uniform plane wave whose fields at any position along z are infinite in extent over the transverse plane.

There are many uses of circularly polarized waves. Perhaps the most obvious advantage is that reception of a wave having circular polarization does not depend on the antenna orientation in the plane normal to the propagation direction. Dipole antennas, for example, are required to be oriented along the electric field direction of the signal they receive. If circularly polarized signals are transmitted, the receiver orientation requirements are relaxed considerably. In optics, circularly polarized light can be passed through a polarizer of any orientation, thus yielding linearly polarized light in any direction (although one loses half the original power this way). Other uses involve treating linearly polarized light as a superposition of circularly polarized waves, to be described next.

Circularly polarized light can be generated using an *anisotropic* medium—a material whose permittivity is a function of electric field direction. Many crystals have this property. A crystal orientation can be found such that along one direction (say, the x axis), the permittivity is lowest, while along the orthogonal direction (y axis), the permittivity is highest. The strategy is to input a linearly polarized wave with its field vector at 45 degrees to the x and y axes of the crystal. It will thus have equal-amplitude x and y components in the crystal, and these will now propagate in the z direction at different speeds. A phase difference (or *retardation*) accumulates between the components as they propagate, which can reach $\pi/2$ if the crystal is long enough. The wave at the output thus becomes circularly polarized. Such a crystal, cut to the right length and used in this manner, is called a *quarter-wave plate,* since it introduces a relative phase shift of $\pi/2$ between E_x and E_y, which is equivalent to $\lambda/4$.

It is useful to express circularly polarized waves in phasor form. To do this, we note that (96) can be expressed as

$$\mathbf{E}(z,t) = \mathcal{R}e\left\{E_0 e^{j\omega t} e^{-j\beta z}\left[\mathbf{a}_x + e^{\pm j\pi/2}\mathbf{a}_y\right]\right\}$$

Using the fact that $e^{\pm j\pi/2} = \pm j$, we identify the phasor form as:

$$\boxed{\mathbf{E}_s = E_0(\mathbf{a}_x \pm j\mathbf{a}_y)\,e^{-j\beta z}} \tag{100}$$

where the plus sign is used for left circular polarization and the minus sign for right circular polarization. If the wave propagates in the negative z direction, we have

$$\boxed{\mathbf{E}_s = E_0(\mathbf{a}_x \pm j\mathbf{a}_y)e^{+j\beta z}} \tag{101}$$

where in this case the positive sign applies to right circular polarization and the minus sign to left circular polarization. The student is encouraged to verify this.

EXAMPLE 11.7

Let us consider the result of superimposing left and right circularly polarized fields of the same amplitude, frequency, and propagation direction, but where a phase shift of δ radians exists between the two.

Solution. Taking the waves to propagate in the $+z$ direction, and introducing a relative phase, δ, the total phasor field is found, using (100):

$$\mathbf{E}_{sT} = \mathbf{E}_{sR} + \mathbf{E}_{sL} = E_0[\mathbf{a}_x - j\mathbf{a}_y]e^{-j\beta z} + E_0[\mathbf{a}_x + j\mathbf{a}_y]e^{-j\beta z}e^{j\delta}$$

Grouping components together, this becomes

$$\mathbf{E}_{sT} = E_0[(1 + e^{j\delta})\,\mathbf{a}_x - j(1 - e^{j\delta})\mathbf{a}_y]e^{-j\beta z}$$

Factoring out an overall phase term, $e^{j\delta/2}$, we obtain

$$\mathbf{E}_{sT} = E_0 e^{j\delta/2}\left[(e^{-j\delta/2} + e^{j\delta/2})\mathbf{a}_x - j(e^{-j\delta/2} - e^{j\delta/2})\mathbf{a}_y\right]e^{-j\beta z}$$

From Euler's identity, we find that $e^{j\delta/2} + e^{-j\delta/2} = 2\cos\delta/2$, and $e^{j\delta/2} - e^{-j\delta/2} = 2j\sin\delta/2$. Using these relations, we obtain

$$\mathbf{E}_{sT} = 2E_0[\cos(\delta/2)\mathbf{a}_x + \sin(\delta/2)\mathbf{a}_y]e^{-j(\beta z - \delta/2)} \tag{102}$$

We recognize (102) as the electric field of a *linearly polarized* wave, whose field vector is oriented at angle $\delta/2$ from the x axis.

Example 11.7 shows that any linearly polarized wave can be expressed as the sum of two circularly polarized waves of opposite handedness, where the linear polarization direction is determined by the relative phase difference between the two waves. Such a representation is convenient (and necessary) when considering, for example, the propagation of linearly polarized light through media which contain organic molecules. These often exhibit spiral structures having left- or right-handed pitch, and they will thus interact differently with left- or right-hand circular polarization. As a result, the left circular component can propagate at a different speed than the right circular component, and so the two waves will accumulate a phase difference as they propagate. As a result, the direction of the linearly polarized field vector at the output of the material will differ from the direction that it had at the input. The extent of this rotation can be used as a measurement tool to aid in material studies.

Polarization issues will become extremely important when we consider wave reflection in Chapter 12.

REFERENCES

1. Balanis, C. A. *Advanced Engineering Electromagnetics*. New York: John Wiley & Sons, 1989.
2. International Telephone and Telegraph Co., Inc. *Reference Data for Radio Engineers*. 7th ed. Indianapolis, Ind.: Howard W. Sams & Co., 1985. This handbook has some excellent data on the properties of dielectric and insulating materials.
3. Jackson, J. D. *Classical Electrodynamics*. 3d ed. New York: John Wiley & Sons, 1999.
4. Ramo, S., J. R. Whinnery, and T. Van Duzer. *Fields and Waves in Communication Electronics*. 3d ed. New York: John Wiley & Sons, 1994.

CHAPTER 11 PROBLEMS

Quizzes

11.1 Show that $E_{xs} = Ae^{j(k_0 z + \phi)}$ is a solution of the vector Helmholtz equation, Eq. (30), for $k_0 = \omega\sqrt{\mu_0\epsilon_0}$ and any ϕ and A.

11.2 A 20 GHz uniform plane wave propagates in the forward z direction in a lossless medium for which $\epsilon_r = \mu_r = 3$. Find (a) v_p; (b) β; (c) λ; (d) $\mathbf{E}_s$; (e) $\mathbf{H}_s$; (f) $\langle\mathbf{S}\rangle$.

11.3 An **H** field in free space is given as $\mathcal{H}(x, t) = 10\cos(10^8 t - \beta x)\mathbf{a}_y$ A/m. Find (a) β; (b) λ; (c) $\mathcal{E}(x, t)$ at $P(0.1, 0.2, 0.3)$ at $t = 1$ ns.

11.4 Small antennas have low efficiencies (as will be seen in Chapter 14), and the efficiency increases with size up to the point at which a critical dimension of the antenna is an appreciable fraction of a wavelength, say $\lambda/8$. (a) An antenna that is 12 cm long is operated in air at 1 MHz. What fraction of a wavelength long is it? (b) The same antenna is embedded in a ferrite material for which $\epsilon_r = 20$ and $\mu_r = 2000$. What fraction of a wavelength is it now?

11.5 Consider two x-polarized waves that counter-propagate along the z axis. The wave traveling in the forward z direction is of frequency ω_2 ; the backward z propagating wave is at frequency ω_1, where $\omega_1 < \omega_2$. Both frequencies are very slightly detuned on either side of their mean frequency, ω_0, such that $\omega_0 - \omega_1 = \omega_2 - \omega_0 << \omega_0$. Using the complex field forms, construct the expression for the total electric field, and from this, find the power density distribution (proportional to EE*). Your answer should be in the form of a "standing wave" that in fact moves slowly along the z axis. Find an expression for the velocity of the wave pattern in terms of given or known parameters. Does the pattern move in the forward or backward z direction?

11.6 A uniform plane wave has electric field $\mathbf{E}_s = (E_{y0}\,\mathbf{a}_y - E_{z0}\,\mathbf{a}_z)e^{-\alpha x}e^{-j\beta x}$ V/m. The intrinsic impedance of the medium is given as $\eta = |\eta|\,e^{j\phi}$, where ϕ is a constant phase. (a) Describe the wave polarization and state the direction of propagation. (b) Find $\mathbf{H}_s$. (c) Find $\varepsilon(x, t)$ and $\mathcal{H}(x, t)$. (d) Find $<\mathbf{S}>$ in W/m^2. (e) Find the time-average power in watts that is intercepted by an antenna of rectangular cross-section, having width w and height h, suspended parallel to the yz plane, and at a distance d from the wave source.

11.7 Express in terms of the penetration depth, δ, the distance into a lossy medium at which wave power has attenuated by (a) 1 dB; (b) 3 dB; (c) 30 dB.

11.8 An electric field in free space is given in spherical coordinates as $\mathbf{E}_s(r) = E_0(r)e^{-jkr}\,\mathbf{a}_\theta$ V/m. (a) Find $\mathbf{H}_s(r)$ assuming uniform plane wave behavior. (b) Find $<\mathbf{S}>$. (c) Express the average outward power in watts through a closed spherical shell of radius r, centered at the origin. (d) Establish the required functional form of $E_0(r)$ that will enable the power flow in part c to be independent of radius. With this condition met, the given field becomes that of an *isotropic radiator* in a lossless medium (radiating equal power density in all directions).

11.9 An example of a *nonuniform* plane wave is a surface or *evanescent* wave, an example of which, in phasor form, is shown here, exhibiting diminishing amplitude with x while propagating in the forward z direction:

$$\mathbf{E}_s(z) = E_0 e^{-\alpha x}e^{-j\beta z}\mathbf{a_y}$$

Such fields form part of the mode structure in dielectric waveguides, as will be explored in Chapter 13. (a) Assuming free-space conditions, use Eq. (23) to find the associated magnetic field, $\mathbf{H}_s(z)$ (note that there will be two components). (b) Find α as a function of β, ω, ϵ_0, and μ_0, such that all of Maxwell's equations are satisfied by the electric and magnetic fields.

11.10 In a medium characterized by intrinsic impedance $\eta = |\eta|e^{j\phi}$, a linearly polarized plane wave propagates, with magnetic field given as $\mathbf{H}_s = (H_{0y}\mathbf{a}_y + H_{0z}\mathbf{a}_z)e^{-\alpha x}e^{-j\beta x}$. Find (a) $\mathbf{E}_s$; (b) $\mathscr{E}(x, t)$; (c) $\mathcal{H}(x, t)$; (d) $\langle\mathbf{S}\rangle$.

11.11 A uniform plane wave at frequency $f = 100$ MHz propagates in a material having conductivity $\sigma = 3.0$ S/m and dielectric constant $\epsilon_r' = 8.00$. The wave carries electric field amplitude $E_0 = 100$ V/m. (a) Calculate the loss tangent and determine whether the medium would qualify as a good dielectric or a good conductor. (b) Calculate α, β, and η. (c) Write the electric field in phasor form, assuming x polarization and forward z travel. (d) Write the magnetic field in phasor form. (e) Write the time-average Poynting vector, $\mathbf{S}$. (f) Find the 6-dB material thickness at which the wave power drops to 25% of its value on entry.

11.12 Repeat Problem 11.11, except the wave now propagates in fresh water, having conductivity $\sigma = 10^{-3}$ S/m, dielectric constant $\epsilon_r' = 80.0$, and permeability μ_0.

11.13 Let $jk = 0.2 + j1.5$ m^{-1} and $\eta = 450 + j60$ Ω for a uniform plane propagating in the $\mathbf{a}_z$ direction. If $\omega = 300$ Mrad/s, find μ, ϵ', and ϵ'' for the medium.

11.14 Describe how the attenuation coefficient of a liquid medium, assumed to be a good conductor, could be determined through measurement of wavelength in the liquid at a known frequency. What restrictions apply? Could this method be used to find the conductivity as well?

11.15 A 10-GHz radar signal may be represented as a uniform plane wave in a sufficiently small region. Calculate the wavelength in centimeters and the attenuation in nepers per meter if the wave is propagating in a nonmagnetic material for which (a) $\epsilon_r' = 1$ and $\epsilon_r'' = 0$; (b) $\epsilon_r' = 1.04$ and $\epsilon_r'' = 9.00 \times 10^{-4}$; (c) $\epsilon_r' = 2.5$ and $\epsilon_r'' = 7.2$.

11.16 Consider the power dissipation term, $\int \mathbf{E} \cdot \mathbf{J} dv$, in Poynting's theorem [Eq. (70)]. This gives the power lost to heat within a volume into which electromagnetic waves enter. The term $p_d = \mathbf{E} \cdot \mathbf{J}$ is thus the power dissipation per unit volume in W/m^3. Following the same reasoning that resulted in Eq. (77), the time-average power dissipation per volume will be $\langle p_d \rangle = (1/2)\mathscr{Re}\{\mathbf{E}_s \cdot \mathbf{J}_s^*\}$. (a) Show that in a conducting medium, through which a uniform plane wave of amplitude E_0 propagates in the forward z direction, $\langle p_d \rangle = (\sigma/2)|E_0|^2 e^{-2\alpha z}$. (b) Confirm this result for the special case of a good conductor by using the left-hand side of Eq. (70), and consider a very small volume.

11.17 Consider a long solid cylindrical wire having uniform conductivity σ. The wire is oriented along the z axis and has radius a and length L. DC voltage V_0 is applied between the two ends. (a) Write the electric field intensity, **E**, in terms of V_0 and L. (*b*) Using Ampere's circuital law, find **H** inside the wire. (*c*) Find the Poynting vector, **S**. (*d*) Evaluate the left side of Poynting's theorem by integrating the **S** over the wire surface. (*e*) Evaluate the right side of Poynting's theorem (specialized for this non-time-varying case) and thus verify that the theorem is satisfied for this situation.

11.18 Given a 100-MHz uniform plane wave in a medium known to be a good dielectric, the phasor electric field is $\mathscr{E}_s = 4e^{-0.5z}e^{-j20z}\mathbf{a}_x$ V/m. Determine (*a*) ϵ'; (*b*) ϵ''; (*c*) η; (*d*) $\mathbf{H}_s$; (*e*) $\langle\mathbf{S}\rangle$; (*f*) the power in watts that is incident on a rectangular surface measuring 20 m × 30 m at $z = 10$ m.

11.19 Perfectly conducting cylinders with radii of 8 mm and 20 mm are coaxial. The region between the cylinders is filled with a perfect dielectric for which $\epsilon = 10^{-9}/4\pi$ F/m and $\mu_r = 1$. If ϵ in this region is $(500/\rho)\cos(\omega t - 4z)\mathbf{a}_\rho$ V/m, find (*a*) ω, with the help of Maxwell's equations in cylindrical coordinates; (*b*) $\mathcal{H}(\rho, z, t)$; (*c*) $\langle\mathbf{S}(\rho, z, t)\rangle$; (*d*) the average power passing through every cross section $8 < \rho < 20$ mm, $0 < \phi < 2\pi$.

11.20 Voltage breakdown in air at standard temperature and pressure occurs at an electric field strength of approximately 3×10^6 V/m. This becomes an issue in some high-power optical experiments, in which tight focusing of light may be necessary. Estimate the lightwave power in watts that can be focused into a cylindrical beam of 10 μm radius before breakdown occurs. Assume uniform plane wave behavior (although this assumption will produce an answer that is higher than the actual number by as much as a factor of 2, depending on the actual beam shape).

11.21 The cylindrical shell, 1 cm $< \rho <$ 1.2 cm, is composed of a conducting material for which $\sigma = 10^6$ S/m. The external and internal regions are nonconducting. Let $H_\phi = 2000$ A/m at $\rho = 1.2$ cm. Find (*a*) **H** everywhere; (*b*) **E** everywhere; (*c*) $\langle\mathbf{S}\rangle$ everywhere.

11.22 The inner and outer dimensions of a coaxial copper transmission line are 2 and 7 mm, respectively. Both conductors have thicknesses much greater than δ. The dielectric is lossless and the operating frequency is 400 MHz. Calculate the resistance per meter length of the (*a*) inner conductor; (*b*) outer conductor; (*c*) transmission line.

11.23 A hollow tubular conductor is constructed from a type of brass having a conductivity of 1.2×10^7 S/m. The inner and outer radii are 9 and 10 mm, respectively. Calculate the resistance per meter length at a frequency of (*a*) dc; (*b*) 20 MHz; (*c*) 2 GHz.

11.24 (*a*) Most microwave ovens operate at 2.45 GHz. Assume that $\sigma = 1.2 \times 10^6$ S/m and $\mu_r = 500$ for the stainless steel interior, and find the depth of penetration. (*b*) Let $E_s = 50\angle 0°$ V/m at the surface of the conductor, and plot a curve of the amplitude of E_s versus the angle of E_s as the field propagates into the stainless steel.

11.25 A good conductor is planar in form, and it carries a uniform plane wave that has a wavelength of 0.3 mm and a velocity of 3×10^5 m/s. Assuming the conductor is nonmagnetic, determine the frequency and the conductivity.

11.26 The dimensions of a certain coaxial transmission line are $a = 0.8$ mm and $b = 4$ mm. The outer conductor thickness is 0.6 mm, and all conductors have $\sigma = 1.6 \times 10^7$ S/m. (a) Find R, the resistance per unit length at an operating frequency of 2.4 GHz. (b) Use information from Sections 6.3 and 8.10 to find C and L, the capacitance and inductance per unit length, respectively. The coax is air-filled. (c) Find α and β if $\alpha + j\beta = \sqrt{j\omega C(R + j\omega L)}$.

11.27 The planar surface $z = 0$ is a brass–Teflon interface. Use data available in Appendix C to evaluate the following ratios for a uniform plane wave having $\omega = 4 \times 10^{10}$ rad/s: (a) $\alpha_{Tef}/\alpha_{brass}$; (b) $\lambda_{Tef}/\lambda_{brass}$; (c) v_{Tef}/v_{brass}.

11.28 A uniform plane wave in free space has electric field vector given by $\mathbf{E}_s = 10e^{-j\beta x}\mathbf{a}_z + 15e^{-j\beta x}\mathbf{a}_y$ V/m. (a) Describe the wave polarization. (b) Find $\mathbf{H}_s$. (c) Determine the average power density in the wave in W/m^2.

11.29 Consider a left circularly polarized wave in free space that propagates in the forward z direction. The electric field is given by the appropriate form of Eq. (100). Determine (a) the magnetic field phasor, $\mathbf{H}_s$; (b) an expression for the average power density in the wave in W/m^2 by direct application of Eq. (77).

11.30 In an *anisotropic* medium, permittivity varies with electric field *direction*, and is a property seen in most crystals. Consider a uniform plane wave propagating in the z direction in such a medium, and which enters the material with equal field components along the x and y axes. The field phasor will take the form:

$$\mathbf{E}_s(z) = E_0(\mathbf{a}_x + \mathbf{a}_y\, e^{j\Delta\beta z})e^{-j\beta z}$$

where $\Delta\beta = \beta_x - \beta_y$ is the difference in phase constants for waves that are linearly polarized in the x and y directions. Find distances into the material (in terms of $\Delta\beta$) at which the field is (a) linearly polarized and (b) circularly polarized. (c) Assume intrinsic impedance η that is approximately constant with field orientation and find $\mathbf{H}_s$ and $< \mathbf{S} >$.

11.31 A linearly polarized uniform plane wave, propagating in the forward z direction, is input to a lossless anisotropic material, in which the dielectric constant encountered by waves polarized along $y(\epsilon_{ry})$ differs from that seen by waves polarized along $x(\epsilon_{rx})$. Suppose $\epsilon_{rx} = 2.15$, $\epsilon_{ry} = 2.10$, and the wave electric field at input is polarized at 45° to the positive x and y axes. (a) Determine, in terms of the free-space wavelength λ, the shortest length of the material, such that the wave, as it emerges from the output, is circularly polarized. (b) Will the output wave be right or left circularly polarized? Problem 11.30 is good background.

11.32 Suppose that the length of the medium of Problem 11.31 is made to be *twice* that determined in the problem. Describe the polarization of the output wave in this case.

11.33 Given a wave for which $\mathbf{E}_s = 15e^{-j\beta z}\mathbf{a}_x + 18e^{-j\beta z}e^{j\phi}\mathbf{a}_y$ V/m in a medium characterized by complex intrinsic impedance $\eta(a)$, find $\mathbf{H}_s$. (b) Determine the average power density in W/m².

11.34 Given a general elliptically polarized wave as per Eq. (93):

$$\mathbf{E}_s = [E_{x0}\mathbf{a}_x + E_{y0}e^{j\phi}\mathbf{a}_y]e^{-j\beta z}$$

(a) Show, using methods similar to those of Example 11.7, that a linearly polarized wave results when superimposing the given field and a phase-shifted field of the form:

$$\mathbf{E}_s = [E_{x0}\mathbf{a}_x + E_{y0}e^{-j\phi}\mathbf{a}_y]e^{-j\beta z}e^{j\delta}$$

where δ is a constant. (b) Find δ in terms of ϕ such that the resultant wave is linearly polarized along x.

CHAPTER 12 Plane Wave Reflection and Dispersion

术语

dispersion 色散。

incidence 入射。

transmission coefficient 透射或传输系数。

intrinsic impedance 固有阻抗。

harmonic oscillator 简谐振荡器。

resonance 共振 / 谐振。

pulse broadening 脉冲变宽。

In Chapter 11, we learned how to mathematically represent uniform plane waves as functions of frequency, medium properties, and electric field orientation. We also learned how to calculate the wave velocity, attenuation, and power. In this chapter we consider wave reflection and transmission at planar boundaries between different media. Our study will allow any orientation between the wave and boundary and will also include the important cases of multiple boundaries. We will also study the practical case of waves that carry power over a finite band of frequencies, as would occur, for example, in a modulated carrier. We will consider such waves in *dispersive* media, in which some parameter that affects propagation (permittivity for example) varies with frequency. The effect of a dispersive medium on a signal is of great importance because the signal envelope will change its shape as it propagates. As a result, detection and faithful representation of the original signal at the receiving end become problematic. Consequently, dispersion and attenuation must both be evaluated when establishing maximum allowable transmission distances. ■

12.1 REFLECTION OF UNIFORM PLANE WAVES AT NORMAL INCIDENCE

We first consider the phenomenon of reflection which occurs when a uniform plane wave is incident on the boundary between regions composed of two different materials. The treatment is specialized to the case of *normal incidence*—in which the wave propagation direction is perpendicular to the boundary. In later sections, we remove this restriction. Expressions will be found for the wave that is reflected from the interface and for that which is transmitted from one region into the other. These results are directly related to impedance-matching problems in ordinary transmission lines, as we have already encountered in Chapter 10. They are also applicable to waveguides, which we will study in Chapter 13.

导读

1. 电磁波在传播过程中遇到参数不同的媒质分界面时，会发生反射与透射或折射，反射与透射量以及折射角取决于媒质的参数或波阻抗；在分界面上场量要满足界面条件，由此约束反射与透射波场量的大小；在分界面两侧波速在界面上的切向分量相等，由此约束折射角与反射角。

2. 电磁波在理想电介质或无损媒质中传播时，波的转播速度仅与材料参数有关，而与频率无关，即不同频率的电磁波的传播波速相同。

3. 实际上没有理想电介质，材料的介电常数都是频率的函数。因此，与材料的介电常数有关的波的传播特性都与频率有关，包括波速、界面上的折射角等。电磁波的波速随频率而改变的现象称为电磁波的色散效应。

4. 我们可认为电磁波的色散一词来源于光的色散，白光通过棱镜分离出不同颜色的光称为角色散效应，这是由于不同频率的光分量在棱镜界面上的折射角不同所致。电磁波的色散效应不仅会导致不同频率的电磁波在媒质交界面上的折射角不同，而且由于一个脉冲电磁波包含不同频率的分量，不同频率的分量到达观测点的时间不同，故使得脉冲波形在时间轴上会变宽。

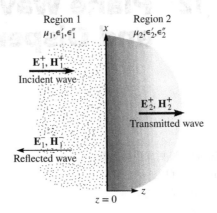

Figure 12.1 A plane wave incident on a boundary establishes reflected and transmitted waves having the indicated propagation directions. All fields are parallel to the boundary, with electric fields along x and magnetic fields along y.

12.1.1 Reflected and Transmitted Waves at a Boundary

We begin by considering the electric field intensity in a wave that propagates toward a boundary between two media. Referring to Figure 12.1, we define region 1 (ϵ_1, μ_1) as the half-space for which $z < 0$; region 2 (ϵ_2, μ_2) is the half-space for which $z > 0$. Initially we establish the incident wave in region 1, traveling in the $+z$ direction, and linearly polarized along x.

$$\mathscr{E}_{x1}^+(z,t) = E_{x10}^+ e^{-\alpha_1 z}\cos(\omega t - \beta_1 z)$$

In phasor form, this is

$$E_{xs1}^+(z) = E_{x10}^+ e^{-jk_1 z} \qquad (1)$$

where we take E_{x10}^+ as real. The subscript 1 identifies the region, and the superscript + indicates a positively traveling wave. Associated with $E_{xs1}^+(z)$ is a magnetic field in the y direction,

$$H_{ys1}^+(z) = \frac{1}{\eta_1} E_{x10}^+ e^{-jk_1 z} \qquad (2)$$

where k_1 and η_1 are complex unless ϵ_1'' (or σ_1) is zero. This uniform plane wave in region 1 that is traveling toward the boundary surface at $z = 0$ is called the *incident* wave. Since the direction of propagation of the incident wave is perpendicular to the boundary plane, we describe it as normal incidence.

We now recognize that energy may be transmitted across the boundary surface at $z = 0$ into region 2 by providing a wave moving in the $+z$ direction in that medium. The phasor electric and magnetic fields for this wave are

$$E_{xs2}^+(z) = E_{x20}^+ e^{-jk_2 z} \qquad (3)$$

$$H_{ys2}^+(z) = \frac{1}{\eta_2} E_{x20}^+ e^{-jk_2 z} \qquad (4)$$

This wave, which moves away from the boundary surface into region 2, is called the *transmitted* wave. Note the use of the different propagation constant k_2 and intrinsic impedance η_2.

The boundary conditions at $z = 0$ must be satisfied with these assumed fields. With **E** polarized along x, the field is tangent to the interface, and therefore the **E** fields in regions 1 and 2 must be equal at $z = 0$. Setting $z = 0$ in (1) and (3) would require that $E_{x10}^+ = E_{x20}^+$. **H**, being y-directed, is also a tangential field, and must be continuous across the boundary (no current sheets are present in real media). When we let $z = 0$ in (2) and (4), we find that we must have $E_{x10}^+/\eta_1 = E_{x20}^+/\eta_2$. Since $E_{x10}^+ = E_{x20}^+$, then $\eta_1 = \eta_2$. But this is a very special condition that does not fit the facts in general, and we are therefore unable to satisfy the boundary conditions with only an incident and a transmitted wave. We require a wave traveling away from the boundary in region 1, as shown in Figure 12.1; this is the *reflected* wave,

$$E_{xs1}^-(z) = E_{x10}^- e^{jk_1 z} \tag{5}$$

$$H_{xs1}^-(z) = -\frac{E_{x10}^-}{\eta_1} e^{jk_1 z} \tag{6}$$

where E_{x10}^- may be a complex quantity. Because this field is traveling in the $-z$ direction, $E_{xs1}^- = -\eta_1 H_{ys1}^-$ for the Poynting vector shows that $\mathbf{E}_1^- \times \mathbf{H}_1^-$ must be in the $-\mathbf{a}_z$ direction.

12.1.2 Reflection and Transmission Coefficients

With the presence of reflected and transmitted fields, the boundary conditions will be satisfied, and in the process the amplitudes of the transmitted and reflected waves may be found in terms of E_{x10}^+. The total electric field intensity is continuous at $z = 0$. The boundary conditions for tangential continuity of **E** and **H** are used in the following manner: First, the electric field must satisfy

$$E_{xs1} = E_{xs2} \qquad (z = 0)$$

or

$$E_{xs1}^+ + E_{xs1}^- = E_{xs2}^+ \qquad (z = 0)$$

Therefore

$$\boxed{E_{x10}^+ + E_{x10}^- = E_{x20}^+} \tag{7}$$

Then, magnetic field must satisfy

$$H_{ys1} = H_{ys2} \qquad (z = 0)$$

or

$$H_{ys1}^+ + H_{ys1}^- = H_{ys2}^+ \qquad (z = 0)$$

and therefore

$$\boxed{\frac{E_{x10}^+}{\eta_1} - \frac{E_{x10}^-}{\eta_1} = \frac{E_{x20}^+}{\eta_2}} \tag{8}$$

Solving (8) for E_{x20}^+ and substituting into (7), we find

$$E_{x10}^+ + E_{x10}^- = \frac{\eta_2}{\eta_1} E_{x10}^+ - \frac{\eta_2}{\eta_1} E_{x10}^-$$

or

$$E_{x10}^- = E_{x10}^+ \frac{\eta_2 - \eta_1}{\eta_2 + \eta_1}$$

The ratio of the amplitudes of the reflected and incident electric fields defines the *reflection coefficient*, designated by Γ,

$$\Gamma = \frac{E_{x10}^-}{E_{x10}^+} = \frac{\eta_2 - \eta_1}{\eta_2 + \eta_1} = |\Gamma| e^{j\phi_r} \qquad (9)$$

It is evident that as η_1 or η_2 may be complex, Γ will also be complex, and so we include a reflective phase shift, ϕ_r. The interpretation of Eq. (9) is identical to that used with transmission lines [Eq. (73), Chapter 10].

The relative amplitude of the transmitted electric field intensity is found by combining (9) and (7) to yield the *transmission coefficient, τ,*

$$\tau = \frac{E_{x20}^+}{E_{x10}^+} = \frac{2\eta_2}{\eta_1 + \eta_2} = 1 + \Gamma = |\tau| e^{j\phi_t} \qquad (10)$$

whose form and interpretation are consistent with the usage in transmission lines [Eq. (75), Chapter 10].

12.1.3 Total Reflection: Standing Wave Ratio

These results may be applied to a few special cases. We first let region 1 be a perfect dielectric and region 2 be a perfect conductor. Then we apply Eq. (48), Chapter 11, with $\epsilon_2'' = \sigma_2/\omega$, obtaining

$$\eta_2 = \sqrt{\frac{j\omega\mu_2}{\sigma_2 + j\omega\epsilon_2'}} = 0$$

in which zero is obtained since $\sigma_2 \to \infty$. Therefore, from (10),

$$E_{x20}^+ = 0$$

No time-varying fields can exist in the perfect conductor. An alternate way of looking at this is to note that the skin depth is zero.

Because $\eta_2 = 0$, Eq. (9) shows that

$$\Gamma = -1$$

and

$$E_{x10}^+ = -E_{x10}^-$$

The incident and reflected fields are of equal amplitude, and so all the incident energy is reflected by the perfect conductor. The fact that the two fields are of opposite sign indicates that at the boundary (or at the moment of reflection), the reflected field is shifted in phase by 180° relative to the incident field. The total **E** field in region 1 is

$$E_{xs1} = E_{xs1}^+ + E_{xs1}^-$$
$$= E_{x10}^+ e^{-j\beta_1 z} - E_{x10}^+ e^{j\beta_1 z}$$

where we have let $jk_1 = 0 + j\beta_1$ in the perfect dielectric. These terms may be combined and simplified,

$$E_{xs1} = (e^{-j\beta_1 z} - e^{j\beta_1 z}) E_{x10}^+$$
$$= -j2 \sin(\beta_1 z) E_{x10}^+ \qquad (11)$$

Multiplying (11) by $e^{j\omega t}$ and taking the real part, we obtain the real instantaneous form:

$$\mathscr{E}_{x1}(z,t) = 2 E_{x10}^+ \sin(\beta_1 z) \sin(\omega t) \qquad (12)$$

We recognize this total field in region 1 as a standing wave, obtained by combining two waves of equal amplitude traveling in opposite directions. We first encountered standing waves in transmission lines, but in the form of counterpropagating voltage waves (see Example 10.1).

Again, we compare the form of (12) to that of the incident wave,

$$\mathscr{E}_{x1}(z,t) = E_{x10}^+ \cos(\omega t - \beta_1 z) \qquad (13)$$

Here we see the term $\omega t - \beta_1 z$ or $\omega(t - z/v_{p1})$, which characterizes a wave traveling in the $+z$ direction at a velocity $v_{p1} = \omega/\beta_1$. In (12), however, the factors involving time and distance are separate trigonometric terms. Whenever $\omega t = m\pi$, $\mathscr{E}_{x1}$ is zero at all positions. On the other hand, spatial nulls in the standing wave pattern occur for all times wherever $\beta_1 z = m\pi$, which in turn occurs when $m = (0, \pm1, \pm2, \ldots)$. In such cases,

$$\frac{2\pi}{\lambda_1} z = m\pi$$

and the null locations occur at

$$z = m\frac{\lambda_1}{2}$$

Thus $E_{x1} = 0$ at the boundary $z = 0$ and at every half-wavelength from the boundary in region 1, $z < 0$, as illustrated in Figure 12.2.

Because $E_{xs1}^+ = \eta_1 H_{ys1}^+$ and $E_{xs1}^- = -\eta_1 H_{ys1}^-$, the magnetic field is

$$H_{ys1} = \frac{E_{x10}^+}{\eta_1} (e^{-j\beta_1 z} + e^{j\beta_1 z})$$

or

$$H_{y1}(z,t) = 2\frac{E_{x10}^+}{\eta_1} \cos(\beta_1 z) \cos(\omega t) \qquad (14)$$

This is also a standing wave, but it shows a maximum amplitude at the positions where $E_{x1} = 0$. It is also 90° out of time phase with E_{x1} everywhere. As a result, the average power as determined through the Poynting vector [Eq. (77), Chapter 11] is zero in the forward and backward directions.

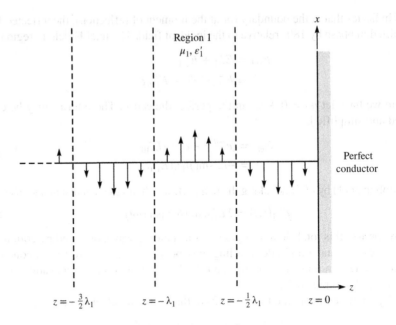

Figure 12.2   The instantaneous values of the total field E_{x1} are shown at $\omega t = \pi/2$. $E_{x1} = 0$ for all time at multiples of one half-wavelength from the conducting surface.

12.1.4   Partial Reflection and Power Reflectivity

Now suppose that perfect dielectrics exist in both regions 1 and 2, so that η_1 and η_2 are both real positive quantities and $\alpha_1 = \alpha_2 = 0$. Equation (9) enables us to calculate the reflection coefficient and find E_{x1}^- in terms of the incident field E_{x1}^+. Knowing E_{x1}^+ and E_{x1}^-, we then find H_{y1}^+ and H_{y1}^-. In region 2, E_{x2}^+ is found from (10), and this then determines H_{y2}^+.

EXAMPLE 12.1

As a numerical example we select

$$\eta_1 = 100 \ \Omega$$
$$\eta_2 = 300 \ \Omega$$
$$E_{x10}^+ = 100 \ \text{V/m}$$

and calculate values for the incident, reflected, and transmitted waves.

Solution. The reflection coefficient is

$$\Gamma = \frac{300 - 100}{300 + 100} = 0.5$$

and thus

$$E_{x10}^- = 50 \ \text{V/m}$$

The magnetic field intensities are

$$H_{y10}^+ = \frac{100}{100} = 1.00 \text{ A/m}$$

$$H_{y10}^- = -\frac{50}{100} = -0.50 \text{ A/m}$$

Using Eq. (77) from Chapter 11, we find that the magnitude of the average incident power density is

$$\langle S_{1i} \rangle = \left| \frac{1}{2} \mathcal{R}e\{ \mathbf{E}_s \times \mathbf{H}_s^* \} \right| = \frac{1}{2} E_{x10}^+ H_{y10}^+ = 50 \text{ W/m}^2$$

The average reflected power density is

$$\langle S_{1r} \rangle = -\frac{1}{2} E_{x10}^- H_{y10}^- = 12.5 \text{ W/m}^2$$

In region 2, using (10),

$$E_{x20}^+ = \tau E_{x10}^+ = 150 \text{ V/m}$$

and

$$H_{y20}^+ = \frac{150}{300} = 0.500 \text{ A/m}$$

Therefore, the average power density that is transmitted through the boundary into region 2 is

$$\langle S_2 \rangle = \frac{1}{2} E_{x20}^+ H_{y20}^+ = 37.5 \text{ W/m}^2$$

We may check and confirm the power conservation requirement:

$$\langle S_{1i} \rangle = \langle S_{1r} \rangle + \langle S_2 \rangle$$

A general rule on the transfer of power through reflection and transmission can be formulated. We consider the same field vector and interface orientations as before, but allow for the case of complex impedances. For the incident power density, we have

$$\langle S_{1i} \rangle = \frac{1}{2} \mathcal{R}e\{ E_{xs1}^+ H_{ys1}^{+*} \} = \frac{1}{2} \mathcal{R}e\left\{ E_{x10}^+ \frac{1}{\eta_1^*} E_{x10}^{+*} \right\} = \frac{1}{2} \mathcal{R}e\left\{ \frac{1}{\eta_1^*} \right\} |E_{x10}^+|^2$$

The reflected power density is then

$$\langle S_{1r} \rangle = -\frac{1}{2} \mathcal{R}e\{ E_{xs1}^- H_{ys1}^{-*} \} = \frac{1}{2} \mathcal{R}e\left\{ \Gamma E_{x10}^+ \frac{1}{\eta_1^*} \Gamma^* E_{x10}^{+*} \right\} = \frac{1}{2} \mathcal{R}e\left\{ \frac{1}{\eta_1^*} \right\} |E_{x10}^+|^2 |\Gamma|^2$$

We thus find the general relation between the reflected and incident power:

$$\boxed{\langle S_{1r} \rangle = |\Gamma|^2 \langle S_{1i} \rangle} \qquad (15)$$

In a similar way, we find the transmitted power density:

$$\langle S_2 \rangle = \frac{1}{2} \mathcal{R}e\{ E_{xs2}^+ H_{ys2}^{+*} \} = \frac{1}{2} \mathcal{R}e\left\{ \tau E_{x10}^+ \frac{1}{\eta_2^*} \tau^* E_{x10}^{+*} \right\} = \frac{1}{2} \mathcal{R}e\left\{ \frac{1}{\eta_2^*} \right\} |E_{x10}^+|^2 |\tau|^2$$

and so we see that the incident and transmitted power densities are related through

$$\langle S_2 \rangle = \frac{\mathscr{R}e\{1/\eta_2^*\}}{\mathscr{R}e\{1/\eta_1^*\}} |\tau|^2 \langle S_{1i} \rangle = \left|\frac{\eta_1}{\eta_2}\right|^2 \left(\frac{\eta_2 + \eta_2^*}{\eta_1 + \eta_1^*}\right) |\tau|^2 \langle S_{1i} \rangle \tag{16}$$

Equation (16) is a relatively complicated way to calculate the transmitted power, unless the impedances are real. It is easier to take advantage of energy conservation by noting that whatever power is not reflected must be transmitted. Eq. (15) can be used to find

$$\boxed{\langle S_2 \rangle = (1 - |\Gamma|^2)\langle S_{1i} \rangle} \tag{17}$$

As would be expected (and which must be true), Eq. (17) can also be derived from Eq. (16).

D12.1. A 1-MHz uniform plane wave is normally incident onto a fresh water lake ($\epsilon_r' = 78$, $\epsilon_r'' = 0$, $\mu_r = 1$). Determine the fraction of the incident power that is (a) reflected and (b) transmitted. (c) Determine the amplitude of the electric field that is transmitted into the lake.

Ans. (a) 0.63; (b) 0.37; (c) 0.20 V/m

12.2 STANDING WAVE RATIO

In cases where $|\Gamma| < 1$, some energy is transmitted into the second region and some is reflected. Region 1 therefore supports a field that is composed of both a traveling wave and a standing wave. We encountered this situation previously in transmission lines, in which partial reflection occurs at the load. Measurements of the voltage standing wave ratio and the locations of voltage minima or maxima enabled the determination of an unknown load impedance or established the extent to which the load impedance was matched to that of the line (Section 10.10). Similar measurements can be performed on the field amplitudes in plane wave reflection.

Using the same fields investigated in the previous section, we combine the incident and reflected electric field intensities. Medium 1 is assumed to be a perfect dielectric ($\alpha_1 = 0$), but region 2 may be any material. The total electric field phasor in region 1 will be

$$E_{x1T} = E_{x1}^+ + E_{x1}^- = E_{x10}^+ e^{-j\beta_1 z} + \Gamma E_{x10}^+ e^{j\beta_1 z} \tag{18}$$

where the reflection coefficient is as expressed in (9):

$$\Gamma = \frac{\eta_2 - \eta_1}{\eta_2 + \eta_1} = |\Gamma|e^{j\phi}$$

We allow for the possibility of a complex reflection coefficient by including its phase, ϕ. This is necessary because although η_1 is real and positive for a lossless medium, η_2

will in general be complex. Additionally, if region 2 is a perfect conductor, η_2 is zero, and so ϕ is equal to π; if η_2 is real and less than η_1, ϕ is also equal to π; and if η_2 is real and greater than η_1, ϕ is zero.

Incorporating the phase of Γ into (18), the total field in region 1 becomes

$$E_{x1T} = \left(e^{-j\beta_1 z} + |\Gamma|e^{j(\beta_1 z + \phi)}\right)E_{x10}^+ \tag{19}$$

The maximum and minimum field amplitudes in (19) are z-dependent and are subject to measurement. Their ratio, as found for voltage amplitudes in transmission lines (Section 10.10), is the *standing wave ratio,* denoted by s. We have a maximum when each term in the larger parentheses in (19) has the same phase angle; so, for E_{x10}^+ positive and real,

$$|E_{x1T}|_{\max} = (1 + |\Gamma|)E_{x10}^+ \tag{20}$$

and this occurs where

$$-\beta_1 z = \beta_1 z + \phi + 2m\pi \qquad (m = 0, \pm 1, \pm 2, \ldots) \tag{21}$$

Therefore

$$\boxed{z_{\max} = -\frac{1}{2\beta_1}(\phi + 2m\pi)} \tag{22}$$

Note that an electric field maximum is located at the boundary plane ($z = 0$) if $\phi = 0$; moreover, $\phi = 0$ when Γ is real and positive. This occurs for real η_1 and η_2 when $\eta_2 > \eta_1$. Thus there is a field maximum at the boundary surface when the intrinsic impedance of region 2 is greater than that of region 1 and both impedances are real. With $\phi = 0$, maxima also occur at $z_{\max} = -m\pi/\beta_1 = -m\lambda_1/2$.

For the perfect conductor $\phi = \pi$, and these maxima are found at $z_{\max} = -\pi/(2\beta_1)$, $-3\pi/(2\beta_1)$, or $z_{\max} = -\lambda_1/4, -3\lambda_1/4$, and so forth.

The minima must occur where the phase angles of the two terms in the larger parentheses in (19) differ by $180°$, thus

$$|E_{x1T}|_{\min} = (1 - |\Gamma|)E_{x10}^+ \tag{23}$$

and this occurs where

$$-\beta_1 z = \beta_1 z + \phi + \pi + 2m\pi \qquad (m = 0, \pm 1, \pm 2, \ldots) \tag{24}$$

or

$$\boxed{z_{\min} = -\frac{1}{2\beta_1}(\phi + (2m + 1)\pi)} \tag{25}$$

The minima are separated by multiples of one half-wavelength (as are the maxima), and for the perfect conductor the first minimum occurs when $-\beta_1 z = 0$, or at the conducting surface. In general, an electric field minimum is found at $z = 0$ whenever $\phi = \pi$; this occurs if $\eta_2 < \eta_1$ and both are real. The results are mathematically identical to those found for the transmission line study in Section 10.10. Figure 10.6 in that chapter provides a visualization.

Further insights can be obtained by working with Eq. (19) and rewriting it in real instantaneous form. The steps are identical to those taken in Chapter 10, Eqs. (81) through (84). We find the total field in region 1 to be

$$\mathscr{E}_{x1T}(z,t) = \underbrace{(1 - |\Gamma|)E_{x10}^+ \cos(\omega t - \beta_1 z)}_{\text{traveling wave}}$$

$$+ \underbrace{2|\Gamma|E_{x10}^+ \cos(\beta_1 z + \phi/2)\cos(\omega t + \phi/2)}_{\text{standing wave}} \tag{26}$$

The field expressed in Eq. (26) is the sum of a traveling wave of amplitude $(1 - |\Gamma|)E_{x10}^+$ and a standing wave having amplitude $2|\Gamma|E_{x10}^+$. The portion of the incident wave that reflects and back-propagates in region 1 interferes with an equivalent portion of the incident wave to form a standing wave. The rest of the incident wave (that does not interfere) is the traveling wave part of (26). The maximum amplitude observed in region 1 is found where the amplitudes of the two terms in (26) add directly to give $(1 + |\Gamma|)E_{x10}^+$. The minimum amplitude is found where the standing wave achieves a null, leaving only the traveling wave amplitude of $(1 - |\Gamma|)E_{x10}^+$. The fact that the two terms in (26) combine in this way with the proper phasing can be confirmed by substituting $z_{\max}$ and $z_{\min}$, as given by (22) and (25).

<hr>

EXAMPLE 12.2

To illustrate some of these results, let us consider a 100-V/m, 3-GHz wave that is propagating in a material having $\epsilon_{r1}' = 4$, $\mu_{r1} = 1$. The wave is normally incident on another perfect dielectric in region 2, $z > 0$, where $\epsilon_{r2}' = 9$ and $\mu_{r2} = 1$ (Figure 12.3). We seek the locations of the maxima and minima of **E**.

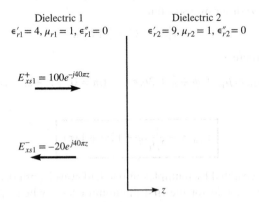

Dielectric 1 $\quad\quad$ Dielectric 2
$\epsilon_{r1}' = 4, \mu_{r1} = 1, \epsilon_{r1}'' = 0 \quad\quad \epsilon_{r2}' = 9, \mu_{r2} = 1, \epsilon_{r2}'' = 0$

$E_{xs1}^+ = 100e^{-j40\pi z}$

$E_{xs1}^- = -20e^{j40\pi z}$

z

Figure 12.3 An incident wave, $E_{xs1}^+ = 100e^{-j40\pi z}$ V/m, is reflected with a reflection coefficient $\Gamma = -0.2$. Dielectric 2 is infinitely thick.

Solution. We calculate $\omega = 6\pi \times 10^9$ rad/s, $\beta_1 = \omega\sqrt{\mu_1\epsilon_1} = 40\pi$ rad/m, and $\beta_2 = \omega\sqrt{\mu_2\epsilon_2} = 60\pi$ rad/m. Although the wavelength would be 10 cm in air, we find here that $\lambda_1 = 2\pi/\beta_1 = 5$ cm, $\lambda_2 = 2\pi/\beta_2 = 3.33$ cm, $\eta_1 = 60\pi$ Ω, $\eta_2 = 40\pi$ Ω, and $\Gamma = (\eta_2 - \eta_1)/(\eta_2 + \eta_1) = -0.2$. Because Γ is real and negative ($\eta_2 < \eta_1$), there will be a minimum of the electric field at the boundary, and it will be repeated at half-wavelength (2.5 cm) intervals in dielectric 1. From (23), we see that $|E_{x1T}|_{min} = 80$ V/m.

Maxima of **E** are found at distances of 1.25, 3.75, 6.25, ... cm from $z = 0$. These maxima all have amplitudes of 120 V/m, as predicted by (20).

There are no maxima or minima in region 2 because there is no reflected wave there.

The ratio of the maximum to minimum amplitudes is the standing wave ratio:

$$s = \frac{|E_{x1T}|\text{max}}{|E_{x1T}|\text{min}} = \frac{1 + |\Gamma|}{1 - |\Gamma|} \tag{27}$$

Because $|\Gamma| < 1$, s is always positive and greater than or equal to unity. For the preceding example,

$$s = \frac{1 + |-0.2|}{1 - |-0.2|} = \frac{1.2}{0.8} = 1.5$$

If $|\Gamma| = 1$, the reflected and incident amplitudes are equal, all the incident energy is reflected, and s is infinite. Planes separated by multiples of $\lambda_1/2$ can be found on which E_{x1} is zero at all times. Midway between these planes, E_{x1} has a maximum amplitude twice that of the incident wave.

If $\eta_2 = \eta_1$, then $\Gamma = 0$, no energy is reflected, and $s = 1$; the maximum and minimum amplitudes are equal.

If one-half the incident power is reflected, $|\Gamma|^2 = 0.5$, $|\Gamma| = 0.707$, and $s = 5.83$.

D12.2. What value of s results when $\Gamma = \pm 1/2$?

Ans. 3

Because the standing wave ratio is a ratio of amplitudes, the relative amplitudes, as measured by a probe, permit its use to determine s experimentally.

EXAMPLE 12.3

A uniform plane wave in air partially reflects from the surface of a material whose properties are unknown. Measurements of the electric field in the region in front of the interface yield a 1.5-m spacing between maxima, with the first maximum occurring 0.75 m from the interface. A standing wave ratio of 5 is measured. Determine the intrinsic impedance, η_u, of the unknown material.

Solution. The 1.5 m spacing between maxima is $\lambda/2$, which implies that a wavelength is 3.0 m, or $f = 100$ MHz. The first maximum at 0.75 m is thus at a distance of $\lambda/4$ from the interface, which means that a field minimum occurs at the boundary. Thus Γ will be real and negative. We use (27) to write

$$|\Gamma| = \frac{s-1}{s+1} = \frac{5-1}{5+1} = \frac{2}{3}$$

So

$$\Gamma = -\frac{2}{3} = \frac{\eta_u - \eta_0}{\eta_u + \eta_0}$$

which we solve for η_u to obtain

$$\eta_u = \frac{1}{5}\eta_0 = \frac{377}{5} = 75.4 \ \Omega$$

12.3 WAVE REFLECTION FROM MULTIPLE INTERFACES

So far we have treated the reflection of waves at the single boundary that occurs between semi-infinite media. In this section, we consider wave reflection from materials that are finite in extent, such that we must consider the effect of the front and back surfaces. Such a two-interface problem would occur, for example, for light incident on a flat piece of glass. Additional interfaces are present if the glass is coated with one or more layers of dielectric material for the purpose (as we will see) of reducing reflections. Such problems in which more than one interface is involved are often encountered; single-interface problems are in fact more the exception than the rule.

12.3.1 The Two-Interface Problem

Consider the general situation shown in Figure 12.4, in which a uniform plane wave propagating in the forward z direction is normally incident from the left onto the interface between regions 1 and 2; these have intrinsic impedances η_1 and η_2. A third region of impedance η_3 lies beyond region 2, and so a second interface exists between regions 2 and 3. We let the second interface location occur at $z = 0$, and so all positions to the left will be described by values of z that are negative. The width of the second region is l, so the first interface will occur at position $z = -l$.

When the incident wave reaches the first interface, events occur as follows: A portion of the wave reflects, while the remainder is transmitted, to propagate toward the second interface. There, a portion is transmitted into region 3, while the rest reflects and returns to the first interface; there it is again partially reflected. This reflected wave then combines with additional transmitted energy from region 1, and the process repeats. We thus have a complicated sequence of multiple reflections that occur within region 2, with partial transmission at each bounce. To analyze the situation in this way would involve keeping track of a very large number of reflections; this would be necessary when studying the *transient* phase of the process, where the incident wave first encounters the interfaces.

If the incident wave is left on for all time, however, a *steady-state* situation is eventually reached, in which (1) an overall fraction of the incident wave is reflected

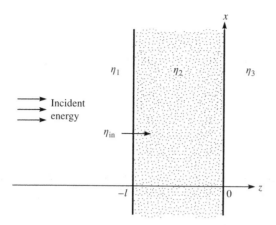

Figure 12.4   Basic two-interface problem, in which the impedances of regions 2 and 3, along with the finite thickness of region 2, are accounted for in the input impedance at the front surface, η_{in}.

from the two-interface configuration and back-propagates in region 1 with a definite amplitude and phase; (2) an overall fraction of the incident wave is transmitted through the two interfaces and forward-propagates in the third region; (3) a net backward wave exists in region 2, consisting of all reflected waves from the second interface; and (4) a net forward wave exists in region 2, which is the superposition of the transmitted wave through the first interface and all waves in region 2 that have reflected from the first interface and are now forward-propagating. The effect of combining many co-propagating waves in this way is to establish a single wave which has a definite amplitude and phase, determined through the sums of the amplitudes and phases of all the component waves. In steady state, we thus have a total of five waves to consider. These are the incident and net reflected waves in region 1, the net transmitted wave in region 3, and the two counterpropagating waves in region 2.

12.3.2   Wave Impedance

The two-interface situation is analyzed in the same manner as that used in the analysis of finite-length transmission lines (Section 10.11). We assume that all regions are composed of lossless media, and consider the two waves in region 2. If we take these as x-polarized, their electric fields combine to yield

$$E_{xs2} = E_{x20}^{+}e^{-j\beta_2 z} + E_{x20}^{-}e^{j\beta_2 z} \tag{28a}$$

where $\beta_2 = \omega\sqrt{\epsilon_{r2}}/c$, and where the amplitudes, E_{x20}^{+} and E_{x20}^{-}, are complex. The y-polarized magnetic field is similarly written, using complex amplitudes:

$$H_{ys2} = H_{y20}^{+}e^{-j\beta_2 z} + H_{y20}^{-}e^{j\beta_2 z} \tag{28b}$$

We now note that the forward and backward electric field amplitudes in region 2 are related through the reflection coefficient at the second interface, Γ_{23}, where

$$\Gamma_{23} = \frac{\eta_3 - \eta_2}{\eta_3 + \eta_2} \tag{29}$$

We thus have

$$E_{x20}^- = \Gamma_{23} E_{x20}^+ \tag{30}$$

We then write the magnetic field amplitudes in terms of electric field amplitudes through

$$H_{y20}^+ = \frac{1}{\eta_2} E_{x20}^+ \tag{31a}$$

and

$$H_{y20}^- = -\frac{1}{\eta_2} E_{x20}^- = -\frac{1}{\eta_2} \Gamma_{23} E_{x20}^+ \tag{31b}$$

We now define the *wave impedance*, η_w, as the z-dependent ratio of the total electric field to the total magnetic field. In region 2, this becomes, using (28a) and (28b),

$$\eta_w(z) = \frac{E_{xs2}}{H_{ys2}} = \frac{E_{x20}^+ e^{-j\beta_2 z} + E_{x20}^- e^{j\beta_2 z}}{H_{y20}^+ e^{-j\beta_2 z} + H_{y20}^- e^{j\beta_2 z}}$$

Then, using (30), (31a), and (31b), we obtain

$$\eta_w(z) = \eta_2 \left[\frac{e^{-j\beta_2 z} + \Gamma_{23} e^{j\beta_2 z}}{e^{-j\beta_2 z} - \Gamma_{23} e^{j\beta_2 z}} \right]$$

Now, using (29) and Euler's identity, we have

$$\eta_w(z) = \eta_2 \times \frac{(\eta_3 + \eta_2)(\cos\beta_2 z - j\sin\beta_2 z) + (\eta_3 - \eta_2)(\cos\beta_2 z + j\sin\beta_2 z)}{(\eta_3 + \eta_2)(\cos\beta_2 z - j\sin\beta_2 z) - (\eta_3 - \eta_2)(\cos\beta_2 z + j\sin\beta_2 z)}$$

This is easily simplified to yield

$$\eta_w(z) = \eta_2 \frac{\eta_3 \cos\beta_2 z - j\eta_2 \sin\beta_2 z}{\eta_2 \cos\beta_2 z - j\eta_3 \sin\beta_2 z} \tag{32}$$

We now use the wave impedance in region 2 to solve our reflection problem. Of interest to us is the net reflected wave amplitude at the first interface. Since tangential **E** and **H** are continuous across the boundary, we have

$$E_{xs1}^+ + E_{xs1}^- = E_{xs2} \qquad (z = -l) \tag{33a}$$

and

$$H_{ys1}^+ + H_{ys1}^- = H_{ys2} \qquad (z = -l) \tag{33b}$$

Then, in analogy to (7) and (8), we may write

$$E_{x10}^+ + E_{x10}^- = E_{xs2}(z = -l) \tag{34a}$$

and

$$\frac{E_{x10}^+}{\eta_1} - \frac{E_{x10}^-}{\eta_1} = \frac{E_{xs2}(z = -l)}{\eta_w(-l)} \tag{34b}$$

where E_{x10}^+ and E_{x10}^- are the amplitudes of the incident and reflected fields. We call $\eta_{w(-l)}$ the *input impedance*, η_{in}, to the two-interface combination. We now solve (34a) and (34*b*) together, eliminating E_{xs2}, to obtain

$$\boxed{\frac{E_{x10}^-}{E_{x10}^+} = \Gamma = \frac{\eta_{in} - \eta_1}{\eta_{in} + \eta_1}}$$ (35)

To find the input impedance, we evaluate (32) at $z = -l$, resulting in

$$\boxed{\eta_{in} = \eta_2 \frac{\eta_3 \cos\beta_2 l + j\eta_2 \sin\beta_2 l}{\eta_2 \cos\beta_2 l + j\eta_3 \sin\beta_2 l}}$$ (36)

Equations (35) and (36) are general results that enable us to calculate the net reflected wave amplitude and phase from two parallel interfaces between lossless media.[1] Note the dependence on the interface spacing, *l*, and on the wavelength as measured in region 2, characterized by β_2. Of immediate importance to us is the fraction of the incident power that reflects from the dual interface and back-propagates in region 1. As we found earlier, this fraction will be $|\Gamma|^2$. Also of interest is the transmitted power, which propagates away from the second interface in region 3. It is simply the remaining power fraction, which is $1 - |\Gamma|^2$. The power in region 2 stays constant in steady state; power leaves that region to form the reflected and transmitted waves, but is immediately replenished by the incident wave. We have already encountered an analogous situation involving cascaded transmission lines, which culminated in Eq. (101) in Chapter 10.

12.3.3 Special Cases: Half-Wave and Quarter-Wave Layers

An important result of situations involving two interfaces is that it is possible to achieve total transmission in certain cases. From (35), we see that total transmission occurs when $\Gamma = 0$, or when $\eta_{in} = \eta_1$. In this case, as in transmission lines, we say that the input impedance is *matched* to that of the incident medium. There are a few methods of accomplishing this.

As a start, suppose that $\eta_3 = \eta_1$, and region 2 is of such thickness that $\beta_2 l = m\pi$, where *m* is an integer. Now $\beta_2 = 2\pi/\lambda_2$, where λ_2 is the wavelength *as measured in region 2*. Therefore

$$\frac{2\pi}{\lambda_2} l = m\pi$$

or

$$l = m\frac{\lambda_2}{2}$$ (37)

With $\beta_2 l = m\pi$, the second region thickness is an integer multiple of half-wavelengths as measured in that medium. Equation (36) now reduces to $\eta_{in} = \eta_3$. Thus the general effect of a multiple half-wave thickness is to render the second region immaterial to

[1] For convenience, (34*a*) and (34*b*) have been written for a specific time at which the incident wave amplitude, E_{x10}^+, occurs at $z = -l$. This establishes a zero-phase reference at the front interface for the incident wave, and so it is from this reference that the reflected wave phase is determined. Equivalently, we have repositioned the $z = 0$ point at the front interface. Eq. (36) allows this because it is only a function of the interface spacing, *l*.

the results on reflection and transmission. Equivalently, we have a single-interface problem involving η_1 and η_3. Now, with $\eta_3 = \eta_1$, we have a matched input impedance, and there is no net reflected wave. This method of choosing the region 2 thickness is known as *half-wave matching*. Its applications include, for example, antenna housings on airplanes known as *radomes,* which form a part of the fuselage. The antenna, inside the aircraft, can transmit and receive through this layer, which can be shaped to enable good aerodynamic characteristics. Note that the half-wave matching condition no longer applies as we deviate from the wavelength that satisfies it. When this is done, the device reflectivity increases (with increased wavelength deviation), so it ultimately acts as a bandpass filter.

Often, it is convenient to express the dielectric constant of the medium through the *refractive index* (or just index), *n*, defined as

$$n = \sqrt{\epsilon_r} \qquad (38)$$

Characterizing materials by their refractive indices is primarily done at optical frequencies (on the order of 10^{14} Hz), whereas at much lower frequencies, a dielectric constant is traditionally specified. Since ϵ_r is complex in lossy media, the index will also be complex. Rather than complicate the situation in this way, we will restrict our use of the refractive index to cases involving lossless media, having $\mathscr{E}_r'' = 0$, and $\mu_r = 1$. Under lossless conditions, we may write the plane wave phase constant and the material intrinsic impedance in terms of the index through

$$\beta = k = \omega \sqrt{\mu_0 \epsilon_0} \sqrt{\epsilon_r} = \frac{n\omega}{c} \qquad (39)$$

and

$$\eta = \frac{1}{\sqrt{\epsilon_r}} \sqrt{\frac{\mu_0}{\epsilon_0}} = \frac{\eta_0}{n} \qquad (40)$$

Finally, the phase velocity and wavelength in a material of index *n* are

$$v_p = \frac{c}{n} \qquad (41)$$

and

$$\lambda = \frac{v_p}{f} = \frac{\lambda_0}{n} \qquad (42)$$

where λ_0 is the wavelength in free space. It is obviously important not to confuse the index *n* with the similar-appearing Greek η (intrinsic impedance), which has an entirely different meaning.

Another application, typically seen in optics, is the *Fabry-Perot interferometer*. This, in its simplest form, consists of a single block of glass or other transparent material of index n, whose thickness, l, is set to transmit wavelengths which satisfy the condition $\lambda = \lambda_0/n = 2l/m$. Often we want to transmit only one wavelength, not several, as (37) would allow. We would therefore like to assure that adjacent wavelengths that are passed through the device are separated as far as possible, so that only one will lie within the input power spectrum. In terms of wavelength as measured in the material, this separation is in general given by

$$\lambda_{m-1} - \lambda_m = \Delta\lambda_f = \frac{2l}{m-1} - \frac{2l}{m} = \frac{2l}{m(m-1)} \doteq \frac{2l}{m^2}$$

Note that m is the number of half-wavelengths in region 2, or $m = 2l/\lambda = 2nl/\lambda_0$, where λ_0 is the desired free-space wavelength for transmission. Thus

$$\Delta\lambda_f \doteq \frac{\lambda_2^2}{2l} \qquad (43a)$$

In terms of wavelength measured in free space, this becomes

$$\Delta\lambda_{f0} = n\Delta\lambda_f \doteq \frac{\lambda_0^2}{2nl} \qquad (43b)$$

$\Delta\lambda_{f0}$ is known as the *free spectral range* of the Fabry-Perot interferometer in terms of free-space wavelength separation. The interferometer can be used as a narrowband filter (transmitting a desired wavelength and a narrow spectrum around this wavelength) if the spectrum to be filtered is narrower than the free spectral range.

EXAMPLE 12.4

Suppose we wish to filter an optical spectrum of full width $\Delta\lambda_{s0} = 50$ nm (measured in free space), whose center wavelength, λ_0, is in the red part of the visible spectrum at 600 nm, where one nm (nanometer) is 10^{-9} m. A Fabry-Perot filter is to be used, consisting of a lossless glass plate in air, having refractive index $n = 1.45$. We need to find the required range of glass thicknesses such that multiple wavelength orders will not be transmitted.

Solution. We require that the free spectral range be greater than the optical spectral width, or $\Delta\lambda_{f0} > \Delta\lambda_s$. Using (43b)

$$l < \frac{\lambda_0^2}{2n\Delta\lambda_{s0}}$$

So

$$l < \frac{600^2}{2(1.45)(50)} = 2.5 \times 10^3 \, \text{nm} = 2.5 \, \mu\text{m}$$

where 1 μm (micrometer) $= 10^{-6}$ m. Fabricating a glass plate of this thickness or less is somewhat ridiculous to contemplate. Instead, what is often used is an airspace of thickness on this order, between two thick plates whose surfaces on the sides opposite the airspace are antireflection coated. This is in fact a more versatile configuration because the wavelength to be transmitted (and the free spectral range) can be adjusted by varying the plate separation.

Next, we remove the restriction $\eta_1 = \eta_3$ and look for a way to produce zero reflection. Returning to Eq. (36), suppose we set $\beta_2 l = (2m-1)\pi/2$, or an odd multiple of $\pi/2$. This means that

$$\frac{2\pi}{\lambda_2} l = (2m-1)\frac{\pi}{2} \qquad (m = 1, 2, 3, \ldots)$$

or

$$\boxed{l = (2m-1)\frac{\lambda_2}{4}} \tag{44}$$

The thickness is an odd multiple of a quarter-wavelength as measured in region 2. Under this condition, (36) reduces to

$$\eta_{in} = \frac{\eta_2^2}{\eta_3} \tag{45}$$

Typically, we choose the second region impedance to allow matching between given impedances η_1 and η_3. To achieve total transmission, we require that $\eta_{in} = \eta_1$, so that the required second region impedance becomes

$$\boxed{\eta_2 = \sqrt{\eta_1 \eta_3}} \tag{46}$$

With the conditions given by (44) and (46) satisfied, we have performed *quarter-wave matching*. The design of antireflective coatings for optical devices is based on this principle.

<hr>

EXAMPLE 12.5

We wish to coat a glass surface with an appropriate dielectric layer to provide total transmission from air to the glass at a free-space wavelength of 570 nm. The glass has refractive index $n_3 = 1.45$. Determine the required index for the coating and its minimum thickness.

Solution. The known impedances are $\eta_1 = 377\ \Omega$ and $\eta_3 = 377/1.45 = 260\ \Omega$. Using (46) we have

$$\eta_2 = \sqrt{(377)(260)} = 313\ \Omega$$

The index of region 2 will then be

$$n_2 = \left(\frac{377}{313}\right) = 1.20$$

The wavelength in region 2 will be

$$\lambda_2 = \frac{570}{1.20} = 475\ \text{nm}$$

The minimum thickness of the dielectric layer is then

$$l = \frac{\lambda_2}{4} = 119\ \text{nm} = 0.119\ \mu\text{m}$$

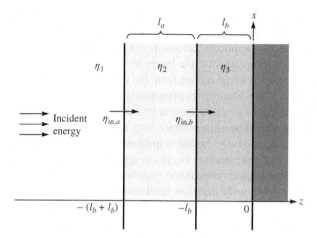

Figure 12.5 A three-interface problem in which input impedance $\eta_{in,a}$ is transformed back to the front interface to form input impedance $\eta_{in,b}$.

12.3.4 The Multilayer Problem: Impedance Transformation

The procedure for evaluating wave reflection from two interfaces has involved calculating an effective impedance at the first interface, η_{in}, which is expressed in terms of the impedances that lie beyond the front surface. This process of *impedance transformation* is more apparent when we consider problems involving more than two interfaces.

For example, consider the three-interface situation shown in Figure 12.5, where a wave is incident from the left in region 1. We wish to determine the fraction of the incident power that is reflected and back-propagates in region 1 and the fraction of the incident power that is transmitted into region 4. To do this, we need to find the input impedance at the front surface (the interface between regions 1 and 2). We start by transforming the impedance of region 4 to form the input impedance at the boundary between regions 2 and 3. This is shown as $\eta_{in,b}$ in Figure 12.5. Using (36), we have

$$\eta_{in,b} = \eta_3 \frac{\eta_4 \cos\beta_3 l_b + j\eta_3 \sin\beta_3 l_b}{\eta_3 \cos\beta_3 l_b + j\eta_4 \sin\beta_3 l_b} \tag{47}$$

We have now effectively reduced the situation to a two-interface problem in which $\eta_{in,b}$ is the impedance of all that lies beyond the second interface. The input impedance at the front interface, $\eta_{in,a}$, is now found by transforming $\eta_{in,b}$ as follows:

$$\eta_{in,a} = \eta_2 \frac{\eta_{in,b} \cos\beta_2 l_a + j\eta_2 \sin\beta_2 l_a}{\eta_2 \cos\beta_2 l_a + j\eta_{in,b} \sin\beta_2 l_a} \tag{48}$$

The reflected power fraction is now $|\Gamma|^2$, where

$$\Gamma = \frac{\eta_{in,a} - \eta_1}{\eta_{in,a} + \eta_1}$$

The fraction of the power transmitted into region 4 is, as before, $1-|\Gamma|^2$. The method of impedance transformation can be applied in this manner to any number of interfaces. The process, although tedious, is easily handled by a computer.

The motivation for using multiple layers to reduce reflection is that the resulting structure is less sensitive to deviations from the design wavelength if the impedances (or refractive indices) are arranged to progressively increase or decrease from layer to layer. For multiple layers to antireflection coat a camera lens, for example, the layer on the lens surface would be of impedance very close to that of the glass. Subsequent layers are given progressively higher impedances. With a large number of layers fabricated in this way, the situation begins to approach (but never reaches) the ideal case, in which the top layer impedance matches that of air, while the impedances of deeper layers continuously decrease until reaching the value of the glass surface. With this continuously varying impedance, there is no surface from which to reflect, and so light of any wavelength is totally transmitted. Multilayer coatings designed in this way produce excellent broadband transmission characteristics.

D12.3. A uniform plane wave in air is normally incident on a dielectric slab of thickness $\lambda_2/4$ and intrinsic impedance $\eta_2 = 260 \ \Omega$. Determine the magnitude and phase of the reflection coefficient.

Ans. 0.356; 180°

12.4 PLANE WAVE PROPAGATION IN GENERAL DIRECTIONS

In this section, we will learn how to mathematically describe uniform plane waves that propagate in any direction. Our motivation for doing this is our need to address the problem of incident waves on boundaries that are not perpendicular to the propagation direction. Such problems of *oblique incidence* generally occur, with normal incidence being a special case. Addressing such problems requires (as always) that we establish an appropriate coordinate system. With the boundary positioned in the x, y plane, for example, the incident wave will propagate in a direction that could involve all three coordinate axes, whereas with normal incidence, we were only concerned with propagation along z. We need a mathematical formalism that will allow for the general direction case.

To begin, consider a wave that propagates in a lossless medium, with propagation constant $\beta = k = \omega\sqrt{\mu\epsilon}$. For simplicity, we consider a two-dimensional case, where the wave travels in a direction between the x and z axes. The first step is to consider the propagation constant as a *vector,* $\mathbf{k}$, indicated in Figure 12.6. The direction of $\mathbf{k}$ is the propagation direction, which is the same as the direction of the Poynting vector in our case.[2] The magnitude of $\mathbf{k}$ is the phase shift per unit distance *along that direction.* Part of the process of characterizing a wave involves specifying its phase at any

---

[2] We assume here that the wave is in an isotropic medium, where the permittivity and permeability do not change with field orientation. In anisotropic media (where ϵ and/or μ depend on field orientation), the directions of the Poynting vector and $\mathbf{k}$ may differ.

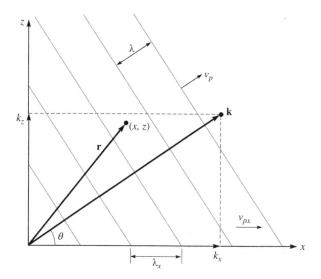

Figure 12.6   Representation of a uniform plane wave with wavevector **k** at angle θ to the x axis. The phase at point (x, z) is given by **k · r**. Planes of constant phase (shown as lines perpendicular to **k**) are spaced by wavelength λ but have wider spacing when measured along the x or z axis.

spatial location. For the waves we have considered that propagate along the z axis, this was accomplished by the factor $e^{\pm jkz}$ in the phasor form. To specify the phase in our two-dimensional problem, we make use of the vector nature of **k** and consider the phase at a general location (x, z) described through the position vector **r**. The phase at that location, referenced to the origin, is given by the projection of **k** along **r** times the magnitude of **r**, or just **k · r**. If the electric field is of magnitude E_0, we can thus write down the phasor form of the wave in Figure 12.6 as

$$\boxed{\mathbf{E}_s = \mathbf{E}_0 e^{-j\mathbf{k}\cdot\mathbf{r}}} \qquad (49)$$

The minus sign in the exponent indicates that the phase along **r** moves in time in the direction of increasing **r**. Again, the wave power flow in an isotropic medium occurs in the direction along which the phase shift per unit distance is maximum—or along **k**. The vector **r** serves as a means to measure phase at any point using **k**. This construction is easily extended to three dimensions by allowing **k** and **r** to each have three components.

In our two-dimensional case of Figure 12.6, we can express **k** in terms of its x and z components:

$$\mathbf{k} = k_x \mathbf{a}_x + k_z \mathbf{a}_z$$

The position vector, $\mathbf{r}$, can be similarly expressed:

$$\boxed{\mathbf{r} = x\mathbf{a}_x + z\mathbf{a}_z}$$

so that

$$\mathbf{k} \cdot \mathbf{r} = k_x x + k_z z$$

Equation (49) now becomes

$$\mathbf{E}_s = \mathbf{E}_0 e^{-j(k_x x + k_z z)} \tag{50}$$

Whereas Eq. (49) provided the general form of the wave, Eq. (50) is the form that is specific to the situation. Given a wave expressed by (50), the angle of propagation from the x axis is readily found through

$$\theta = \tan^{-1}\left(\frac{k_z}{k_x}\right)$$

The wavelength and phase velocity depend on the direction one is considering. In the direction of $\mathbf{k}$, these will be

$$\lambda = \frac{2\pi}{k} = \frac{2\pi}{\left(k_x^2 + k_z^2\right)^{1/2}}$$

and

$$v_p = \frac{\omega}{k} = \frac{\omega}{\left(k_x^2 + k_z^2\right)^{1/2}}$$

If, for example, we consider the x direction, these quantities will be

$$\lambda_x = \frac{2\pi}{k_x}$$

and

$$v_{px} = \frac{\omega}{k_x}$$

Note that both λ_x and v_{px} are greater than their counterparts along the direction of $\mathbf{k}$. This result, at first surprising, can be understood through the geometry of Figure 12.6. The diagram shows a series of phase fronts (planes of constant phase) which intersect $\mathbf{k}$ at right angles. The phase shift between adjacent fronts is set at 2π in the figure; this corresponds to a spatial separation along the $\mathbf{k}$ direction of one wavelength, as shown. The phase fronts intersect the x axis, and we see that *along* x the front separation is greater than it was along $\mathbf{k}$. λ_x is the spacing between fronts along x and is indicated on the figure. The phase velocity along x is the velocity of the intersection points between the phase fronts and the x axis. Again, from the geometry, we see that this velocity must be faster than the velocity along **k** and will, of course, exceed the speed of light in the medium. This does not constitute a violation of special relativity, however, since the energy in the wave flows in the direction of **k** and not along x or z. The wave frequency is $f = \omega/2\pi$ and is invariant with direction. Note, for example, that in the directions we have considered,

$$f = \frac{v_p}{\lambda} = \frac{v_{px}}{\lambda_x} = \frac{\omega}{2\pi}$$

EXAMPLE 12.6

Consider a 50-MHz uniform plane wave having electric field amplitude 10 V/m. The medium is lossless, having $\epsilon_r = \epsilon_r' = 9.0$ and $\mu_r = 1.0$. The wave propagates in the x, y plane at a 30° angle to the x axis and is linearly polarized along z. Write down the phasor expression for the electric field.

Solution. The propagation constant magnitude is

$$k = \omega\sqrt{\mu\epsilon} = \frac{\omega\sqrt{\epsilon_r}}{c} = \frac{2\pi \times 50 \times 10^6 (3)}{3 \times 10^8} = 3.2 \text{ m}^{-1}$$

The vector **k** is now

$$\mathbf{k} = 3.2(\cos 30 \mathbf{a}_x + \sin 30 \mathbf{a}_y) = 2.8\mathbf{a}_x + 1.6\mathbf{a}_y \text{ m}^{-1}$$

Then

$$\mathbf{r} = x\mathbf{a}_x + y\mathbf{a}_y$$

With the electric field directed along z, the phasor form becomes

$$\mathbf{E}_s = E_0 e^{-j\mathbf{k} \cdot \mathbf{r}} \mathbf{a}_z = 10 e^{-j(2.8x + 1.6y)} \mathbf{a}_z$$

D12.4. For Example 12.6, calculate λ_x, λ_y, v_{px}, and v_{py}.

Ans. 2.2 m; 3.9 m; 1.1×10^8 m/s; 2.0×10^8 m/s

12.5 PLANE WAVE REFLECTION AT OBLIQUE INCIDENCE ANGLES

We now consider the problem of wave reflection from plane interfaces, in which the incident wave propagates at some angle to the surface. Our objectives are (1) to determine the relation between incident, reflected, and transmitted angles, and (2) to derive reflection and transmission coefficients that are functions of the incident angle and wave polarization.

We will also show that cases exist in which total reflection or total transmission may occur at the interface between two dielectrics if the angle of incidence and the polarization are appropriately chosen.

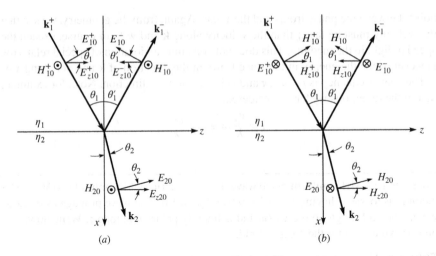

Figure 12.7 Geometries for plane wave incidence at angle θ_1 onto an interface between dielectrics having intrinsic impedances η_1 and η_2. The two polarization cases are shown: (*a*) p-polarization (or TM), with **E** in the plane of incidence; (*b*) s-polarization (or TE), with **E** perpendicular to the plane of incidence.

The situation is illustrated in Figure 12.7, in which the incident wave direction and position-dependent phase are characterized by wavevector $\mathbf{k}_1^+$. The angle of incidence is the angle between $\mathbf{k}_1^+$ and a line that is normal to the surface (the x axis in this case). The incidence angle is shown as θ_1. The reflected wave, characterized by wavevector $\mathbf{k}_1^-$, will propagate away from the interface at angle θ_1'. Finally, the transmitted wave, characterized by $\mathbf{k}_2$, will propagate into the second region at angle θ_2 as shown. One would suspect (from previous experience) that the incident and reflected angles are equal ($\theta_1 = \theta_1'$), which is correct. We need to show this, however, to be complete.

The two media are lossless dielectrics, characterized by intrinsic impedances η_1 and η_2. We will assume, as before, that the materials are nonmagnetic, and thus have permeability μ_0. Consequently, the materials are adequately described by specifying their dielectric constants, ϵ_{r1} and ϵ_{r2}, or their refractive indices, $n_1 = \sqrt{\epsilon_{r1}}$ and $n_2 = \sqrt{\epsilon_{r2}}$.

In Figure 12.7, two cases are shown that differ by the choice of electric field orientation. In Figure 12.7*a*, the **E** field is polarized in the plane of the page, with **H** therefore perpendicular to the page and pointing outward. In this illustration, the plane of the page is also the *plane of incidence*, which is more precisely defined as the plane spanned by the incident **k** vector and the normal to the surface. With **E** lying in the plane of incidence, the wave is said to have *parallel polarization* or to be *p-polarized* (**E** is parallel to the incidence plane). Note that although **H** is perpendicular to the incidence plane, it lies parallel (or transverse) to the interface. Consequently, another name for this type of polarization is *transverse magnetic,* or TM polarization.

Figure 12.7*b* shows the situation in which the field directions have been rotated by 90°. Now **H** lies in the plane of incidence, whereas **E** is perpendicular to the plane.

Because **E** is used to define polarization, the configuration is called *perpendicular polarization,* or is said to be *s-polarized*.[3] **E** is also parallel to the interface, and so the case is also called *transverse electric,* or TE polarization. We will find that the reflection and transmission coefficients will differ for the two polarization types, but that reflection and transmission angles will not depend on polarization. We only need to consider s- and p-polarizations because any other field direction can be constructed as some combination of s and p waves.

Our desired knowledge of reflection and transmission coefficients, as well as how the angles relate, can be found through the field boundary conditions at the interface. Specifically, we require that the transverse components of **E** and **H** be continuous across the interface. These were the conditions we used to find Γ and τ for normal incidence ($\theta_1 = 0$), which is in fact a special case of our current problem. We will consider the case of p-polarization (Figure 12.7*a*) first. To begin, we write down the incident, reflected, and transmitted fields in phasor form, using the notation developed in Section 12.4:

$$\mathbf{E}_{s1}^+ = \mathbf{E}_{10}^+ e^{-j\mathbf{k}_1^+ \cdot \mathbf{r}} \tag{51}$$

$$\mathbf{E}_{s1}^- = \mathbf{E}_{10}^- e^{-j\mathbf{k}_1^- \cdot \mathbf{r}} \tag{52}$$

$$\mathbf{E}_{s2} = \mathbf{E}_{20} e^{-j\mathbf{k}_2 \cdot \mathbf{r}} \tag{53}$$

where

$$\mathbf{k}_1^+ = k_1(\cos\theta_1\, \mathbf{a}_x + \sin\theta_1\, \mathbf{a}_z) \tag{54}$$

$$\mathbf{k}_1^- = k_1(-\cos\theta_1'\, \mathbf{a}_x + \sin\theta_1'\, \mathbf{a}_z) \tag{55}$$

$$\mathbf{k}_2 = k_2(\cos\theta_2\, \mathbf{a}_x + \sin\theta_2\, \mathbf{a}_z) \tag{56}$$

and where

$$\mathbf{r} = x\mathbf{a}_x + z\mathbf{a}_z \tag{57}$$

The wavevector magnitudes are $k_1 = \omega\sqrt{\epsilon_{r1}}/c = n_1\omega/c$ and $k_2 = \omega\sqrt{\epsilon_{r2}}/c = n_2\omega/c$.

Now, to evaluate the boundary condition that requires continuous tangential electric field, we need to find the components of the electric fields (z components) that are parallel to the interface. Projecting all **E** fields in the z direction, and using (51) through (57), we find

$$E_{zs1}^+ = E_{z10}^+ e^{-j\mathbf{k}_1^+ \cdot \mathbf{r}} = E_{10}^+ \cos\theta_1 e^{-jk_1(x\cos\theta_1 + z\sin\theta_1)} \tag{58}$$

$$E_{zs1}^- = E_{z10}^- e^{-j\mathbf{k}_1^- \cdot \mathbf{r}} = E_{10}^- \cos\theta_1' e^{jk_1(x\cos\theta_1' - z\sin\theta_1')} \tag{59}$$

$$E_{zs2} = E_{z20} e^{-j\mathbf{k}_2 \cdot \mathbf{r}} = E_{20} \cos\theta_2 e^{-jk_2(x\cos\theta_2 + z\sin\theta_2)} \tag{60}$$

---

[3] The *s* designation is an abbreviation for the German *senkrecht,* meaning *perpendicular.* The *p* in *p-polarized* is an abbreviation for the German word for parallel, which is *parallel.*

The boundary condition for a continuous tangential electric field now reads:

$$E_{zs1}^+ + E_{zs1}^- = E_{zs2} \quad (\text{at } x = 0)$$

We now substitute Eqs. (58) through (60) into (61) and evaluate the result at $x = 0$ to obtain

$$E_{10}^+ \cos\theta_1 \, e^{-jk_1 z \sin\theta_1} + E_{10}^- \cos\theta_1' e^{-jk_1 z \sin\theta_1'} = E_{20} \cos\theta_2 e^{-jk_2 z \sin\theta_2} \tag{61}$$

Note that E_{10}^+, E_{10}^-, and E_{20} are all constants (independent of z). Further, we require that (61) hold for all values of z (everywhere on the interface). For this to occur, it must follow that all the phase terms appearing in (61) are equal. Specifically,

$$k_1 z \sin\theta_1 = k_1 z \sin\theta_1' = k_2 z \sin\theta_2$$

From this, we see immediately that $\theta_1' = \theta_1$, or the angle of reflection is equal to the angle of incidence. We also find that

$$\boxed{k_1 \sin\theta_1 = k_2 \sin\theta_2} \tag{62}$$

Equation (62) is known as *Snell's law of refraction*. Because, in general, $k = n\omega/c$, we can rewrite (62) in terms of the refractive indices:

$$\boxed{n_1 \sin\theta_1 = n_2 \sin\theta_2} \tag{63}$$

Equation (63) is the form of Snell's law that is most readily used for our present case of nonmagnetic dielectrics. Equation (62) is a more general form which would apply, for example, to cases involving materials with different permeabilities as well as different permittivities. In general, we would have $k_1 = (\omega/c)\sqrt{\mu_{r1}\epsilon_{r1}}$ and $k_2 = (\omega/c)\sqrt{\mu_{r2}\epsilon_{r2}}$.

Having found the relations between angles, we next turn to our second objective, which is to determine the relations between the amplitudes, E_{10}^+, E_{10}^-, and E_{20}. To accomplish this, we need to consider the other boundary condition, requiring tangential continuity of $\mathbf{H}$ at $x = 0$. The magnetic field vectors for the p-polarized wave are all negative y-directed. At the boundary, the field amplitudes are related through

$$H_{10}^+ + H_{10}^- = H_{20} \tag{64}$$

Then, when we use the fact that $\theta_1' = \theta_1$ and invoke Snell's law, (61) becomes

$$E_{10}^+ \cos\theta_1 + E_{10}^- \cos\theta_1 = E_{20}\cos\theta_2 \tag{65}$$

Using the medium intrinsic impedances, we know, for example, that $E_{10}^+/H_{10}^+ = \eta_1$ and $E_{20}^+/H_{20}^+ = \eta_2$. Eq. (64) can be written as follows:

$$\frac{E_{10}^+ \cos\theta_1}{\eta_{1p}} - \frac{E_{10}^- \cos\theta_1}{\eta_{1p}} = \frac{E_{20}^+ \cos\theta_2}{\eta_{2p}} \tag{66}$$

Note the minus sign in front of the second term in (66), which results from the fact that $E_{10}^- \cos\theta_1$ is negative (from Figure 12.7a), whereas H_{10}^- is positive (again from the figure). When we write Eq. (66), *effective impedances*, valid for p-polarization, are defined through

$$\boxed{\eta_{1p} = \eta_1 \cos \theta_1} \tag{67}$$

and

$$\boxed{\eta_{2p} = \eta_2 \cos \theta_2} \tag{68}$$

Using this representation, Eqs. (65) and (66) are now in a form that enables them to be solved together for the ratios E_{10}^-/E_{10}^+ and E_{20}/E_{10}^+. Performing analogous procedures to those used in solving (7) and (8), we find the reflection and transmission coefficients:

$$\boxed{\Gamma_p = \frac{E_{10}^-}{E_{10}^+} = \frac{\eta_{2p} - \eta_{1p}}{\eta_{2p} + \eta_{1p}}} \tag{69}$$

$$\boxed{\tau_p = \frac{E_{20}}{E_{10}^+} = \frac{2\eta_{2p}}{\eta_{2p} + \eta_{1p}} \left(\frac{\cos \theta_1}{\cos \theta_2} \right)} \tag{70}$$

A similar procedure can be carried out for s-polarization, referring to Figure 12.7b. The details are left as an exercise; the results are

$$\boxed{\Gamma_s = \frac{E_{y10}^-}{E_{y10}^+} = \frac{\eta_{2s} - \eta_{1s}}{\eta_{2s} + \eta_{1s}}} \tag{71}$$

$$\boxed{\tau_s = \frac{E_{y20}}{E_{y10}^+} = \frac{2\eta_{2s}}{\eta_{2s} + \eta_{1s}}} \tag{72}$$

where the effective impedances for s-polarization are

$$\boxed{\eta_{1s} = \eta_1 \sec \theta_1} \tag{73}$$

and

$$\boxed{\eta_{2s} = \eta_2 \sec \theta_2} \tag{74}$$

Equations (67) through (74) are what we need to calculate wave reflection and transmission for either polarization, and at any incident angle.

EXAMPLE 12.7

A uniform plane wave is incident from air onto glass at an angle from the normal of 30°. Determine the fraction of the incident power that is reflected and transmitted for (a) p-polarization and (b) s-polarization. Glass has refractive index $n_2 = 1.45$.

Solution. First, we apply Snell's law to find the transmission angle. Using $n_1 = 1$ for air, we use (63) to find

$$\theta_2 = \sin^{-1}\left(\frac{\sin 30}{1.45}\right) = 20.2°$$

Now, for p-polarization:

$$\eta_{1p} = \eta_1 \cos 30 = (377)(.866) = 326\ \Omega$$

$$\eta_{2p} = \eta_2 \cos 20.2 = \frac{377}{1.45}(.938) = 244\ \Omega$$

Then, using (69), we find

$$\Gamma_p = \frac{244 - 326}{244 + 326} = -0.144$$

The fraction of the incident power that is reflected is

$$\frac{P_r}{P_{\text{inc}}} = |\Gamma_p|^2 = .021$$

The transmitted fraction is then

$$\frac{P_t}{P_{\text{inc}}} = 1 - |\Gamma_p|^2 = .979$$

For s-polarization, we have

$$\eta_{1s} = \eta_1 \sec 30 = 377/.866 = 435\ \Omega$$

$$\eta_{2s} = \eta_2 \sec 20.2 = \frac{377}{1.45(.938)} = 277\ \Omega$$

Then, using (71):

$$\Gamma_s = \frac{277 - 435}{277 + 435} = -.222$$

The reflected power fraction is thus

$$|\Gamma_s|^2 = .049$$

The fraction of the incident power that is transmitted is

$$1 - |\Gamma_s|^2 = .951$$

In Example 12.7, reflection coefficient values for the two polarizations were found to be negative. The meaning of a negative reflection coefficient is that the component of the reflected electric field that is parallel to the interface will be directed opposite the incident field component when both are evaluated at the boundary.

This effect is also observed when the second medium is a perfect conductor. In this case, we know that the electric field inside the conductor must be zero. Consequently, $\eta_2 = E_{20}/H_{20} = 0$, and the reflection coefficients will be $\Gamma_p = \Gamma_s = -1$. Total reflection occurs, regardless of the incident angle or polarization.

12.6 TOTAL REFLECTION AND TOTAL TRANSMISSION OF OBLIQUELY INCIDENT WAVES

Now that we have methods available to us for solving problems involving oblique incidence reflection and transmission, we can explore the special cases of *total reflection* and *total transmission*. We look for special combinations of media, incidence angles, and polarizations that produce these properties. To begin, we identify the necessary condition for total reflection. We want total *power* reflection, so that $|\Gamma|^2 = \Gamma\Gamma^* = 1$, where Γ is either Γ_p or Γ_s. The fact that this condition involves the possibility of a complex Γ allows some flexibility. For the incident medium, we note that η_{1p} and η_{1s} will always be real and positive. On the other hand, when we consider the second medium, η_{2p} and η_{2s} involve factors of $\cos\theta_2$ or $1/\cos\theta_2$, where

$$\cos\theta_2 = \left[1 - \sin^2\theta_2\right]^{1/2} = \left[1 - \left(\frac{n_1}{n_2}\right)^2 \sin^2\theta_1\right]^{1/2} \tag{75}$$

where Snell's law has been used. We observe that $\cos\theta_2$, and hence η_{2p} and η_{2s}, become imaginary whenever $\sin\theta_1 > n_2/n_1$. Let us consider parallel polarization, for example. Under conditions of imaginary η_{2p}, (69) becomes

$$\Gamma_p = \frac{j|\eta_{2p}| - \eta_{1p}}{j|\eta_{2p}| + \eta_{1p}} = -\frac{\eta_{1p} - j|\eta_{2p}|}{\eta_{1p} + j|\eta_{2p}|} = -\frac{Z}{Z^*}$$

where $Z = \eta_{1p} - j|\eta_{2p}|$. *We can therefore see that $\Gamma_p\Gamma_p^* = 1$, meaning total power reflection, whenever η_{2p} is imaginary.* The same will be true whenever η_{2p} is zero, which will occur when $\sin\theta_1 = n_2/n_1$. We thus have our condition for total reflection, which is

$$\boxed{\sin\theta_1 \geq \frac{n_2}{n_1}} \tag{76}$$

From this condition arises the *critical angle* of total reflection, θ_c, defined through

$$\boxed{\sin\theta_c = \frac{n_2}{n_1}} \tag{77}$$

The total reflection condition can thus be more succinctly written as

$$\boxed{\theta_1 \geq \theta_c \quad \text{(for total reflection)}} \tag{78}$$

Note that for (76) and (77) to make sense, it must be true that $n_2 < n_1$, or the wave must be incident from a medium of higher refractive index than that of the medium beyond the boundary. For this reason, the total reflection condition is sometimes called total *internal* reflection; it is often seen (and applied) in optical devices such as beam-steering prisms, where light within the glass structure totally reflects from glass–air interfaces.

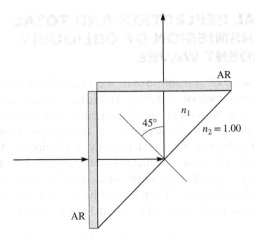

Figure 12.8 Beam-steering prism for Example 12.8.

EXAMPLE 12.8

A prism is to be used to turn a beam of light by 90°, as shown in Figure 12.8. Light enters and exits the prism through two antireflective (AR-coated) surfaces. Total reflection is to occur at the back surface, where the incident angle is 45° to the normal. Determine the minimum required refractive index of the prism material if the surrounding region is air.

Solution. Considering the back surface, the medium beyond the interface is air, with $n_2 = 1.00$. Because $\theta_1 = 45°$, (76) is used to obtain

$$n_1 \geq \frac{n_2}{\sin 45} = \sqrt{2} = 1.41$$

Because fused silica glass has refractive index $n_g = 1.45$, it is a suitable material for this application and is in fact widely used.

Another important application of total reflection is in *optical waveguides*. These, in their simplest form, are constructed of three layers of glass, in which the middle layer has a slightly higher refractive index than the outer two. Figure 12.9 shows the basic structure. Light, propagating from left to right, is confined to the middle layer by total reflection at the two interfaces, as shown. Optical fiber waveguides are constructed on this principle, in which a cylindrical glass core region of small radius is surrounded coaxially by a lower-index cladding glass material of larger radius. Basic waveguiding principles as applied to metallic and dielectric structures will be presented in Chapter 13.

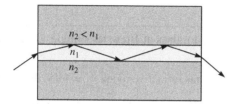

Figure 12.9 A dielectric slab waveguide (symmetric case), showing light confinement to the center material by total reflection.

We next consider the possibility of *total transmission*. In this case, the requirement is simply that $\Gamma = 0$. We investigate this possibility for the two polarizations. First, we consider s-polarization. If $\Gamma_s = 0$, then from (71) we require that $\eta_{2s} = \eta_{1s}$, or

$$\eta_2 \sec \theta_2 = \eta_1 \sec \theta_1$$

Using Snell's law to write θ_2 in terms of θ_1, the preceding equation becomes

$$\eta_2 \left[1 - \left(\frac{n_1}{n_2}\right)^2 \sin^2 \theta_1\right]^{-1/2} = \eta_1 \left[1 - \sin^2 \theta_1\right]^{-1/2}$$

There is no value of θ_1 that will satisfy this, so we turn instead to p-polarization. Using (67), (68), and (69), with Snell's law, we find that the condition for $\Gamma_p = 0$ is

$$\eta_2 \left[1 - \left(\frac{n_1}{n_2}\right)^2 \sin^2 \theta_1\right]^{1/2} = \eta_1 \left[1 - \sin^2 \theta_1\right]^{1/2}$$

This equation does have a solution, which is

$$\boxed{\sin \theta_1 = \sin \theta_B = \frac{n_2}{\sqrt{n_1^2 + n_2^2}}} \tag{79}$$

where we have used $\eta_1 = \eta_0/n_1$ and $\eta_2 = \eta_0/n_2$. We call this special angle θ_B, where total transmission occurs, the *Brewster angle* or *polarization angle*. The latter name comes from the fact that if light having both s- and p-polarization components is incident at $\theta_1 = \theta_B$, the p component will be totally transmitted, leaving the partially reflected light entirely s-polarized. At angles that are slightly off the Brewster angle, the reflected light is still predominantly s-polarized. Most reflected light that we see originates from horizontal surfaces (such as the surface of the ocean), and so the light has mostly horizontal polarization. Polaroid sunglasses take advantage of this fact to reduce glare, for they are made to block the transmission of horizontally polarized light while passing light that is vertically polarized.

EXAMPLE 12.9

Light is incident from air to glass at Brewster's angle. Determine the incident and transmitted angles.

Solution. Because glass has refractive index $n_2 = 1.45$, the incident angle will be

$$\theta_1 = \theta_B = \sin^{-1}\left(\frac{n_2}{\sqrt{n_1^2 + n_2^2}}\right) = \sin^{-1}\left(\frac{1.45}{\sqrt{1.45^2 + 1}}\right) = 55.4°$$

The transmitted angle is found from Snell's law, through

$$\theta_2 = \sin^{-1}\left(\frac{n_1}{n_2}\sin\theta_B\right) = \sin^{-1}\left(\frac{n_1}{\sqrt{n_1^2 + n_2^2}}\right) = 34.6°$$

Note from this exercise that $\sin\theta_2 = \cos\theta_B$, which means that the sum of the incident and refracted angles at the Brewster condition is always $90°$.

Many of the results we have seen in this section are summarized in Figure 12.10, in which Γ_p and Γ_s, from (69) and (71), are plotted as functions of the incident angle, θ_1. Curves are shown for selected values of the refractive index ratio, n_1/n_2. For all plots in which $n_1/n_2 > 1$, Γ_s and Γ_p achieve values of ±1 at the critical angle. At larger angles, the reflection coefficients become imaginary (and are not shown) but nevertheless retain magnitudes of unity. The occurrence of the Brewster angle is evident in the curves for Γ_p (Figure 12.10a) because all curves cross the θ_1 axis. This behavior is not seen in the Γ_s functions because Γ_s is positive for all values of θ_1 when $n_1/n_2 > 1$.

D12.5. In Example 12.9, calculate the reflection coefficient for s-polarized light.

Ans. −0.355

12.7 WAVE PROPAGATION IN DISPERSIVE MEDIA

In Chapter 11, we encountered situations in which the complex permittivity of the medium depends on frequency. This is true in all materials through a number of possible mechanisms. One of these, mentioned earlier, is that oscillating bound charges in a material are in fact harmonic oscillators that have resonant frequencies associated with them (see Appendix D). When the frequency of an incoming electromagnetic wave is at or near a bound charge resonance, the wave will induce strong oscillations; these in turn have the effect of depleting energy from the wave in its original form. The wave thus experiences absorption, and it does so to a greater extent than it would at a frequency that is detuned from resonance. A related effect is that the real part of

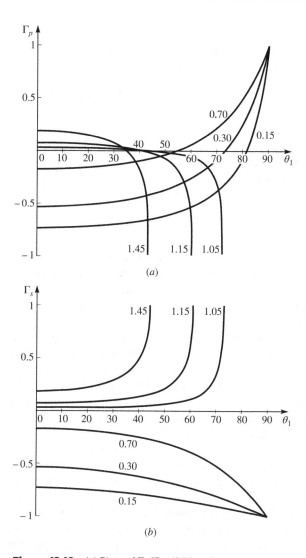

Figure 12.10 (a) Plots of Γ_p [Eq. (69)] as functions of the incident angle, θ_1, as shown in Figure 12.7a. Curves are shown for selected values of the refractive index ratio, n_1/n_2. Both media are lossless and have $\mu_r = 1$. Thus $\eta_1 = \eta_0/n_1$ and $\eta_2 = \eta_0/n_2$. (b) Plots of Γ_s [Eq. (71)] as functions of the incident angle, θ_1, as shown in Figure 12.7b. As in Figure 12.10a, the media are lossless, and curves are shown for selected n_1/n_2.

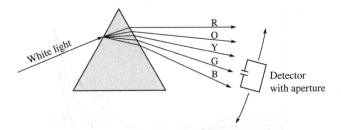

Figure 12.11   The angular dispersion of a prism can be measured using a movable device which measures both wavelength and power. The device senses light through a small aperture, thus improving wavelength resolution.

the dielectric constant will be different at frequencies near resonance than at frequencies far from resonance. In short, resonance effects give rise to values of ϵ' and ϵ'' that will vary continuously with frequency. These in turn will produce a fairly complicated frequency dependence in the attenuation and phase constants as expressed in Eqs. (44) and (45) in Chapter 11.

This section concerns the effect of a frequency-varying dielectric constant (or refractive index) on a wave as it propagates in an otherwise lossless medium. This situation arises quite often because significant refractive index variation can occur at frequencies far away from resonance, where absorptive losses are negligible. A classic example of this is the separation of white light into its component colors by a glass prism. In this case, the frequency-dependent refractive index results in different angles of refraction for the different colors—hence the separation. The color separation effect produced by the prism is known as *angular dispersion,* or more specifically, *chromatic* angular dispersion.

The term *dispersion* implies a *separation* of distinguishable components of a wave. In the case of the prism, the components are the various colors that have been spatially separated. An important point here is that the spectral *power* has been dispersed by the prism. We can illustrate this idea by considering what it would take to measure the difference in refracted angles between, for example, blue and red light. One would need to use a power detector with a very narrow aperture, as shown in Figure 12.11. The detector would be positioned at the locations of the blue and red light from the prism, with the narrow aperture allowing essentially one color at a time (or light over a very narrow spectral range) to pass through to the detector. The detector would then measure the power in what we could call a "spectral packet," or a very narrow slice of the total power spectrum. The smaller the aperture, the narrower the spectral width of the packet, and the greater the precision in the measurement.[4]

---

[4]To perform this experiment, one would need to measure the wavelength as well. To do this, the detector would likely be located at the output of a spectrometer or monochrometer whose input slit performs the function of the bandwidth-limiting aperture.

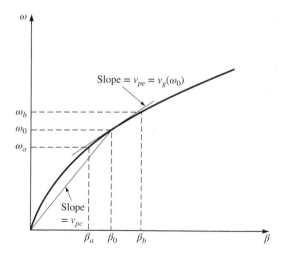

Figure 12.12 ω-β diagram for a material in which the refractive index increases with frequency. The slope of a line tangent to the curve at ω_0 is the group velocity at that frequency. The slope of a line joining the origin to the point on the curve at ω_0 is the phase velocity at ω_0.

It is important for us to think of wave power as subdivided into spectral packets in this way because it will figure prominently in our interpretation of the main topic of this section, which is wave dispersion *in time*.

We now consider a lossless nonmagnetic medium in which the refractive index varies with frequency. The phase constant of a uniform plane wave in this medium will assume the form

$$\beta(\omega) = k = \omega\sqrt{\mu_0 \epsilon(\omega)} = n(\omega)\frac{\omega}{c} \tag{80}$$

If we take $n(\omega)$ to be a monotonically increasing function of frequency (as is usually the case), a plot of ω versus β would look something like the curve shown in Figure 12.12. Such a plot is known as an ω-β *diagram* for the medium. Much can be learned about how waves propagate in the material by considering the shape of the ω-β curve.

Suppose we have two waves at two frequencies, ω_a and ω_b, which are co-propagating in the material and whose amplitudes are equal. The two frequencies are labeled on the curve in Figure 12.12, along with the frequency midway between the two, ω_0. The corresponding phase constants, β_a, β_b, and β_0, are also labeled. The electric fields of the two waves are linearly polarized in the same direction (along x, for example), while both waves propagate in the forward z direction. The waves will thus interfere with each other, producing a resultant wave whose field function can be found simply by adding the **E** fields of the two waves. This addition is done using the complex fields:

$$E_{c,\text{net}}(z,t) = E_0[e^{-j\beta_a z} e^{j\omega_a t} + e^{-j\beta_b z} e^{j\omega_b t}]$$

Note that we must use the full complex forms (with frequency dependence retained) as opposed to the phasor forms, since the waves are at different frequencies. Next, we factor out the term $e^{-j\beta_0 z} e^{j\omega_0 t}$:

$$E_{c,\text{net}}(z,t) = E_0 e^{-j\beta_0 z} e^{j\omega_0 t}\left[e^{j\Delta\beta z} e^{-j\Delta\omega t} + e^{-j\Delta\beta z} e^{j\Delta\omega t}\right]$$

$$= 2E_0 e^{-j\beta_0 z} e^{j\omega_0 t}\cos(\Delta\omega t - \Delta\beta z) \tag{81}$$

where

$$\Delta\omega = \omega_0 - \omega_a = \omega_b - \omega_0$$

and

$$\Delta\beta = \beta_0 - \beta_a = \beta_b - \beta_0$$

The preceding expression for $\Delta\beta$ is approximately true as long as $\Delta\omega$ is small. This can be seen from Figure 12.12 by observing how the shape of the curve affects $\Delta\beta$, given uniform frequency spacings.

The real instantaneous form of (81) is found through

$$\varepsilon_{\text{net}}(z,t) = \mathcal{R}e\{E_{c,\text{net}}\} = 2E_0\cos(\Delta\omega t - \Delta\beta z)\cos(\omega_0 t - \beta_0 z) \tag{82}$$

If $\Delta\omega$ is fairly small compared to ω_0, we recognize (82) as a carrier wave at frequency ω_0 that is sinusoidally modulated at frequency $\Delta\omega$. The two original waves are thus "beating" together to form a slow modulation, as one would hear when the same note is played by two slightly out-of-tune musical instruments. The resultant wave is shown in Figure 12.13.

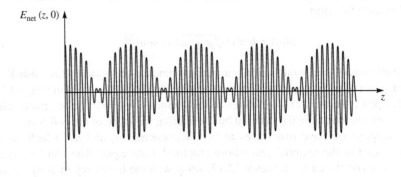

Figure 12.13 Plot of the total electric field strength as a function of z (with $t = 0$) of two co-propagating waves having different frequencies, ω_a and ω_b, as per Eq. (81). The rapid oscillations are associated with the carrier frequency, $\omega_0 = (\omega_a + \omega_b)/2$. The slower modulation is associated with the envelope or "beat" frequency, $\Delta\omega = (\omega_b - \omega_a)/2$.

Of interest to us are the phase velocities of the carrier wave and the modulation envelope. From (82), we can immediately write these down as:

$$v_{pc} = \frac{\omega_0}{\beta_0} \quad \text{(carrier velocity)} \tag{83}$$

$$v_{pe} = \frac{\Delta\omega}{\Delta\beta} \quad \text{(envelope velocity)} \tag{84}$$

Referring to the ω-β diagram, Figure 12.12, we recognize the carrier phase velocity as the slope of the straight line that joins the origin to the point on the curve whose coordinates are ω_0 and β_0. We recognize the envelope velocity as a quantity that approximates the slope of the ω-β curve at the location of an operation point specified by (ω_0, β_0). The envelope velocity in this case is thus somewhat less than the carrier velocity. As $\Delta\omega$ becomes vanishingly small, the envelope velocity is exactly the slope of the curve at ω_0. We can therefore state the following for our example:

$$\boxed{\lim_{\Delta\omega \to 0} \frac{\Delta\omega}{\Delta\beta} = \left.\frac{d\omega}{d\beta}\right|_{\omega_0} = v_g(\omega_0)} \tag{85}$$

The quantity $d\omega/d\beta$ is called the *group velocity* function for the material, $v_g(\omega)$. When evaluated at a specified frequency ω_0, it represents the velocity of a group of frequencies within a spectral packet of vanishingly small width, centered at frequency ω_0. In stating this, we have extended our two-frequency example to include waves that have a continuous frequency spectrum. Each frequency component (or packet) is associated with a group velocity at which the energy in that packet propagates. Since the slope of the ω-β curve changes with frequency, group velocity will obviously be a function of frequency. The *group velocity dispersion* of the medium is, to the first order, the rate at which the slope of the ω-β curve changes with frequency. It is this behavior that is of critical practical importance to the propagation of modulated waves within dispersive media and to understanding the extent to which the modulation envelope may degrade with propagation distance.

EXAMPLE 12.10

Consider a medium in which the refractive index varies linearly with frequency over a certain range:

$$n(\omega) = n_0 \frac{\omega}{\omega_0}$$

Determine the group velocity and the phase velocity of a wave at frequency ω_0.

Solution. First, the phase constant will be

$$\beta(\omega) = n(\omega)\frac{\omega}{c} = \frac{n_0\omega^2}{\omega_0 c}$$

Now

$$\frac{d\beta}{d\omega} = \frac{2n_0\omega}{\omega_0 c}$$

so that

$$v_g = \frac{d\omega}{d\beta} = \frac{\omega_0 c}{2 n_0 \omega}$$

The group velocity at ω_0 is

$$v_g(\omega_0) = \frac{c}{2 n_0}$$

The phase velocity at ω_0 will be

$$v_p(\omega_0) = \frac{\omega}{\beta(\omega_0)} = \frac{c}{n_0}$$

12.8 PULSE BROADENING IN DISPERSIVE MEDIA

To see how a dispersive medium affects a modulated wave, let us consider the propagation of an electromagnetic pulse. Pulses are used in digital signals, where the presence or absence of a pulse in a given time slot corresponds to a digital "one" or "zero." The effect of the dispersive medium on a pulse is to broaden it in time. To see how this happens, we consider the pulse *spectrum*, which is found through the Fourier transform of the pulse in time domain. In particular, suppose the pulse shape in time is Gaussian, and has electric field given at position $z = 0$ by

$$E(0,t) = E_0 e^{-\frac{1}{2}(t/T)^2} e^{j\omega_0 t} \tag{86}$$

where E_0 is a constant, ω_0 is the carrier frequency, and T is the characteristic half-width of the pulse envelope; this is the time at which the pulse *intensity*, or magnitude of the Poynting vector, falls to $1/e$ of its maximum value (note that intensity is proportional to the square of the electric field). The frequency spectrum of the pulse is the Fourier transform of (86), which is

$$E(0,\omega) = \frac{E_0 T}{\sqrt{2\pi}} e^{-\frac{1}{2}T^2(\omega - \omega_0)^2} \tag{87}$$

Note from (87) that the frequency displacement from ω_0 at which the spectral *intensity* (proportional to $|E(0,\omega)|^2$) falls to $1/e$ of its maximum is $\Delta\omega = \omega - \omega_0 = 1/T$.

Figure 12.14a shows the Gaussian intensity spectrum of the pulse, centered at ω_0, where the frequencies corresponding to the $1/e$ spectral intensity positions, ω_a and ω_b, are indicated. Figure 12.14b shows the same three frequencies marked on the ω-β curve for the medium. Three lines are drawn that are tangent to the curve at the three frequency locations. The slopes of the lines indicate the group velocities at ω_a, ω_b, and ω_0, indicated as v_{ga}, v_{gb}, and v_{g0}. We can think of the pulse spreading in time as resulting from the differences in propagation times of the spectral energy packets that make up the pulse spectrum. Since the pulse spectral energy is highest at the center frequency, ω_0, we can use this as a reference point about which further

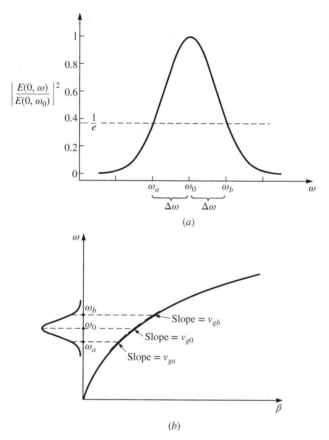

Figure 12.14 (a) Normalized power spectrum of a Gaussian pulse, as determined from Eq. (86). The spectrum is centered at carrier frequency ω_0 and has 1/e half-width, $\Delta\omega$. Frequencies ω_a and ω_b correspond to the 1/e positions on the spectrum. (b) The spectrum of Figure 12.14a as shown on the ω-β diagram for the medium. The three frequencies specified in Figure 12.14a are associated with three different slopes on the curve, resulting in different group delays for the spectral components.

spreading of the energy will occur. For example, let us consider the difference in arrival times (group delays) between the frequency components, ω_0 and ω_b, after propagating through a distance z of the medium:

$$\Delta\tau = z\left(\frac{1}{v_{gb}} - \frac{1}{v_{g0}}\right) = z\left(\left.\frac{d\beta}{d\omega}\right|_{\omega_b} - \left.\frac{d\beta}{d\omega}\right|_{\omega_0}\right) \qquad (88)$$

The essential point is that the medium is acting as what could be called a *temporal prism*. Instead of spreading out the spectral energy packets spatially, it is spreading them out in time. In this process, a new temporal pulse envelope is constructed whose width is based fundamentally on the spread of propagation delays of the different spectral components. By determining the delay difference between the peak spectral component and the component at the spectral half-width, we construct an expression for the new *temporal* half-width. This assumes, of course, that the initial pulse width is negligible in comparison, but if not, we can account for that also, as will be shown later on.

To evaluate (88), we need more information about the ω-β curve. If we assume that the curve is smooth and has fairly uniform curvature, we can express $\beta(\omega)$ as the first three terms of a Taylor series expansion about the carrier frequency, ω_0:

$$\beta(\omega) \doteq \beta(\omega_0) + (\omega - \omega_0)\beta_1 + \frac{1}{2}(\omega - \omega_0)^2\beta_2 \tag{89}$$

where

$$\beta_0 = \beta(\omega_0)$$
$$\beta_1 = \frac{d\beta}{d\omega}\bigg|_{\omega_0} \tag{90}$$

and

$$\beta_2 = \frac{d^2\beta}{d\omega^2}\bigg|_{\omega_0} \tag{91}$$

Note that if the ω-β curve were a straight line, then the first two terms in (89) would precisely describe $\beta(\omega)$. It is the third term in (89), involving β_2, that describes the curvature and ultimately the dispersion.

Noting that β_0, β_1, and β_2 are constants, we take the first derivative of (89) with respect to ω to find

$$\frac{d\beta}{d\omega} = \beta_1 + (\omega - \omega_0)\beta_2 \tag{92}$$

We now substitute (92) into (88) to obtain

$$\Delta\tau = [\beta_1 + (\omega_b - \omega_0)\beta_2]z - [\beta_1 + (\omega_0 - \omega_0)\beta_2]z = \Delta\omega\beta_2 z = \frac{\beta_2 z}{T} \tag{93}$$

where $\Delta\omega = (\omega_b - \omega_0) = 1/T$. β_2, as defined in Eq. (91), is the *dispersion parameter*. Its units are in general time^2/distance, that is, pulse spread in time per unit spectral bandwidth, per unit distance. In optical fibers, for example, the units most commonly used are picoseconds^2/kilometer (psec^2/km). β_2 can be determined when we know how β varies with frequency, or it can be measured.

If the initial pulse width is very short compared to $\Delta\tau$, then the broadened pulse width at location z will be simply $\Delta\tau$. If the initial pulse width is comparable to $\Delta\tau$, then the pulse width at z can be found through the convolution of the initial Gaussian pulse envelope of width T with a Gaussian envelope whose width is $\Delta\tau$. Thus, in general, the pulse width at location z will be

$$T' = \sqrt{T^2 + (\Delta\tau)^2} \tag{94}$$

EXAMPLE 12.11

An optical fiber link is known to have dispersion $\beta_2 = 20$ ps^2/km. A Gaussian light pulse at the input of the fiber is of initial width $T = 10$ ps. Determine the width of the pulse at the fiber output if the fiber is 15 km long.

Solution. The pulse spread will be

$$\Delta\tau = \frac{\beta_2 z}{T} = \frac{(20)(15)}{10} = 30 \text{ ps}$$

So the output pulse width is

$$T' = \sqrt{(10)^2 + (30)^2} = 32 \text{ ps}$$

An interesting by-product of pulse broadening through chromatic dispersion is that the broadened pulse is *chirped*. This means that the instantaneous frequency of the pulse varies monotonically (either increases or decreases) with time over the pulse envelope. This again is just a manifestation of the broadening mechanism, in which the spectral components at different frequencies are spread out in time as they propagate at different group velocities. We can quantify the effect by calculating the group delay, τ_g, as a function of frequency, using (92). We obtain:

$$\tau_g = \frac{z}{v_g} = z\frac{d\beta}{d\omega} = (\beta_1 + (\omega - \omega_0)\beta_2)z \tag{95}$$

This equation tells us that the group delay will be a linear function of frequency and that higher frequencies will arrive at later times if β_2 is positive. We refer to the chirp as positive if the lower frequencies lead the higher frequencies in time [requiring a positive β_2 in (95)]; chirp is negative if the higher frequencies lead in time (negative β_2). Figure 12.15 shows the broadening effect and illustrates the chirping phenomenon.

D12.6. For the fiber link of Example 12.11, a 20-ps pulse is input instead of the 10-ps pulse in the example. Determine the output pulsewidth.

Ans. 25 ps

As a final point, we note that the pulse bandwidth, $\Delta\omega$, was found to be $1/T$. This is true as long as the Fourier transform of the pulse *envelope* is taken, as was done with (86) to obtain (87). In that case, E_0 was taken to be a constant, and so the only time variation arose from the carrier wave and the Gaussian envelope. Such a

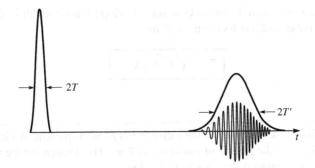

Figure 12.15 Gaussian pulse intensities as functions of time (smooth curves) before and after propagation through a dispersive medium, as exemplified by the ω-β diagram of Figure 12.14b. The electric field oscillations are shown under the second trace to demonstrate the chirping effect as the pulse broadens. Note the reduced amplitude of the broadened pulse, which occurs because the pulse energy (the area under the intensity envelope) is constant.

pulse, whose frequency spectrum is obtained only from the pulse envelope, is known as *transform-limited*. In general, however, additional frequency bandwidth may be present since E_0 may vary with time for one reason or another (such as phase noise that could be present on the carrier). In these cases, pulse broadening is found from the more general expression

$$\Delta\tau = \Delta\omega\beta_2 z \tag{96}$$

where $\Delta\omega$ is the net spectral bandwidth arising from all sources. Clearly, transform-limited pulses are preferred in order to minimize broadening because these will have the smallest spectral width for a given pulse width.

REFERENCES

1. DuBroff, R. E., S. V. Marshall, and G. G. Skitek. *Electromagnetic Concepts and Applications*. 4th ed. Englewood Cliffs, N. J.: Prentice-Hall, 1996. Chapter 9 of this text develops the concepts presented here, with additional examples and applications.

2. Iskander, M. F. *Electromagnetic Fields and Waves*. Englewood Cliffs, N. J.: Prentice-Hall, 1992. The multiple interface treatment in Chapter 5 of this text is particularly good.

3. Harrington, R. F. *Time-Harmonic Electromagnetic Fields*. New York: McGraw-Hill, 1961. This advanced text provides a good overview of general wave reflection concepts in Chapter 2.

4. Marcuse, D. *Light Transmission Optics*. New York: Van Nostrand Reinhold, 1982. This intermediate-level text provides detailed coverage of optical waveguides and pulse propagation in dispersive media.

CHAPTER 12 PROBLEMS

12.1 A uniform plane wave in air, $E_{x1}^{+} = E_{x10}^{+}\cos(10^{10}t - \beta z)$ V/m, is normally incident on a copper surface at $z = 0$. What percentage of the incident power density is transmitted into the copper?

12.2 The plane $z = 0$ defines the boundary between two dielectrics. For $z < 0$, $\epsilon_{r1}' = 9$, $\epsilon_{r1}'' = 0$, and $\mu_1 = \mu_0$. For $z > 0$, $\epsilon_{r2}' = 3$, $\epsilon_{r2}'' = 0$, and $\mu_2 = \mu_0$. Let $E_{x1}^{+} = 10\cos(\omega t - 15z)$V/m and find (a) ω; (b)$\langle \mathbf{S}_1^{+} \rangle$; (c)$\langle \mathbf{S}_1^{-} \rangle$; (d)$\langle \mathbf{S}_2^{+} \rangle$.

12.3 A uniform plane wave in region 1 is normally incident on the planar boundary separating regions 1 and 2. If $\epsilon_1'' = \epsilon_2'' = 0$, while $\epsilon_{r1}' = \mu_{r1}^3$ and $\epsilon_{r2}' = \mu_{r2}^3$, find the ratio $\epsilon_{r2}'/\epsilon_{r1}'$ if 20 percent of the energy in the incident wave is reflected at the boundary. There are two possible answers.

12.4 A 10-MHz uniform plane wave having an initial average power density of 5 W/m^2 is normally incident from free space onto the surface of a lossy material in which $\epsilon_2''/\epsilon_2' = 0.05$, $\epsilon_{r2}' = 5$, and $\mu_2 = \mu_0$. Calculate the distance into the lossy medium at which the transmitted wave power density is down by 10 dB from the initial 5 W/m^2.

12.5 The region $z < 0$ is characterized by $\epsilon_r' = \mu_r = 1$and $\epsilon_r'' = 0$. The total **E** field here is given as the sum of two uniform plane waves, $\mathbf{E}_s = 150e^{-j10z}\mathbf{a}_x + (50\angle 20°)e^{j10z}\mathbf{a}_x$V/m. (a) What is the operating frequency? (b) Specify the intrinsic impedance of the region $z > 0$ that would provide the appropriate reflected wave. (c) At what value of z, -10 cm $< z < 0$, is the total electric field intensity a maximum amplitude?

12.6 In the beam-steering prism of Example 12.8, suppose the antireflective coatings are removed, leaving bare glass-to-air interfaces. Calculate the ratio of the prism output power to the input power, assuming a single transit.

12.7 The semi-infinite regions $z < 0$ and $z > 1$ m are free space. For $0 < z < 1$ m, $\epsilon_r' = 4$, $\mu_r = 1$, and $\epsilon_r'' = 0$. A uniform plane wave with $\omega = 4 \times 10^8$ rad/s is traveling in the $\mathbf{a}_z$ direction toward the interface at $z = 0$. (a) Find the standing wave ratio in each of the three regions. (b) Find the location of the maximum |**E**| for $z < 0$ that is nearest to $z = 0$.

12.8 A wave starts at point a, propagates 1 m through a lossy dielectric rated at 0.1 dB/cm, reflects at normal incidence at a boundary at which $\Gamma = 0.3 + j0.4$, and then returns to point a. Calculate the ratio of the final power to the incident power after this round trip, and specify the overall loss in decibels.

12.9 Region 1, $z < 0$, and region 2, $z > 0$, are both perfect dielectrics ($\mu = \mu_0$, $\epsilon'' = 0$). A uniform plane wave traveling in the $\mathbf{a}_z$ direction has a radian frequency of 3×10^{10} rad/s. Its wavelengths in the two regions are $\lambda_1 = 5$ cm and $\lambda_2 = 3$ cm. What percentage of the energy incident on the boundary is (a) reflected; (b) transmitted? (c) What is the standing wave ratio in region 1?

12.10 In Figure 12.1, let region 2 be free space, while $\mu_{r1} = 1$, $\epsilon_{r1}'' = 0$, and ϵ_{r1}' is unknown. Find ϵ_{r1}' if (a) the amplitude of $\mathbf{E}_1^-$ is one-half that of $\mathbf{E}_1^+$; (b) $\langle \mathbf{S}_1^- \rangle$ is one-half of $\langle \mathbf{S}_1^+ \rangle$; (c) $|\mathbf{E}_1|_{min}$ is one-half of $|\mathbf{E}_1|_{max}$.

12.11 A 150-MHz uniform plane wave is normally incident from air onto a material whose intrinsic impedance is unknown. Measurements yield a standing wave ratio of 3 and the appearance of an electric field minimum at 0.3 wavelengths in front of the interface. Determine the impedance of the unknown material.

12.12 A 50-MHz uniform plane wave is normally incident from air onto the surface of a calm ocean. For seawater, $\sigma = 4$ S/m, and $\epsilon_r' = 78$. (a) Determine the fractions of the incident power that are reflected and transmitted. (b) Qualitatively, how (if at all) will these answers change as the frequency is increased?

12.13 A right-circularly polarized plane wave is normally incident from air onto a semi-infinite slab of plexiglas ($\epsilon_r' = 3.45$, $\epsilon_r'' = 0$). Calculate the fractions of the incident power that are reflected and transmitted. Also, describe the polarizations of the reflected and transmitted waves.

12.14 A left-circularly polarized plane wave is normally incident onto the surface of a perfect conductor. (a) Construct the superposition of the incident and reflected waves in phasor form. (b) Determine the real instantaneous form of the result of part a. (c) Describe the wave that is formed.

12.15 Sulfur hexafluoride (SF_6) is a high-density gas that has refractive index $n_s = 1.8$ at a specified pressure, temperature, and wavelength. Consider the retro-reflecting prism shown in Fig. 12.16 that is immersed in SF_6. Light enters through a quarter-wave antireflective coating and then totally reflects from the back surfaces of the glass. In principle, the beam should experience zero loss at the design wavelength ($P_{out} = P_{in}$). (a) Determine the minimum required value of the glass refractive index, n_g, so that the interior beam will totally reflect. (b) Knowing n_g, find the required refractive index of the quarter-wave film, n_f. (c) With the SF_6 gas evacuated from the chamber, and with the glass and film values as previously found, find the ratio P_{out}/P_{in}.

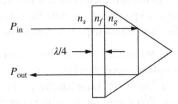

Figure 12.16   See Problem 12.15.

Assume very slight misalignment so that the long beam path through the prism is not retraced by reflected waves.

12.16 In Figure 12.5, let regions 2 and 3 both be of quarter-wave thickness. Region 4 is glass, having refractive index $n_4 = 1.45$; region 1 is air. (a) Find $\eta_{in,b}$. (b) Find $\eta_{in,a}$. (c) Specify a relation between the four intrinsic impedances that will enable total transmission of waves incident from the left into region 4. (d) Specify refractive index values for regions 2 and 3 that will accomplish the condition of part c. (e) Find the fraction of incident power transmitted if the two layers were of half-wave thickness instead of quarter-wave.

12.17 A uniform plane wave in free space is normally incident onto a dense dielectric plate of thickness $\lambda/4$, having refractive index n. Find the required value of n such that exactly half the incident power is reflected (and half transmitted). Remember that $n > 1$.

12.18 A uniform plane wave is normally incident onto a slab of glass ($n = 1.45$) whose back surface is in contact with a perfect conductor. Determine the reflective phase shift at the front surface of the glass if the glass thickness is (a) $\lambda/2$; (b) $\lambda/4$; (c) $\lambda/8$.

12.19 You are given four slabs of lossless dielectric, all with the same intrinsic impedance, η, known to be different from that of free space. The thickness of each slab is $\lambda/4$, where λ is the wavelength as measured in the slab material. The slabs are to be positioned parallel to one another, and the combination lies in the path of a uniform plane wave, normally incident. The slabs are to be arranged such that the airspaces between them are either zero, one-quarter wavelength, or one-half wavelength in thickness. Specify an arrangement of slabs and airspaces such that (a) the wave is totally transmitted through the stack, and (b) the stack presents the highest reflectivity to the incident wave. Several answers may exist.

12.20 The 50-MHz plane wave of Problem 12.12 is incident onto the ocean surface at an angle to the normal of 60°. Determine the fractions of the incident power that are reflected and transmitted for (a) s-polarization, and (b) p-polarization.

12.21 A right-circularly polarized plane wave in air is incident at Brewster's angle onto a semi-infinite slab of plexiglas ($\epsilon_r' = 3.45$, $\epsilon_r'' = 0$). (a) Determine the fractions of the incident power that are reflected and transmitted. (b) Describe the polarizations of the reflected and transmitted waves.

12.22 A dielectric waveguide is shown in Figure 12.17 with refractive indices as labeled. Incident light enters the guide at angle ϕ from the front surface normal as shown. Once inside, the light totally reflects at the upper $n_1 - n_2$ interface, where $n_1 > n_2$. All subsequent reflections from the upper and

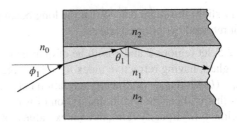

Figure 12.17 See Problems 12.22 and 12.23.

lower boundaries will be total as well, and so the light is confined to the guide. Express, in terms of n_1 and n_2, the maximum value of ϕ such that total confinement will occur, with $n_0 = 1$. The quantity $\sin \phi$ is known as the *numerical aperture* of the guide.

12.23 Suppose that ϕ in Figure 12.17 is Brewster's angle, and that θ_1 is the critical angle. Find n_0 in terms of n_1 and n_2.

12.24 A *Brewster prism* is designed to pass p-polarized light without any reflective loss. The prism of Figure 12.18 is made of glass ($n = 1.45$) and is in air. Considering the light path shown, determine the vertex angle α.

12.25 In the Brewster prism of Figure 12.18, determine for s-polarized light the fraction of the incident power that is transmitted through the prism, and from this specify the dB *insertion loss*, defined as $10\log_{10}$ of that number.

12.26 Show how a single block of glass can be used to turn a p-polarized beam of light through 180°, with the light suffering (in principle) zero reflective loss. The light is incident from air, and the returning beam (also in air) may be displaced sideways from the incident beam. Specify all pertinent angles and use $n = 1.45$ for glass. More than one design is possible here.

12.27 Using Eq. (79) in Chapter 11 as a starting point, determine the ratio of the group and phase velocities of an electromagnetic wave in a good conductor. Assume conductivity does not vary with frequency.

12.28 Over a small wavelength range, the refractive index of a certain material varies approximately linearly with wavelength as $n(\lambda) \doteq n_a + n_b(\lambda - \lambda_a)$, where n_a, n_b and λ_a are constants, and where λ is the free-space wavelength.

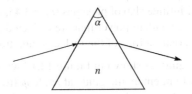

Figure 12.18 See Problems 12.24 and 12.25.

(*a*) Show that $d/d\omega = -(2\pi c/\omega^2)d/d\lambda$. (*b*) Using $\beta(\lambda) = 2\pi n/\lambda$, determine the wavelength-dependent (or independent) group delay over a unit distance. (*c*) Determine β_2 from your result of part *b*. (*d*) Discuss the implications of these results, if any, on pulse broadening.

12.29 A $T = 5$ ps transform-limited pulse propagates in a dispersive medium for which $\beta_2 = 10$ ps^2/km. Over what distance will the pulse spread to twice its initial width?

12.30 A $T = 20$ ps transform-limited pulse propagates through 10 km of a dispersive medium for which $\beta_2 = 12$ ps^2/km. The pulse then propagates through a second 10-km medium for which $\beta_2 = -12$ ps^2/km. Describe the pulse at the output of the second medium and give a physical explanation for what happened.

CHAPTER 13   Guided Waves

术语

waveguide   波导。
cutoff frequency   截止频率。
mode number   模数。
optical fiber   光纤。
TE wave   横电波。
TM wave   横磁波。

I n this chapter, we investigate several structures for guiding electromagnetic waves, and we explore the principles by which these operate. Included are transmission lines, which we first explored from the viewpoint of their currents and voltages in Chapter 10, and which we now revisit from a fields point of view. We then broaden the discussion to include several waveguiding devices. Broadly defined, a waveguide is a structure through which electromagnetic waves can be transmitted from point to point and within which the fields are confined to a certain extent. A transmission line fits this description, but it is a special case that employs two conductors, and it propagates a purely TEM field configuration. Waveguides in general depart from these restrictions and may employ any number of conductors and dielectrics—or as we will see, dielectrics alone and no conductors.

The chapter begins with a presentation of several transmission line structures, with emphasis on obtaining expressions for the primary constants, L, C, G, and R, for high- and low-frequency operating regimes. Next, we begin our study of waveguides by first taking a broad view of waveguide devices to obtain a physical understanding of how they work and the conditions under which they are used. We then explore the simple parallel-plate structure and distinguish between its operation as a transmission line and as a waveguide. In this device, the concept of waveguide modes is developed, as are the conditions under which these will occur. We will study the electric and magnetic field configurations of the guided modes using simple plane wave models and the wave equation. We will then study more complicated structures, including rectangular waveguides, dielectric slab waveguides, and optical fibers. ∎

13.1 TRANSMISSION LINE FIELDS AND PRIMARY CONSTANTS

We begin by establishing the equivalence between transmission line operations when considering voltage and current, from the point of view of the fields within the line. Consider, for example, the parallel-plate line shown in Figure 13.1. In the line, we

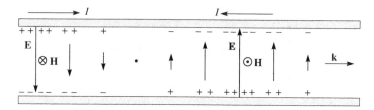

Figure 13.1   A transmission-line wave represented by voltage and current distributions along the length is associated with transverse electric and magnetic fields, forming a TEM wave.

1. 采取双导线、同轴电缆与波导可以实现电磁波的定向传输。双线方式在传输高频电磁波时导线的欧姆损耗、特别是辐射损失会加剧；同轴电缆的导线欧姆损耗也随频率的增高而增大；其结果是不仅传输效率低还可能产生信号失真。所以，对于高频电磁波的定向传输要采用波导。

2. 电磁波在中空的金属波导管中传输，是经管壁多次反射曲折地向前传播。对于介质波导，较大入射角度的电磁波可以在波导管内发生全反射而向前传播，这就是光纤传输光信号的原理。

3. 对于一种结构尺寸的金属波导，存在一个截止频率，只有当电磁波的频率大于该值时，电磁波才能通过波导进行传输。

assume that the plate spacing, d, is much less than the line width, b (into the page), so electric and magnetic fields can be assumed to be uniform within any transverse plane. Lossless propagation is also assumed. Figure 13.1 shows the side view, which includes the propagation axis z. The fields, along with the voltage and current, are shown at an instant in time.

The voltage and current in phasor form are:

$$V_s(z) = V_0 e^{-j\beta z} \tag{1a}$$

$$I_s(z) = \frac{V_0}{Z_0} e^{-j\beta z} \tag{1b}$$

where $Z_0 = \sqrt{L/C}$. The electric field in a given transverse plane at location z is just the parallel-plate capacitor field:

$$E_{sx}(z) = \frac{V_s}{d} = \frac{V_0}{d} e^{-j\beta z} \tag{2a}$$

The magnetic field is equal to the surface current density, assumed uniform, on either plate [Eq. (12), Chapter 7]:

$$H_{sy}(z) = K_{sz} = \frac{I_s}{b} = \frac{V_0}{bZ_0} e^{-j\beta z} \tag{2b}$$

The two fields, both uniform, orthogonal, and lying in the transverse plane, are identical in form to those of a uniform plane wave. As such, they are transverse electromagnetic (TEM) fields, also known simply as transmission-line fields. They differ from the fields of the uniform plane wave only in that they exist within the interior of the line, and nowhere else.

The power flow down the line is found through the time-average Poynting vector, integrated over the line cross section. Using (2a) and (2b), we find:

$$P_z = \int_0^b \int_0^d \frac{1}{2} \mathcal{R}e\{E_{xs}H_{ys}^*\}dxdy = \frac{1}{2}\frac{V_0}{d}\frac{V_0^*}{bZ_0^*}(bd) = \frac{|V_0|^2}{2Z_0^*} = \frac{1}{2}\mathcal{R}e\{V_s I_s^*\} \tag{3}$$

The power transmitted by the line is one of the most important quantities that we wish to know from a practical standpoint. Eq. (3) shows that this can be obtained consistently through the line fields, or through the voltage and current. As would be

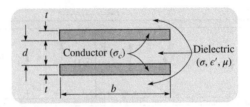

Figure 13.2 The geometry of the parallel-plate transmission line.

expected, this consistency is maintained when losses are included. The fields picture is in fact advantageous, and is generally preferred, since it is easy to incorporate dielectric loss mechanisms (other than conductivity) in addition to the dispersive properties of the dielectric. The transmission-line fields are also needed to produce the primary constants, as we now demonstrate for the parallel-plate line and other selected line geometries.

We assume the line is filled with dielectric having permittivity ϵ', conductivity σ, and permeability μ, usually μ_0 (Figure 13.2). The upper and lower plate thickness is t, which, along with the plate width b and plate conductivity σ_c, is used to evaluate the resistance per unit length parameter R under low-frequency conditions. We will, however, consider high-frequency operation, in which the skin effect gives an effective plate thickness or skin depth δ that is much less than t.

First, the capacitance and conductance per unit length are simply those of the parallel-plate structure, assuming static fields. Using Eq. (27) from Chapter 6, we find

$$C = \frac{\epsilon' b}{d} \tag{4}$$

The value of permittivity used should be appropriate for the range of operating frequencies considered.

The conductance per unit length may be determined easily from the capacitance expression by use of the simple relation between capacitance and resistance [Eq. (45), Chapter 6]:

$$G = \frac{\sigma}{\epsilon'} C = \frac{\sigma b}{d} \tag{5}$$

The evaluation of L and R involves the assumption of a well-developed skin effect such that $\delta \ll t$. Consequently, the inductance is primarily external because the magnetic flux within either conductor is negligible compared to that between conductors. Therefore,

$$L \doteq L_{\text{ext}} = \frac{\mu d}{b} \tag{6}$$

Note that $L_{\text{ext}} C = \mu \epsilon' = 1/v_p^2$, and we are therefore able to evaluate the external inductance for any transmission line for which we know the capacitance and insulator characteristics.

The last of the four parameters that we need is the resistance R per unit length. If the frequency is very high and the skin depth δ is very small, then we obtain an appropriate expression for R by distributing the total current uniformly throughout a depth δ. The skin effect resistance (through both conductors in series over a unit length) is

$$R = \frac{2}{\sigma_c \delta b} \tag{7}$$

Finally, it is convenient to include the common expression for the characteristic impedance of the line here with the parameter formulas:

$$Z_0 = \sqrt{\frac{L_{\text{ext}}}{C}} = \sqrt{\frac{\mu}{\epsilon'}} \frac{d}{b} \tag{8}$$

If necessary, a more accurate value may be obtained from Eq. (47), Chapter 10. Note that when substituting (8) into (2b), and using (2a), we obtain the expected relation for a TEM wave, $E_{xs} = \eta H_{ys}$, where $\eta = \sqrt{\mu/\epsilon'}$.

D13.1. Parameters for the planar transmission line shown in Figure 13.2 are $b = 6$ mm, $d = 0.25$ mm, $t = 25$ mm, $\sigma_c = 5.5 \times 10^7$ S/m, $\epsilon' = 25$ pF/m, $\mu = \mu_0$, and $\sigma/\omega\epsilon' = 0.03$. If the operating frequency is 750 MHz, calculate: (a) α; (b) β; (c) Z_0.

Ans. (a) 0.47 Np/m; (b) 26 rad/m; (c) $9.3\angle 0.7° \, \Omega$

13.1.1 Coaxial (High Frequencies)

We next consider a coaxial cable in which the dielectric has an inner radius a and outer radius b (Figure 13.3). The capacitance per unit length, obtained as Eq. (5) of Section 6.3, is

$$C = \frac{2\pi\epsilon'}{\ln(b/a)} \tag{9}$$

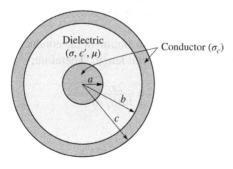

Figure 13.3 Coaxial transmission-line geometry.

Now, using the relation $RC = \epsilon/\sigma$ (see Problem 6.16), the conductance is

$$G = \frac{2\pi\sigma}{\ln(b/a)} \qquad (10)$$

where σ is the conductivity of the dielectric between the conductors at the operating frequency.

The inductance per unit length was computed for the coaxial cable as Eq. (50) in Section 8.10,

$$L_{\text{ext}} = \frac{\mu}{2\pi}\ln(b/a) \qquad (11)$$

Again, this is an external inductance, for the small skin depth precludes any appreciable magnetic flux within the conductors.

For a circular conductor of radius a and conductivity σ_c, we let Eq. (90) of Section 11.4 apply to a unit length, obtaining

$$R_{\text{inner}} = \frac{1}{2\pi a \delta \sigma_c}$$

There is also a resistance for the outer conductor, which has an inner radius b. We assume the same conductivity σ_c and the same value of skin depth δ, leading to

$$R_{\text{outer}} = \frac{1}{2\pi b \delta \sigma_c}$$

Because the line current flows through these two resistances in series, the total resistance is the sum:

$$R = \frac{1}{2\pi\delta\sigma_c}\left(\frac{1}{a} + \frac{1}{b}\right) \qquad (12)$$

Finally, the characteristic impedance, assuming low losses, is

$$Z_0 = \sqrt{\frac{L_{\text{ext}}}{C}} = \frac{1}{2\pi}\sqrt{\frac{\mu}{\epsilon'}}\ln\frac{b}{a} \qquad (13)$$

13.1.2 Coaxial (Low Frequencies)

We now obtain the coaxial line parameter values at very low frequencies where there is no appreciable skin effect and the current is assumed to be distributed uniformly throughout the conductor cross sections.

We first note that the current distribution in the conductor does not affect either the capacitance or conductance per unit length. Therefore,

$$C = \frac{2\pi\epsilon'}{\ln(b/a)} \qquad (14)$$

and

$$G = \frac{2\pi\sigma}{\ln(b/a)} \qquad (15)$$

The resistance per unit length may be calculated by dc methods, $R = l/(\sigma_c S)$, where $l = 1$ m and σ_c is the conductivity of the outer and inner conductors. The area of the center conductor is πa^2 and that of the outer is $\pi(c^2 - b^2)$. Adding the two resistance values, we have

$$R = \frac{1}{\sigma_c \pi}\left(\frac{1}{a^2} + \frac{1}{c^2 - b^2}\right) \tag{16}$$

Only one of the four parameter values remains to be found, the inductance per unit length. The external inductance that we calculated at high frequencies is the greatest part of the total inductance. To it, however, must be added smaller terms representing the internal inductances of the inner and outer conductors.

At very low frequencies where the current distribution is uniform, the internal inductance of the center conductor is the subject of Problem 43 in Chapter 8; the relationship is also given as Eq. (62) in Section 8.10:

$$L_{a,\text{int}} = \frac{\mu}{8\pi} \tag{17}$$

The determination of the internal inductance of the outer shell is a more difficult problem, and most of the work was requested in Problem 36 in Chapter 8. There, we found that the energy stored per unit length in an outer cylindrical shell of inner radius b and outer radius c with uniform current distribution is

$$W_H = \frac{\mu I^2}{16\pi(c^2 - b^2)}\left(b^2 - 3c^2 + \frac{4c^4}{c^2 - b^2}\ln\frac{c}{b}\right)$$

Thus the internal inductance of the outer conductor at very low frequencies is

$$L_{bc,\text{int}} = \frac{\mu}{8\pi(c^2 - b^2)}\left(b^2 - 3c^2 + \frac{4c^4}{c^2 - b^2}\ln\frac{c}{b}\right) \tag{18}$$

At low frequencies the total inductance is obtained by adding (11), (17), and (18):

$$L = \frac{\mu}{2\pi}\left[\ln\frac{b}{a} + \frac{1}{4} + \frac{1}{4(c^2 - b^2)}\left(b^2 - 3c^2 + \frac{4c^4}{c^2 - b^2}\ln\frac{c}{b}\right)\right] \tag{19}$$

13.1.3 Coaxial (Intermediate Frequencies)

There still remains the frequency interval where the skin depth is neither very much larger than nor very much smaller than the radius. In this case, the current distribution is governed by Bessel functions, and both the resistance and internal inductance are complicated expressions. Values are tabulated in the handbooks, and it is necessary to use them for very small conductor sizes at high frequencies and for larger conductor sizes used in power transmission at low frequencies.[1]

---

[1] Bessel functions are discussed within the context of optical fiber in Section 13.7. The current distribution, internal inductance, and internal resistance of round wires is discussed (with numerical examples) in Weeks, pp. 35–44. See the References at the end of this chapter.

D13.2. The dimensions of a coaxial transmission line are $a = 4$ mm, $b = 17.5$ mm, and $c = 20$ mm. The conductivity of the inner and outer conductors is 2×10^7 S/m, and the dielectric properties are $\mu_r = 1$, $\epsilon_r' = 3$, and $\sigma/\omega\epsilon' = 0.025$. Assume that the loss tangent is constant with frequency. Determine: (a) L, C, R, G, and Z_0 at 150 MHz; (b) L and R at 60 Hz.

Ans. (a) $0.30 \, \mu$H/m, 113 pF/m, $0.27 \, \Omega$/m, 2.7 mS/m, 51 Ω; (b) $0.36 \, \mu$H/m, 1.16 mΩ/m

13.1.4 Two-Wire (High Frequencies)

For the two-wire transmission line of Figure 13.4 with conductors of radius a and conductivity σ_c with center-to-center separation d in a medium of permeability μ, permittivity ϵ', and conductivity σ_c, the capacitance per unit length is found using the results of Section 6.4:

$$C = \frac{\pi\epsilon'}{\cosh^{-1}(d/2a)} \tag{20}$$

or

$$C \doteq \frac{\pi\epsilon'}{\ln(d/a)} \quad (a \ll d)$$

The external inductance may be found from $L_{ext}C = \mu\epsilon'$. It is

$$L_{ext} = \frac{\mu}{\pi}\cosh^{-1}(d/2a) \tag{21}$$

or

$$L_{ext} \doteq \frac{\mu}{\pi}\ln(d/a) \quad (a \ll d)$$

The conductance per unit length may be written immediately from an inspection of the capacitance expression, and using the relation $RC = \epsilon/\sigma$:

$$G = \frac{\pi\sigma}{\cosh^{-1}(d/2a)} \tag{22}$$

The resistance per unit length is twice that of the center conductor of the coax,

$$R = \frac{1}{\pi a\delta\sigma_c} \tag{23}$$

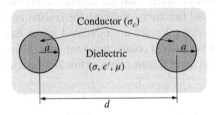

Figure 13.4 The geometry of the two-wire transmission line.

Finally, using the capacitance and the external inductance expressions, we obtain a value for the characteristic impedance,

$$Z_0 = \sqrt{\frac{L_{\text{ext}}}{C}} = \frac{1}{\pi}\sqrt{\frac{\mu}{\epsilon}}\cosh^{-1}(d/2a) \tag{24}$$

13.1.5 Two-Wire (Low Frequencies)

At low frequencies where a uniform current distribution may be assumed, we again must modify the L and R expressions, but not those for C and G. The latter two are again expressed by (20) and (22):

$$C = \frac{\pi\epsilon'}{\cosh^{-1}(d/2a)}$$

$$G = \frac{\pi\sigma}{\cosh^{-1}(d/2a)}$$

The inductance per unit length must be increased by twice the internal inductance of a straight round wire,

$$L = \frac{\mu}{\pi}\left[\frac{1}{4} + \cosh^{-1}(d/2a)\right] \tag{25}$$

The resistance becomes twice the dc resistance of a wire of radius a, conductivity σ_c, and unit length:

$$R = \frac{2}{\pi a^2 \sigma_c} \tag{26}$$

D13.3. The conductors of a two-wire transmission line each have a radius of 0.8 mm and a conductivity of 3×10^7 S/m. They are separated by a center-to-center distance of 0.8 cm in a medium for which $\epsilon_r' = 2.5$, $\mu_r = 1$, and $\sigma = 4 \times 10^{-9}$ S/m. If the line operates at 60 Hz, find: (a) δ; (b) C; (c) G; (d) L; (e) R.

Ans. (a) 1.2 cm; (b) 30 pF/m; (c) 5.5 nS/m; (d) 1.02 μH/m; (e) 0.033 Ω/m

---

13.1.6 Microstrip Line (Low Frequencies)

Microstrip line is one example of a class of configurations involving planar conductors of finite widths on or within dielectric substrates; they are usually employed as device interconnects for microelectronic circuitry. The microstrip configuration, shown in Figure 13.5, consists of a dielectric (assumed lossless) of thickness d and of permittivity $\epsilon' = \epsilon_r \epsilon_0$, sandwiched between a conducting ground plane and a narrow conducting strip of width w. The region above the top strip is air (assumed here) or a dielectric of lower permittivity.

The structure approaches the case of the parallel-plate line if $w \gg d$. In a microstrip, such an assumption is generally not valid, and so significant charge densities exist on both surfaces of the upper conductor. The resulting electric field, originating at the top conductor and terminating on the bottom conductor, will exist within both

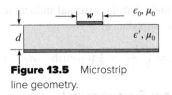

Figure 13.5 Microstrip line geometry.

substrate *and* air regions. The same is true for the magnetic field, which circulates around the top conductor. This electromagnetic field configuration cannot propagate as a purely TEM wave because wave velocities within the two media will differ. Instead, waves having z components of **E** and **H** occur, with the z component magnitudes established so that the air and dielectric fields do achieve equal phase velocities (the reasoning behind this will be explained in Section 13.6). Analyzing the structure while allowing for the special fields is complicated, but it is usually permissible to approach the problem under the assumption of negligible z components. This is the *quasi TEM* approximation, in which the static fields (obtainable through numerical solution of Laplace's equation, for example) are used to evaluate the primary constants. Accurate results are obtained at low frequencies (below 1 or 2 GHz). At higher frequencies, results obtained through the static fields can still be used but in conjunction with appropriate modifying functions. We will consider the simple case of low-frequency operation and assume lossless propagation.[2]

To begin, it is useful to consider the microstrip line characteristics when the dielectric is *not* present. Assuming that both conductors have very small thicknesses, the internal inductance will be negligible, and so the phase velocity within the air-filled line, v_{p0}, will be

$$v_{p0} = \frac{1}{\sqrt{L_{ext}C_0}} = \frac{1}{\sqrt{\mu_0 \epsilon_0}} = c \qquad (27a)$$

where C_0 is the capacitance of the air-filled line (obtained from the electric field for that case), and c is the velocity of light. With the dielectric in place, the capacitance changes, *but the inductance does not,* provided the dielectric permeability is μ_0. Using (27a), the phase velocity now becomes

$$v_p = \frac{1}{\sqrt{L_{ext}C}} = c\sqrt{\frac{C_0}{C}} = \frac{c}{\sqrt{\epsilon_{r,eff}}} \qquad (27b)$$

where the *effective dielectric constant* for the microstrip line is

$$\epsilon_{r,eff} = \frac{C}{C_0} = \left(\frac{c}{v_p}\right)^2 \qquad (28)$$

It is implied from (28) that the microstrip capacitance C would result if both the air and substrate regions were filled homogeneously with material having dielectric constant $\epsilon_{r,eff}$. The effective dielectric constant is a convenient parameter

---

[2]The high-frequency case is treated in detail in Edwards (Reference 2).

to use because it provides a way of unifying the effects of the dielectric and the conductor geometry. To see this, consider the two extreme cases involving large and small width-to-height ratios, w/d. If w/d is very large, then the line resembles the parallel-plate line, in which nearly all of the electric field exists within the dielectric. In this case $\epsilon_{r,\,\text{eff}} \doteq \epsilon_r$. On the other hand, for a very narrow top strip, or small w/d, the dielectric and air regions contain roughly equal amounts of electric flux. In that case, the effective dielectric constant approaches its minimum, given by the average of the two dielectric constants. We therefore obtain the range of allowed values of $\epsilon_{r,\,\text{eff}}$:

$$\frac{1}{2}(\epsilon_r + 1) < \epsilon_{r,\text{eff}} < \epsilon_r \tag{29}$$

The physical interpretation of $\epsilon_{r,\text{eff}}$ is that it is a *weighted average* of the dielectric constants of the substrate and air regions, with the weighting determined by the extent to which the electric field fills either region. We may thus write the effective dielectric constant in terms of a *field filling factor, q,* for the substrate:

$$\epsilon_{r,\text{eff}} = 1 + q(\epsilon_r - 1) \tag{30}$$

where $0.5 < q < 1$. With large w/d, $q \to 1$; with small w/d, $q \to 0.5$.

Now, the characteristic impedances of the air-filled line and the line with dielectric substrate are, respectively, $Z_0^{\text{air}} = \sqrt{L_{\text{ext}}/C_0}$ and $Z_0 = \sqrt{L_{\text{ext}}C}$. Then, using (28), we find

$$Z_0 = \frac{Z_0^{\text{air}}}{\sqrt{\epsilon_{r,\text{eff}}}} \tag{31}$$

A procedure for obtaining the characteristic impedance would be to first evaluate the air-filled impedance for a given w/d. Then, knowing the effective dielectric constant, determine the actual impedance using (31). Another problem would be to determine the required ratio w/d for a given substrate material in order to achieve a desired characteristic impedance.

Detailed analyses have led to numerous approximation formulas for the evaluation of $\epsilon_{r,\text{eff}}$, Z_0^{air}, and Z_0 within different regimes (again, see Reference 2 and the references therein). For example, with dimensions restricted such that $1.3 < w/d < 3.3$, applicable formulas include:

$$Z_0^{\text{air}} \doteq 60 \ln\left[4\left(\frac{d}{w}\right) + \sqrt{16\left(\frac{d}{w}\right)^2 + 2}\right] \qquad \frac{w}{d} < 3.3 \tag{32}$$

and

$$\epsilon_{r,\text{eff}} \doteq \frac{\epsilon_r + 1}{2} + \frac{\epsilon_r - 1}{2}\left(1 + 10\frac{d}{w}\right)^{-0.555} \qquad \frac{w}{d} > 1.3 \tag{33}$$

Or, if a line is to be fabricated having a desired value of Z_0, the effective dielectric constant (from which the required w/d can be obtained) is found through:

$$\epsilon_{r,\text{eff}} \doteq \epsilon_r[0.96 + \epsilon_r(0.109 - 0.004\epsilon_r)(\log_{10}(10 + Z_0) - 1)]^{-1} \qquad \frac{w}{d} > 1.3 \tag{34}$$

D13.4. A microstrip line is fabricated on a lithium niobate substrate ($\epsilon_r = 4.8$) of 1 mm thickness. If the top conductor is 2 mm wide, find (*a*) $\epsilon_{r, \text{eff}}$; (*b*) Z_0; (*c*) v_p.

Ans. (*a*) 3.6; (*b*) 47 Ω; (*c*) 1.6×10^8 m/s

13.2 BASIC WAVEGUIDE OPERATION

Waveguides assume many different forms that depend on the purpose of the guide and on the frequency of the waves to be transmitted. The simplest form (in terms of analysis) is the parallel-plate guide shown in Figure 13.6. Other forms are the hollow-pipe guides, including the rectangular waveguide of Figure 13.7, and the cylindrical guide, shown in Figure 13.8. Dielectric waveguides, used primarily at optical frequencies, include the slab waveguide of Figure 13.9 and the optical fiber, shown in Figure 13.10. Each of these structures has certain advantages over the others, depending on the application and the frequency of the waves to be transmitted. All guides, however, exhibit the same basic operating principles, which we will explore in this section.

To develop an understanding of waveguide behavior, we consider the parallel-plate waveguide of Figure 13.6. At first, we recognize this as one of the transmission-line structures that we investigated in Section 13.1. So the first question that arises is: how does a waveguide differ from a transmission line to begin with? The difference lies in the form of the electric and magnetic fields within the line. To see this, consider again Figure 13.1, which shows the fields when the line operates as a transmission line. As we saw earlier, a sinusoidal voltage wave, with voltage applied between conductors, leads to an electric field that is directed vertically between the conductors as shown. Because current flows only in the z direction, magnetic field will be oriented in and out of the page (in the y direction). The interior fields comprise a plane electromagnetic wave which propagates in the z direction (as the Poynting vector will show), since both fields lie in the transverse plane. We refer to

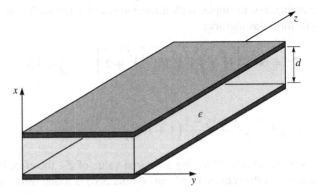

Figure 13.6 Parallel-plate waveguide, with metal plates at $x = 0, d$. Between the plates is a dielectric of permittivity ϵ.

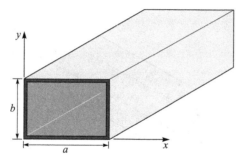

Figure 13.7 Rectangular waveguide.

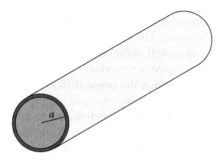

Figure 13.8 Cylindrical waveguide.

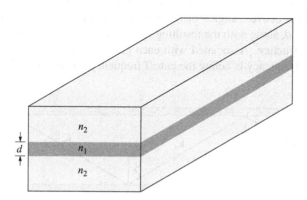

Figure 13.9 Symmetric dielectric slab waveguide, with slab region (refractive index n_1) surrounded by two dielectrics of index $n_2 < n_1$.

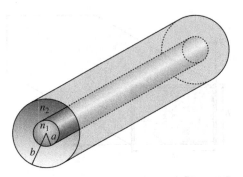

Figure 13.10 Optical fiber waveguide, with the core dielectric ($\rho < a$) of refractive index n_1. The cladding dielectric ($a < \rho < b$) is of index $n_2 < n_1$.

this as a transmission-line wave, which, as discussed in Section 13.1, is a transverse electromagnetic, or TEM, wave. The wavevector **k**, shown in Figure 13.1, indicates the direction of wave travel as well as the direction of power flow.

As the frequency is increased, a remarkable change occurs in the way the fields progagate down the line. Although the original field configuration of Figure 13.1 may still be present, another possibility emerges, which is shown in Figure 13.11. Again, a plane wave is guided in the z direction, but by means of a progression of zig-zag reflections at the upper and lower plates. Wavevectors $\mathbf{k}_u$ and $\mathbf{k}_d$ are associated with the upward-and downward-propagating waves, respectively, and these have identical magnitudes,

$$|\mathbf{k}_u| = |\mathbf{k}_d| = k = \omega\sqrt{\mu\epsilon}$$

For such a wave to propagate, all upward-propagating waves must be *in phase* (as must be true of all downward-propagating waves). This condition can only be satisfied at certain discrete angles of incidence, shown as θ in Figure 13.11. An allowed value of θ, along with the resulting field configuration, comprise a *waveguide mode* of the structure. Associated with each guided mode is a *cutoff frequency*. If the operating frequency is below the cutoff frequency, the mode will not propagate.

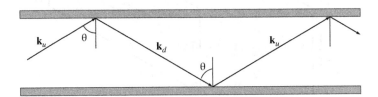

Figure 13.11 In a parallel-plate waveguide, plane waves can propagate by oblique reflection from the conducting walls. This produces a waveguide mode that is not TEM.

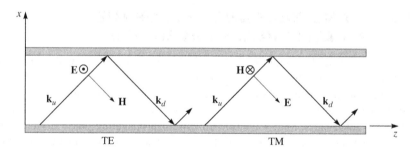

Figure 13.12 Plane wave representation of TE and TM modes in a parallel-plate guide.

If above cutoff, the mode propagates. The TEM mode, however, has no cutoff; it will be supported at any frequency. At a given frequency, the guide may support several modes, the quantity of which depends on the plate separation and on the dielectric constant of the interior medium, as will be shown. The number of modes increases as the frequency is raised.

So to answer our initial question on the distinction between transmission lines and waveguides, we can state the following: Transmission lines consist of two or more conductors and as a rule will support TEM waves (or something which could approximate such a wave). A waveguide may consist of one or more conductors, or no conductors at all, and will support waveguide modes of forms similar to those just described. Waveguides may or may not support TEM waves, depending on the design.

In the parallel-plate guide, two types of waveguide modes can be supported. These are shown in Figure 13.12 as arising from the s and p orientations of the plane wave polarizations. In a manner consistent with our previous discussions on oblique reflection (Section 12.5), we identify a *transverse electric* or *TE* mode when **E** is perpendicular to the plane of incidence (s-polarized); this positions **E** parallel to the transverse plane of the waveguide, as well as to the boundaries. Similarly, a *transverse magnetic* or *TM* mode results with a p polarized wave; the entire **H** field is in the y direction and is thus within the transverse plane of the guide. Both possibilities are illustrated in Figure 13.12. Note, for example, that with **E** in the y direction (TE mode), **H** will have x and z components. Likewise, a TM mode will have x and z components of **E**.[3] In any event, the reader can verify from the geometry of Figure 13.12 that it is not possible to achieve a purely TEM mode for values of θ other than 90°. Other wave polarizations are possible that lie between the TE and TM cases, but these can always be expressed as superpositions of TE and TM modes.

---

[3] Other types of modes can exist in other structures (not the parallel-plate guide) in which *both* **E** and **H** have z components. These are known as *hybrid* modes, and they typically occur in guides with cylindrical cross sections, such as the optical fiber.

13.3 PLANE WAVE ANALYSIS OF THE PARALLEL-PLATE WAVEGUIDE

We now investigate the conditions under which waveguide modes will occur, using our plane wave model for the mode fields. In Figure 13.13*a*, a zig-zag path is again shown, but this time phase fronts are drawn that are associated with two of the upward-propagating waves. The first wave has reflected twice (at the top and bottom surfaces) to form the second wave (the downward-propagating phase fronts are not shown). Note that the phase fronts of the second wave do not coincide with those of the first wave, and so the two waves are out of phase. In Figure 13.13*b*, the wave angle has been adjusted so that the two waves are now in phase. Having satisfied this condition for the two waves, we will find that *all* upward-propagating waves will have coincident phase fronts. The same condition will automatically occur for all downward-propagating waves. This is the requirement to establish a guided mode.

13.3.1 Wave Geometry

In Figure 13.14 we show the wavevector, $\mathbf{k}_u$, and its components, along with a series of phase fronts. A drawing of this kind for $\mathbf{k}_d$ would be the same, except the

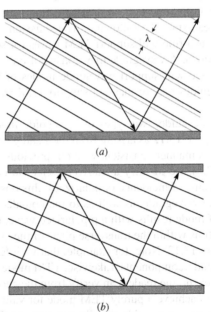

(*a*)

(*b*)

Figure 13.13 (*a*) Plane wave propagation in a parallel-plate guide in which the wave angle is such that the upward-propagating waves are not in phase. (*b*) The wave angle has been adjusted so that the upward waves are in phase, resulting in a guided mode.

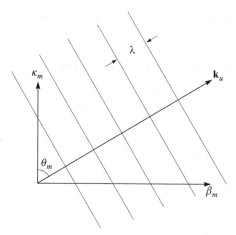

Figure 13.14 The components of the upward wavevector are κ_m and β_m, the transverse and axial phase constants. To form the downward wavevector, $\mathbf{k}_d$, the direction of κ_m is reversed.

x component, κ_m, would be reversed. In Section 12.4, we measured the phase shift per unit distance along the x and z directions by the components, k_x and k_z, which varied continuously as the direction of $\mathbf{k}$ changed. In our discussion of waveguides, we introduce a different notation, where κ_m and β_m are used for k_x and k_z. The subscript m is an integer indicating the *mode number*. This provides a subtle hint that β_m and κ_m will assume only certain discrete values that correspond to certain allowed directions of $\mathbf{k}_u$ and $\mathbf{k}_d$, such that our coincident phase front requirement is satisfied.[4] From the geometry we see that for any value of m,

$$\boxed{\beta_m = \sqrt{k^2 - \kappa_m^2}} \tag{35}$$

Use of the symbol β_m for the z components of $\mathbf{k}_u$ and $\mathbf{k}_d$ is appropriate because β_m will ultimately be the phase constant for the mth waveguide mode, measuring phase shift per distance down the guide; it is also used to determine the phase velocity of the mode, ω/β_m, and the group velocity, $d\omega/d\beta_m$.

Throughout the discussion, we will assume that the medium within the guide is lossless and nonmagnetic, so that

$$\boxed{k = \omega\sqrt{\mu_0\epsilon'} = \frac{\omega\sqrt{\epsilon_r'}}{c} = \frac{\omega n}{c}} \tag{36}$$

---

[4] Subscripts (m) are not shown on $\mathbf{k}_u$ and $\mathbf{k}_d$ but are understood. Changing m does not affect the magnitudes of these vectors, only their directions.

which we express either in terms of the dielectric constant, ϵ_r', or the refractive index, n, of the medium.

13.3.2 Transverse Resonance and Cutoff

It is κ_m, the x component of $\mathbf{k}_u$ and $\mathbf{k}_d$, that will be useful to us in quantifying our requirement on coincident phase fronts through a condition known as *transverse resonance*. This condition states that the net phase shift measured during a round trip over the full transverse dimension of the guide must be an integer multiple of 2π radians. This is another way of stating that all upward- (or downward-) propagating plane waves must have coincident phases. The various segments of this round trip are illustrated in Figure 13.15. We assume for this exercise that the waves are frozen in time and that an observer moves vertically over the round trip, measuring phase shift along the way. In the first segment (Figure 13.15a), the observer starts at a position just above the lower conductor and moves vertically to the top conductor through distance d. The measured phase shift over this distance is $\kappa_m d$ rad. On reaching the top surface, the observer will note a possible phase shift on reflection (Figure 13.15b). This will be π if the wave is TE polarized and will be zero if the wave is TM polarized (see Figure 13.16 for a demonstration of this). Next, the observer moves along the reflected wave phases down to the lower conductor and again measures a phase shift of $\kappa_m d$ (Figure 13.15c). Finally, after including the phase shift on reflection at the bottom conductor, the observer is back at the original starting point and is noting the phase of the next upward-propagating wave.

The total phase shift over the round trip is required to be an integer multiple of 2π:

$$\kappa_m d + \phi + \kappa_m d + \phi = 2m\pi \qquad (37)$$

where ϕ is the phase shift on reflection at each boundary. Note that with $\phi = \pi$ (TE waves) or 0 (TM waves), the net reflective phase shift over a round trip is 2π or 0, regardless of the angle of incidence. Thus the reflective phase shift has no bearing on the current problem, and we may simplify (37) to read

$$\kappa_m = \frac{m\pi}{d} \qquad (38)$$

which is valid for *both* TE and TM modes. Note from Figure 13.14 that $\kappa_m = k \cos\theta_m$. Thus the wave angles for the allowed modes are readily found from (38) with (36):

$$\theta_m = \cos^{-1}\left(\frac{m\pi}{kd}\right) = \cos^{-1}\left(\frac{m\pi c}{\omega n d}\right) = \cos^{-1}\left(\frac{m\lambda}{2nd}\right) \qquad (39)$$

where λ is the wavelength in free space.

We can next solve for the phase constant for each mode, using (35) with (38):

$$\beta_m = \sqrt{k^2 - \kappa_m^2} = k\sqrt{1 - \left(\frac{m\pi}{kd}\right)^2} = k\sqrt{1 - \left(\frac{m\pi c}{\omega n d}\right)^2} \qquad (40)$$

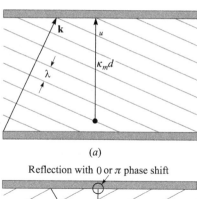

(a)

Reflection with 0 or π phase shift

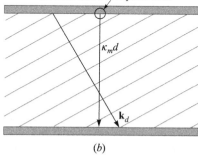

(b)

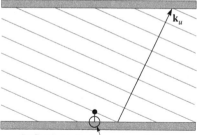

Reflection with 0 or π phase shift

(c)

Figure 13.15   The net phase shift over a round trip in the parallel-plate guide is found by first measuring the transverse phase shift between plates of the initial upward wave (*a*); next, the transverse phase shift in the reflected (downward) wave is measured, while accounting for the reflective phase shift at the top plate (*b*); finally, the phase shift on reflection at the bottom plate is added, thus returning to the starting position, but with a new upward wave (*c*). Transverse resonance occurs if the phase at the final point is the same as that at the starting point (the two upward waves are in phase).

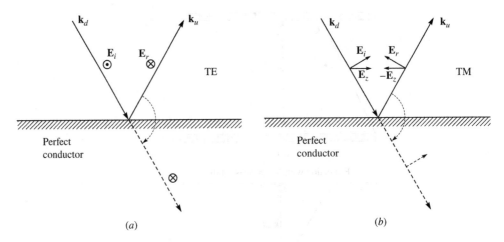

Figure 13.16 The phase shift of a wave on reflection from a perfectly conducting surface depends on whether the incident wave is TE (s-polarized) or TM (p-polarized). In both drawings, electric fields are shown as they would appear immediately adjacent to the conducting boundary. In (*a*) the field of a TE wave reverses direction upon reflection to establish a zero net field at the boundary. This constitutes a π phase shift, as is evident by considering a fictitious transmitted wave (dashed line) formed by a simple rotation of the reflected wave into alignment with the incident wave. In (*b*) an incident TM wave experiences a reversal of the *z component* of its electric field. The resultant field of the reflected wave, however, has not been phase-shifted; rotating the reflected wave into alignment with the incident wave (dashed line) shows this.

We define the radian *cutoff frequency* for mode *m* as

$$\omega_{cm} = \frac{m\pi c}{nd} \tag{41}$$

so that (40) becomes

$$\beta_m = \frac{n\omega}{c}\sqrt{1 - \left(\frac{\omega_{cm}}{\omega}\right)^2} \tag{42}$$

The significance of the cutoff frequency is readily seen from (42): If the operating frequency ω is greater than the cutoff frequency for mode *m*, then that mode will have phase constant β_m that is real-valued, and so the mode will propagate. For $\omega < \omega_{cm}$, β_m is imaginary, and the mode does not propagate.

Associated with the cutoff frequency is the *cutoff wavelength*, λ_{cm}, defined as the free-space wavelength at which cutoff for mode *m* occurs. This will be

$$\lambda_{cm} = \frac{2\pi c}{\omega_{cm}} = \frac{2nd}{m} \tag{43}$$

Note, for example, that in an air-filled guide ($n = 1$) the wavelength at which the lowest-order mode first starts to propagate is $\lambda_{c1} = 2d$, or the plate separation is one-half wavelength. Mode m will propagate whenever $\omega > \omega_{cm}$, or equivalently whenever $\lambda < \lambda_{cm}$. Use of the cutoff wavelength enables us to construct a second useful form of Eq. (42):

$$\boxed{\beta_m = \frac{2\pi n}{\lambda}\sqrt{1 - \left(\frac{\lambda}{\lambda_{cm}}\right)^2}} \qquad (44)$$

EXAMPLE 13.1

A parallel-plate waveguide has plate separation $d = 1$ cm and is filled with Teflon having dielectric constant $\epsilon_r' = 2.1$. Determine the maximum operating frequency such that only the TEM mode will propagate. Also find the range of frequencies over which the TE$_1$ and TM$_1$ ($m = 1$) modes, and no higher-order modes, will propagate.

Solution. Using (41), the cutoff frequency for the first waveguide mode ($m = 1$) will be

$$f_{c1} = \frac{\omega_{c1}}{2\pi} = \frac{2.99 \times 10^{10}}{2\sqrt{2.1}} = 1.03 \times 10^{10} \text{ Hz} = 10.3 \text{ GHz}$$

To propagate only TEM waves, we must have $f < 10.3$ GHz. To allow TE$_1$ and TM$_1$ (along with TEM) only, the frequency range must be $\omega_{c1} < \omega < \omega_{c2}$, where $\omega_{c2} = 2\omega_{c1}$, from (41). Thus, the frequencies at which we will have the $m = 1$ modes and TEM will be 10.3 GHz $< f <$ 20.6 GHz.

EXAMPLE 13.2

In the parallel-plate guide of Example 13.1, the operating wavelength is $\lambda = 2$ mm. How many waveguide modes will propagate?

Solution. For mode m to propagate, the requirement is $\lambda < \lambda_{cm}$. For the given waveguide and wavelength, the inequality becomes, using (43),

$$2 \text{ mm} < \frac{2\sqrt{2.1} \, (10 \text{ mm})}{m}$$

from which

$$m < \frac{2\sqrt{2.1} \, (10 \text{ mm})}{2 \text{ mm}} = 14.5$$

Thus the guide will support modes at the given wavelength up to order $m = 14$. Since there will be a TE and a TM mode for each value of m, this gives, not including the TEM mode, a total of 28 guided modes that are above cutoff.

13.3.3 Plane Wave Superposition, Phase and Group Velocities

The field configuration for a given mode can be found through the superposition of the fields of all the reflected waves. We can do this for the TE waves, for example, by writing the electric field phasor in the guide in terms of incident and reflected fields through

$$E_{ys} = E_0 e^{-j\mathbf{k}_u \cdot \mathbf{r}} - E_0 e^{-j\mathbf{k}_d \cdot \mathbf{r}} \tag{45}$$

where the wavevectors, $\mathbf{k}_u$ and $\mathbf{k}_d$, are indicated in Figure 13.12. The minus sign in front of the second term arises from the π phase shift on reflection. From the geometry depicted in Figure 13.14, we write

$$\mathbf{k}_u = \kappa_m \mathbf{a}_x + \beta_m \mathbf{a}_z \tag{46}$$

and

$$\mathbf{k}_d = -\kappa_m \mathbf{a}_x + \beta_m \mathbf{a}_z \tag{47}$$

Then, using

$$\mathbf{r} = x\mathbf{a}_x + z\mathbf{a}_z$$

Eq. (45) becomes

$$E_{ys} = E_0(e^{-j\kappa_m x} - e^{j\kappa_m x})e^{-j\beta_m z} = 2j\,E_0 \sin(\kappa_m x)e^{-j\beta_m z} = E_0' \sin(\kappa_m x)e^{-j\beta_m z} \tag{48}$$

where the plane wave amplitude, E_0, and the overall phase are absorbed into E_0'. In real instantaneous form, (48) becomes

$$E_y(z, t) = \mathcal{R}e\left(E_{ys}e^{j\omega t}\right) = E_0' \sin(\kappa_m x)\cos(\omega t - \beta_m z) \quad \text{(TE mode above cutoff)} \tag{49}$$

We interpret this as a wave that propagates in the positive z direction (down the guide) while having a field profile that varies with x.[5] The TE mode field is the *interference pattern* resulting from the superposition of the upward and downward plane waves. Note that if $\omega < \omega_{cm}$, then (42) yields an imaginary value for β_m, which we may write as $-j|\beta_m| = -j\alpha_m$. Eqs. (48) and (49) then become

$$E_{ys} = E_0' \sin(\kappa_m x)e^{-\alpha_m z} \tag{50}$$

$$E(z, t) = E_0' \sin(\kappa_m x)e^{-\alpha_m z}\cos(\omega t) \quad \text{(TE mode below cutoff)} \tag{51}$$

This mode does not propagate, but simply oscillates at frequency ω, while exhibiting a field pattern that decreases in strength with increasing z. The attenuation coefficient, α_m, is found from (42) with $\omega < \omega_{cm}$:

$$\alpha_m = \frac{n\omega_{cm}}{c}\sqrt{1 - \left(\frac{\omega}{\omega_{cm}}\right)^2} = \frac{2\pi n}{\lambda_{cm}}\sqrt{1 - \left(\frac{\lambda_{cm}}{\lambda}\right)^2} \tag{52}$$

---

[5] We can also interpret this field as that of a standing wave in x while it is a traveling wave in z.

We note from (39) and (41) that the plane wave angle is related to the cutoff frequency and cutoff wavelength through

$$\cos\theta_m = \frac{\omega_{cm}}{\omega} = \frac{\lambda}{\lambda_{cm}} \qquad (53)$$

So we see that at cutoff ($\omega = \omega_{cm}$), $\theta_m = 0$, and the plane waves are just reflecting back and forth over the cross section; they are making no forward progress down the guide. As ω is increased beyond cutoff (or λ is decreased), the wave angle increases, approaching $90°$ as ω approaches infinity (or as λ approaches zero). From Figure 13.14, we have

$$\beta_m = k\sin\theta_m = \frac{n\omega}{c}\sin\theta_m \qquad (54)$$

and so the phase velocity of mode m will be

$$v_{pm} = \frac{\omega}{\beta_m} = \frac{c}{n\sin\theta_m} \qquad (55)$$

The velocity minimizes at c/n for all modes, approaching this value at frequencies far above cutoff; v_{pm} approaches infinity as the frequency is reduced to approach the cutoff frequency. Again, phase velocity is the speed of the phases in the z direction, and the fact that this velocity may exceed the speed of light in the medium is not a violation of relativistic principles, as discussed in Section 12.7.

The energy will propagate at the group velocity, $v_g = d\omega/d\beta$. Using (42), we have

$$v_{gm}^{-1} = \frac{d\beta_m}{d\omega} = \frac{d}{d\omega}\left[\frac{n\omega}{c}\sqrt{1 - \left(\frac{\omega_{cm}}{\omega}\right)^2}\right] \qquad (56)$$

The derivative is straightforward. Carrying it out and taking the reciprocal of the result yields:

$$v_{gm} = \frac{c}{n}\sqrt{1 - \left(\frac{\omega_{cm}}{\omega}\right)^2} = \frac{c}{n}\sin\theta_m \qquad (57)$$

Group velocity is thus identified as the projection of the velocity associated with $\mathbf{k}_u$ or $\mathbf{k}_d$ into the z direction. This will be less than or equal to the velocity of light in the medium, c/n, as expected.

EXAMPLE 13.3

In the guide of Example 13.1, the operating frequency is 25 GHz. Consequently, modes for which $m = 1$ and $m = 2$ will be above cutoff. Determine the *group delay difference* between these two modes over a distance of 1 cm. This is the difference in propagation times between the two modes when energy in each propagates over the 1-cm distance.

Solution. The group delay difference is expressed as

$$\Delta t = \left(\frac{1}{v_{g2}} - \frac{1}{v_{g1}}\right)(\text{s/cm})$$

From (57), along with the results of Example 13.1, we have

$$v_{g1} = \frac{c}{\sqrt{2.1}}\sqrt{1 - \left(\frac{10.3}{25}\right)^2} = 0.63c$$

$$v_{g2} = \frac{c}{\sqrt{2.1}}\sqrt{1 - \left(\frac{20.6}{25}\right)^2} = 0.39c$$

Then

$$\Delta t = \frac{1}{c}\left[\frac{1}{.39} - \frac{1}{.63}\right] = 3.3 \times 10^{-11}\,\text{s/cm} = 33\,\text{ps/cm}$$

This computation gives a rough measure of the *modal dispersion* in the guide, applying to the case of having only two modes propagating. A pulse, for example, whose center frequency is 25 GHz would have its energy divided between the two modes. The pulse would broaden by approximately 33 ps/cm of propagation distance as the energy in the modes separates. If, however, we include the TEM mode (as we really must), then the broadening will be even greater. The group velocity for TEM will be $c/\sqrt{2.1}$. The group delay difference of interest will then be between the TEM mode and the $m = 2$ mode (TE or TM). We would therefore have

$$\Delta t_{\text{net}} = \frac{1}{c}\left[\frac{1}{.39} - 1\right] = 52\,\text{ps/cm}$$

D13.5. Determine the wave angles θ_m for the first four modes ($m = 1, 2, 3, 4$) in a parallel-plate guide with $d = 2$ cm, $\epsilon_r' = 1$, and $f = 30$ GHz.

Ans. 76°; 60°; 41°; 0°

D13.6. A parallel-plate guide has plate spacing $d = 5$ mm and is filled with glass ($n = 1.45$). What is the maximum frequency at which the guide will operate in the TEM mode only?

Ans. 20.7 GHz

D13.7. A parallel-plate guide having $d = 1$ cm is filled with air. Find the cutoff wavelength for the $m = 2$ mode (TE or TM).

Ans. 1 cm

13.4 PARALLEL-PLATE GUIDE ANALYSIS USING THE WAVE EQUATION

The most direct approach in the analysis of any waveguide is through the wave equation, which we solve subject to the boundary conditions at the conducting walls. The form of the equation that we will use is that of Eq. (28) in Section 11.1, which was written for the case of free-space propagation. We account for the dielectric properties in the waveguide by replacing k_0 in that equation with k to obtain:

$$\boxed{\nabla^2 \mathbf{E}_s = -k^2 \mathbf{E}_s} \tag{58}$$

where $k = n\omega/c$ as before.

We can use the results of the last section to help us visualize the process of solving the wave equation. For example, we may consider TE modes first, in which there will be only a y component of $\mathbf{E}$. The wave equation becomes:

$$\frac{\partial^2 E_{ys}}{\partial x^2} + \frac{\partial^2 E_{ys}}{\partial y^2} + \frac{\partial^2 E_{ys}}{\partial z^2} + k^2 E_{ys} = 0 \tag{59}$$

We assume that the width of the guide (in the y direction) is very large compared to the plate separation d. Therefore we can assume no y variation in the fields (fringing fields are ignored), and so $\partial^2 E_{ys}/\partial y^2 = 0$. We also know that the z variation will be of the form $e^{-j\beta_m z}$. The form of the field solution will thus be

$$\boxed{E_{ys} = E_0 f_m(x) e^{-j\beta_m z}} \tag{60}$$

where E_0 is a constant, and where $f_m(x)$ is a normalized function to be determined (whose maximum value is unity). We have included subscript m on β, κ, and $f(x)$, since we anticipate several solutions that correspond to discrete modes, to which we associate mode number m. We now substitute (60) into (59) to obtain

$$\boxed{\frac{d^2 f_m(x)}{dx^2} + (k^2 - \beta_m^2) f_m(x) = 0} \tag{61}$$

where E_0 and $e^{-j\beta_m z}$ have divided out, and where we have used the fact that

$$\frac{d^2}{dz^2} e^{-j\beta_m z} = -\beta_m^2 e^{-j\beta_m z}$$

Note also that we have written (61) using the total derivative d^2/dx^2, as f_m is a function only of x. We next make use of the geometry of Figure 13.14, and we note that

$k^2 - \beta_m^2 = \kappa_m^2$. Using this in (61) we obtain

$$\frac{d^2 f_m(x)}{dx^2} + \kappa_m^2 f_m(x) = 0 \tag{62}$$

The general solution of (62) will be

$$f_m(x) = \cos(\kappa_m x) + \sin(\kappa_m x) \tag{63}$$

We next apply the appropriate boundary conditions in our problem to evaluate κ_m. From Figure 13.6, conducting boundaries appear at $x = 0$ and $x = d$, at which the tangential electric field (E_y) must be zero. In Eq. (63), only the $\sin(\kappa_m x)$ term will allow the boundary conditions to be satisfied, so we retain it and drop the cosine term. The $x = 0$ condition is automatically satisfied by the sine function. The $x = d$ condition is met when we choose the value of κ_m such that

$$\kappa_m = \frac{m\pi}{d} \tag{64}$$

We recognize Eq. (64) as the same result that we obtained using the transverse resonance condition of Section 13.3. The final form of E_{ys} is obtained by substituting $f_m(x)$ as expressed through (63) and (64) into (60), yielding a result that is consistent with the one expressed in Eq. (48):

$$E_{ys} = E_0 \sin\left(\frac{m\pi x}{d}\right) e^{-j\beta_m z} \tag{65}$$

An additional significance of the mode number m is seen when considering the form of the electric field of (65). Specifically, m is the number of spatial half-cycles of electric field that occur over the distance d in the transverse plane. This can be understood physically by considering the behavior of the guide at cutoff. As we learned in the last section, the plane wave angle of incidence in the guide at cutoff is zero, meaning that the wave simply bounces up and down between the conducting walls. The wave must be resonant in the structure, which means that the net round trip phase shift is $2m\pi$. With the plane waves oriented vertically, $\beta_m = 0$, and so $\kappa_m = k = 2n\pi/\lambda_{cm}$. So at cutoff,

$$\frac{m\pi}{d} = \frac{2n\pi}{\lambda_{cm}} \tag{66}$$

which leads to

$$d = \frac{m\lambda_{cm}}{2n} \qquad \text{at cutoff} \tag{67}$$

Eq. (65) at cutoff then becomes

$$E_{ys} = E_0 \sin\left(\frac{m\pi x}{d}\right) = E_0 \sin\left(\frac{2n\pi x}{\lambda_{cm}}\right) \tag{68}$$

The waveguide is simply a one-dimensional *resonant cavity*, in which a wave can oscillate in the x direction if its wavelength as measured in the medium is an integer multiple of $2d$ where the integer is m.

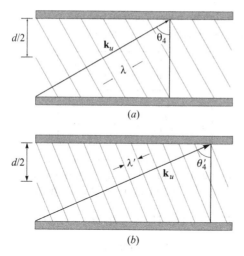

Figure 13.17 (*a*) A plane wave associated with an $m = 4$ mode, showing a net phase shift of 4π (two wavelengths measured in x) occurring over distance d in the transverse plane. (*b*) As frequency increases, an increase in wave angle is required to maintain the 4π transverse phase shift.

Now, as the frequency increases, wavelength will decrease, and so the requirement of wavelength equaling an integer multiple of $2d$ is no longer met. The response of the mode is to establish z *components* of $\mathbf{k}_u$ and $\mathbf{k}_d$, which results in the decreased wavelength being compensated by an increase in wavelength *as measured in the x direction*. Figure 13.17 shows this effect for the $m = 4$ mode, in which the wave angle, θ_4, steadily increases with increasing frequency. Thus, the mode retains precisely the functional form of its field in the x direction, but it establishes an increasing value of β_m as the frequency is raised. This invariance in the transverse spatial pattern means that the mode will retain its identity at all frequencies. Group velocity, expressed in (57), is changing as well, meaning that the changing wave angle with frequency is a mechanism for group velocity dispersion, known simply as *waveguide dispersion*. Pulses, for example, that propagate in a single waveguide mode will thus experience broadening in the manner considered in Section 12.8.

Having found the electric field, we can find the magnetic field using Maxwell's equations. We note from our plane wave model that we expect to obtain x and z components of $\mathbf{H}_s$ for a TE mode. We use the Maxwell equation

$$\nabla \times \mathbf{E}_s = -j\omega\mu\mathbf{H}_s \tag{69}$$

where, in the present case of having only a y component of $\mathbf{E}_s$, we have

$$\nabla \times \mathbf{E}_s = \frac{\partial E_{ys}}{\partial x}\mathbf{a}_z - \frac{\partial E_{ys}}{\partial z}\mathbf{a}_x = \kappa_m E_0 \cos(\kappa_m x)e^{-j\beta_m z}\mathbf{a}_z + j\beta_m E_0 \sin(\kappa_m x)e^{-j\beta_m z}\mathbf{a}_x \tag{70}$$

We solve for $\mathbf{H}_s$ by dividing both sides of (69) by $-j\omega\mu$. Performing this operation on (70), we obtain the two magnetic field components:

$$H_{xs} = -\frac{\beta_m}{\omega\mu}E_0\sin(\kappa_m x)e^{-j\beta_m z} \tag{71}$$

$$H_{zs} = j\frac{\kappa_m}{\omega\mu}E_0\cos(\kappa_m x)e^{-j\beta_m z} \tag{72}$$

Together, these two components form closed-loop patterns for $\mathbf{H}_s$ in the x, z plane, as can be verified using the streamline plotting methods developed in Section 2.6.

It is interesting to consider the magnitude of $\mathbf{H}_s$, which is found through

$$|\mathbf{H}_s| = \sqrt{\mathbf{H}_s \cdot \mathbf{H}_s^*} = \sqrt{H_{xs}H_{xs}^* + H_{zs}H_{zs}^*} \tag{73}$$

Carrying this out using (71) and (72) results in

$$|\mathbf{H}_s| = \frac{E_0}{\omega\mu}\left(\kappa_m^2 + \beta_m^2\right)^{1/2}\left(\sin^2(\kappa_m x) + \cos^2(\kappa_m x)\right)^{1/2} \tag{74}$$

Using the fact that $\kappa_m^2 + \beta_m^2 = k^2$ and using the identity $\sin^2(\kappa_m x) + \cos^2(\kappa_m x) = 1$, (74) becomes

$$|\mathbf{H}_s| = \frac{k}{\omega\mu}E_0 = \frac{\omega\sqrt{\mu\epsilon}}{\omega\mu} = \frac{E_0}{\eta} \tag{75}$$

where $\eta = \sqrt{\mu/\epsilon}$. This result is consistent with our understanding of waveguide modes based on the superposition of plane waves, in which the relation between $\mathbf{E}_s$ and $\mathbf{H}_s$ is through the medium intrinsic impedance, η.

D13.8. Determine the group velocity of the $m = 1$ (TE or TM) mode in an air-filled parallel-plate guide with $d = 0.5$ cm at $f =$ (a) 30 GHz, (b) 60 GHz, and (c) 100 GHz.

Ans. (a) 0; (b) 2.6×10^8 m/s; (c) 2.9×10^8 m/s

D13.9. A TE mode in a parallel-plate guide is observed to have three maxima in its electric field pattern between $x = 0$ and $x = d$. What is the value of m?

Ans. 3

13.5 RECTANGULAR WAVEGUIDES

In this section we consider the rectangular waveguide, a structure that is usually used in the microwave region of the electromagnetic spectrum. The guide is shown in Figure 13.7. As always, the propagation direction is along the z axis. The guide is of width a along x and height b along y. We can relate the geometry to that of the parallel-plate guide of previous sections by thinking of the rectangular guide as two parallel-plate guides of orthogonal orientation that are assembled to form one unit. We have a pair of horizontal conducting walls (along the x direction) and a pair of vertical walls (along y), all of which form one continuous boundary. The wave equation in its full three-dimensional form [Eq. (59)] must now be solved, for in general we may have field variations in all three coordinate directions.

In the parallel-plate guide, we found that the TEM mode can exist, along with TE and TM modes. The rectangular guide will support the TE and TM modes, but it *will not support a TEM mode*. This is because, in contrast to the parallel-plate guide, we now have a conducting boundary that completely surrounds the transverse plane. The nonexistence of TEM can be understood by remembering that any electric field must have a zero tangential component at the boundary. This means that it is impossible to set up an electric field that will not exhibit the sideways variation that is necessary to satisfy this boundary condition. Because **E** varies in the transverse plane, the computation of **H** through $\nabla \times \mathbf{E} = -j\omega\mu\mathbf{H}$ *must* lead to a z component of **H**, and so we cannot have a TEM mode. We cannot find any other orientation of a completely transverse **E** in the guide that will allow a completely transverse **H**.

13.5.1 Using Maxwell's Equations to Relate Field Components

With the modes dividing into TE and TM types, the standard approach is to first solve the wave equation for the z components. By definition, $E_z = 0$ in a TE mode, and $H_z = 0$ in a TM mode. Therefore, we will find the TE mode solution by solving the wave equation for H_z, and we will obtain the TM mode solution by solving for E_z. Using these results, all transverse field components are obtained directly through Maxwell's equations. This procedure may sound a little tedious (which it is), but we can be certain to find all the modes this way. First, we will handle the problem of finding transverse components in terms of the z components.

To begin the process, we assume that the phasor electric and magnetic fields will be forward-z propagating functions that exhibit spatial variation in the xy plane; the only z variation is that of a forward-propagating wave:

$$\mathbf{E}_s(x,y,z) = \mathbf{E}_s(x,y,0)e^{-j\beta z} \tag{76a}$$

$$\mathbf{H}_s(x,y,z) = \mathbf{H}_s(x,y,0)e^{-j\beta z} \tag{76b}$$

We can then obtain expressions for the transverse components of the phasor fields by evaluating the x and y components of the Maxwell curl equations in sourceless media. In evaluating the curl, it is evident from (76) that $\partial/\partial z = -j\beta$. The result is

$$\nabla \times \mathbf{E}_s = -j\omega\mu\,\mathbf{H}_s \rightarrow \begin{cases} \partial E_{zs}/\partial y + j\beta\,E_{ys} = -j\omega\mu\,H_{xs} & (x \text{ component}) \quad (77a) \\ j\beta\,E_{xs} + \partial E_{zs}/\partial x = j\omega\mu\,H_{ys} & (y \text{ component}) \quad (77b) \end{cases}$$

$$\nabla \times \mathbf{H}_s = j\omega\epsilon\,\mathbf{E}_s \rightarrow \begin{cases} \partial H_{zs}/\partial y + j\beta\,H_{ys} = j\omega\epsilon\,E_{xs} & (x \text{ component}) \quad (78a) \\ j\beta\,H_{xs} + \partial H_{zs}/\partial x = -j\omega\epsilon\,E_{ys} & (y \text{ component}) \quad (78b) \end{cases}$$

Now, pairs of the preceding equations can be solved together in order to express the individual transverse field components in terms of derivatives of the z components of **E** and **H**. For example, (77a) and (78b) can be combined, eliminating E_{ys}, to give

$$H_{xs} = \frac{-j}{\kappa^2}\left[\beta\frac{\partial H_{zs}}{\partial x} - \omega\epsilon\frac{\partial E_{zs}}{\partial y}\right] \qquad (79a)$$

Then, using (76b) and (77a), eliminate E_{xs} between them to obtain

$$H_{ys} = \frac{-j}{\kappa^2}\left[\beta\frac{\partial H_{zs}}{\partial y} + \omega\epsilon\frac{\partial E_{zs}}{\partial x}\right] \qquad (79b)$$

Using the same equation pairs, the transverse electric field components are then found:

$$E_{xs} = \frac{-j}{\kappa^2}\left[\beta\frac{\partial E_{zs}}{\partial x} + \omega\mu\frac{\partial H_{zs}}{\partial y}\right] \qquad (79c)$$

$$E_{ys} = \frac{-j}{\kappa^2}\left[\beta\frac{\partial E_{zs}}{\partial y} - \omega\mu\frac{\partial H_{zs}}{\partial x}\right] \qquad (79d)$$

κ is defined in the same manner as in the parallel-plate guide [Eq. (35)]:

$$\kappa = \sqrt{k^2 - \beta^2} \qquad (80)$$

where $k = \omega\sqrt{\mu\epsilon}$. In the parallel-plate geometry, we found that discrete values of κ and β resulted from the analysis, which we then subscripted with the integer mode number, m (κ_m and β_m). The interpretation of m was the number of field maxima that occurred between plates (in the x direction). In the rectangular guide, field variations will occur in both x and y, and so we will find it necessary to assign *two* integer subscripts to κ and β, thus leading to

$$\kappa_{mp} = \sqrt{k^2 - \beta_{mp}^2} \qquad (81)$$

where m and p indicate the number of field variations in the x and y directions. The form of Eq. (81) suggests that plane wave (ray) theory could be used to obtain the mode fields in the rectangular guide, as was accomplished in Section 13.3 for the parallel-plate guide. This is, in fact, the case, and is readily accomplished for cases in which plane wave reflections occur between only two opposing boundaries (either top to bottom or side to side), and this would be true only for certain TE modes. The method becomes complicated when reflections occur at all four surfaces; but in any case, the interpretation of κ_{mp} is the transverse (xy plane) component of the plane wave-vector k, while β_{mp} is the z component, as before.

The next step is to solve the wave equation for the z components of **E** and **H**, from which we will find the fields of the TM and TE modes.

13.5.2 TM Modes

Finding the TM modes begins with the wave equation [Eq. (59)], in which derivatives with respect to z are equivalent to multiplying by $j\beta$. We write the equation for the z component of $\mathbf{E}_s$:

$$\frac{\partial^2 E_{zs}}{\partial x^2} + \frac{\partial^2 E_{zs}}{\partial y^2} + (k^2 - \beta_{mp}^2)E_{zs} = 0 \tag{82}$$

The solution of (82) can be written as a sum of terms, each of which involves the product of three functions that exhibit individual variation with x, y, and z:

$$\boxed{E_{zs}(x,y,z) = \sum_{m,p} F_m(x)\, G_p(y) \exp(-j\beta_{mp}\, z)} \tag{83}$$

where the functions $F_m(x)$ and $G_p(y)$ (not normalized) are to be determined. Each term in (83) corresponds to one mode of the guide, and will by itself be a solution to (82). To determine the functions, a single term in (83) is substituted into (82). Noting that all derivatives are applied to functions of a single variable (and thus partial derivatives become total derivatives), and using (81), the result is

$$G_p(y)\frac{d^2 F_m}{dx^2} + F_m(x)\frac{d^2 G_p}{dy^2} + \kappa_{mp}^2\, F_m(x)\, G_p(y) = 0 \tag{84}$$

in which the $\exp(-j\beta_{mp}\, z)$ term has divided out. Rearranging (84), we get

$$\underbrace{\frac{1}{F_m}\frac{d^2 F_m}{dx^2}}_{-\kappa_m^2} + \underbrace{\frac{1}{G_p}\frac{d^2 G_p}{dy^2}}_{-\kappa_p^2} + \kappa_{mp}^2 = 0 \tag{85}$$

Terms in (85) are grouped such that all of the x variation is in the first term, which varies *only* with x, and all y variation is in the second term, which varies *only* with y. Now, consider what would happen if x is allowed to vary while holding y fixed. The second and third terms would be fixed, and Eq. (85) must always hold. Therefore, the x-varying first term must be a constant. This constant is denoted $-\kappa_m^2$, as indicated in (85). The same is true for the second term, which must also turn out to be a constant if y is allowed to vary while x is fixed. We assign the second term the constant value $-\kappa_p^2$ as indicated. Eq. (85) then states that

$$\boxed{\kappa_{mp}^2 = \kappa_m^2 + \kappa_p^2} \tag{86}$$

which suggests an immediate geometrical interpretation: As κ_{mp} is the transverse plane component of the wavevector k, κ_m and κ_p are clearly the x and y components of κ_{mp} (and of k)—again if one thinks in terms of plane waves and how they would bounce around in the waveguide to form the overall mode. Also indicated in (86) is the fact that κ_m and κ_p will be functions, respectively, of the integers m and p, as we will find.

Under the preceding conditions, Eq. (85) will now separate into two equations, one in each variable:

$$\frac{d^2 F_m}{dx^2} + \kappa_m^2\, F_m = 0 \tag{87a}$$

$$\frac{d^2 G_p}{dy^2} + \kappa_p^2\, G_p = 0 \tag{87b}$$

Equation (87) is now easily solved. We obtain

$$F_m(x) = A_m \cos(\kappa_m x) + B_m \sin(\kappa_m x) \tag{88a}$$

$$G_p(y) = C_p \cos(\kappa_p y) + D_p \sin(\kappa_p y) \tag{88b}$$

Using these, along with (83), the general solution for z component of $\mathbf{E}_s$ for a single TM mode can be constructed:

$$E_{zs} = [A_m \cos(\kappa_m x) + B_m \sin(\kappa_m x)][C_p \cos(\kappa_p y) + D_p \sin(\kappa_p y)]\exp(-j\beta_{mp} z) \tag{89}$$

The constants in (89) can be evaluated by applying the boundary conditions of the field on all four surfaces. Specifically, as E_{zs} is tangent to all the conducting surfaces, it must vanish on all of them. Referring to Figure 13.7, the boundary conditions are

$$E_{zs} = 0 \text{ at } x = 0, y = 0, x = a, \text{ and } y = b$$

Obtaining zero field at $x = 0$ and $y = 0$ is accomplished by dropping the cosine terms in (89) (setting $A_m = C_p = 0$). The values of κ_m and κ_p that appear in the remaining sine terms are then set to the following, in order to assure zero field at $x = a$ and $y = b$:

$$\kappa_m = \frac{m\pi}{a} \tag{90a}$$

$$\kappa_p = \frac{p\pi}{b} \tag{90b}$$

Using these results, and defining $B = B_m D_p$, Eq. (89) becomes:

$$\boxed{E_{zs} = B \, \sin(\kappa_m x)\sin(\kappa_p y)\exp(-j\beta_{mp} z)} \tag{91a}$$

Now, to find the remaining (transverse) field components, we substitute Eq. (91a) into Eqs. (79) to obtain:

$$\boxed{E_{xs} = -j\beta_{mp}\frac{\kappa_m}{\kappa_{mp}^2}B \cos(\kappa_m x) \sin(\kappa_p y)\exp(-j\beta_{mp} z)} \tag{91b}$$

$$\boxed{E_{ys} = -j\beta_{mp}\frac{\kappa_p}{\kappa_{mp}^2}B \sin(\kappa_m x)\cos(\kappa_p y)\exp(-j\beta_{mp} z)} \tag{91c}$$

$$\boxed{H_{xs} = j\omega\epsilon\frac{\kappa_p}{\kappa_{mp}^2}B \, \sin(\kappa_m x)\cos(\kappa_p y)\exp(-j\beta_{mp} z)} \tag{91d}$$

$$\boxed{H_{ys} = -j\omega\epsilon\frac{\kappa_m}{\kappa_{mp}^2}B \cos(\kappa_m x)\sin(\kappa_p y)\exp(-j\beta_{mp} z)} \tag{91e}$$

The preceding field components pertain to modes designated TM$_{mp}$. Note that for these modes, both m and p must be greater than or equal to 1. A zero value for either integer will zero all fields.

13.5.3 TE Modes

To obtain the TE mode fields, we solve the wave equation for the z component of **H** and then use Eq. (79) as before to find the transverse components. The wave equation is now the same as (82), except that E_{zs} is replaced by H_{zs}:

$$\frac{\partial^2 H_{zs}}{\partial x^2} + \frac{\partial^2 H_{zs}}{\partial y^2} + (k^2 - \beta_{mp}^2)H_{zs} = 0 \qquad (92)$$

and the solution is of the form:

$$\boxed{H_{zs}(x,y,z) = \sum_{m,p} F_m'(x)\, G_p'(y)\exp(-j\beta_{mp}\,z)} \qquad (93)$$

The procedure from here is identical to that involving TM modes, and the general solution will be

$$H_{zs} = [A_m' \cos(\kappa_m x) + B_m' \sin(\kappa_m x)][C_p' \cos(\kappa_p y) + D_p' \sin(\kappa_p y)]\exp(-j\beta_{mp}z) \quad (94)$$

Again, the expression is simplified by using the appropriate boundary conditions. We know that tangential electric field must vanish on all conducting boundaries. When we relate the electric field to magnetic field derivatives using (79c) and (79d), the following conditions develop:

$$E_{xs}\Big|_{y=0,b} = 0 \Rightarrow \frac{\partial H_{zs}}{\partial y}\Big|_{y=0,b} = 0 \qquad (95a)$$

$$E_{ys}\Big|_{x=0,a} = 0 \Rightarrow \frac{\partial H_{zs}}{\partial x}\Big|_{x=0,a} = 0 \qquad (95b)$$

The boundary conditions are now applied to Eq. (94), giving, for Eq. (95a)

$$\frac{\partial H_{zs}}{\partial y} = [A_m' \cos(\kappa_m x) + B_m' \sin(\kappa_m x)]$$
$$\times \left[-\kappa_p C_p' \sin(\kappa_p y) + \kappa_p D_p' \cos(\kappa_p y) \right] \exp(-j\beta_{mp}\,z)$$

The underlined terms are those that were modified by the partial differentiation. Requiring this result to be zero at $y = 0$ and $y = b$ leads to dropping the $\cos(\kappa_p y)$ term (set $D_p' = 0$) and requiring that $\kappa_p = p\pi/b$ as before. Applying Eq. (95b) to (94) results in

$$\frac{\partial H_{zs}}{\partial x} = [-\kappa_m A_m' \sin(\kappa_m x) + \kappa_m B_m' \cos(\kappa_m x)]$$
$$\times \left[C_p' \cos(\kappa_p y) + D_p' \sin(\kappa_p y) \right] \exp(-j\beta_{mp}\,z)$$

where again, the underlined term has been modified by partial differentiation with respect to x. Setting this result to zero at $x = 0$ and $x = a$ leads to dropping the $\cos(\kappa_m x)$ term (setting $B_m' = 0$) and requiring that $\kappa_m = m\pi/a$ as before. With all the above boundary conditions applied, the final expression for H_{zs} is now

$$\boxed{H_{zs} = A\cos(\kappa_m x)\cos(\kappa_p y)\exp(-j\beta_{mp}\,z)} \qquad (96a)$$

where we define $A = A'_m C'_p$. Applying Eqs. (79a) through (79d) to (96a) gives the transverse field components:

$$H_{xs} = j\beta_{mp}\frac{\kappa_m}{\kappa_{mp}^2}A\sin(\kappa_m x)\cos(\kappa_p y)\exp(-j\beta_{mp} z) \tag{96b}$$

$$H_{ys} = j\beta_{mp}\frac{\kappa_p}{\kappa_{mp}^2}A\cos(\kappa_m x)\sin(\kappa_p y)\exp(-j\beta_{mp} z) \tag{96c}$$

$$E_{xs} = j\omega\mu\frac{\kappa_p}{\kappa_{mp}^2}A\cos(\kappa_m x)\sin(\kappa_p y)\exp(-j\beta_{mp} z) \tag{96d}$$

$$E_{ys} = -j\omega\mu\frac{\kappa_m}{\kappa_{mp}^2}A\sin(\kappa_m x)\cos(\kappa_p y)\exp(-j\beta_{mp} z) \tag{96e}$$

These field components pertain to modes designated TE$_{mp}$. For these modes, either m or p may be zero, thus allowing for the possibility of the important TE$_{m0}$ or TE$_{0p}$ cases, as will be discussed later. Some very good illustrations of TE and TM modes are presented in Reference 3.

13.5.4 Cutoff Conditions

The phase constant for a given mode can be expressed using Eq. (81):

$$\beta_{mp} = \sqrt{k^2 - \kappa_{mp}^2} \tag{97}$$

Then, using (86), along with (90a) and (90b), we have

$$\beta_{mp} = \sqrt{k^2 - \left(\frac{m\pi}{a}\right)^2 - \left(\frac{p\pi}{b}\right)^2} \tag{98}$$

This result can be written in a manner consistent with Eq. (42) by using $k = \omega\sqrt{\mu\epsilon}$ and defining a radian cutoff frequency, $\omega_{c_{mp}}$, appropriate for the rectangular guide. We obtain:

$$\beta_{mp} = \omega\sqrt{\mu\epsilon}\sqrt{1 - \left(\frac{\omega_{Cmp}}{\omega}\right)^2} \tag{99}$$

where

$$\omega_{Cmp} = \frac{1}{\sqrt{\mu\epsilon}}\left[\left(\frac{m\pi}{a}\right)^2 + \left(\frac{p\pi}{b}\right)^2\right]^{1/2} \tag{100}$$

As discussed for the parallel-plate guide, it is again clear from (99) that the operating frequency, ω, must exceed the cutoff frequency, ω_{Cmp}, to achieve a real value for β_{mp} (and thus enabling mode mp to propagate). Equation (100) applies to both TE and TM modes, and thus some combination of both mode types may be present (or above cutoff) at a given frequency. It is evident that the choice of guide dimensions, a and

b, along with the material properties, ϵ_r and μ_r, will determine the number of modes that will propagate. For the typical case in which $\mu_r = 1$, using $n = \sqrt{\epsilon_r}$, and identifying the speed of light, $c = 1/\sqrt{\mu_0\epsilon_0}$, we may re-write (100) in a manner consistent with Eq. (41):

$$\omega_{Cmp} = \frac{c}{n}\left[\left(\frac{m\pi}{a}\right)^2 + \left(\frac{p\pi}{b}\right)^2\right]^{1/2} \tag{101}$$

This would lead to an expression for the cutoff wavelength, λ_{Cmp}, in a manner consistent with Eq. (43):

$$\lambda_{Cmp} = \frac{2\pi c}{\omega_{Cmp}} = 2n\left[\left(\frac{m}{a}\right)^2 + \left(\frac{p}{b}\right)^2\right]^{-1/2} \tag{102}$$

λ_{Cmp} is the *free space* wavelength at cutoff. If measured in the medium that fills the waveguide, the cutoff wavelength would be given by Eq. (102) divided by n.

Now, in a manner consistent with Eq. (44), Eq. (99) becomes

$$\beta_{mp} = \frac{2\pi n}{\lambda}\sqrt{1 - \left(\frac{\lambda}{\lambda_{Cmp}}\right)^2} \tag{103}$$

where λ is the free space wavelength. As we saw before, a TE_{mp} or TM_{mp} mode can propagate if its operating wavelength, λ, is less than λ_{Cmp}.

13.5.5 Special Cases: TE_{m0} and TE_{0p} Modes

The most important mode in the rectangular guide is the one that can propagate by itself. As we know, this will be the mode that has the lowest cutoff frequency (or the highest cutoff wavelength), so that over a certain range of frequencies, this mode will be above cutoff, while all others are below cutoff. By inspecting Eq. (101), and noting that $a > b$, the lowest cutoff frequency will occur for the mode in which $m = 1$ and $p = 0$, which will be the TE_{10} mode (remember that a TM_{10} mode does not exist, as can be shown in (91)). It turns out that this mode, and those of the same general type, are of the same form as those of the parallel-plate structure.

The specific fields for the TE_{m0} family of modes are obtained from (96a) through (96e) by setting $p = 0$, which means, using (86) and (90), that

$$\kappa_m = \kappa_{mp}\Big|_{p=0} = \frac{m\pi}{a} \tag{104}$$

and $\kappa_p = 0$. Under these conditions, the only surviving field components in (91) will be E_{ys}, H_{xs}, and H_{zs}. It is convenient to define the field equations in terms of an electric field amplitude, E_0, which is composed of all the amplitude terms in Eq. (96e). Specifically, define

$$E_0 = -j\omega\mu\frac{\kappa_m}{\kappa_{m0}^2}A = -j\frac{\omega\mu}{\kappa_m}A \tag{105}$$

Substituting (104) and (105) into Eqs. (96e), (96c), and (96a) leads to the following expressions for the TE_{m0} mode fields:

$$E_{ys} = E_0 \sin(\kappa_m x) e^{-j\beta_{m0} z} \tag{106}$$

$$H_{xs} = -\frac{\beta_{m0}}{\omega\mu} E_0 \sin(\kappa_m x) e^{-j\beta_{m0} z} \tag{107}$$

$$H_{zs} = j\frac{\kappa_m}{\omega\mu} E_0 \cos(\kappa_m x) e^{-j\beta_{m0} z} \tag{108}$$

It can be seen that these expressions are identical to the parallel-plate fields, Eqs. (65), (71), and (72). For TE_{m0}, we again note that the subscripts indicate that there are m half cycles of the electric field over the x dimension and there is zero variation in y. The cutoff frequency for the TE_{m0} mode is given by (101), appropriately modified:

$$\omega_{Cm0} = \frac{m\pi c}{na} \tag{109}$$

Using (109) in (99), the phase constant is

$$\beta_{m0} = \frac{n\omega}{c}\sqrt{1 - \left(\frac{m\pi c}{\omega n a}\right)^2} \tag{110}$$

All of the implications on mode behavior above and below cutoff are exactly the same as we found for the parallel-plate guide. The plane wave analysis is also carried out in the same manner. TE_{m0} modes can be modeled as plane waves that propagate down the guide by reflecting between the vertical side walls.

The electric field of the fundamental (TE_{10}) mode is, from (106):

$$E_{ys} = E_0 \sin\left(\frac{\pi x}{a}\right) e^{-j\beta_{10} z} \tag{111}$$

This field is plotted in Figure 13.18a. The field is vertically polarized, terminates on the top and bottom plates, and becomes zero at the two vertical walls, as is required from our boundary condition on a tangential electric field at a conducting surface. Its cutoff wavelength is found from (102) to be

$$\lambda_{C10} = 2na \tag{112}$$

which means that cutoff for this mode is achieved when the guide horizontal dimension, a, is equal to a half-wavelength (as measured in the medium).

Another possibility is the TE_{0p} field configuration, which consists of a horizontally polarized electric field. Figure 13.18b shows the field for TE_{01}. The specific fields for the TE_{0p} family are obtained from Eqs. (96a) through (96e) by setting $m = 0$,

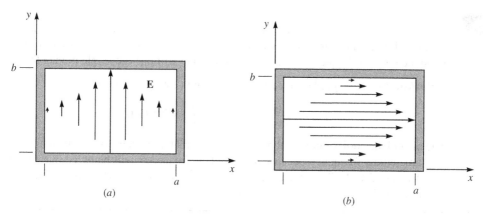

Figure 13.18   (a) TE$_{10}$ and (b) TE$_{01}$ mode electric field configurations in a rectangular waveguide.

which means, using (86) and (90), that

$$\kappa_p = \kappa_{mp}\big|_{m=0} = \frac{p\pi}{b} \tag{113}$$

and $\kappa_m = 0$. Now, the surviving field components in Eqs. (91a) through (91e) will be E_{xs}, H_{ys}, and H_{zs}. Now, define the electric field amplitude, E_0', which is composed of all the amplitude terms in Eq. (96d):

$$E_0' = j\omega\mu\frac{\kappa_p}{\kappa_{0p}^2}A = j\frac{\omega\mu}{\kappa_p}A \tag{114}$$

Using (113) and (114) in Eqs. (96d), (96b), and (96a) leads to the following expressions for the TE$_{0p}$ mode fields:

$$E_{xs} = E_0 \sin\left(\kappa_p y\right)e^{-j\beta_{0p}z} \tag{115}$$

$$H_{ys} = \frac{\beta_{0p}}{\omega\mu}E_0 \sin\left(\kappa_p y\right)e^{-j\beta_{0p}z} \tag{116}$$

$$H_{zs} = -j\frac{\kappa_p}{\omega\mu}E_0 \cos(\kappa_p y)e^{-j\beta_{0p}z} \tag{117}$$

where the cutoff frequency will be

$$\omega_{C0p} = \frac{p\pi c}{nb} \tag{118}$$

EXAMPLE 13.4

An air-filled rectangular waveguide has dimensions $a = 2$ cm and $b = 1$ cm. Determine the range of frequencies over which the guide will operate single mode (TE_{10}).

Solution. Since the guide is air-filled, $n = 1$, and (109) gives, for $m = 1$:

$$f_{C10} = \frac{\omega_{C10}}{2\pi} = \frac{c}{2a} = \frac{3 \times 10^{10}}{2(2)} = 7.5 \text{ GHz}$$

The next higher-order mode will be either TE_{20} or TE_{01}, which from (100) will have the same cutoff frequency because $a = 2b$. This frequency will be twice that found for TE_{10}, or 15 GHz. Thus the operating frequency range over which the guide will be single mode is 7.5 GHz $< f <$ 15 GHz.

13.5.6   The Need for Rectangular Waveguides

Having seen how rectangular waveguides work, questions arise: Why are they used, and when are they useful? Let us consider for a moment the operation of a transmission line at frequencies high enough such that waveguide modes can occur. The onset of guided modes in a transmission line, known as *moding*, is in fact a problem that needs to be avoided, because signal distortion may result. A signal that is input to such a line will find its power divided in some proportions among the various modes. The signal power in each mode propagates at a group velocity unique to that mode. With the power thus distributed, distortion will occur over sufficient distances, as the signal components among the modes lose synchronization with each other, owing to the different delay times (group delays) associated with the different modes. We encountered this concept in Example 13.3.

The problem of *modal dispersion* in transmission lines is avoided by ensuring that only the TEM mode propagates and that all waveguide modes are below cutoff. This is accomplished either by using line dimensions that are smaller than one-half the signal wavelength or by assuring an upper limit to the operating frequency in a given line. But it is more complicated than this.

In Section 13.1, we saw that increasing the frequency increases the line loss as a result of the skin effect. This is manifested through the increase in the series resistance per unit length, R. One can compensate by increasing one or more dimensions in the line cross section, as shown, for example, in Eqs. (7) and (12), but only to the point at which moding may occur. Typically, the increasing loss with increasing frequency will render the transmission line useless before the onset of moding, but one still cannot increase the line dimensions to reduce losses without considering the possibility of moding. This limitation on dimensions also limits the power-handling capability of the line, as the voltage at which dielectric breakdown occurs decreases with decreasing conductor separation. Consequently, the use of transmission lines, as frequencies are increased beyond a certain point, becomes undesirable, as losses will become excessive, and as the limitation on dimensions will limit the power-handling capability. Instead, we look to other guiding structures, among which is the rectangular guide.

Because the rectangular guide will not support a TEM mode, it will not operate until the frequency exceeds the cutoff frequency of the lowest-order guided mode of the structure. Thus, the guide must be constructed large enough to accomplish this for a given frequency; the required transverse dimensions will consequently be larger than those of a transmission line that is designed to support only the TEM mode. The increased size, coupled with the fact that there is more conductor surface area than in a transmission line of equal volume, means that losses will be substantially lower in the rectangular waveguide structure. Additionally, the guides will support more power at a given electric field strength than a transmission line, as the rectangular guide will have a higher cross-sectional area.

Still, hollow pipe guides must operate in a single mode in order to avoid the signal distortion problems arising from multimode transmission. This means that the guides must be of dimensions such that they operate above the cutoff frequency of the lowest-order mode, but below the cutoff frequency of the next higher-order mode, as demonstrated in Example 13.4. Increasing the operating frequency again means that the guide transverse dimensions must be decreased to maintain single-mode operation. This can be accomplished to a point at which skin effect losses again become problematic (remember that the skin depth is decreasing with increasing frequency, in addition to the decrease in metal surface area with diminishing guide size). In addition, the guides become too difficult to fabricate, with machining tolerances becoming more stringent. So again, as frequencies are further increased, we look for another type of structure.

D13.10 Specify the minimum width, a, and the maximum height, b, of an air-filled rectangular guide so that it will operate in a single mode over the frequency range 15 GHz $< f <$ 20 GHz.

Ans. 1 cm; 0.75 cm

13.6 PLANAR DIELECTRIC WAVEGUIDES

When skin effect losses become excessive, a good way to remove them is to remove the metal in the structure entirely and use interfaces between dielectrics for the confining surfaces. We thus obtain a *dielectric waveguide;* a basic form, the *symmetric slab waveguide,* is shown in Figure 13.19. The structure is so named because of its vertical symmetry about the z axis. The guide is assumed to have width in y much greater than the slab thickness d, so the problem becomes two-dimensional, with fields presumed to vary with x and z while being independent of y. The slab waveguide works in very much the same way as the parallel-plate waveguide, except wave reflections occur at the interfaces between dielectrics, having different refractive indices, n_1 for the slab and n_2 for the surrounding regions above and below. In the dielectric guide, total reflection is needed, so the incident angle must exceed the critical angle. Consequently, as discussed in Section 12.6, the slab index, n_1, must be greater than that of the surrounding materials, n_2. Dielectric guides differ from conducting guides in that power is not completely confined to the slab but resides partially above and below.

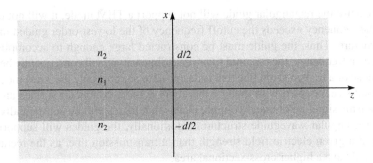

Figure 13.19 Symmetric dielectric slab waveguide structure, in which waves propagate along z. The guide is assumed to be infinite in the y direction, thus making the problem two-dimensional.

Dielectric guides are used primarily at optical frequencies (on the order of 10^{14} Hz). Again, guide transverse dimensions must be kept on the order of a wavelength to achieve operation in a single mode. A number of fabrication methods can be used to accomplish this. For example, a glass plate can be doped with materials that will raise the refractive index. The doping process allows materials to be introduced only within a thin layer adjacent to the surface that is a few micrometers thick.

13.6.1 Plane Wave Superposition Model

To understand how the guide operates, consider Figure 13.20, which shows a wave propagating through the slab by multiple reflections, but where *partial transmission* into the upper and lower regions occurs at each bounce. Wavevectors are shown in the middle and upper regions, along with their components in the x and z directions. As we found in Chapter 12, the z components (β) of all wavevectors are equal, as must be true if the field boundary conditions at the interfaces are to be satisfied for all positions and times. Partial transmission at the boundaries is, of course, an undesirable situation, since power in the slab will eventually leak away. We thus have a *leaky wave* propagating in the structure, whereas we need to have a guided mode. Note that in either case, we still have the two possibilities of wave polarization, and the resulting mode designation—either TE or TM.

Total power reflection at the boundaries for TE or TM waves implies, respectively, that $|\Gamma_s|^2$ or $|\Gamma_p|^2$ is unity, where the reflection coefficients are given in Eqs. (71) and (69) in Chapter 12.

$$\Gamma_s = \frac{\eta_{2s} - \eta_{1s}}{\eta_{2s} + \eta_{1s}} \qquad (119)$$

and

$$\Gamma_p = \frac{\eta_{2p} - \eta_{1p}}{\eta_{2p} + \eta_{1p}} \qquad (120)$$

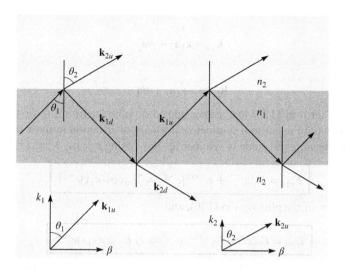

Figure 13.20 Plane wave geometry of a leaky wave in a symmetric slab waveguide. For a guided mode, total reflection occurs in the interior, and the x components of $\mathbf{k}_{2u}$ and $\mathbf{k}_{2d}$ are imaginary.

As discussed in Section 12.6, we require that the effective impedances, η_{2s} or η_{2p}, be purely imaginary, zero, or infinite if (119) or (120) is to have unity magnitude. Knowing that

$$\eta_{2s} = \frac{\eta_2}{\cos\theta_2} \tag{121}$$

and

$$\eta_{2p} = \eta_2 \cos\theta_2 \tag{122}$$

the requirement is that $\cos\theta_2$ be zero or imaginary, where, from Eq. (75), Section 12.6,

$$\cos\theta_2 = \left[1 - \sin^2\theta_2\right]^{1/2} = \left[1 - \left(\frac{n_1}{n_2}\right)^2 \sin^2\theta_1\right]^{1/2} \tag{123}$$

As a result, we require that

$$\theta_1 \geq \theta_c \tag{124}$$

where the critical angle is defined through

$$\sin\theta_c = \frac{n_2}{n_1} \tag{125}$$

Now, from the geometry of Figure 13.20, we can construct the field distribution of a TE wave in the guide using plane wave superposition. In the slab region ($-d/2 < x < d/2$), we have

$$E_{y1s} = E_0 e^{-j\mathbf{k}_{1u} \cdot \mathbf{r}} \pm E_0 e^{-j\mathbf{k}_{1d} \cdot \mathbf{r}} \qquad \left(-\frac{d}{2} < x < \frac{d}{2}\right) \tag{126}$$

where

$$\mathbf{k}_{1u} = \kappa_1 \mathbf{a}_x + \beta \mathbf{a}_z \tag{127}$$

and

$$\mathbf{k}_{1d} = -\kappa_1 \mathbf{a}_x + \beta \mathbf{a}_z \tag{128}$$

The second term in (126) may either add to or subtract from the first term, since either case would result in a symmetric intensity distribution in the x direction. We expect this because the guide is symmetric. Now, using $\mathbf{r} = x\mathbf{a}_x + z\mathbf{a}_z$, (126) becomes

$$E_{y1s} = E_0[e^{j\kappa_1 x} + e^{-j\kappa_1 x}]e^{-j\beta z} = 2E_0 \cos(\kappa_1 x)e^{-j\beta z} \tag{129}$$

for the choice of the plus sign in (126), and

$$E_{y1s} = E_0[e^{j\kappa_1 x} - e^{-j\kappa_1 x}]e^{-j\beta z} = 2j\,E_0 \sin(\kappa_1 x)e^{-j\beta z} \tag{130}$$

if the minus sign is chosen. Because $\kappa_1 = n_1 k_0 \cos\theta_1$, we see that larger values of κ_1 imply smaller values of θ_1 at a given frequency. In addition, larger κ_1 values result in a greater number of spatial oscillations of the electric field over the transverse dimension, as (129) and (130) show. We found similar behavior in the parallel-plate guide. In the slab waveguide, as with the parallel-plate guide, we associate higher-order modes with increasing values of κ_1.[6]

13.6.2 Surface Waves

In the regions above and below the slab, waves propagate according to wavevectors $\mathbf{k}_{2u}$ and $\mathbf{k}_{2d}$ as shown in Figure 13.20. Above the slab, for example ($x > d/2$), the TE electric field will be of the form

$$E_{y2s} = E_{02}e^{-j\mathbf{k}_2 \cdot \mathbf{r}} = E_{02}e^{-j\kappa_2 x}e^{-j\beta z} \tag{131}$$

However, $\kappa_2 = n_2 k_0 \cos\theta_2$, where $\cos\theta_2$, given in (123), is imaginary. We may therefore write

$$\kappa_2 = -j\gamma_2 \tag{132}$$

where γ_2 is real and is given by [using (123)]

$$\gamma_2 = j\kappa_2 = jn_2 k_0 \cos\theta_2 = jn_2 k_0 (-j)\left[\left(\frac{n_1}{n_2}\right)^2 \sin^2\theta_1 - 1\right]^{1/2} \tag{133}$$

Equation (131) now becomes

$$E_{y2s} = E_{02}e^{-\gamma_2(x-d/2)}e^{-j\beta z} \qquad \left(x > \frac{d}{2}\right) \tag{134}$$

---

[6] It would be appropriate to add the mode number subscript, m, to κ_1, κ_2, β, and θ_1, because, as was true with the metal guides, we will obtain discrete values of these quantities. To keep notation simple, the m subscript is suppressed, and we will assume it to be understood. Again, subscripts 1 and 2 in this section indicate, respectively, the slab and surrounding *regions*, and have nothing to do with mode number.

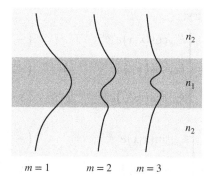

$$m = 1 \qquad m = 2 \qquad m = 3$$

Figure 13.21 Electric field amplitude distributions over the transverse plane for the first three TE modes in a symmetric slab waveguide.

where the x variable in (131) has been replaced by $x - (d/2)$ to position the field magnitude, E_{02}, at the boundary. Using similar reasoning, the field in the region below the lower interface, where x is negative, and where $\mathbf{k}_{2d}$ is involved, will be

$$E_{y2s} = E_{02}e^{\gamma_2(x+d/2)}e^{-j\beta z} \qquad \left(x < -\frac{d}{2}\right)$$

(135)

The fields expressed in (134) and (135) are those of *surface* waves. Note that they propagate in the z direction only, according to $e^{-j\beta z}$, but simply reduce in amplitude with increasing |x|, according to the $e^{-\gamma_2(x-d/2)}$ term in (134) and the $e^{\gamma_2(x+d/2)}$ term in (135). These waves represent a certain fraction of the total power in the mode, and so we see an important fundamental difference between dielectric waveguides and metal waveguides: in the dielectric guide, the fields (and guided power) exist over a cross section that extends beyond the confining boundaries, and in principle they exist over an infinite cross section. In practical situations, the exponential decay of the fields above and below the boundaries is typically sufficient to render the fields negligible within a few slab thicknesses from each boundary.

13.6.3 Mode Fields

The total electric field distribution is composed of the field in all three regions and is sketched in Figure 13.21 for the first few modes. Within the slab, the field is oscillatory and is of a similar form to that of the parallel-plate waveguide. The difference is that the fields in the slab waveguide do not reach zero at the boundaries but connect to the evanescent fields above and below the slab. The restriction is that the TE fields on either side of a boundary (being tangent to the interface) must match at the boundary. Specifically,

$$E_{y1s}|_{x=\pm d/2} = E_{y2s}|_{x=\pm d/2}$$

(136)

Applying this condition to (129), (130), (134), and (135) results in the final expressions for the TE electric field in the symmetric slab waveguide, for the cases of even and odd symmetry:

$$
E_{se}(\text{even TE}) = \begin{cases} E_{0e}\cos\left(\kappa_1 x\right)e^{-j\beta z} & \left(-\dfrac{d}{2} < x < \dfrac{d}{2}\right) \\[2mm] E_{0e}\cos\left(\kappa_1\dfrac{d}{2}\right)e^{-\gamma_2(x-d/2)}e^{-j\beta z} & \left(x > \dfrac{d}{2}\right) \\[2mm] E_{0e}\cos\left(\kappa_1\dfrac{d}{2}\right)e^{\gamma_2(x+d/2)}e^{-j\beta z} & \left(x < -\dfrac{d}{2}\right) \end{cases} \tag{137}
$$

$$
E_{so}(\text{odd TE}) = \begin{cases} E_{0o}\sin\left(\kappa_1 x\right)e^{-j\beta z} & \left(-\dfrac{d}{2} < x < \dfrac{d}{2}\right) \\[2mm] E_{0o}\sin\left(\kappa_1\dfrac{d}{2}\right)e^{-\gamma_2(x-d/2)}e^{-j\beta z} & \left(x > \dfrac{d}{2}\right) \\[2mm] -E_{0o}\sin\left(\kappa_1\dfrac{d}{2}\right)e^{\gamma_2(x+d/2)}e^{-j\beta z} & \left(x < -\dfrac{d}{2}\right) \end{cases} \tag{138}
$$

Solution of the wave equation yields (as it must) results identical to these. The reader is referred to References 2 and 3 for the details. The magnetic field for the TE modes will consist of x and z components, as was true for the parallel-plate guide. Finally, the TM mode fields will be nearly the same in form as those of TE modes, but with a simple rotation in polarization of the plane wave components by 90°. Thus, in TM modes, H_y will result, and it will have the same form as E_y for TE, as presented in (137) and (138).

Apart from the differences in the field structures, the dielectric slab waveguide operates in a manner that is qualitatively similar to the parallel-plate guide. Again, a finite number of discrete modes will be allowed at a given frequency, and this number increases as frequency increases. Higher-order modes are characterized by successively smaller values of θ_1.

An important difference in the slab waveguide occurs at cutoff for any mode. We know that $\theta = 0$ at cutoff in the metal guides. In the dielectric guide at cutoff, the wave angle, θ_1, is equal to the *critical angle, θ_c*. Then, as the frequency of a given mode is raised, its θ_1 value increases beyond θ_c in order to maintain transverse resonance, while maintaining the same number of field oscillations in the transverse plane.

As wave angle increases, however, the character of the evanescent fields changes significantly. This can be understood by considering the wave angle dependence on evanescent decay coefficient γ_2, as given by (133). Note in that equation that as θ_1 increases (as frequency goes up), γ_2 also increases, leading to a more rapid falloff of the fields with increasing distance above and below the slab. The mode therefore becomes more tightly confined to the slab as frequency is raised. Also, at a given frequency, lower-order modes, having smaller wave angles, will have lower values of γ_2 as (133) indicates. Consequently, when considering several modes propagating together at a single frequency, the higher-order modes will carry a greater percentage of their power in the upper and lower regions surrounding the slab than will modes of lower order.

13.6.4 Transverse Resonance and Cutoff Conditions

One can determine the conditions under which modes will propagate by using the transverse resonance condition, as we did with the parallel-plate guide. We perform the transverse round trip analysis in the slab region in the same manner that was done in Section 13.3 and obtain an equation similar to (37):

$$\boxed{\kappa_1 d + \phi_{TE} + \kappa_1 d + \phi_{TE} = 2m\pi} \qquad (139)$$

for TE waves and

$$\boxed{\kappa_1 d + \phi_{TM} + \kappa_1 d + \phi_{TM} = 2m\pi} \qquad (140)$$

for the TM case. Eqs. (139) and (140) are called the *eigenvalue equations* for the symmetric dielectric slab waveguide. The phase shifts on reflection, ϕ_{TE} and ϕ_{TM}, are the phases of the reflection coefficients, Γ_s and Γ_p, given in (119) and (120). These are readily found, but they turn out to be functions of θ_1. As we know, κ_1 also depends on θ_1, but in a different way than ϕ_{TE} and ϕ_{TM}. Consequently, (139) and (140) are *transcendental* in θ_1, and they cannot be solved in closed form. Instead, numerical or graphical methods must be used (see References 4 or 5). Emerging from this solution process, however, is a fairly simple cutoff condition for any TE or TM mode:

$$\boxed{k_0 d\sqrt{n_1^2 - n_2^2} \geq (m-1)\pi \qquad (m = 1,2,3,\dots)} \qquad (141)$$

For mode m to propagate, (141) must hold. The physical interpretation of the mode number m is again the number of half-cycles of the electric field (for TE modes) or magnetic field (for TM modes) that occur over the transverse dimension. The lowest-order mode ($m = 1$) is seen to have no cutoff—it will propagate from zero frequency on up. We will thus achieve single-mode operation (actually a single pair of TE and TM modes) if we can assure that the $m = 2$ modes are below cutoff. Using (141), our single-mode condition will thus be:

$$\boxed{k_0 d\sqrt{n_1^2 - n_2^2} < \pi} \qquad (142)$$

Using $k_0 = 2\pi/\lambda$, the wavelength range over which single-mode operation occurs is

$$\boxed{\lambda > 2d\sqrt{n_1^2 - n_2^2}} \qquad (143)$$

EXAMPLE 13.5

A symmetric dielectric slab waveguide is to guide light at wavelength $\lambda = 1.30\ \mu$m. The slab thickness is to be $d = 5.00\ \mu$m, and the refractive index of the surrounding material is $n_2 = 1.450$. Determine the maximum allowable refractive index of the slab material that will allow single TE and TM mode operation.

Solution. Equation (143) can be rewritten in the form

$$n_1 < \sqrt{\left(\frac{\lambda}{2d}\right)^2 + n_2^2}$$

Thus

$$n_1 < \sqrt{\left(\frac{1.30}{2(5.00)}\right)^2 + (1.450)^2} = 1.456$$

Clearly, fabrication tolerances are very exacting when constructing dielectric guides for single-mode operation!

D13.11 A 0.5-mm-thick slab of glass ($n_1 = 1.45$) is surrounded by air ($n_2 = 1$). The slab waveguides infrared light at wavelength $\lambda = 1.0$ μm. How many TE and TM modes will propagate?

Ans. 2102

13.7 OPTICAL FIBER

Optical fiber works on the same principle as the dielectric slab waveguide, except of course for the round cross section. A *step index* fiber is shown in Figure 13.10, in which a high-index *core* of radius a is surrounded by a lower-index *cladding* of radius b. Light is confined to the core through the mechanism of total reflection, but again some fraction of the power resides in the cladding as well. As we found in the slab waveguide, the cladding power again moves in toward the core as frequency is raised. Additionally, as is true in the slab waveguide, the fiber supports a mode that has no cutoff.

Analysis of the optical fiber is complicated. This is mainly because of the round cross section, along with the fact that it is generally a three-dimensional problem; the slab waveguide had only two dimensions to be concerned about. It is possible to analyze the fiber using rays within the core that reflect from the cladding boundary as light progresses down the fiber. We did this with the slab waveguide and obtained results fairly quickly. The method is difficult in fiber, however, because ray paths are complicated. There are two types of rays in the core: (1) those that pass through the fiber axis (z axis), known as *meridional rays,* and (2) those that avoid the axis but progress in a spiral-like path as they propagate down the guide. These are known as *skew rays;* their analysis, although possible, is tedious. Fiber modes are developed that can be associated with the individual ray types, or with combinations thereof, but it is easier to obtain these by solving the wave equation directly. Our purpose in this section is to provide a first exposure to the optical fiber problem (and to avoid an excessively long treatment). To accomplish this, we will solve the simplest case in the quickest way.

13.7.1 Mode Fields in Weakly Guiding Fiber

The simplest fiber configuration is that of a step index, but with the core and cladding indices of values that are very close, that is $n_1 \doteq n_2$. This is the *weak-guidance* condition, whose simplifying effect on the analysis is significant. We already saw how core and cladding indices in the slab waveguide need to be very close in value in order to achieve single-mode or few-mode operation. Fiber manufacturers have taken this result to heart, such that the weak-guidance condition is in fact satisfied by most commercial fibers today. Typical dimensions of a single-mode fiber are between 5 and 10 μm for the core diameter, with the cladding diameter usually 125 μm. Refractive index differences between core and cladding are typically a small fraction of a percent.

The main result of the weak-guidance condition is that a set of modes appears in which each mode is *linearly polarized.* This means that light having *x*-polarization, for example, will enter the fiber and establish itself in a mode or in a set of modes that preserve the *x*-polarization. Magnetic field is essentially orthogonal to **E**, and so it would in that case lie in the *y* direction. The *z* components of both fields, although present, are too weak to be of significance; the nearly equal core and cladding indices lead to ray paths that are essentially parallel to the guide axis—deviating only slightly. In fact, we may write for a given mode, $E_x \doteq \eta H_y$, when η is approximated as the intrinsic impedance of the cladding. Therefore, in the weak-guidance approximation, the fiber mode fields are treated as plane waves (nonuniform, of course). The designation for these modes is LP$_{\ell m}$, meaning linearly polarized, with integer order parameters ℓ and m. The latter express the numbers of variations over the two dimensions in the circular transverse plane. Specifically, ℓ, the *azimuthal mode number,* is one-half the number of power density maxima (or minima) that occur at a given radius as ϕ varies from 0 to 2π. The *radial mode number, m,* expresses the number of maxima that occur along a radial line (at constant ϕ) that extends from zero to infinity.

Although we may assume a linearly polarized field in a rectangular coordinate system, we are obliged to work in cylindrical coordinates for obvious reasons. In a manner that reminds us of the rectangular waveguide, it is possible to write the *x*-polarized phasor electric field within a weakly guiding cylindrical fiber as a product of three functions, each of which varies with one of the coordinate variables, ρ, ϕ, and z:

$$E_{xs}(\rho,\phi,z) = \sum_i R_i(\rho)\Phi_i(\phi)\exp(-j\beta_i z) \qquad (144)$$

Each term in the summation is an individual mode of the fiber. Note that the *z* function is just the propagation term, $e^{-j\beta z}$, since we are assuming an infinitely long lossless fiber.

The wave equation is Eq. (58), which we may write for the assumed *x* component of $\mathbf{E}_s$, but in which the Laplacian operator is written in cylindrical coordinates:

$$\frac{1}{\rho}\frac{\partial}{\partial\rho}\left(\rho\frac{\partial^2 E_{xs}}{\partial\rho}\right) + \frac{1}{\rho^2}\frac{\partial^2 E_{xs}}{\partial\phi^2} + (k^2 - \beta^2)E_{xs} = 0 \qquad (145)$$

where we recognize that the $\partial^2/\partial z^2$ operation, when applied to (144), leads to a factor of $-\beta^2$. We now substitute a single term of (144) into (145) [since each term in (144) should alone satisfy the wave equation]. Dropping the subscript *i*, expanding the radial derivative, and rearranging terms, we obtain:

$$\underbrace{\frac{\rho^2}{R}\frac{d^2 R}{d\rho^2} + \frac{\rho}{R}\frac{dR}{d\rho} + \rho^2(k^2 - \beta^2)}_{\ell^2} = \underbrace{-\frac{1}{\Phi}\frac{d^2\Phi}{d\phi^2}}_{\ell^2} \qquad (146)$$

We note that the left-hand side of (146) varies only with ρ, whereas the right-hand side varies only with ϕ. Since the two variables are independent, it must follow that each side of the equation must be equal to a constant. Calling this constant ℓ^2, as shown, we may write separate equations for each side; the variables are now separated:

$$\frac{d^2\Phi}{d\phi^2} + \ell^2\Phi = 0 \tag{147a}$$

$$\frac{d^2R}{d\rho^2} + \frac{1}{\rho}\frac{dR}{d\rho} + \left[k^2 - \beta^2 - \frac{\ell^2}{\rho^2}\right]R = 0 \tag{147b}$$

The solution of (147a) is of the form of the sine or cosine of ϕ:

$$\Phi(\phi) = \begin{cases} \cos(\ell\phi + \alpha) \\ \sin(\ell\phi + \alpha) \end{cases} \tag{148}$$

where α is a constant. The form of (148) dictates that ℓ must be an integer, since the same mode field must occur in the transverse plane as ϕ is changed by 2π radians. Since the fiber is round, the orientation of the x and y axes in the transverse plane is immaterial, so we may choose the cosine function and set $\alpha = 0$. We will thus use $\Phi(\phi) = \cos(\ell\phi)$.

The solution of (147b) to obtain the radial function is more complicated. Eq. (147b) is a form of Bessel's equation, whose solutions are Bessel functions of various forms. The key parameter is the function $\beta_t = (k^2 - \beta^2)^{1/2}$, the square of which appears in (147b). Note that β_t will differ in the two regions: Within the core $(\rho < a), \beta_t = \beta_{t1} = (n_1^2k_0^2 - \beta^2)^{1/2}$; within the cladding $(\rho > a)$, we have $\beta_t = \beta_{t2} = (n_2^2k_0^2 - \beta^2)^{1/2}$. Depending on the relative magnitudes of k and β, β_t may be real or imaginary. These possibilities lead to two solution forms of (147b):

$$R(\rho) = \begin{cases} AJ_\ell(\beta_t\rho) & \beta_t \text{ real} \\ BK_\ell(|\beta_t|\rho) & \beta_t \text{ imaginary} \end{cases} \tag{149}$$

where A and B are constants. $J_\ell(\beta_t\rho)$ is the ordinary Bessel function of the first kind, of order ℓ and of argument $\beta_t\rho$. $K_\ell(|\beta_t|\rho)$ is the modified Bessel function of the second kind, of order ℓ, and having argument $|\beta_t|\rho$. The first two orders of each of these functions are illustrated in Figures 13.22a and b. In our study, it is necessary to know the precise zero crossings of the J_0 and J_1 functions. Those shown in Figure 13.22a are as follows: For J_0, the zeros are 2.405, 5.520, 8.654, 11.792, and 14.931. For J_1, the zeros are 0, 3.832, 7.016, 10.173, and 13.324. Other Bessel function types would contribute to the solutions in Eq. (149), but these exhibit nonphysical behavior with radius and are not included.

We next need to determine which of the two solutions is appropriate for each region. Within the core $(\rho < a)$ we expect to get an oscillatory solution for the field—much in the same manner as we found in the slab waveguide. Therefore, we assign the ordinary Bessel function solutions to that region by requiring that $\beta_{t1} = (n_1^2k_0^2 - \beta^2)^{1/2}$ is real. In the cladding $(\rho > a)$, we expect surface waves that decrease in amplitude with increasing radius away from the core/cladding boundary.

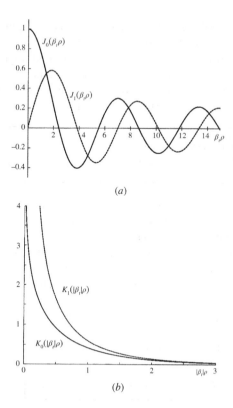

(a)

(b)

Figure 13.22 (a) Ordinary Bessel functions of the first kind, of orders 0 and 1, and of argument $\beta_t\rho$, where β_t is real. (b) Modified Bessel functions of the second kind, of orders 0 and 1, and of argument $|\beta_t|\rho$, where β_t is imaginary.

The Bessel K functions provide this behavior and will apply if β_{t2} is imaginary. Requiring this, we may therefore write $|\beta_{t2}| = (\beta^2 - n_2^2 k_0^2)^{1/2}$. The diminishing field amplitude with increasing radius within the cladding allows us to neglect the effect of the outer cladding boundary (at $\rho = b$), as fields there are presumed to be too weak for this boundary to have any effect on the mode field.

Because β_{t1} and β_{t2} are in units of m^{-1}, it is convenient to normalize these quantities (while making them dimensionless) by multiplying both by the core radius, a. Our new normalized parameters become

$$u \equiv a\beta_{t1} = a\sqrt{n_1^2 k_0^2 - \beta^2}$$ (150a)

$$w \equiv a|\beta_{t2}| = a\sqrt{\beta^2 - n_2^2 k_0^2}$$ (150b)

u and w are in direct analogy with the quantities κ_{1d} and κ_{2d} in the slab waveguide. As in those parameters, β is the z component of *both* $n_1 k_0$ and $n_2 k_0$ and is the phase constant of the guided mode. β must be the same in both regions so that the field boundary conditions will be satisfied at $\rho = a$ for all z and t.

We may now construct the total solution for E_{xs} for a single guided mode, using (144) along with (148), (149), (150a), and (150b):

$$E_{xs} = \begin{cases} E_0 J_\ell(u\rho/a)\cos(\ell\phi)e^{-j\beta z} & \rho \leq a \\ E_0 [J_\ell(u)/K_\ell(w)]K_\ell(w\rho/a)\cos(\ell\phi)e^{-j\beta z} & \rho \geq a \end{cases} \tag{151}$$

Note that we have let the coefficient A in (149) equal E_0, and $B = E_0[J_\ell(u)/K_\ell(w)]$. These choices ensure that the expressions for E_{xs} in the two regions become equal at $\rho = a$, a condition approximately true as long as $n_1 \doteq n_2$ (the weak-guidance approximation).

Again, the weak-guidance condition also allows the approximation $H \doteq E/\eta$, with η taken as the intrinsic impedance of the cladding. Having $\mathbf{E}_s$ and $\mathbf{H}_s$ enables us to find the $\text{LP}_{\ell m}$ mode average power density (or light intensity) through

$$|\langle \mathbf{S} \rangle| = \left| \frac{1}{2} \mathcal{R}e\{\mathbf{E}_s \times \mathbf{H}_s^*\} \right| = \frac{1}{2}\mathcal{R}e\, E_{xs} H_{ys}^* = \frac{1}{2\eta}|E_{xs}|^2 \tag{152}$$

Using (151) in (152), the mode intensity in W/m^2 becomes

$$I_{\ell m} = I_0 J_\ell^2\left(\frac{u\rho}{a}\right)\cos^2(\ell\phi) \qquad \rho \leq a \tag{153a}$$

$$I_{\ell m} = I_0 \left(\frac{J_\ell(u)}{K_\ell(w)}\right)^2 K_\ell^2\left(\frac{w\rho}{a}\right)\cos^2(\ell\phi) \qquad \rho \geq a \tag{153b}$$

where I_0 is the peak intensity value. The significance of the azimuthal mode number ℓ, as evident in (153a) and (153b), is that twice its value gives the number of intensity variations around the circle, $0 < \phi < 2\pi$; it also determines the order of the Bessel functions that are used. The influence of the radial mode number, m, is not immediately apparent in (153a) and (153b). Briefly stated, m determines the range of allowed values of u that occur in the Bessel function, $J(u\rho/a)$. The greater the value of m, the greater the allowed values of u; with larger u, the Bessel function goes through more oscillations over the range $0 < \rho < a$, and so more radial intensity variations occur with larger m. In the slab waveguide, the mode number (also m) determines the allowed ranges of κ_1. As we saw in Section 13.6, increasing κ_1 at a given frequency means that the slab ray propagates closer to the normal (smaller θ_1), and so more spatial oscillations of the field occur in the transverse direction (larger m).

13.7.2 Mode Parameters and Cutoff Conditions

The final step in the analysis is to obtain an equation from which values of mode parameters (u, w, and β, for example) can be determined for a given operating frequency and fiber construction. In the slab waveguide, two equations, (139) and (140), were found using transverse resonance arguments, and these were associated with TE and TM waves in the slab. In our fiber, we do not apply transverse resonance directly, but rather *implicitly,* by requiring that all fields satisfy the boundary conditions at the core/cladding interface, $\rho = a$.[7] We have already applied conditions on the transverse fields to obtain Eq. (151). The remaining condition is continuity of the z components of **E** and **H**. In the weak-guidance approximation, we have neglected all z components, but we will consider them now for this last exercise. Using Faraday's law in point form, continuity of H_{zs} at $\rho = a$ is the same as the continuity of the z component of $\nabla \times \mathbf{E}_s$, provided that $\mu = \mu_0$ (or is the same value) in both regions. Specifically

$$(\nabla \times \mathbf{E}_{s1})_z|_{\rho=a} = (\nabla \times \mathbf{E}_{s2})_z|_{\rho=a} \tag{154}$$

The procedure begins by expressing the electric field in (151) in terms of ρ and ϕ components and then applying (154). This is a lengthy procedure and is left as an exercise (or may be found in Reference 5). The result is the eigenvalue equation for LP modes in the weakly guiding step index fiber:

$$\frac{J_{\ell-1}(u)}{J_\ell(u)} = -\frac{w}{u}\,\frac{K_{\ell-1}(w)}{K_\ell(w)} \tag{155}$$

This equation, like (139) and (140), is transcendental, and it must be solved for u and w numerically or graphically. This exercise in all of its aspects is beyond the scope of our treatment. Instead, we will obtain from (155) the conditions for cutoff for a given mode and some properties of the most important mode—that which has no cutoff, and which is therefore the mode that is present in single-mode fiber.

The solution of (155) is facilitated by noting that u and w can be combined to give a new parameter that is independent of β and depends only on the fiber construction and on the operating frequency. This new parameter, called the *normalized frequency,* or V number, is found using (150a) and (150b):

$$V \equiv \sqrt{u^2 + w^2} = ak_0\sqrt{n_1^2 - n_2^2} \tag{156}$$

We note that an increase in V is accomplished through an increase in core radius, frequency, or index difference.

The cutoff condition for a given mode can now be found from (155) in conjunction with (156). To do this, we note that cutoff in a dielectric guide means that total reflection at the core/cladding boundary just ceases, and power just begins to propagate radially, away from the core. The effect on the electric field of Eq. (151) is to produce a cladding field that no longer diminishes with increasing radius. This occurs in the modified Bessel function, $K(w\rho/a)$, when $w = 0$. This is our general

---

[7] Recall that the equations for reflection coefficient (119) and (120), from which the phase shift on reflection used in transverse resonance is determined, originally came from the application of the field boundary conditions.

cutoff condition, which we now apply to (155), whose right-hand side becomes zero when $w = 0$. This leads to cutoff values of u and V (u_c and V_c), and, by (156), $u_c = V_c$. Eq. (155) at cutoff now becomes:

$$\boxed{J_{\ell-1}(V_c) = 0} \qquad (157)$$

Finding the cutoff condition for a given mode is now a matter of finding the appropriate zero of the relevant ordinary Bessel function, as determined by (157). This gives the value of V at cutoff for that mode.

For example, the lowest-order mode is the simplest in structure; therefore it has no variations in ϕ and one variation (one maximum) in ρ. The designation for this mode is therefore LP_{01}, and with $\ell = 0$, (157) gives the cutoff condition as $J_{-1}(V_c) = 0$. Because $J_{-1} = J_1$ (true only for the J_1 Bessel function), we take the first zero of J_1, which is $V_c(01) = 0$. The LP_{01} mode therefore has no cutoff and will propagate at the exclusion of all other modes provided V for the fiber is greater than zero but less than V_c for the next-higher-order mode. By inspecting Figure 13.22a, we see that the next Bessel function zero is 2.405 (for the J_0 function). Therefore, $\ell - 1 = 0$ in (156), and so $\ell = 1$ for the next-higher-order mode. Also, we use the lowest value of m_ℓ ($m = 1$), and the mode is therefore identified as LP_{11}. Its cutoff V is $V_c(11) = 2.405$. If $m = 2$ were to be chosen instead, we would obtain the cutoff V number for the LP_{12} mode. We use the next zero of the J_0 function, which is 5.520, or $V_c(12) = 5.520$. In this way, the radial mode number, m, numbers the zeros of the Bessel function of order $\ell - 1$, taken in order of increasing value.

13.7.3 Single-Mode Fiber

When we follow the reasoning described above, the condition for single-mode operation in a step index fiber is found to be

$$\boxed{V < V_c(11) = 2.405} \qquad (158)$$

Then, using (156) along with $k_0 = 2\pi/\lambda$, we find

$$\boxed{\lambda > \lambda_c = \frac{2\pi a}{2.405}\sqrt{n_1^2 - n_2^2}} \qquad (159)$$

as the requirement on free-space wavelength to achieve single-mode operation in a step index fiber. The similarity to the single-mode condition in the slab waveguide [Eq. (143)] is apparent. The *cutoff wavelength*, λ_c, is that for the LP_{11} mode. Its value is quoted as a specification of most commercial single-mode fiber.

EXAMPLE 13.6

The cutoff wavelength of a step index fiber is quoted as $\lambda_c = 1.20~\mu$m. If the fiber is operated at wavelength $\lambda = 1.55~\mu$m, what is V?

Solution. Using (156) and (159), we find

$$V = 2.405\,\frac{\lambda_c}{\lambda} = 2.405\left(\frac{1.20}{1.55}\right) = 1.86$$

The intensity profiles of the first two modes can be found using (153a) and (153b), having determined u and w values for each mode from (155). For LP_{01}, we find

$$I_{01} = \begin{cases} I_0 J_0^2\left(u_{01}\rho/a\right) & \rho \leq a \\ I_0\left(\dfrac{J_0(u_{01})}{K_0(w_{01})}\right)^2 K_0^2(w_{01}\rho/a) & \rho \geq a \end{cases} \tag{160}$$

and for LP_{11} we find

$$I_{11} = \begin{cases} I_0 J_1^2(u_{11}\rho/a)\cos^2\phi & \rho \leq a \\ I_0\left(\dfrac{J_1(u_{11})}{K_1(w_{11})}\right)^2 K_1^2(w_{11}\rho/a)\cos^2\phi & \rho \geq a \end{cases} \tag{161}$$

The two intensities for a single V value are plotted as functions of radius at $\phi = 0$ in Figure 13.23. We again note the lower confinement of the higher-order mode to the core, as was true in the slab waveguide.

As V increases (accomplished by increasing the frequency, for example), existing modes become more tightly confined to the core, while new modes of higher order may begin to propagate. The behavior of the lowest-order mode with changing V is depicted in Figure 13.24, where we again note that the mode becomes more tightly confined as V increases. In determining the intensities, Eq. (155) must in general be solved numerically to obtain u and w. Various analytic approximations to the exact numerical solution exist, the best of which is the Rudolf-Neumann formula for the LP_{01} mode, valid over the range $1.3 < V < 3.5$:

$$w_{01} \doteq 1.1428V - 0.9960 \tag{162}$$

Having w_{01}, u_{01} can be found from (156), knowing V.

Another important simplification for the LP_{01} mode is the approximation of its intensity profile by a Gaussian function. An inspection of any of the intensity plots of Figure 13.24 shows a resemblance to a Gaussian, which would be expressed as

$$I_{01} \approx I_0 e^{-2\rho^2/\rho_0^2} \tag{163}$$

where ρ_0, termed the *mode field radius,* is defined as the radius from the fiber axis at which the mode intensity falls to $1/e^2$ times its on-axis value. This radius depends on frequency, and most generally on V. A similar approximation can be made for the fundamental symmetric slab waveguide mode intensity. In step index fiber, the best fit between the Gaussian approximation and the actual mode intensity as given in (160) is given by the Marcuse formula:

$$\frac{\rho_0}{a} \approx 0.65 + \frac{1.619}{V^{3/2}} + \frac{2.879}{V^6} \tag{164}$$

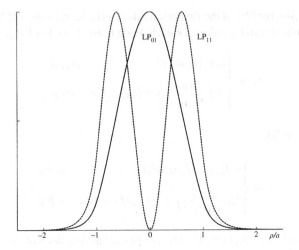

Figure 13.23 Intensity plots from Eqs. (160) and (161) of the first two LP modes in a weakly guiding step index fiber, as functions of normalized radius, ρ/a. Both functions were evaluated at the same operating frequency; the relatively weak confinement of the LP_{11} mode compared to that of LP_{01} is evident.

The mode field radius (at a quoted wavelength) is another important specification (along with the cutoff wavelength) of commercial single-mode fiber. It is important to know for several reasons: First, in splicing or connecting two single-mode fibers together, the lowest connection loss will be attained if both fibers have the same mode field radius, and if the fiber axes are precisely aligned. Different radii or displaced axes result in increased loss, but this can be calculated and compared with measurement. Alignment tolerance (allowable deviation from precise axis alignment) is relaxed somewhat if the fibers have larger mode field radii. Second, a smaller mode field radius means that the fiber is less likely to suffer loss as a result of bending. A loosely confined mode tends to radiate away more as the fiber is bent. Finally, mode field radius is directly related to the mode phase constant, β, since if u and w are known (found from ρ_0), β can be found from (150a) or (150b). Therefore, knowledge of how β changes with frequency (leading to the quantification of dispersion) can be found by measuring the change in mode field radius with frequency. Again, References 4 and 5 (and references therein) provide more detail.

D13.12 For the fiber of Example 13.6, the core radius is given as $a = 5.0 \ \mu$m. Find the mode field radius at wavelengths (a) 1.55 μm; (b) 1.30 μm.

Ans. (a) 6.78 μm; (b) 5.82 μm

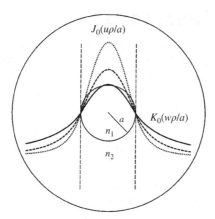

Figure 13.24 Intensity plots for the LP_{01} mode in a weakly guiding step index fiber. Traces are shown for $V = 1.0$ (solid), $V = 1.2$ (dashed), and $V = 1.5$ (dotted), corresponding to increases in frequency in those proportions. Dashed vertical lines indicate the core/cladding boundary, at which for all three cases, the J_0 radial dependence in the core connects to the k_0 radial dependence in the cladding, as demonstrated in Eq. (160). The migration of mode power toward the fiber axis as frequency increases is evident.

REFERENCES

1. Weeks, W. L. *Transmission and Distribution of Electrical Energy*. New York: Harper and Row, 1981. Line parameters for various configurations of power transmission and distribution systems are discussed in Chapter 2, along with typical parameter values.

2. Edwards, T. C. *Foundations for Microstrip Circuit Design*. Chichester, N.Y.: Wiley-Interscience, 1981. Chapters 3 and 4 provide an excellent treatment of microstrip lines, with many design formulas.

3. Ramo, S., J. R. Whinnery, and T. Van Duzer. *Fields and Waves in Communication Electronics*. 3d ed. New York: John Wiley & Sons, 1990. In-depth treatment of parallel-plate and rectangular waveguides is presented in Chapter 8.

4. Marcuse, D. *Theory of Dielectric Optical Waveguides*. 2d ed. New York: Academic Press, 1990. This book provides a very general and complete discussion of dielectric slab waveguides, plus other types.

5. Buck, J. A. *Fundamentals of Optical Fibers*. 2d ed. New York: Wiley-Interscience, 2004. Symmetric slab dielectric guides and weakly guiding fibers are emphasized in this book by one of the coauthors.

CHAPTER 13   PROBLEMS

Quizzes

13.1 The conductors of a coaxial transmission line are copper ($\sigma_c = 5.8 \times 10^7$ S/m), and the dielectric is polyethylene ($\epsilon'_r = 2.26$, $\sigma/\omega\epsilon' = 0.0002$). If the inner radius of the outer conductor is 4 mm, find the radius of the inner conductor so that (a) $Z_0 = 50\ \Omega$; (b) $C = 100$ pF/m; (c) $L = 0.2\ \mu$H/m. A lossless line can be assumed.

13.2 Find R, L, C, and G for a coaxial cable with $a = 0.25$ mm, $b = 2.50$ mm, $c = 3.30$ mm, $\epsilon_r = 2.0$, $\mu_r = 1$, $\sigma_c = 1.0 \times 10^7$ S/m, $\sigma = 1.0 \times 10^{-5}$ S/m, and $f = 300$ MHz.

13.3 Two aluminum-clad steel conductors are used to construct a two-wire transmission line. Let $\sigma_{Al} = 3.8 \times 10^7$ S/m, $\sigma_{St} = 5 \times 10^6$ S/m, and $\mu_{St} = 100\ \mu$H/m. The radius of the steel wire is 0.5 in., and the aluminum coating is 0.05 in. thick. The dielectric is air, and the center-to-center wire separation is 4 in. Find C, L, G, and R for the line at 10 MHz.

13.4 Find R, L, C, and G for a two-wire transmission line in polyethylene at $f = 800$ MHz. Assume copper conductors of radius 0.50 mm and separation 0.80 cm. Use $\epsilon_r = 2.26$ and $\sigma/(\omega\epsilon') = 4.0 \times 10^{-4}$.

13.5 Each conductor of a two-wire transmission line has a radius of 0.5 mm; their center-to-center separation is 0.8 cm. Let $f = 150$ MHz, and assume σ and σ_c are zero. Find the dielectric constant of the insulating medium if (a) $Z_0 = 300\ \Omega$; (b) $C = 20$ pF/m; (c) $v_p = 2.6 \times 10^8$ m/s.

13.6 Consider an air-filled coaxial transmission line having inner and outer radii a and b. (a) Show that skin effect loss in the line is minimized when $b/a = 3.6$, yielding characteristic impedance $Z_0 = 77\ \Omega$. (b) Show that the maximum power the line can carry before air breakdown occurs is achieved when $b/a = 1.65$, yielding characteristic impedance $Z_0 = 30$ ohms. Assume low loss, such that the attenuation coefficient is given by Eq. (54a) in Chapter 10.

13.7 Pertinent dimensions for the transmission line shown in Figure 13.2 are $b = 3$ mm and $d = 0.2$ mm. The conductors and the dielectric are nonmagnetic. (a) If the characteristic impedance of the line is $15\ \Omega$, find ϵ'_r. Assume a low-loss dielectric. (b) Assume copper conductors and operation at 2×10^8 rad/s. If $RC = GL$, determine the loss tangent of the dielectric.

13.8 A transmission line constructed from perfect conductors and an air dielectric is to have a maximum dimension of 8 mm for its cross section. The line is to be used at high frequencies. Specify the dimensions if it is (a) a two-wire line with $Z_0 = 300\ \Omega$; (b) a planar line with $Z_0 = 15\ \Omega$; (c) a $72\ \Omega$ coax having a zero-thickness outer conductor.

13.9 A microstrip line is to be constructed using a lossless dielectric for which $\epsilon'_r = 7.0$. If the line is to have a $50\ \Omega$ characteristic impedance, determine (a) $\epsilon_{r,\text{eff}}$; (b) w/d.

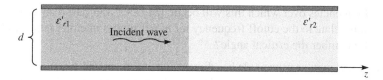

Figure 13.25 See Problems 13.17 and 13.18.

13.10 Two microstrip lines are fabricated end-to-end on a 2-mm-thick wafer of lithium niobate ($\epsilon'_r = 4.8$). Line 1 is of 4 mm width; line 2 (unfortunately) has been fabricated with a 5 mm width. Determine the power loss in dB for waves transmitted through the junction.

13.11 A parallel-plate waveguide is known to have a cutoff wavelength for the $m = 1$ TE and TM modes of $\lambda_{c1} = 4.1$ mm. The guide is operated at wavelength $\lambda = 1.0$ mm. How many modes propagate?

13.12 A parallel-plate guide is to be constructed for operation in the TEM mode only over the frequency range $0 < f < 3$ GHz. The dielectric between plates is to be Teflon ($\epsilon'_r = 2.1$). Determine the maximum allowable plate separation, d.

13.13 A lossless parallel-plate waveguide is known to propagate the $m = 2$ TE and TM modes at frequencies as low as 10 GHz. If the plate separation is 1 cm, determine the dielectric constant of the medium between plates.

13.14 A $d = 1$ cm parallel-plate guide is made with glass ($n = 1.45$) between plates. If the operating frequency is 32 GHz, which modes will propagate?

13.15 For the guide of Problem 13.14, and at the 32 GHz frequency, determine the difference between the group delays of the highest-order mode (TE or TM) and the TEM mode. Assume a propagation distance of 10 cm.

13.16 The cutoff frequency of the $m = 1$ TE and TM modes in an air-filled parallel-plate guide is known to be $f_{c1} = 7.5$ GHz. The guide is used at wavelength $\lambda = 1.5$ cm. Find the group velocity of the $m = 2$ TE and TM modes.

13.17 A parallel-plate guide is partially filled with two lossless dielectrics (Figure 13.25) where $\epsilon'_{r1} = 4.0$, $\epsilon'_{r2} = 2.1$, and $d = 1$ cm. At a certain frequency, it is found that the TM_1 mode propagates through the guide without suffering any reflective loss at the dielectric interface. (a) Find this frequency. (b) Is the guide operating at a single TM mode at the frequency found in part a? *Hint*: Remember Brewster's angle?

13.18 In the guide of Figure 13.25, it is found that $m = 1$ modes propagating from left to right totally reflect at the interface, so that no power is transmitted into the region of dielectric constant ϵ'_{r2}. (a) Determine the range of frequencies over which this will occur. (*b*) Does your part *a* answer in any way relate to the cutoff frequency for $m = 1$ modes in either region? *Hint:* Remember the critical angle?

13.19 A rectangular waveguide has dimensions $a = 6$ cm and $b = 4$ cm. (*a*) Over what range of frequencies will the guide operate single mode? (*b*) Over what frequency range will the guide support *both* TE_{10} and TE_{01} modes and no others?

13.20 Two rectangular waveguides are joined end-to-end. The guides have identical dimensions, where $a = 2b$. One guide is air-filled; the other is filled with a lossless dielectric characterized by ϵ'_r. (*a*) Determine the maximum allowable value of ϵ'_r such that single-mode operation can be simultaneously assured in *both* guides at some frequency. (*b*) Write an expression for the frequency range over which single-mode operation will occur in both guides; your answer should be in terms of ϵ'_r, guide dimensions as needed, and other known constants.

13.21 An air-filled rectangular waveguide is to be constructed for single-mode operation at 15 GHz. Specify the guide dimensions, *a* and *b*, such that the design frequency is 10 percent higher than the cutoff frequency for the TE_{10} mode, while being 10 percent lower than the cutoff frequency for the next-higher-order mode.

13.22 Consider the TE_{11} mode in a rectangular guide having width *a* and height *b*. Using Eqs. (96d) and (96e), along with methods described in Chapter 2, Section 2.6, show that the equation for the streamlines for this field (at $z = 0$) is

$$y = \frac{b}{\pi} \cos^{-1}\left[\frac{C}{\cos(\pi x/a)}\right] \quad \left(0 < x < a, 0 < y < b\right)$$

where *C* is a constant that identifies a given streamline.

13.23 (*a*) Using the relation $\langle \mathbf{S} \rangle = \frac{1}{2} \mathcal{R}e\{\mathbf{E}_s \times \mathbf{H}_s^*\}$ and Eqs. (106) through (108), show that the average power density in the TE_{10} mode in a rectangular waveguide is given by

$$\langle \mathbf{S} \rangle = \frac{\beta_{10}}{2\omega\mu}E_0^2 \sin^2(\kappa_{10}x)\mathbf{a}_z \text{ W/m}^2$$

(*b*) Integrate the result of part *a* over the guide cross section, $0 < x < a, 0 < y < b$, to show that the average power in watts transmitted down the guide is given as

$$P_{av} = \frac{\beta_{10}ab}{4\omega\mu}E_0^2 = \frac{ab}{4\eta}E_0^2 \sin\theta_{10} \text{ W}$$

where $\eta = \sqrt{\mu/\epsilon}$ and θ_{10} is the wave angle associated with the TE_{10} mode. Interpret.

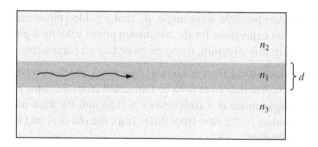

Figure 13.26   See Problem 13.29.

13.24 Show that the group dispersion parameter, $d^2\beta/d\omega^2$, for a given mode in a parallel-plate or rectangular waveguide is given by

$$\frac{d^2\beta}{d\omega^2} = -\frac{n}{\omega c}\left(\frac{\omega_c}{\omega}\right)^2\left[1 - \left(\frac{\omega_c}{\omega}\right)^2\right]^{-3/2}$$

where ω_c is the radian cutoff frequency for the mode in question [note that the first derivative form was already found, resulting in Eq. (57)].

13.25 Consider a transform-limited pulse of center frequency $f = 10$ GHz, and of full-width $2T = 1.0$ ns. The pulse propagates in a lossless single-mode rectangular guide which is air-filled and in which the 10 GHz operating frequency is 1.1 times the cutoff frequency of the TE_{10} mode. Using the result of Problem 13.24, determine the length of guide over which the pulse broadens to twice its initial width. What simple step can be taken to reduce the amount of pulse broadening in this guide, while maintaining the same initial pulse width? Additional background for this problem is found in Section 12.6.

13.26 A symmetric dielectric slab waveguide has a slab thickness $d = 10\ \mu m$, with $n_1 = 1.48$ and $n_2 = 1.45$. If the operating wavelength is $\lambda = 1.3\ \mu m$, what modes will propagate?

13.27 A symmetric slab waveguide is known to support only a single pair of TE and TM modes at wavelength $\lambda = 1.55\ \mu m$. If the slab thickness is 5 μm, what is the maximum value of n_1 if $n_2 = 3.30$?

13.28 In a symmetric slab waveguide, $n_1 = 1.50$, $n_2 = 1.45$, and $d = 10\ \mu m$.
(a) What is the phase velocity of the $m = 1$ TE or TM mode at cutoff?
(b) How will your part a result change for higher-order modes (if at all)?

13.29 An *asymmetric* slab waveguide is shown in Figure 13.26. In this case, the regions above and below the slab have unequal refractive indices, where $n_1 > n_3 > n_2$. (a) Write, in terms of the appropriate indices, an expression for the minimum possible wave angle, θ_1, that a guided mode may have. (*b*) Write an expression for the maximum phase velocity a guided mode may have in this structure, using given or known parameters.

13.30 A step index optical fiber is known to be single mode at wavelengths $\lambda > 1.2$ μm. Another fiber is to be fabricated from the same materials, but it is to be single mode at wavelengths $\lambda > 0.63$ μm. By what percentage must the core radius of the new fiber differ from the old one, and should it be larger or smaller?

13.31 Is the mode field radius greater than or less than the fiber core radius in single-mode step index fiber?

13.32 The mode field radius of a step index fiber is measured as 4.5 μm at free-space wavelength $\lambda = 1.30$ μm. If the cutoff wavelength is specified as $\lambda_c = 1.20$ μm, find the expected mode field radius at $\lambda = 1.55$ μm.

CHAPTER 14 Electromagnetic Radiation and Antennas

术语

radiation　辐射。

hertzian dipole　赫兹偶极子。

far-field　远场。

solid angle　立体角。

steradian(sr)　球面度。表征立体角的单位，简写为sr。

normalize　归一化，指将函数值的最大值变为1，其他值成比例减小。

monopole　单极子天线。

reciprocity theorem　互易定理。

We are used to the idea that loss mechanisms in electrical devices, including transmission lines and waveguides, are associated with resistive effects in which electrical power is transformed into heat. We have also assumed that time-varying electric and magnetic fields are totally confined to a waveguide or circuit. In fact, confinement is rarely complete, and electromagnetic power will *radiate* away from the device to some degree. Radiation may generally be an unwanted effect, as it represents an additional power loss mechanism, or a device may receive unwanted signals from the surrounding region. On the other hand, a well-designed antenna provides an efficient interface between guided waves and free-space waves for purposes of intentionally radiating or receiving electromagnetic power. In either case, it is important to understand the radiation phenomenon so that it can either be used most effectively or be reduced to a minimum. In this chapter, our goal is to establish such an understanding and to explore several practical examples of antenna design. ∎

14.1 BASIC RADIATION PRINCIPLES: THE HERTZIAN DIPOLE

The essential point of this chapter is that *any* time-varying current distribution will radiate electromagnetic power. So our first task is to find the fields that radiate from a specific time-varying source. This problem is different from any that we have explored. In our treatment of waves and fields in bulk media and in waveguides, only the wave motion in the medium was investigated, and the sources of the fields were not considered. Earlier in Chapter 11, we found the current distribution in a conductor by relating it to assumed electric and magnetic field intensities at the conductor surface. Although this would relate the current source to the field, it is not practical for our purposes because the conductors were considered infinite in size in at least one dimension.

We begin by studying a current filament of infinitesimally small cross-section, positioned within an infinite lossless medium that is specified by permeability μ and

导读

1. 在远离天线的远场区域才呈现出电磁波的特性，即电场与磁场相互为源、相互激励而使能量向外辐射的特性；在距离天线较近的近场区域，除了电场与磁场相互为源之外，天线上电流产生的磁场与电荷产生的电场所占比例较大，甚至是主要分量。

2. 电偶极子（或赫兹偶极子）天线是一段导线，末端开路不会有传导电流流出，即传导电流为零，而在天线中却有电流，这是时变场中才有的现象。可以将天线端部看作电容的极板，天线中的电流用来在端部（或电容极板上）积累或释放电荷，即电荷的积累与释放形成天线中的电流。

3. 天线的辐射电阻表征辐射能力的大小。电偶极子天线越长、信号频率越高，则辐射功率越大。

4. 利用适当排布的多个天线构成的天线阵列可以实现电磁波的定向辐射。

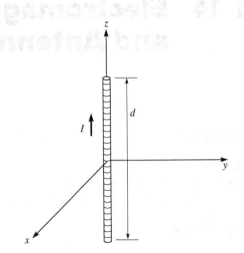

Figure 14.1   A differential current filament of length d carries a current $I = I_0 \cos \omega t$.

permittivity ϵ (both real). The filament is specified as having a differential length, but we will later extend the results easily to larger dimensions that are on the order of a wavelength. The filament is positioned with its center at the origin and is oriented along the z axis as shown in Figure 14.1. The positive sense of the current is taken in the $\mathbf{a}_z$ direction. A uniform current $I(t) = I_0 \cos \omega t$ is assumed to flow in this short length d. The existence of such a current would imply the existence of time-varying charges of equal and opposite instantaneous amplitude on each end of the wire. For this reason, the wire is termed an *elemental* or *Hertzian dipole*. This is distinct in meaning from the more general definition of a dipole antenna that we will use later in this chapter.

14.1.1   Retarded Vector Potential for the Hertzian Dipole

The first step is the application of the retarded vector magnetic potential expression, as presented in Section 9.5,

$$A = \int \frac{\mu\, I[t - R/v]d\mathbf{L}}{4\pi R} \tag{1}$$

where I is a function of the retarded time $t - R/v$.

When a single frequency is used to drive the antenna, v is the phase velocity of a wave at that frequency in the medium around the current element and is given by $v = 1/\sqrt{\mu\epsilon}$. Since no integration is required for the short filament, we have

$$\boxed{\mathbf{A} = \frac{\mu\, I[t - R/v]d}{4\pi R}\,\mathbf{a}_z} \tag{2}$$

Only the z component of $\mathbf{A}$ is present, for the current is only in the $\mathbf{a}_z$ direction. At any point P at distance R from the origin, the vector potential is retarded by R/v and so we use

$$I[t - R/v] = I_0 \cos\left[\omega\left(t - \frac{R}{v}\right)\right] = I_0 \cos[\omega t - kR] \tag{3}$$

where the wavenumber in the lossless medium is $k = \omega/v = \omega \sqrt{\mu\epsilon}$. In phasor form, Eq. (3) becomes

$$I_s = I_0 \ e^{-jkR} \tag{4}$$

where the current amplitude, I_0, is assumed to be real (as it will be throughout this chapter). Incorporating (4) into (2), we find the phasor retarded potential:

$$\boxed{\mathbf{A}_s = A_{zs}\,\mathbf{a}_z = \frac{\mu I_0 d}{4\pi R} e^{-jkR}\,\mathbf{a}_z} \tag{5}$$

Using a mixed coordinate system for the moment, we now replace R with the small r of the spherical coordinate system and then determine which spherical components are represented by A_{zs}. Using the projections as illustrated in Figure 14.2, we find

$$A_{rs} = A_{zs}\cos\theta \tag{6a}$$

$$A_{\theta s} = -A_{zs}\sin\theta \tag{6b}$$

and therefore

$$A_{rs} = \frac{\mu I_0 d}{4\pi r}\cos\theta \ e^{-jkr} \tag{7a}$$

$$A_{\theta s} = -\frac{\mu I_0 d}{4\pi r}\sin\theta \ e^{-jkr} \tag{7b}$$

14.1.2 Obtaining the General Electric and Magnetic Fields

From the preceding two components of the vector magnetic potential at P we can now find $\mathbf{B}_s$ or $\mathbf{H}_s$ from the definition of $\mathbf{A}_s$,

$$\mathbf{B}_s = \mu\mathbf{H}_s = \nabla \times \mathbf{A}_s \tag{8}$$

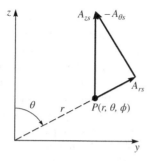

Figure 14.2 The resolution of A_{zs} at $P(r, \theta, \phi)$ into the two spherical components A_{rs} and $A_{\theta s}$. The sketch is arbitrarily drawn in the $\phi = 90°$ plane.

Taking the indicated partial derivatives as specified by the curl operator in spherical coordinates, we can separate Eq. (8) into its three spherical components, of which only the ϕ component is nonzero:

$$H_{\phi s} = \frac{1}{\mu r} \frac{\partial}{\partial r}(r A_{\theta s}) - \frac{1}{\mu r} \frac{\partial A_{rs}}{\partial \theta} \tag{9}$$

Now, substituting (7a) and (7b) into (9), we find the magnetic field:

$$\boxed{H_{\phi s} = \frac{I_0 d}{4\pi} \sin\theta \, e^{-jkr}\left(j\frac{k}{r} + \frac{1}{r^2}\right)} \tag{10}$$

The electric field that is associated with Eq. (10) is found from one of Maxwell's equations—specifically the point form of Ampère's circuital law as applied to the surrounding region (where conduction and convection current are absent). In phasor form, this is Eq. (23) in Chapter 11, except that in the present case we allow for a lossless medium having permittivity ϵ:

$$\nabla \times \mathbf{H}_s = j\omega\epsilon \, \mathbf{E}_s \tag{11}$$

Using (11), we expand the curl in spherical coordinates, assuming the existence of only a ϕ component for $\mathbf{H}_s$. The resulting electric field components are:

$$E_{rs} = \frac{1}{j\omega\epsilon} \frac{1}{r \sin\theta} \frac{\partial}{\partial \theta}(H_{\phi s} \sin\theta) \tag{12a}$$

$$E_{\theta s} = \frac{1}{j\omega\epsilon}\left(-\frac{1}{r}\right)\frac{\partial}{\partial r}(r H_{\phi s}) \tag{12b}$$

Then on substituting (10) into (12a) and (12b) we find:

$$\boxed{E_{rs} = \frac{I_0 d}{2\pi} \eta \cos\theta \, e^{-jkr}\left(\frac{1}{r^2} + \frac{1}{jkr^3}\right)} \tag{13a}$$

$$\boxed{E_{\theta s} = \frac{I_0 d}{4\pi} \eta \sin\theta \, e^{-jkr}\left(\frac{jk}{r} + \frac{1}{r^2} + \frac{1}{jkr^3}\right)} \tag{13b}$$

where the intrinsic impedance is, as always, $\eta = \sqrt{\mu/\epsilon}$.

Equations (10), (13a), and (13b) are the fields that we set out to find. The next step is to interpret them. We first notice the e^{-jkr} factor appearing with each component. By itself, this term describes a spherical wave, propagating outward from the origin in the positive r direction with a phase constant $k = 2\pi/\lambda$. λ is the wavelength as measured in the medium. Matters are complicated by the complex r-dependent terms in parentheses that appear in all three equations. These terms can be expressed in polar form (magnitude and phase), leading to the following modified versions of the three field equations for the Hertzian dipole:

$$\boxed{H_{\phi s} = \frac{I_0 k d}{4\pi r}\left[1 + \frac{1}{(kr)^2}\right]^{1/2} \sin\theta \, \exp\left[-j(kr - \delta_\phi)\right]} \tag{14}$$

$$E_{rs} = \frac{I_0 d}{2\pi r^2}\eta \left[1 + \frac{1}{(kr)^2}\right]^{1/2} \cos\theta \exp\left[-j(kr - \delta_r)\right] \qquad (15)$$

$$E_{\theta s} = \frac{I_0 k d}{4\pi r}\eta \left[1 - \frac{1}{(kr)^2} + \frac{1}{(kr)^4}\right]^{1/2} \sin\theta \exp\left[-j(kr - \delta_\theta)\right] \qquad (16)$$

where the additional phase terms are

$$\delta_\phi = \tan^{-1}[kr] \qquad (17a)$$

$$\delta_r = \tan^{-1}[kr] - \frac{\pi}{2} \qquad (17b)$$

and

$$\delta_\theta = \tan^{-1}\left[kr\left(1 - \frac{1}{(kr)^2}\right)\right] \qquad (18)$$

In (17) and (18), the principal value is always taken when evaluating the inverse tangent. This means that the phases as expressed in (17) and (18) will occur within the range $\pm\pi/2$ as kr varies between zero and infinity. Suppose a single frequency (k value) is chosen, and the fields are observed at a fixed instant in time. Consider observing the field along a path in the direction of increasing r, in which spatial oscillations will be seen as r varies. As a result of the phase terms in (17) and (18), the oscillation period will change with increasing r. We may demonstrate this by considering the H_ϕ component as a function of r under the following conditions:

$$I_0 d = 4\pi \qquad \theta = 90° \qquad t = 0$$

Using $k = 2\pi/\lambda$, Eq. (14) becomes

$$H_{\phi s} = \frac{2\pi}{\lambda r}\left[1 + \left(\frac{\lambda}{2\pi r}\right)^2\right]^{1/2} \exp\left\{-j\left[\frac{2\pi r}{\lambda} - \tan^{-1}\left(\frac{2\pi r}{\lambda}\right)\right]\right\} \qquad (19)$$

The real part of (19) gives the real instantaneous field at $t = 0$:

$$\mathscr{H}_\phi(r,0) = \frac{2\pi}{\lambda r}\left[1 + \left(\frac{\lambda}{2\pi r}\right)^2\right]^{1/2} \cos\left[\tan^{-1}\left(\frac{2\pi r}{\lambda}\right) - \frac{2\pi r}{\lambda}\right] \qquad (20)$$

We next use the identity, $\cos(a - b) = \cos a \cos b + \sin a \sin b$, in addition to $\cos(\tan^{-1}x) = 1/\sqrt{1 + x^2}$ and $\sin(\tan^{-1}x) = x/\sqrt{1 + x^2}$. With these, Eq. (20) simplifies to

$$\mathscr{H}_\phi = \frac{1}{r^2}\left[\cos\left(\frac{2\pi r}{\lambda}\right) + \frac{2\pi r}{\lambda}\sin\left(\frac{2\pi r}{\lambda}\right)\right] \qquad (21)$$

A few important points emerge when studying Eq. (21). First, at distances r that are on the order of a wavelength, the expression consists of two sinusoidal functions having the same period but in which the amplitude of the second one increases with increasing r. This leads to significant nonsinusoidal behavior in that the field as a function of r/λ will oscillate, but with nonuniform periodicity, and with positive and negative amplitudes that differ in each cycle. Second, at distances r that are much greater than a wavelength, the second term in (21) dominates, and the field variation with r approaches that of a pure sinusoid. We may therefore say that, for all practical purposes, the wave at large distances, where $r \gg \lambda$, is a uniform plane wave having a sinusoidal variation with distance (and time, of course) and a well-defined wavelength. This wave evidently carries power away from the differential antenna.

We should now take a more careful look at the expressions containing terms varying as $1/r^3$, $1/r^2$, and $1/r$ in Eqs. (10), (13a), and (13b). At points very close to the current element, the $1/r^3$ term must be dominant. In the numerical example we have used, the relative values of the terms in $1/r^3$, $1/r^2$, and $1/r$ in the $E_{\theta s}$ expression are about 250, 16, and 1, respectively, when r is 1 cm. The variation of an electric field as $1/r^3$ should remind us of the *electrostatic* field of the dipole (Chapter 4). The development of this concept is the subject of Problem 14.4. The near-field terms represent energy stored in a reactive (capacitive) field, and they do not contribute to the radiated power. The inverse-square term in the $H_{\phi s}$ expression is similarly important only in the region very near to the current element. It corresponds to the *induction* field of the dc element, as found through the Biot-Savart law (Problem 14.5).

14.1.3   Fields in the Far Zone

At distances corresponding to, say, 10 or more wavelengths from the current element, the product $kr = 2\pi r/\lambda > 20\pi$, and the fields dramatically simplify. In Eqs. (14)–(16), the terms within brackets involving $1/(kr)^2$ and $1/(kr)^4$ can be considered much less than unity and can be neglected. In addition, the phases [Eqs. (17) and (18)] all approach $\pi/2$. The effect is also seen in Eqs. (10), (13a), and (13b), in which all terms except the inverse-distance ($1/r$) term may be neglected. At such distances, at which $kr \gg 1$ (equivalently $r \gg \lambda$), we are said to be in the *far field* or *far zone*. The remaining field terms that have the $1/r$ dependence are the *radiation* fields. This produces an approximately zero E_{rs} field, leaving only $E_{\theta s}$ and $H_{\phi s}$. Thus, in the far zone:

$$
\begin{aligned}
E_{rs} &\doteq 0 \\[2mm]
E_{\theta s} &= j\frac{I_0 kd}{4\pi r}\eta \ \sin\theta \ e^{-jkr} \quad\quad (22) \\[2mm]
H_{\phi s} &= j\frac{I_0 kd}{4\pi r}\sin\theta \ e^{-jkr} \quad\quad\quad (23)
\end{aligned}
$$

The relation between these fields is evidently the same as that of a uniform plane wave, which an expanding spherical wave approximates at large radii, and over regions in which $1/r$ is approximately constant. Specifically,

$$
E_{\theta s} = \eta H_{\phi s} \quad\quad (kr \gg 1 \text{ or } r \gg \lambda) \quad\quad (24)
$$

The variation of both radiation fields with the polar angle θ is the same; the fields maximize in the equatorial plane (xy plane) of the current element and vanish off the ends of the element. The variation with angle may be shown by plotting a vertical, or *E-plane* pattern (assuming a vertical orientation of the current element). The E plane

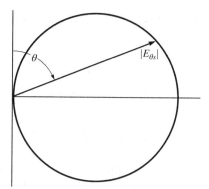

Figure 14.3 The polar plot of the
E-plane pattern of a vertical current element. The crest amplitude of $E_{\theta s}$ is plotted
as a function of the polar angle θ at a
constant distance r. The locus is a circle.

is simply the coordinate plane that contains the electric field, which in our present
case is any surface of constant ϕ in the spherical coordinate system. Figure 14.3 shows
an *E*-plane plot of Eq. (22) in polar coordinates, in which the relative magnitude of
$E_{\theta s}$ is plotted against θ for a constant r. The length of the vector shown in the figure
represents the magnitude of E_θ, normalized to unity at $\theta = 90°$; the vector length is
just $|\sin \theta|$, and so as θ varies, the tip of the vector traces out a circle as shown.

A horizontal, or *H-plane* pattern may also be plotted for this or more complicated antenna systems. In the present case, this would show the variation of field
intensity with ϕ. The *H*-plane of the current element (the plane that contains the
magnetic field) is any plane that is normal to the z axis. As E_θ is not a function of ϕ,
the *H*-plane plot would be simply a circle centered at the origin.

> **D14.1.** A short antenna with a uniform current distribution in air has $I_0 d = 3 \times$
> 10^{-4} A $\cdot$ m and $\lambda = 10$ cm. Find $|E_{\theta s}|$ at $\theta = 90°$, $\phi = 0°$, and $r =:$ (*a*) 1 cm;
> (*b*) 2 cm; (*c*) 20 cm; (*d*) 200 cm; (*e*) 2 m.
>
> **Ans.** (*a*) 125 V/m; (*b*) 25 V/m; (*c*) 2.8 V/m; (*d*) 0.28 V/m; (*e*) 0.028 V/m

14.2 ANTENNA SPECIFICATIONS

It is important to fully describe and quantify the radiation from a general antenna. To
do this, we need to be aware of a few new concepts and definitions.

14.2.1 Emitted Power and Radiation Resistance

In order to evaluate the radiated power, the time-average Poynting vector must be
found (Eq. (77), Chapter 11). In the present case, this will become

$$< \mathbf{S} > = \frac{1}{2} \mathscr{R}e\{E_{\theta s} \, H_{\phi s}^*\}\mathbf{a}_r \quad \text{W/m}^2 \tag{25}$$

Substituting (22) and (23) into (25), we obtain the time-average Poynting vector magnitude:

$$|<\mathbf{S}>| = S_r = \frac{1}{2}\left(\frac{I_0 kd}{4\pi r}\right)^2 \eta \sin^2\theta \tag{26}$$

From this we find the time-average power that crosses the surface of a sphere of radius r, centered at the antenna:

$$P_r = \int_{\phi=0}^{2\pi}\int_{\theta=0}^{\pi} S_r \, r^2\sin\theta \, d\theta \, d\phi = 2\pi\left(\frac{1}{2}\right)\left(\frac{I_0 kd}{4\pi}\right)^2 \eta \int_0^{\pi}\sin^3\theta \, d\theta \tag{27}$$

The integral is evaluated, and we substitute $k = 2\pi/\lambda$. We will also assume that the medium is free space, where $\eta = \eta_0 \doteq 120\pi$. We finally obtain:

$$P_r = 40\pi^2\left(\frac{I_0 d}{\lambda}\right)^2 \quad \text{W} \tag{28}$$

This is the same average power that would be dissipated in a resistance R_{rad} by sinusoidal current of amplitude I_0 in the absence of any radiation, where

$$P_r = \frac{1}{2}I_0^2 \, R_{rad} \tag{29}$$

We call this effective resistance the *radiation resistance* of the antenna. For the differential antenna, this becomes

$$R_{rad} = \frac{2P_r}{I_0^2} = 80\pi^2\left(\frac{d}{\lambda}\right)^2 \tag{30}$$

If, for example, the differential length is 0.01λ, then R_{rad} is about 0.08 Ω. This small resistance is probably comparable to the *ohmic* resistance of a practical antenna (providing a measure of the power dissipated through heat), and thus the efficiency of the antenna is likely to be too low. Effective matching to the source also becomes very difficult to achieve, for the input reactance of an electrically short antenna is much greater in magnitude than the input resistance R_{rad}.

14.2.2 Solid Angle

Evaluating the net power from the antenna, as carried out in (27), involved the integration of the Poynting vector over a spherical shell of presumed large radius, such that the antenna appeared as a point source at the sphere center. In view of this, a new concept of power density can be introduced; this involves the power that is carried within a very thin cone with its vertex at the antenna location. The axis of the cone extends along a line of radius, and thus the cone intersects the spherical surface over which the integral in (27) is taken. That portion of the sphere area that the cone intersects will have area A. We define the *solid angle* of the cone in the following manner: If $A = r^2$, where r is the sphere radius, then the cone is defined as having a solid angle, Ω, equal to one *steradian* (sr).[1] As the total sphere area is $4\pi r^2$, we see that the total solid angle contained within a sphere is 4π steradians.

As a consequence of this definition, differential area on the sphere surface can be expressed in terms of a differential solid angle through:

$$dA = r^2 \, d\Omega \qquad (31)$$

The total sphere area can then be expressed as an integral over solid angle, or equivalently by an integral using spherical coordinates:

$$A_{\text{net}} = 4\pi r^2 = \int_0^{4\pi} r^2 \, d\Omega = \int_0^{2\pi} \int_0^{\pi} r^2 \sin\theta \, d\theta \, d\phi \qquad (32)$$

from which we identify the differential solid angle as expressed in spherical coordinates:

$$\boxed{d\Omega = \sin\theta \, d\theta \, d\phi} \qquad (33)$$

The preceding relations are depicted in Fig. 14.4.

D14.2. A cone is centered on the positive z axis with its vertex at the origin. The cone angle in spherical coordinates is θ_1. (*a*) If the cone subtends 1 sr of solid angle, determine θ_1. (*b*) If $\theta_1 = 45°$, find the solid angle subtended.

Ans. (*a*) 32.8°; (*b*) $\pi \sqrt{2}$

14.2.3 Radiation Intensity

We can now express the Poynting vector magnitude as found in Eq. (26) in units of power per unit solid angle. To do this, we multiply the W/m^2 power density in (26) by the sphere area encompassed in one steradian—which is r^2. The result, known as the *radiation intensity*, is

$$\boxed{K(\theta, \phi) = r^2 S_r \quad \text{W/sr}} \qquad (34)$$

For the Hertzian dipole, the intensity is independent of ϕ, and we would have (using (26)):

$$K(\theta) = \frac{1}{2} \left(\frac{I_0 k d}{4\pi} \right)^2 \eta \, \sin^2\theta \quad \text{W/sr} \qquad (35)$$

In the general case, the total radiated power is then

$$\boxed{P_r = \int_0^{4\pi} K \, d\Omega = \int_0^{2\pi} \int_0^{\pi} K(\theta, \phi) \sin\theta \, d\theta \, d\phi \quad \text{W}} \qquad (36)$$

which, for the Hertzian dipole, gives the same result as found in (28).

---

[1] This definition is related to that of the radian, in which the arc length on a circle traced out by a change of angle of one radian is the circle radius, r.

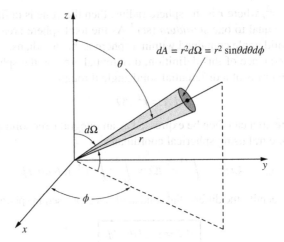

Figure 14.4 A cone having differential solid angle $d\Omega$ subtends a (shaded) differential area on the surface of a sphere of radius r. This area, given by $dA = r^2\, d\Omega$, can also be expressed in our more familiar spherical coordinate system as $dA = r^2\sin\theta\, d\theta\, d\phi$.

The advantage of using the radiation intensity for power density is that this quantity is independent of the radius. This is true, however, only if the original power density exhibits a $1/r^2$ dependence. In fact, *all* antennas have this functional dependence on radius in the far zone, in that when far enough away, the antenna appears as a point source of power. Assuming the surrounding medium does not absorb any power, the integral of the Poynting vector over a closed sphere of *any* radius must give the same result. This fact demands an inverse-square dependence on radius for the power density. With the radial dependence removed, one can concentrate on the angular dependence of the power density as expressed by K, and this will differ significantly among different antennas.

14.2.4 Directivity

A special case of a power source is an *isotropic radiator*, defined as having a *constant* radiation intensity (i.e., $K = K_{\text{iso}}$ is independent of θ and ϕ). This gives a simple relation between K and the total radiated power:

$$P_r = \int_0^{4\pi} K_{\text{iso}}\, d\Omega = 4\pi\, K_{\text{iso}} \Rightarrow K_{\text{iso}} = P_r/4\pi \quad \text{(isotropic radiator)} \qquad (37)$$

Generally, K *will* vary with angle, giving more intensity in some directions than in others. It is useful to compare the radiation intensity in a given direction to that which would occur if the antenna were to radiate *the same net power* isotropically.

The *directivity* function, $D(\theta, \phi)$, does this.[2] Using (36) and (37), we can write the directivity:

$$D(\theta,\phi) = \frac{K(\theta,\phi)}{K_{iso}} = \frac{K(\theta,\phi)}{P_r/4\pi} = \frac{4\pi K(\theta,\phi)}{\oint K\, d\Omega} \qquad (38)$$

Of particular interest in most cases is the maximum value of the directivity, D_{max}, which is sometimes called simply D (without the θ and ϕ dependence indicated):

$$D = D_{max} = \frac{4\pi K_{max}}{\oint K\, d\Omega} \qquad (39)$$

in which the maximum radiation intensity, K_{max}, will usually occur at more than one set of values of θ and ϕ. Typically, the directivity is quoted in decibels, according to the definition

$$D_{dB} = 10\, \log_{10}(D_{max})\ dB \qquad (40)$$

EXAMPLE 14.1

Evaluate the directivity of the Hertzian dipole.

Solution. Use Eqs. (35) and (28) with $k = 2\pi/\lambda$ and $\eta = \eta_0 = 120\pi$ in the expression:

$$D(\theta,\phi) = \frac{4\pi K(\theta,\phi)}{P_r} = \frac{2\pi \left(\frac{I_0 d}{2\lambda}\right)^2 120\pi\, \sin^2\theta}{40\pi^2 \left(\frac{I_0 d}{\lambda}\right)^2} = \frac{3}{2}\sin^2\theta$$

The maximum of this result, occurring at $\theta = \pi/2$, is:

$$D_{max} = \frac{3}{2} \quad \text{Or, in decibels:} \quad D_{dB} = 10\, \log_{10}\left(\frac{3}{2}\right) = 1.76\ dB$$

D14.3. What is the directivity in dB of a power source at the origin that radiates: (a) uniformly into the upper half-space, but nothing into the lower half-space; (b) into all space with a $\cos^2\theta$ power density dependence; (c) into all space with a $|\cos^n\theta|$ dependence?

Ans. (a) 3; (b) 4.77; (c) $10\log_{10}(n+1)$

Usually, one would like to have a much higher directivity than what we just found for the Hertzian dipole. One implication of a low directivity (and a problem with the

[2] In earlier times (and in older texts), the directivity function was called the *directive gain*. The latter term has since been discarded by the Antenna Standards Committee of the IEEE Antennas and Propagation Society, in favor of the term "directivity." Details are found in IEEE Std 145-1993.

short antenna) is that power is radiated over a broad angular range in the E plane. In most cases, we wish to confine the power to a narrow range, or small *beamwidth*, thus increasing the directivity. The *3-dB beamwidth* is defined as the separation between the two angles at which the directivity falls to one-half its maximum value. For the Hertzian dipole, and using the $D(\theta, \phi)$ result from the previous example, the beamwidth will be the span between the two θ values on either side of $90°$ at which $\sin^2 \theta = 1/2$, or $|\sin\theta| = 1/\sqrt{2} = 0.707$. These two values are $45°$ and $135°$, representing a 3-dB beamwidth of $135° - 45° = 90°$. We will see that using a longer antenna leads to both a narrower beamwidth and an increased radiation resistance. In the H plane, radiation is uniform at all values of ϕ, no matter what length is used. It is necessary to use multiple antennas in an *array* in order to narrow the beam in the H plane.

14.2.5 Antenna Gain and Radiation Efficiency

We have based several definitions on the total average power that is radiated by the antenna, P_r. It is desirable, however, to distinguish the radiated power from the *input* power that is *supplied* to the antenna, P_{in}. It is likely that P_{in} will be somewhat greater than P_r because of resistive losses in the conducting materials that make up the antenna. To overcome this resistance, a greater input voltage amplitude would be necessary to generate a given current amplitude, I_0, on which all of our power computations have been based. The antenna *gain* is defined in such a way as to accommodate this difference.[3]

Specifically, suppose that the antenna in question were to isotropically radiate *all* of the electrical power that is supplied to it, which is P_{in}. The radiation intensity would simply be $K_s = P_{in}/4\pi$. Gain is defined as the ratio of the actual radiation intensity in a specified direction, to K_s:

$$G(\theta, \phi) = \frac{K(\theta, \phi)}{K_s} = \frac{4\pi\, K(\theta, \phi)}{P_{in}} \tag{41}$$

Note that the term $4\pi\, K(\theta, \phi)$ would be the radiated power of an isotropic antenna whose (in that case, constant) radiation intensity is $K(\theta, \phi)$. The gain thus expresses the ratio of the radiated power of an antenna to the input power *as if the antenna were to radiate isotropically* with constant K as evaluated at a selected θ and ϕ. Using (38), we can relate directivity to gain:

$$D(\theta, \phi) = \frac{4\pi\, K(\theta, \phi)}{P_r} = \frac{P_{in}}{P_r} G(\theta, \phi) = \frac{1}{\eta_r} G(\theta, \phi) \tag{42}$$

where η_r is the *radiation efficiency* of the antenna, defined as the ratio of the radiated power to the input power. Other ways of writing this are:

$$\eta_r = \frac{P_r}{P_{in}} = \frac{G(\theta, \phi)}{D(\theta, \phi)} = \frac{G_{max}}{D_{max}} \tag{43}$$

which expresses η_r as the maximum gain divided by the maximum directivity.

---

[3] The antenna gain defined in this way is sometimes called the power gain.

14.3 MAGNETIC DIPOLE

An interesting device that is closely related to the Hertzian dipole is the *magnetic dipole* antenna. Shown in Figure 14.5, the antenna consists of a circular current loop of radius a, centered at the origin, and in the xy plane. The loop current is sinusoidal and is given by $I(t) = I_0 \cos \omega t$, as was the case in the Hertzian dipole. Although it is possible to work out the fields for this antenna, beginning with the retarded potentials as in the previous section, there is a much faster way.

We first note that the circulating current implies the existence of a circulating electric field that overlaps the wire and that has the same time dependence. So one could simply replace the wire with a circular electric field that we could designate as $\mathbf{E}(a, t) = E_0(a) \cos(\omega t) \mathbf{a}_\phi$. Such a change would replace conduction current with displacement current, which will have no effect on the surrounding field solutions for $\mathbf{E}$ and $\mathbf{H}$. Next, suppose that we could replace the electric field with a magnetic field, again of the form $\mathbf{H}(a, t) = H_0 \cos(\omega t) \mathbf{a}_\phi$. This is the magnetic field that would be generated by the Hertzian dipole at radius a in the xy plane, and it enables us to obtain the solution for the current loop field through the following method:

We begin with Maxwell's equations in a sourceless medium ($\rho_v = \mathbf{J} = 0$):

$$\nabla \times \mathbf{H} = \epsilon \frac{\partial \mathbf{E}}{\partial t} \qquad (44a)$$

$$\nabla \times \mathbf{E} = -\mu \frac{\partial \mathbf{H}}{\partial t} \qquad (44b)$$

$$\nabla \cdot \mathbf{E} = 0 \qquad (44c)$$

$$\nabla \cdot \mathbf{H} = 0 \qquad (44d)$$

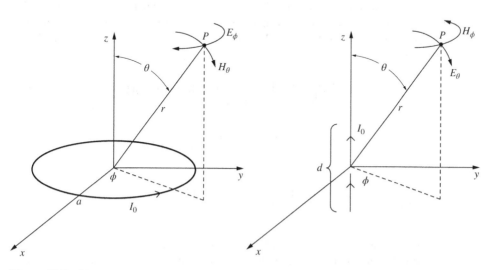

Figure 14.5 Magnetic (left) and electric dipole antennas are dual structures, producing identical field patterns but with the roles of **E** and **H** interchanged.

By inspection, we see that the equations would be unchanged if we replaced **E** with **H**, **H** with −**E**, ϵ with μ, and μ with ϵ. This illustrates the concept of *duality* in electromagnetics. The fact that the current loop *electric* field will have the same functional form as the electric dipole *magnetic* field means that with the above substitutions, we can construct the current loop fields directly from the electric dipole results. It is because of this duality between field solutions of the two devices that the name, *magnetic dipole antenna*, is applied to the current loop device.

Before making the substitutions, we must relate the currents and geometries of the two devices. To do this, consider first the static electric dipole result of Chapter 4 [Eq. (35)]. We can specialize this result by finding the electric field on the z axis ($\theta = 0$). We find

$$\mathbf{E}|_{\theta=0} = \frac{Qd}{2\pi\epsilon z^3}\mathbf{a}_z \tag{45}$$

We can next study the current loop magnetic field as found on the z axis, in which a steady current I_0 is present. This result can be obtained using the Biot–Savart law:

$$\mathbf{H}|_{\theta=0} = \frac{\pi a^2 I_0}{2\pi z^3}\mathbf{a}_z \tag{46}$$

Now the current associated with a harmonically time-varying charge on the electric dipole, $Q(t)$, is

$$I_0 = \frac{dQ}{dt} = j\omega Q \quad\Rightarrow\quad Q = \frac{I_0}{j\omega} \tag{47}$$

If we substitute Eq. (47) into Eq. (45) and replace d with $j\omega\epsilon(\pi a^2)$, we find that Eq. (45) is transformed to Eq. (46). We now perform these substitutions, along with the replacements, **H** for **E**, −**E** for **H**, ϵ for μ, and μ for ϵ, on Eqs. (14), (15), and (16). The results are

$$E_{\phi s} = -j\frac{\omega\mu(\pi a^2)I_0 k}{4\pi r}\left[1 + \frac{1}{(kr)^2}\right]^{1/2}\sin\theta\,\exp[-j(kr-\delta_\phi)] \tag{48}$$

$$H_{rs} = j\frac{\omega\mu(\pi a^2)I_0}{2\pi r^2}\frac{1}{\eta}\left[1 + \frac{1}{(kr)^2}\right]^{1/2}\cos\theta\,\exp[-j(kr-\delta_r)] \tag{49}$$

$$H_{\theta s} = j\frac{\omega\mu(\pi a^2)I_0 k}{4\pi r}\frac{1}{\eta}\left[1 - \frac{1}{(kr)^2} + \frac{1}{(kr)^4}\right]^{1/2}\sin\theta\,\exp[-j(kr-\delta_\theta)] \tag{50}$$

where δ_r, δ_θ, and δ_ϕ are as defined in Eqs. (17) and (18). In the far field ($kr \gg 1$), $E_{\phi s}$ and $H_{\theta s}$ survive, and these simplify to compare closely with (22) and (23). This process of exploiting duality in electromagnetics is a very powerful method that can be applied in many situations.

14.4 THIN WIRE ANTENNAS

In addition to giving insights on radiation fundamentals, the Hertzian dipole results provide us with a basis from which the fields associated with more complicated antennas can be derived. In this section this methodology is applied to the more practical problem of straight thin wire antennas of any length. We will find that for a given wavelength, changes in antenna length lead to dramatic variations in (and control of) the radiation pattern. We will also note improvement in directivity and efficiency when using certain antenna lengths.

The basic arrangement is shown in Figure 14.6. In a simplistic way, it is possible to think of the antenna as having been formed by bending the two wires of an open-ended transmission line down and up by 90°. The midpoint at which the bends occur is known as the *feed* point. The current, originally present, persists and is instantaneously flowing in the same direction in the lower and upper sections of the antenna. If the current is sinusoidal, a standing wave is set up in the antenna wires, with zeros occurring at the wire ends at $z = \pm \ell$. A symmetric antenna of this type is called a *dipole*.

14.4.1 Current Distribution on a Dipole

The actual current distribution on a very thin wire antenna is very nearly sinusoidal. With zero current at the ends, maxima occur one-quarter wavelength from each end, and the current continues to vary in this manner toward the feed point. The current at the feed will be small for an antenna whose overall length, 2ℓ, is an integral number of wavelengths; but it will be equal to the maximum found at any point on the antenna if the antenna length is an odd multiple of a half-wavelength.

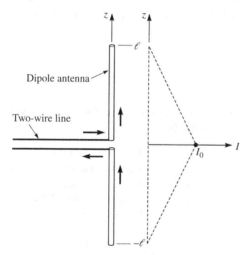

Figure 14.6 A thin dipole antenna driven sinusoidally by a two-wire line. The current amplitude distribution, shown in the adjacent plot, is approximately linear if the overall length is sufficiently less than a half-wavelength. Current amplitude maximizes at the center (feed) point.

On a short antenna, where 2ℓ is significantly less than a half-wavelength, we see only the first portion of the sine wave; the amplitude of the current increases in an approximately linear manner, from zero at the ends to a maximum value at the feed, as indicated in Figure 14.6. The gap at the feed point is small and has negligible effects. The short antenna approximation (in which a linear current variation along the length can be assumed) is reasonable for antennas having an overall length that is less than about one-tenth of a wavelength.

A simple extension of the Hertzian dipole results can be performed in the short antenna regime ($\ell < \lambda/20$). If this is the case, then retardation effects may be neglected. That is, signals arriving at any field point P from the two ends of the antenna are approximately in phase. The average current along the antenna is $I_0/2$, where I_0 is the input current at the feed. The electric and magnetic field intensities will thus be one-half the values given in (22) and (23), and there are no changes in the vertical and horizontal patterns. The power will be one-quarter of its previous value, and thus the radiation resistance will also be one-quarter of the value given by (30). Matters improve as the antenna length is increased, but retardation effects must then be included.

For longer lengths, the current distribution is treated the same as that of an open-ended transmission line that propagates a TEM wave. This will be a standing wave in which the current phasor is given by

$$I_s(z) \doteq I_0 \sin(kz) \tag{51}$$

where the open end is located at $z = 0$. Also, for a TEM wave on a transmission line, the phase constant will be $\beta = k = \omega \sqrt{\mu\epsilon}$. When the line is unfolded to form the antenna, the z axis is rotated to the vertical orientation with $z = 0$ occurring at the feed point. The current in (51) is then modified to be

$$I_s(z) \doteq \begin{cases} I_0 \sin k(\ell - z) & (z > 0) \\ I_0 \sin k(\ell + z) & (z < 0) \end{cases} = I_0 \sin k(\ell - |z|) \tag{52}$$

14.4.2 Modeling the Wire Antenna as a Superposition of Hertzian Dipoles

The antenna is now modeled as a stack of Hertzian dipoles, each having length dz (Figure 14.7). The current amplitude in each Hertzian is determined according to its position z along the length, and is given by (52). The far-zone field from each Hertzian can then be written using Eq. (22), suitably modified. We write this as a *differential* field contribution at a far point at distance r' and spherical coordinate angle, θ':

$$dE_{\theta s} = j \frac{I_s(z)\, k\, dz}{4\pi r'} \eta \, \sin\theta' \, e^{-jkr'} \tag{53}$$

The coordinates r' and θ' are, of course, referenced from the center of the Hertzian, which itself is at a position z along the antenna length. We need to reference these local coordinates to the origin, which occurs at the antenna feed point. To do this, we borrow from the methods used to analyze the static electric dipole as presented in Section 4.7. Referring to Figure 14.7, we can write the relation between the

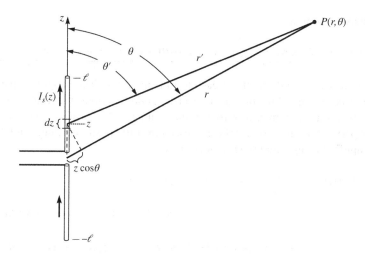

Figure 14.7 A dipole antenna can be represented as a stack of Hertzian dipoles whose individual phasor currents are given by $I_s(z)$. One Hertzian dipole is shown at location z, and has length dz. When the observation point, P, lies in the far zone, distance lines r and r' are approximately parallel, so they differ in length by $z \cos \theta$.

distance r' from the Hertzian at location z and the distance r from the origin to the same point as

$$r' \doteq r - z \cos\theta \tag{54}$$

where, in the far field, $\theta' \doteq \theta$, and distance lines r' and r are approximately parallel. Eq. (53) is then modified to read

$$d E_{\theta s} = j \frac{I_s(z) k \, dz}{4\pi r} \eta \, \sin\theta \, e^{-jk(r - z \cos\theta)} \tag{55}$$

Notice that in obtaining (55) from (53) we have approximated $r' \doteq r$ in the denominator, as the use of Eq. (54) will make little difference when considering *amplitude* variations with z and θ. The exponential term in (55) *does* include (54) because slight variations in z or θ will greatly affect the phase.

Now, the total electric field at the far-zone position (r, θ) will be the sum of all the Hertzian dipole contributions along the antenna length, which becomes the integral:

$$
\begin{aligned}
E_{\theta s}(r, \theta) = \int d E_{\theta s} &= \int_{-\ell}^{\ell} j \frac{I_s(z) k \, dz}{4\pi r} \eta \, \sin\theta \, e^{-jk(r - z \cos\theta)} \\
&= \left[j \frac{I_0 k}{4\pi r} \eta \, \sin\theta \, e^{-jkr} \right] \int_{-\ell}^{\ell} \sin k(\ell - |z|) e^{jkz\cos\theta} \, dz
\end{aligned} \tag{56}
$$

To evaluate the last integral, we first express the complex exponential in terms of sine and cosine terms using the Euler identity. Denoting the bracketed terms outside the integral as A, we write:

$$E_{\theta s}(r,\theta) = A \int_{-\ell}^{\ell} \underbrace{\sin k(\ell - |z|)}_{\text{even}} \underbrace{\cos(kz \cos\theta)}_{\text{even}} + j \underbrace{\sin k(\ell - |z|)}_{\text{even}} \underbrace{\sin(kz \cos\theta)}_{\text{odd}} dz$$

in which the even or odd parity of each term is indicated. The imaginary part of the integrand, consisting of the product of even and odd functions, yields a term with net odd parity; it thus integrates to zero over the symmetric limits of $-\ell$ to ℓ. This leaves the real part, whose integral can be expressed over the positive z range and then further simplified using trigonometric identities:

$$E_{\theta s}(r,\theta) = 2A \int_0^{\ell} \sin k(\ell - z) \cos(kz \cos\theta)\, dz$$

$$= A \int_0^{\ell} \sin[k(\ell - z) + kz \cos\theta] + \sin[k(\ell - z) - kz \cos\theta]dz$$

$$= A \int_0^{\ell} \sin[kz(\cos\theta - 1) + k\ell] - \sin[kz(\cos\theta + 1) - k\ell]dz$$

The last integral is straightforward and evaluates as

$$E_{\theta s}(r,\theta) = 2A \left[\frac{\cos(k\ell \cos\theta) - \cos(k\ell)}{k \sin^2\theta} \right]$$

Now, reincorporating the expression for A gives the final result:

$$E_{\theta s}(r,\theta) = j \frac{I_0 \eta}{2\pi r} e^{-jkr} \left[\frac{\cos(k\ell \cos\theta) - \cos(k\ell)}{\sin\theta} \right] = E_0 F(\theta) \left[\frac{e^{-jkr}}{r} \right] \qquad (57)$$

where we identify the field amplitude

$$E_0 = j \frac{I_0 \eta}{2\pi} \qquad (58)$$

14.4.3   E-Plane Pattern and Radiation Intensity

The terms involving θ and ℓ in Eq. (57) are isolated to form the *pattern function* for the dipole antenna:

$$F(\theta) = \left[\frac{\cos(k\ell \cos\theta) - \cos(k\ell)}{\sin\theta} \right] \qquad (59)$$

This important function, when normalized, is the E-plane pattern of the dipole antenna. It explicitly shows how choices in dipole length affect the θ dependence in the pattern, and it ultimately determines the dependence on ℓ of the directive gain, directivity, and radiated power for a given current.

Plots of the magnitude of $F(\theta)$ in the E-plane are shown in Figure 14.8a and b for selected dipole lengths. In these, the xz plane is chosen, although the results will be the same in any plane that contains the z axis. The plots show a trend toward narrower radiation beams as length increases, but to the point at which secondary maxima, or

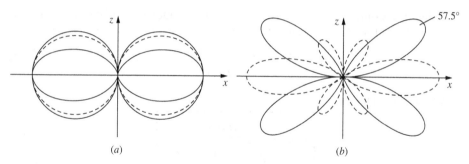

Figure 14.8 *E*-plane plots, normalized to maxima of 1.0, found from $F(\theta)$ for dipole antennas having overall lengths, 2ℓ, of (a) $\lambda/16$ (solid black), $\lambda/2$ (dashed), and λ (blue), and (b) 1.3λ (dashed), and 2λ (blue). In (a), the beam-narrowing trend is evident as length increases (or as wavelength decreases). Note that the $\lambda/16$ curves are nearly circular and thus approximate the Hertzian dipole pattern. At lengths that exceed one wavelength, sidelobes begin to develop, as exhibited in the smaller beams in the 1.3λ pattern in (b). As length increases, the sidelobes grow to form the four symmetrically arranged main lobes of the 2λ antenna, where the lobe in the first quadrant maximizes at $\lambda = 57.5°$. The main lobes along x that were present in the 1.3λ antenna diminish with increasing length, and have vanished completely when the length reaches 2λ.

sidelobes, develop for overall antenna lengths (2ℓ) that exceed one wavelength. The presence of sidelobes is usually not wanted, mainly because they represent radiated power in directions other than that of the main beam ($\theta = \pi/2$). Sidelobe power will therefore likely miss the intended receiver. In addition, the sidelobe directions change with wavelength, and will therefore impart an angular spread to a radiated signal, to an extent which will of course increase with increasing signal bandwidth. These problems are avoided by using antenna lengths that are less than one wavelength.

The radiation intensity can now be found for the dipole antenna by using Eq. (34), along with (25):

$$K(\theta) = r^2 S_r = \frac{1}{2} \mathcal{R}e\{E_{\theta s} H_{\phi s}^*\} r^2$$

where $H_{\phi s} = E_{\theta s}/\eta$. Substituting (57), we obtain

$$K(\theta) = \frac{\eta I_0^2}{8\pi^2}[F(\theta)]^2 = \frac{15\,I_0^2}{\pi}[F(\theta)]^2 \qquad \text{W/sr} \tag{60}$$

where in the last equality free space is assumed, in which $\eta = \eta_0 = 120\pi$. The total radiated power is now the integral of the radiation intensity over all solid angles, or

$$P_r = \int_0^{4\pi} K\,d\Omega = \int_0^{2\pi}\int_0^{\pi} K(\theta)\sin\theta\,d\theta\,d\phi \tag{61}$$

Again assuming free space we find

$$P_r = 30\,I_0^2 \int_0^{\pi} [F(\theta)]^2 \sin\theta\,d\theta \quad \text{W} \tag{62}$$

Using this result, expressions for the directivity and radiation resistance can now be found. From Eq. (42), and using (60) and (62), the directivity in free space is

$$D(\theta) = \frac{4\pi\, K(\theta)}{P_r} = \frac{2\,[F(\theta)]^2}{\int_0^\pi [F(\theta)]^2 \sin\theta\, d\theta} \tag{63}$$

whose maximum value is

$$D_{\max} = \frac{2\,[F(\theta)]_{\max}^2}{\int_0^\pi [F(\theta)]^2 \sin\theta\, d\theta} \tag{64}$$

Finally, the radiation resistance will be

$$R_{\mathrm{rad}} = \frac{2P_r}{I_0^2} = 60 \int_0^\pi [F(\theta)]^2 \sin\theta\, d\theta \tag{65}$$

D14.4. Evaluate the percentage of the maximum power density that is found in the direction $\theta = 45°$ for dipole antennas of overall length (*a*) $\lambda/4$, (*b*) $\lambda/2$, (*c*) λ.

Ans. (*a*) 45.7%; (*b*) 38.6%; (*c*) 3.7%

14.4.4 Half-Wave Dipole

When the antenna length is chosen to be $2\ell = \lambda/2$, we form a "half-wave" dipole; this length choice has several advantages in practice. We begin with an example:

EXAMPLE 14.2

Write the specific pattern function, and evaluate the beamwidth, directivity, and radiation resistance of a half-wave dipole.

Solution. The length is $2\ell = \lambda/2$, or $\ell = \lambda/4$. Therefore, $k\ell = (2\pi/\lambda)(\lambda/4) = \pi/2$, which is now substituted into Eq. (59) to obtain:

$$F(\theta) = \frac{\cos\left(\frac{\pi}{2}\cos\theta\right)}{\sin\theta} \tag{66}$$

The magnitude of this function is plotted as the dashed curve in Figure 14.8*a*. Its maxima (equal to 1) occur at $\theta = \pi/2,\ 3\pi/2$, whereas zeros occur at $\theta = 0$ and π. Beamwidth is found by evaluating the solutions of

$$\frac{\cos\left(\frac{\pi}{2}\cos\theta\right)}{\sin\theta} = \frac{1}{\sqrt{2}}$$

Numerically, it is found that the two angles on either side of the maximum at $\theta = 90°$ that satisfy the preceding equation are $\theta_{1/2} = 51°$ and $129°$. The half-power beamwidth is thus $129° - 51° = 78°$.

Directivity and radiation resistance are then found using (64) and (65), in which the integral of $[F(\theta)]^2$ can be performed numerically. The results are $D_{\max} = 1.64$ (or 2.15 dB), and $R_{\mathrm{rad}} = 73\ \Omega$.

In the half-wave dipole, the standing wave current amplitude maximizes at the feed point, and the antenna is said to be operated *on resonance*. As a result, the driving point impedance, one-quarter wavelength in front of the open ends, would *in principle* be purely real[4] and equal to the 73-Ω radiation resistance, assuming that the antenna is otherwise lossless. This is the primary motivation for using half-wave dipoles, in that they provide a fairly close impedance match to conventional transmission lines (whose characteristic impedances are on the same order).

Actually, because the antenna is essentially an unfolded transmission line, the half-wave dipole does not behave as an ideal quarter-wave transmission line section, as we might suspect considering the discussions in Section 14.1. An appreciable reactive part of the input impedance will likely be present, but the half-wavelength dimension is very close to the length at which the reactance goes to zero. Methods of evaluating the reactance are beyond the scope of our treatment, but are considered in detail in Ref. 1. For a thin lossless dipole of length exactly $\lambda/2$, the input impedance would be $Z_{in} = 73 + jX$, where X is in the vicinity of 40 Ω. The input reactance is extremely sensitive to the antenna length and can be reduced to zero by a very slight reduction in the overall length below $\lambda/2$, leaving the real part essentially unaffected. Similar behavior is seen in dipoles having lengths that are integer multiples of $\lambda/2$, but in these, radiation resistances are considerably higher, thus yielding a poorer impedance match. At dipole lengths between half-wavelength multiples, input reactances can be much higher (in the vicinity of $j600$ Ω) and can become sensitive to the wire thickness, in addition to the length. In practice, when connecting a transmission line feed, the input reactance can be zeroed by length reduction or by using matching techniques such as those discussed in Chapter 10.

Plots of directivity and radiation resistance as functions of antenna length are shown in Figure 14.9. Directivity increases modestly with length, whereas radiation resistance reaches a local maximum at a length between $3\lambda/4$ and λ. At greater lengths, additional peaks in R_{rad} occur at higher levels, but performance is compromised by the presence of sidelobes. Again, half-wave dipoles are typically used because single-lobe behavior is assured over a broad spectral bandwidth, whereas radiation resistance (73 Ω) is close to the impedance of standard transmission lines that are used to feed the antenna.

14.4.5 Monopole Antenna

As a final exercise in wire antennas, we consider the operation of a *monopole* antenna. This is one-half a dipole plus a perfectly conducting plane, as shown in Figure 14.10*a*. The image principle discussed in Section 5.5 provides the image shown in Figure 14.10*b*, so that the monopole and its image form a dipole. Therefore, all field equations that pertain to the dipole apply directly to the upper half-space. The Poynting vector is therefore also the same above the plane, but the integration to find the total power radiated is performed only through the hemisphere that surrounds the upper half-space. So the radiated power and the radiation resistance for the monopole are half the corresponding values for the dipole. As an example, a quarter-wave

---

[4] Think of a half-rotation ($\lambda/4$) around the Smith chart from the open circuit point, toward the generator, where, with loss present, the end position would be somewhere on the negative real axis.

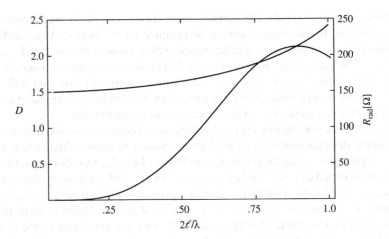

Figure 14.9 Plots of directivity (black) and radiation resistance (blue) as functions of overall antenna length, expressed in wavelengths.

monopole (presenting a half-wave dipole when including the image) yields a radiation resistance of $R_{\text{rad}} = 36.5 \, \Omega$.

Monopole antennas may be driven by a coaxial cable below the plane, having its center conductor connected to the antenna through a small hole, and having its outer conductor connected to the plane. If the region below the plane is inaccessible or inconvenient, the coax may be laid on top of the plane and its outer conductor connected to it. Examples of this type of antenna include AM broadcasting towers and citizens' band antennas.

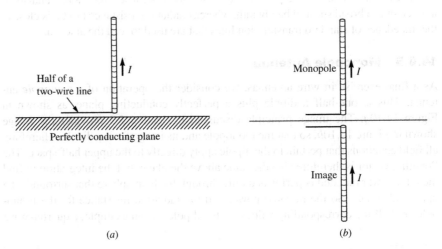

Figure 14.10 (a) An ideal monopole is always associated with a perfectly conducting plane. (b) The monopole plus its image form a dipole.

D14.5. The monopole antenna of Figure 14.10a has a length $d/2 = 0.080$ m and may be assumed to carry a triangular current distribution for which the feed current I_0 is 16.0 A at a frequency of 375 MHz in free space. At point P ($r = 400$ m, $\theta = 60°$, $\phi = 45°$) find (a) $H_{\phi s}$, (b) $E_{\theta s}$, and (c) the amplitude of $\mathcal{P}_r$.

Ans. (a) $j1.7$ mA/m; (b) $j0.65$ V/m; (c) 1.1 W/m^2

14.5 ARRAYS OF TWO ELEMENTS

We next address the problem of establishing better control of the directional properties of antenna radiation. Although some control of directivity is achieved through adjustment of the length of a wire antenna, these results only appear as changes in the E-plane pattern. The H-plane pattern always remains a circle (no ϕ variation), as long as a single vertical wire antenna is used. By using multiple elements in an *array*, significant improvement in directivity as determined in *both E and H* planes can be achieved. Our objective in this section is to lay the groundwork for the analysis of arrays by considering the simple case of using two elements. The resulting methods are readily extendable to multiple element configurations.

14.5.1 Array Fields in the Far Zone

The basic configuration is shown in Figure 14.11. Here, we have our original wire antenna with its feed at the origin, and oriented along the z axis. A second identical antenna, parallel to the first, is positioned at location d on the x axis. The two carry the same current amplitude, I_0 (leading to far-field amplitude E_0), but we allow the second antenna current to exhibit a constant phase difference, ξ, from that of the first. The far-field observation point, P, lies at spherical coordinates (r, θ, ϕ). From this point, the antennas appear close enough together so that (1) the radial lines, r and r_1,

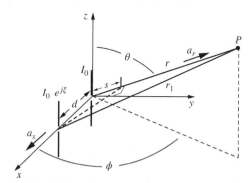

Figure 14.11 The original z-directed wire antenna with its center at the origin is now joined by a second parallel antenna, crossing the x axis at distance d. The second antenna carries the same current amplitude as the first, but with a constant phase shift, ξ. Fields are observed at point P.

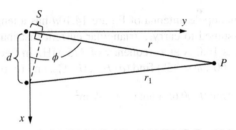

Figure 14.12 Top view of the arrangement of Figure 14.11 (looking down onto the x-y plane). In the far-field approximation, the blue lines are essentially parallel, and $r_1 \doteq r - s$.

are essentially parallel, and (2) the electric field directions at P are essentially the same (along $\mathbf{a}_\theta$). Using Eq. (57), we may therefore write the total field at P, with the understanding that the presence of the second antenna on the x axis will introduce a ϕ dependence in the field that was previously not present:

$$E_{\theta\,P}(r,\theta,\phi) = E_0\, F(\theta) \left[\frac{e^{-jkr}}{r} + \frac{e^{j\xi}e^{-jkr_1}}{r_1} \right] \tag{67}$$

Next, we may express the distance to P from the second antenna, r_1, in terms of the distance to the first antenna, r (also the spherical coordinate radius), by noting that in the far-field approximation we have

$$r_1 \doteq r - s$$

where s is one leg of the right triangle formed by drawing a perpendicular line segment between the second antenna and the line of radius r as shown in Figures 14.11 and 14.12. The length, s, is the projection of the antenna separation, d, onto the radial line, r, and is found through

$$s = d\, \mathbf{a}_x \cdot \mathbf{a}_r = d\, \sin\theta\, \cos\phi \tag{68}$$

Therefore,

$$r_1 \doteq r - d\, \sin\theta\, \cos\phi \tag{69}$$

In the far field, the distance, $d \sin\theta \cos\phi$, is very small compared to r, which allows us to neglect the difference between r and r_1 in the magnitude terms in (67) (so that $1/r_1 \doteq 1/r$). As we know from the dipole studies, the difference cannot be neglected in the phase terms in (67) because phase is very sensitive to slight changes in r. With these considerations in mind, Eq. (67) becomes

$$E_{\theta\,P}(r,\theta,\phi) = \frac{E_0 F(\theta)}{r} \left[e^{-jkr} + e^{j\xi} e^{-jk(r - d\sin\theta\cos\phi)} \right] \tag{70}$$

which simplifies to

$$E_{\theta P}(r,\theta,\phi) = \frac{E_0\,F(\theta)}{r}e^{-jkr}\left[1 + e^{j\psi}\right] \tag{71}$$

where

$$\psi = \xi + kd\,\sin\theta\,\cos\phi \tag{72}$$

ψ is the net phase difference between the two antenna fields that is observed at $P(r,\theta,\phi)$. Equation (71) can be further simplified by factoring out the term $e^{j\psi/2}$ to obtain

$$E_{\theta P}(r,\theta,\phi) = \frac{2E_0\,F(\theta)}{r}e^{-jkr}e^{j\psi/2}\cos(\psi/2) \tag{73}$$

from which we may determine the field amplitude through

$$|E_{\theta P}(r,\theta,\phi)| = \sqrt{E_{\theta P}E_{\theta P}^*} = \frac{2E_0}{r}|F(\theta)||\cos(\psi/2)| \tag{74}$$

14.5.2 Pattern Multiplication Principle

Equation (74) demonstrates the important principle of *pattern multiplication* that applies to arrays of identical antennas. Specifically, the total field magnitude consists of the product of the pattern function magnitude, or *element factor* for the individual antennas, $|F(\theta)|$, and the normalized *array factor* magnitude, given by $|\cos(\psi/2)|$. The array factor is often denoted by

$$A(\theta,\phi) = \cos(\psi/2) = \cos\left[\frac{1}{2}(\xi + kd\,\sin\theta\,\cos\phi)\right] \tag{75}$$

Equation (74) then becomes

$$|E_{\theta P}(r,\theta,\phi)| = \frac{2E_0}{r}|F(\theta)||A(\theta,\phi)| \tag{76}$$

This principle can be extended to arrays of multiple elements by appropriately modifying the array factor, as we will find. The underlying assumption is that the individual array elements are essentially uncoupled; that is, they induce negligible currents in each other. With appreciable coupling, the problem is far more complicated, and pattern multiplication cannot be used.

In the field pattern expressed in (76), the E plane (or θ dependence) is primarily determined by the individual elements, or by $|F(\theta)|$. It is in the H plane where the effect of the array is the strongest. In fact, the main reason for using an array of this configuration is to enable control of the H-plane pattern. In the H plane ($\theta = \pi/2$), Eqs. (75) and (76) give the field dependence on ϕ as

$$E_{\theta P}(r, \pi/2, \phi) \propto A(\pi/2, \phi) = \cos\left[\frac{1}{2}(\xi + kd \cos\phi)\right] \tag{77}$$

The H-plane pattern depends on the choices of the relative current phase, ξ, and the element spacing, d.

EXAMPLE 14.3

Investigate the H-plane pattern when the currents are in phase ($\xi = 0$).

Solution. With $\xi = 0$, Eq. (77) becomes

$$A(\pi/2, \phi) = \cos\left[\frac{kd}{2}\cos\phi\right] = \cos\left[\frac{\pi d}{\lambda}\cos\phi\right]$$

This reaches a maximum at $\phi = \pi/2$ and $3\pi/2$, or along the direction that is normal to the plane of the antennas (the y axis). This occurs regardless of the choice of d, and the array is thus referred to as a *broadside array*. Now, by choosing $d = \lambda/2$, we obtain $A = \cos[(\pi/2) \cos \phi]$, which becomes zero at $\phi = 0$ and π (along the x axis), and we have single main beams along the positive and negative y axis. When d is increased beyond $\lambda/2$, additional maxima (sidelobes) appear as ϕ is varied, but zeros still occur along the x axis if d is set to odd multiples of $\lambda/2$.

The broadside array of the previous example can be regarded as the simplest case. More interesting behavior occurs when a nonzero phase difference exists between the two currents, and adjustments can be performed in the phase and element spacing.

EXAMPLE 14.4

Determine the necessary conditions to establish an *endfire* array, in which the maximum radiation is directed along the x axis.

Solution. Setting $\phi = 0$ or π in Eq. (77) and requiring the equation to achieve a maximum results in the condition:

$$A = \cos\left[\frac{\xi}{2} \pm \frac{\pi d}{\lambda}\right] = \pm 1$$

or

$$\frac{\xi}{2} \pm \frac{\pi d}{\lambda} = m\pi$$

where m is an integer that includes 0, and where the plus sign in the bracket applies for $\phi = 0$, and the minus sign for $\phi = \pi$. One case of practical interest occurs when $m = 0$, $d = \lambda/4$, and $\xi = -\pi/2$, which satisfies the preceding condition when the positive sign is chosen. Equation (77) now becomes

$$A(\pi/2, \phi) = \cos\left[\frac{\pi}{4}(\cos\phi - 1)\right]$$

This function maximizes at $\phi = 0$ and reaches zero at $\phi = \pi$. We have thus created an array that radiates a *single* main lobe along the positive x axis. The way this works can be understood by realizing that the phase lag in current in the element at $x = d$ just compensates for the phase lag that arises from the propagation delay between the element at the origin and the one at $x = d$. The second element radiation is therefore precisely in phase with the radiation from the first element. The two fields, therefore, constructively interfere and propagate together in the forward x direction. In the reverse direction, radiation from the antenna at $x = d$ arrives at the origin to find itself π radians out of phase with the radiation from the $x = 0$ element. The two fields therefore destructively interfere, and no radiation occurs in the negative x direction.

D14.6. In the broadside configuration of Example 14.3, the element spacing is changed to $d = \lambda$. Determine (a) the ratio of the emitted intensities in the $\phi = 0$ and $\phi = 90°$ directions in the H plane, (b) the directions (values of ϕ) of the main beams in the H-plane pattern, and (c) the locations (values of ϕ) of the zeros in the H-plane pattern.

Ans. (a) 1; (b) (0, ±90°, 180°); (c) (± 45°, ± 135°)

D14.7. In the endfire configuration of Example 14.4, determine the directions (values of ϕ) for the main beams in the H plane if the wavelength is shortened from $\lambda = 4d$ to (a) $\lambda = 3d$, (b) $\lambda = 2d$, and (c) $\lambda = d$.

Ans. (a) ± 41.4°; (b) ± 45.0°; (c) ± 75.5°

14.6 UNIFORM LINEAR ARRAYS

We next expand our treatment to arrays of more than two elements. With these, the designer has more options to improve the directivity and possibly to increase the bandwidth of the antenna, for example. As might be imagined, a full treatment of this subject would require an entire book. Here, we consider only the case of the uniform linear array to demonstrate the analysis methods and to present some of the key results.

14.6.1 Far Zone Array Pattern

The uniform linear array configuration is shown in Figure 14.13. The array is linear because the elements are arranged along a straight line (the x axis in this case). The array is uniform because all elements are identical, have equal spacing d, and carry the same current amplitude I_0, and the phase progression in current from element to element is given by a constant value, ξ. The normalized array factor for the two-element array can be expressed using (71) as

$$|A(\theta,\phi)| = |A_2(\theta,\phi)| = |\cos(\psi/2)| = \frac{1}{2}|1 + e^{j\psi}| \tag{78}$$

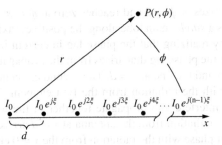

Figure 14.13 *H*-plane diagram of a uniform linear array of *n* dipoles, arranged along *x*, and with individual dipoles oriented along *z* (out of the page). All elements have equal spacing, *d*, and carry equal current amplitudes, I_0. Current phase shift ξ occurs between adjacent elements. Fields are evaluated at far-zone point *P*, from which the dipoles appear to be grouped at the origin.

where the subscript 2 is applied to *A* to indicate that the function applies to two elements. The array factor for a linear array of *n* elements as depicted in Figure 14.13 is a direct extension of (78) and becomes

$$|A_n(\theta,\phi)| = |A_n(\psi)| = \frac{1}{n}\left|1 + e^{j\psi} + e^{j2\psi} + e^{j3\psi} + e^{4\psi} + \ldots + e^{j(n-1)\psi}\right| \tag{79}$$

With the elements arranged along the *x* axis as shown in Figure 14.13, we have $\psi = \xi + kd \sin\theta\cos\phi$, as before. The geometric progression that comprises Eq. (79) can be expressed in closed form to give

$$|A_n(\psi)| = \frac{1}{n}\frac{|1 - e^{jn\psi}|}{|1 - e^{j\psi}|} = \frac{1}{n}\frac{|e^{jn\psi/2}(e^{-jn\psi/2} - e^{jn\psi/2})|}{|e^{j\psi/2}(e^{-j\psi/2} - e^{j\psi/2})|} \tag{80}$$

In the far right side of Eq. (80), we recognize the Euler identities for the sine function in both numerator and denominator, leading finally to

$$|A_n(\psi)| = \frac{1}{n}\frac{|\sin(n\psi/2)|}{|\sin(\psi/2)|} \tag{81}$$

The electric field in the far zone for an array of *n* dipoles can now be written in terms of A_n by extending the result in Eq. (76). Writing $|A_n(\psi)| = |A_n(\theta,\phi)|$, we have

$$|E_{\theta P}(r,\theta,\phi)| = \frac{n\,E_0}{r}|F(\theta)||A_n(\theta,\phi)| \tag{82}$$

demonstrating again the principle of pattern multiplication, in which we now have a new array function that pertains to the linear array.

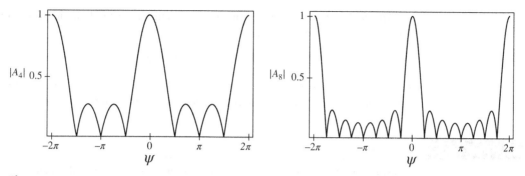

Figure 14.14 $|A_n(\psi)|$ as evaluated from Eq. (81) over the range $-2\pi < \psi < 2\pi$ for cases in which the number of elements, n, is (a) 4, and (b) 8.

14.6.2 Special Cases

Plots of Eq. (81) are shown in Figure 14.14 for the cases in which $n = 4$ and $n = 8$. Note that the functions always maximize to unity when $\psi = 2m\pi$, where m is an integer that includes zero. These principal maxima corrrespond to the main beams of the array pattern. The effect of increasing the number of elements is to narrow the main lobes and to bring in more secondary maxima (sidelobes).

To see how the array pattern is shaped, it is necessary to interpret the array function, Eq. (81), with regard to angular variation in the H plane. In this plane (where $\theta = \pi/2$), we have $\psi = \xi + kd \cos\phi$. Then, knowing that ϕ varies from 0 to 2π radians, $\cos\phi$ varies between ± 1, and we can see that ψ will be within the range

$$\xi - kd \leq \psi \leq \xi + kd \qquad (83)$$

Choices of the current phasing ξ and the antenna spacing d determine the range of ψ values that will appear in the actual array pattern. This could lead, in some cases, to a fairly narrow range in ψ that may or may not include a principal maximum. The current phase determines the central value of ψ, and the antenna spacing determines the maximum variation of ψ that occurs about the central value as the azimuth angle ϕ is varied.

As discussed in Section 14.5, a *broadside array* has main beams that occur normal to the array plane (at $\phi = \pi/2, 3\pi/2$). The condition for this is that the principal maximum, $\psi = 0$, will occur at these angles. We therefore write

$$\psi = 0 = \xi + kd \cos(\pi/2) = \xi$$

and so we would set $\xi = 0$ to obtain a broadside array. In this case, (83) gives $-kd < \psi < kd$. The central value of ψ is thus zero, and so the principal maximum there is included in the pattern. In the H plane, and with $\xi = 0$, we thus have $\psi = kd \cos\phi$. The $\psi = 0$ point will always occur at $\phi = \pi/2$ and $3\pi/2$, and this will be true regardless of the choice of element spacing d. The effect of increasing d is to enlarge the range of ψ that results when ϕ varies over its range of 0 to 2π. Therefore, for a given number of elements, the main beam will get narrower, but more sidelobes will be present in the pattern when the element spacing is increased.

An *endfire array* requires a principal maximum to occur along the *x* axis. In the *H* plane, we may therefore write

$$\psi = 0 = \xi + kd \cos(0) = \xi + kd$$

or $\xi = -kd$ to obtain endfire operation with a maximum occurring along the positive *x* axis. This may or may not result in a main beam occurring along the negative *x* axis as well.

EXAMPLE 14.5

For arrays of 4 and 8 elements, select the current phase and element spacing that will give *unidirectional endfire* operation, in which the main beam exists in the $\phi = 0$ direction, whereas no radiation occurs in the direction of $\phi = \pi$, nor in the broadside directions ($\phi = \pm\pi/2$).

Solution. We want $\psi = 0$ when $\phi = 0$. Therefore, from $\psi = \xi + kd \cos \phi$, we would require that $0 = \xi + kd$, or that $\xi = -kd$. Using 4 or 8 elements, we find either from Eq. (81) or from Figure 14.14 that zeros will occur when $\psi = \pm\pi/2$ and $\pm\pi$. Therefore, if we choose $\xi = -\pi/2$ and $d = \lambda/4$, we obtain $\psi = -\pi/2$ at $\phi = \pi/2$, $3\pi/2$, and $\psi = -\pi$ at $\phi = \pi$. We thus have $\psi = -(\pi/2)(1 - \cos \phi)$. Polar plots of the resulting array functions are shown in Figure 14.15*a* and *b*. Again, the move from 4 to 8 elements has the effect of decreasing the main beamwidth while increasing the sidelobe count from 1 to 3 in this case. If an odd number of elements is used with the above choices in phasing and spacing, a small sidelobe will be present in the $\phi = \pi$ direction.

In general, we may choose current phasing and element spacing to establish the main beam in any direction. Choosing the $\psi = 0$ principal maximum, we may write

$$\psi = 0 = \xi + kd \cos \phi_{max} \quad \Rightarrow \quad \cos \phi_{max} = -\frac{\xi}{kd}$$

so that the main beam direction can be changed by varying the current phasing.

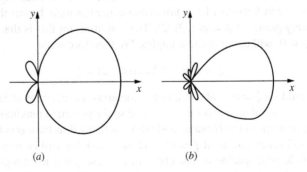

(a)   *(b)*

Figure 14.15 *H*-plane plots of (*a*) 4-element and (*b*) 8-element arrays having element spacing of $d = \lambda/4$ and current phasing $\xi = -\pi/2$.

D14.8. In an endfire linear dipole array in which $\xi = -kd$, what minimum element spacing d in wavelengths results in bidirectional operation, in which equal intensities occur in the H plane at $\phi = 0$ and $\phi = \pi$?

Ans. $d = \lambda/2$

D14.9. For a linear dipole array in which the element spacing is $d = \lambda/4$, what current phase ξ will result in a main beam in the direction of (a) $\phi = 30°$; (b) $\phi = 45°$?

Ans. (a) $-\pi \sqrt{3}/4$; (b) $-\pi \sqrt{2}/4$

14.7 ANTENNAS AS RECEIVERS

We next turn to the other fundamental purpose of an antenna, which is its use as a means to detect, or receive, radiation that originates from a distant source. We will approach this problem through study of a transmit-receive antenna system. This is composed of two antennas, along with their supporting electronics, that play the interchangeable roles of transmitter and detector.

14.7.1 Transmit-Receive Link as a Two-Port Network: Reciprocity

Figure 14.16 shows an example of a transmit-receive arrangement, in which the two coupled antennas together comprise a linear two-port network. Voltage V_1 and current I_1 on the antenna at the left affect the voltage and current (V_2 and I_2) on the antenna at the right—and vice-versa. This coupling is quantified through transimpedance parameters Z_{12} and Z_{21}. The governing equations take the form

$$V_1 = Z_{11} I_1 + Z_{12} I_2 \qquad (84a)$$

$$V_2 = Z_{21} I_1 + Z_{22} I_2 \qquad (84b)$$

Z_{11} and Z_{22} are the input impedances to antennas 1 and 2 when either antenna is isolated and is used as a transmitter, or equivalently, if the two antennas are sufficiently

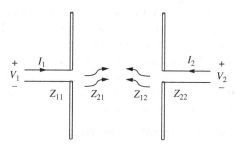

Figure 14.16 A pair of coupled antennas, demonstrating Eqs. (84a) and (84b).

far away from each other. The real parts of Z_{11} and Z_{22} will be the associated radiation resistances, provided ohmic losses in all conductors and all losses to surrounding objects are reduced to zero. We will assume this, in addition to far-zone operation, to be true here. The trans-impedances, Z_{12} and Z_{21}, depend on the spacing and relative orientation between the antennas, as well as on the characteristics of the surrounding medium. A critical property of the transimpedances in a linear medium is that they are equal. This property is the embodiment of the *reciprocity theorem*. Stated simply,

$$\boxed{Z_{12} = Z_{21}} \tag{85}$$

Further insights can be found by inverting (84*a*) and (84*b*) and invoking the admittance parameters, Y_{ij}:

$$I_1 = Y_{11} V_1 + Y_{12} V_2 \tag{86a}$$

$$I_2 = Y_{21} V_1 + Y_{22} V_2 \tag{86b}$$

where, again, the reciprocity theorem tells us that $Y_{12} = Y_{21}$.

Now, suppose that the terminals of antenna 2 are shorted, so that $V_2 = 0$. In this case, Eq. (86*b*) gives $I_2' = Y_{21} V_1'$, where the single prime denotes the condition of a shorted antenna 2. Instead, we could short antenna 1, resulting in $I_1'' = Y_{12} V_2''$ (with the double prime indicating conditions with antenna 1 shorted). Because reciprocity holds, it follows that

$$\boxed{\frac{V_2''}{I_1''} = \frac{V_1'}{I_2'}} \tag{87}$$

Equation (87) applies regardless of the relative positioning and orientation of the two antennas. We know that in a given direction, each antenna will transmit a power density whose value is determined by the antenna radiation pattern. Furthermore, we would expect to see the current that is set up on the receiving antenna depend on that antenna's orientation; that is, there is a *reception pattern* that the receiving antenna presents to the incoming signal. Now, for a fixed relative orientation between the two antennas, with antenna 1 as the transmitter, and antenna 2 shorted, a certain ratio V_1'/I_2' will be obtained. This ratio will depend on the relative orientation, which in turn will depend on the radiation pattern of antenna 1 and on the reception pattern of antenna 2. If roles are reversed such that the transmitter now becomes the receiver, and with antenna 1 shorted, a ratio V_2''/I_1'' will be obtained that by Eq. (87) is the same as before. The conclusion we must come to is that the extent to which the receiving antenna *accepts* power will be determined by its *radiation* pattern. This means, for example, that the main beam direction in the radiation pattern of the receiving antenna corresponds to the direction from which it is most sensitive to incoming signals. *The radiation and receiving patterns of any antenna are the same.*

14.7.2 Received Power and Effective Area

We next consider a more general transmission case, in which the receiving antenna is to deliver power to a load. Antenna 1 (Figure 14.16) serves as the transmitter,

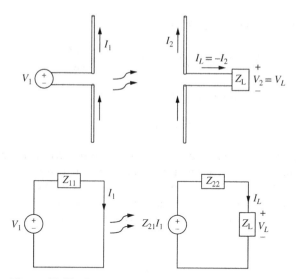

Figure 14.17   Transmitting and receiving antennas, and their equivalent circuits.

while antenna 2 is the receiver, at which the load is attached. A primary assumption is that the antennas are far enough away from each other so that only forward coupling (through Z_{21}) will be appreciable. The large separation distance means that the induced current I_2 is likely to be much less than I_1. Reverse coupling (through Z_{12}) would involve transmission of the *received* signal in antenna 2 back to antenna 1; specifically, the induced current I_2 further induces a (now very weak) additional current I_1' on antenna 1; that antenna would then carry a net current of $I_1 + I_1'$, where $I_1' \ll I_1$. We therefore assume that the product $Z_{12}I_2$ can be neglected, under which Eq. (84a) gives $V_1 = Z_{11}I_1$. A load impedance, Z_L, is connected across the terminals of antenna 2, as shown in the upper part of Figure 14.17. V_2 is the voltage across this load. Current $I_L = -I_2$ now flows through the load. Taking this current as positive, Eq. (84b) becomes

$$V_2 = V_L = Z_{21}I_1 - Z_{22}I_L \qquad (88)$$

This is just the Kirchhoff voltage law equation for the right-hand equivalent circuit shown in the lower part of Figure 14.17. The term $Z_{21}I_1$ is interpreted as the source voltage for this circuit, originating from antenna 1. Using (88), along with $V_L = Z_L I_L$, leads to

$$I_L = \frac{Z_{21}I_1}{Z_{22} + Z_L} \qquad (89)$$

The time-average power dissipated by Z_L is now

$$P_L = \frac{1}{2}\mathcal{R}e\{V_L I_L^*\} = \frac{1}{2}|I_L|^2 \mathcal{R}e\{Z_L\} = \frac{1}{2}|I_1|^2 \left|\frac{Z_{21}}{Z_{22} + Z_L}\right|^2 \mathcal{R}e\{Z_L\} \qquad (90)$$

The maximum power transferred to the load occurs when the load impedance is conjugate-matched to the driving point impedance, or $Z_L = Z_{22}^*$. Making this substitution in (90), and using the fact that $Z_{22} + Z_{22}^* = 2R_{22}$ gives

$$P_L = \frac{1}{2}|I_1|^2 \left|\frac{Z_{21}}{2R_{22}}\right|^2 \mathcal{R}e\{Z_{22}\} = \frac{|I_1|^2|Z_{21}|^2}{8R_{22}} \tag{91}$$

The time-average power transmitted by antenna 1 is

$$P_r = \frac{1}{2}\mathcal{R}e\{V_1 I_1^*\} = \frac{1}{2}R_{11}|I_1|^2 \tag{92}$$

By comparing this result with Eq. (65), we can interpret R_{11} as the radiation resistance of the transmitting antenna if (1) there are no resistive losses and (2) the current amplitude at the driving point is the maximum amplitude I_0. As we found earlier, the latter will occur in a dipole if the overall antenna length is an integer multiple of a half-wavelength. Using (91) and (92), we write the ratio of the received and transmitted powers:

$$\frac{P_L}{P_r} = \frac{|Z_{21}|^2}{4R_{11}R_{22}} \tag{93}$$

At this stage, more understanding is needed of the transimpedance, Z_{21} (or Z_{12}). This quantity will depend on the distance and relative orientations of the two antennas, in addition to other parameters. Figure 14.18 shows two dipole antennas, separated by

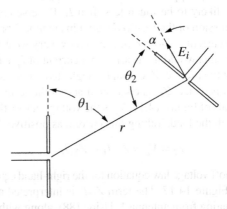

Figure 14.18   A transmit-receive antenna pair, showing relative orientation angles for the case in which the antennas lie in the same plane (in which case the ϕ coordinates are not necessary). Incident electric field E_i from antenna 1 is shown arriving at antenna 2 and presenting angle α to the antenna 2 axis. The field is perpendicular to the distance line r, and thus $\alpha = 90° - \theta_2$. Far-zone operation is assumed, so the two antennas appear as point objects to each other.

radial distance r, and with relative orientations that are specified by values of θ, as measured with respect to *each* antenna axis.[5] With antenna 1 serving as the transmitter and antenna 2 serving as the receiver, the radiation pattern of antenna 1 is given as a function of θ_1 and ϕ_1, while the receiving pattern of antenna 2 (equivalent to *its* radiation pattern) is given as a function of θ_2 and ϕ_2.

A convenient way to express the received power in an antenna is through its *effective area*, denoted $A_e(\theta, \phi)$ and expressed in m^2. Refer to Figure 14.18, and consider the average power density at the receiver (antenna 2) position, originating from the transmitter (antenna 1). As per Eqs. (25) and (26), this will be the magnitude of the Poynting vector at that location, $S_r(r, \theta_1, \phi_1)$ in W/m^2, where a dependence on ϕ is now necessary to describe all possible relative orientations. The effective area of the receiving antenna is defined such that when the power density is multiplied by the effective area, the power dissipated by a matched load at the receiving antenna is obtained. With antenna 2 as the receiver, we write

$$P_{L2} = S_{r1}(r, \theta_1, \phi_1) \times A_{e2}(\theta_2, \phi_2) \quad [\text{W}] \tag{94}$$

But now, using Eqs. (34) and (38), we may write the power density in terms of the directivity of antenna 1:

$$\boxed{S_{r1}(r, \theta_1, \phi_1) = \frac{P_{r1}}{4\pi r^2} D_1(\theta_1, \phi_1)} \tag{95}$$

Combining Eqs. (94) and (95), we obtain the ratio of the power received by antenna 2 to the power radiated by antenna 1:

$$\frac{P_{L2}}{P_{r1}} = \frac{A_{e2}(\theta_2, \phi_2) D_1(\theta_1, \phi_1)}{4\pi r^2} = \frac{|Z_{21}|^2}{4R_{11}R_{22}} \tag{96}$$

where the second equality repeats Eq. (93). We can solve (96) to find

$$\boxed{|Z_{21}|^2 = \frac{R_{11}R_{22}A_{e2}(\theta_2, \phi_2) D_1(\theta_1, \phi_1)}{\pi r^2}} \tag{97a}$$

We next note that if roles are reversed, in which antenna 2 transmits to antenna 1, we would find

$$\boxed{|Z_{12}|^2 = \frac{R_{11}R_{22}A_{e1}(\theta_1, \phi_1) D_2(\theta_2, \phi_2)}{\pi r^2}} \tag{97b}$$

---

[5] One way to express the relative orientations is to define the z axis along the radial distance line, r. Then angles θ_i and ϕ_i ($i = 1, 2$) are used locally to describe the orientations of the antenna axes, in which the origins of the two spherical coordinate systems are located at each antenna feed. The ϕ coordinate would thus be the angle of rotation about the r axis. For example, in Figure 14.18, with both antennas in the plane of the page, both ϕ coordinates could be assigned values of zero. With antenna 2 rotated about r such that it is normal to the page, ϕ_2 would be 90°, and the antennas would be cross-polarized.

The reciprocity theorem states that $Z_{12} = Z_{21}$. By equating Eqs. (97a) and (97b), it therefore follows that

$$\boxed{\frac{D_1(\theta_1,\phi_1)}{A_{e1}(\theta_1,\phi_1)} = \frac{D_2(\theta_2,\phi_2)}{A_{e2}(\theta_2,\phi_2)} = \text{Constant}} \tag{98}$$

That is, the ratio of directivity to effective area for *any* antenna is a universal constant, independent of the antenna type or the direction in which these parameters are evaluated. To evaluate the constant, we only need to look at one case.

EXAMPLE 14.6

Find the effective area of a Hertzian dipole, and determine the general relation between the directivity and effective area of any antenna.

Solution. With the Hertzian dipole as the receiving antenna, and having length d, its load voltage V_L will depend on the electric field that it intercepts from antenna 1. Specifically, we find the projection of the transmitting antenna field along the length of receiving antenna. This projected field, when multiplied by the length of antenna 2, gives the input voltage to the receiving antenna equivalent circuit. Referring to Figure 14.18, the projection angle is α, and thus the voltage that drives the Hertzian dipole will be

$$V_{\text{in}} = E_i \cos\alpha \times d = E_i\, d\, \sin\theta_2$$

The equivalent circuit for the Hertzian dipole is now the same as that of the receiving antenna as shown in Figure 14.17, except that we replace the source voltage, $I_1 Z_{21}$, with V_{in} as given above. Assuming a conjugate-matched load ($Z_L = Z_{22}^*$), the current through the load is now

$$I_L = \frac{E_i\, d\, \sin\theta_2}{Z_{22} + Z_L} = \frac{E_i\, d\, \sin\theta_2}{2R_{22}}$$

The power delivered to the matched load is then

$$P_{L2} = \frac{1}{2}\mathcal{R}e\{V_L I_L^*\} = \frac{1}{2}R_{22}|I_L|^2 = \frac{(E_i\, d)^2 \sin^2\theta_2}{8R_{22}} \tag{99}$$

For the Hertzian dipole, R_{22} is the radiation resistance. This was previously found to be (Eq. (30))

$$R_{22} = R_{\text{rad}} = 80\pi^2 \left(\frac{d}{\lambda}\right)^2$$

Substituting this into (99), we find

$$P_{L2} = \frac{1}{640}\left(\frac{E_i\, \lambda\, \sin\theta_2}{\pi}\right)^2 \text{ [W]} \tag{100}$$

The average power density that is incident on the receiving antenna is now

$$S_{r1}(r,\theta_1,\phi_1) = \frac{E_i(r,\theta_1,\phi_1)^2}{2\eta_0} = \frac{E_i^2}{240\pi} \text{ [W/m}^2] \tag{101}$$

Using (100) and (101), the effective area of the Hertzian is

$$A_{e2}(\theta_2) = \frac{P_{L2}}{S_{r1}} = \frac{3}{8\pi}\lambda^2 \sin^2(\theta_2) \quad [\text{m}^2] \tag{102}$$

The directivity for the Hertzian dipole, derived in Example 14.1, is

$$D_2(\theta_2) = \frac{3}{2}\sin^2(\theta_2) \tag{103}$$

Comparing Eqs. (102) and (103), we find the relation that we are looking for: The effective area and directivity for *any* antenna are related through

$$\boxed{D(\theta,\phi) = \frac{4\pi}{\lambda^2}A_e(\theta,\phi)} \tag{104}$$

We can now return to Eq. (96) and use Eq. (104) to rewrite the ratio of the power delivered to the receiving antenna load to the total power radiated by the transmitting antenna. This yields an expression that involves the simple product of the effective areas, known as the *Friis transmission formula*:

$$\boxed{\frac{P_{L2}}{P_{r1}} = \frac{A_{e2}(\theta_2,\phi_2)\,D_1(\theta_1,\phi_1)}{4\pi r^2} = \frac{A_{e1}(\theta_1,\phi_1)A_{e2}(\theta_2,\phi_2)}{\lambda^2 r^2}} \tag{105}$$

The result can also be expressed in terms of the directivities:

$$\boxed{\frac{P_{L2}}{P_{r1}} = \frac{\lambda^2}{(4\pi r)^2}D_1(\theta_1,\phi_1)D_2(\theta_2,\phi_2)} \tag{106}$$

These results provide an effective summary of what was discussed in this section, by way of giving us a very useful design tool for a free-space communication link. Again, Eq. (105) assumes lossless antennas in the far zones of each other and gives the power dissipated by a load that is conjugate-matched to the receiving antenna impedance.

D14.10. Given: an antenna having a maximum directivity of 6 dB and operating at wavelength $\lambda = 1$ m. What is the maximum effective area of the antenna?

Ans. $1/\pi$ m^2

D14.11. The power of 1 mW is dissipated by the matched load of a receiving antenna of a 1-m^2 effective area. This antenna is positioned at the center of the main beam of the transmitting antenna, located 1.0 km away. What total power is radiated by the transmitter if its directivity is (*a*) 10 dB, (*b*) 7 dB?

Ans. (*a*) 4π kW; (*b*) 8π kW

REFERENCES

1. Balanis, C. *Antenna Theory: Analysis and Design.* 3d ed. Hoboken, N. J.: Wiley, 2005. A widely used text at the advanced senior or graduate level, offering much detail.

2. Silver, S., ed. *Microwave Antenna Theory and Design.* London: Peter Peregrinus, Ltd on behalf of IEE, 1984. This is a reprint of volume 9 of the famous MIT Radiation Laboratory series, originally published by McGraw-Hill in 1949. It contains much information from original sources that later appeared in modern textbooks.

3. Jordan, E. C., and K. G. Balmain. *Electromagnetic Waves and Radiating Systems.* 2d ed. Englewood Cliffs, N. J.: Prentice-Hall, 1968. A classic text, covering waveguides and antennas.

4. Blake, L. V. *Antennas.* New York: Wiley, 1966. A short, well-written, and very readable text at a basic level.

5. Smith, G. S. *An Introduction to Classical Electromagnetic Radiation.* Cambridge, U. K.: Cambridge University Press, 1997. This excellent graduate-level text provides a unique perspective and rigorous treatment of the radiation problem as related to all types of antennas.

Quizzes

CHAPTER 14   PROBLEMS

14.1 A short dipole-carrying current $I_0 \cos \omega t$ in the $\mathbf{a}_z$ direction is located at the origin in free space. (*a*) If $k = 1$ rad/m, $r = 2$ m, $\theta = 45°$, $\phi = 0$, and $t = 0$, give a unit vector in rectangular components that shows the instantaneous direction of $\mathbf{E}$. (*b*) What fraction of the total average power is radiated in the belt, $80° < \theta < 100°$?

14.2 Prepare a curve, r vs. θ in polar coordinates, showing the locus in the $\phi = 0$ plane in which for a Hertzian dipole (*a*) the radiation field $|E_{\theta s}|$ is one-half of its value at $r = 10^4$ m, $\theta = \pi/2$; (*b*) average radiated power density $<S_r>$ is one-half its value at $r = 10^4$ m, $\theta = \pi/2$.

14.3 Two short antennas at the origin in free space carry identical currents of $5 \cos \omega t$ A, one in the $\mathbf{a}_z$ direction, and one in the $\mathbf{a}_y$ direction. Let $\lambda = 2\pi$ m and $d = 0.1$ m. Find $\mathbf{E}_s$ at the distant point where (*a*) $(x = 0, y = 1000, z = 0)$; (*b*) $(0, 0, 1000)$; (*c*) $(1000, 0, 0)$. (*d*) Find $\mathbf{E}$ at $(1000, 0, 0)$ at $t = 0$. (*e*) Find $|\mathbf{E}|$ at $(1000, 0, 0)$ at $t = 0$.

14.4 Write the Hertzian dipole electric field whose components are given in Eqs. (15) and (16) in the near zone in free space where $kr \ll 1$. In this case, only a single term in each of the two equations survives, and the phases, δ_r and δ_θ, simplify to a single value. Construct the resulting electric field vector and compare your result to the static dipole result (Eq. (35) in Chapter 4). What relation must exist between the static dipole charge Q and the current amplitude I_0 so that the two results are identical?

14.5 Consider the term in Eq. (14) (or in Eq. (10)) that gives the $1/r^2$ dependence in the Hertzian dipole magnetic field. Assuming that this term dominates and that $kr \ll 1$, show that the resulting magnetic field is the same as that found by applying the Biot-Savart law (Eq. (2), Chapter 7) to a current element of differential length d, oriented along the z axis, and centered at the origin.

14.6 Evaluate the time-average Poynting vector, $<\mathbf{S}> = \left(\frac{1}{2}\right)\mathscr{R}e\{\mathbf{E}_s \times \mathbf{H}_s^*\}$ for the Hertzian dipole, assuming the general case that involves the field components as given by Eqs. (10), (13a), and (13b). Compare your result to the far-zone case, Eq. (26).

14.7 A short current element has $d = 0.03\lambda$. Calculate the radiation resistance that is obtained for each of the following current distributions: (a) uniform, I_0; (b) linear, $I(z) = I_0(0.5d - |z|)/0.5d$; ($c$) step, I_0 for $0 < |z| < 0.25d$ and $0.5I_0$ for $0.25d < |z| < 0.5d$.

14.8 Evaluate the time-average Poynting vector, $<\mathbf{S}> = (1/2)\mathscr{R}e\{\mathbf{E}_s \times \mathbf{H}_s^*\}$ for the magnetic dipole antenna in the far zone, in which all terms of order $1/r^2$ and $1/r^4$ are neglected in Eqs. (48), (49), and (50). Compare your result to the far-zone power density of the Hertzian dipole, Eq. (26). In this comparison, and assuming equal current amplitudes, what relation between loop radius a and dipole length d would result in equal radiated powers from the two devices?

14.9 A dipole antenna in free space has a linear current distribution with zero current at each end and with peak current I_0 at the center. If the length d is 0.02λ, what value of I_0 is required to (a) provide a radiation-field amplitude of 100 mV/m at a distance of 1 mi, at $\theta = 90°$; (b) radiate a total power of 1 W?

14.10 Show that the chord length in the E-plane plot of Figure 14.4 is equal to $b \sin \theta$, where b is the circle diameter.

14.11 A monopole antenna extends vertically over a perfectly conducting plane and has a linear current distribution. If the length of the antenna is 0.01λ, what value of I_0 is required to (a) provide a radiation-field amplitude of 100 mV/m at a distance of 1 mi, at $\theta = 90°$; (b) radiate a total power of 1 W? Assume free space above the plane.

14.12 Find the zeros in θ for the E-plane pattern of a dipole antenna for which (a) $\ell = \lambda$; (b) $2\ell = 1.3\,\lambda$. Use Figure 14.8 as a guide.

14.13 The radiation field of a certain short vertical current element is $E_{\theta s} = (20/r)$ $\sin \theta\, e^{-j10\pi r}$ V/m if it is located at the origin in free space. (a) Find $E_{\theta s}$ at $P(r = 100, \theta = 90°, \phi = 30°)$. ($b$) Find $E_{\theta s}$ at $P(100, 90°, 30°)$ if the vertical element is located at $A(0.1, 90°, 90°)$. (c) Find $E_{\theta s}$ at $P(100, 90°, 30°)$ if identical vertical elements are located at $A(0.1, 90°, 90°)$ and $B(0.1, 90°, 270°)$.

14.14 For a dipole antenna of overall length $2\ell = \lambda$, evaluate the maximum directivity in decibels, and the half-power beamwidth.

14.15 For a dipole antenna of overall length $2\ell = 1.3\lambda$, determine the locations in θ and the peak intensity of the sidelobes, expressed as a fraction of the main lobe intensity. Express your result as the sidelobe level in decibels, given by $S_s[\text{dB}] = 10\log_{10}(S_{r,\text{main}}/S_{r,\text{sidelobe}})$. Again, use Figure 14.8 as a guide.

14.16 For a dipole antenna of overall length, $2\ell = 1.5\lambda$, (a) evaluate the locations in θ at which the zeros and maxima in the E-plane pattern occur; (b) determine the sidelobe level, as per the definition in Problem 14.14; (c) determine the maximum directivity.

14.17 Consider a lossless half-wave dipole in free space, with radiation resistance, $R_{\text{rad}} = 73$ ohms, and maximum directivity $D_{\max} = 1.64$. If the antenna carries a 1-A current amplitude, (a) how much total power (in watts) is radiated? (b) How much power is intercepted by a 1-m² aperture situated at distance $r = 1$ km away? The aperture is on the equatorial plane and squarely faces the antenna. Assume uniform power density over the aperture.

14.18 Repeat Problem 14.17, but with a full-wave antenna ($2\ell = \lambda$). Numerical integrals may be necessary.

14.19 Design a two-element dipole array that will radiate equal intensities in the $\phi = 0, \pi/2, \pi$, and $3\pi/2$ directions in the H plane. Specify the smallest relative current phasing, ξ, and the smallest element spacing, d.

14.20 A two-element dipole array is configured to provide zero radiation in the broadside ($\phi = \pm 90°$) and endfire ($\phi = 0, 180°$) directions, but with maxima occurring at angles in between. Consider such a setup with $\psi = \pi$ at $\phi = 0$ and $\psi = -3\pi$ at $\phi = \pi$, with both values determined in the H plane. (a) Verify that these values give zero broadside and endfire radiation. (b) Determine the required relative current phase, ξ. (c) Determine the required element spacing, d. (d) Determine the values of ϕ at which maxima in the radiation pattern occur.

14.21 In the two-element endfire array of Example 14.4, consider the effect of varying the operating frequency, f, away from the original design frequency, f_0, while maintaining the original current phasing, $\xi = -\pi/2$. Determine the values of ϕ at which the maxima occur when the frequency is changed to (a) $f = 1.5 f_0$; (b) $f = 2 f_0$.

14.22 Revisit Problem 14.21, but with the current phase allowed to vary with frequency (this will automatically occur if the phase difference is established by a simple time delay between the feed currents). Now, the current phase difference will be $\xi' = \xi f/f_0$, where f_0 is the original (design) frequency. Under this condition, radiation will maximize in the $\phi = 0$ direction regardless of frequency (show this). Backward radiation (along $\phi = \pi$) will develop, however, as the frequency is tuned away from f_0. Derive an expression for the *front-to-back ratio*, defined as the ratio of the radiation intensities at $\phi = 0$ and $\phi = \pi$, expressed in decibels. Express this result as a function of the frequency ratio f/f_0. Evaluate the front-to-back ratio for (a) $f = 1.5 f_0$, (b) $f = 2 f_0$, and (c) $f = 0.75 f_0$.

14.23 A *turnstile* antenna consists of two crossed dipole antennas, positioned in this case in the *xy* plane. The dipoles are identical, lie along the *x* and *y* axes, and are both fed at the origin. Assume that equal currents are supplied to each antenna and that a zero phase reference is applied to the *x*-directed antenna. Determine the relative phase, ξ, of the *y*-directed antenna so that the net radiated electric field as measured on the positive *z* axis is (*a*) left circularly polarized; (*b*) linearly polarized along the 45° axis between *x* and *y*.

14.24 Consider a linear endfire array, designed for maximum radiation intensity at $\phi = 0$, using ξ and *d* values as suggested in Example 14.5. Determine an expression for the front-to-back ratio (defined in Problem 14.22) as a function of the number of elements *n* if *n* is an odd number.

14.25 A six-element linear dipole array has element spacing $d = \lambda/2$. (*a*) Select the appropriate current phasing ξ to achieve maximum radiation along $\phi = \pm 60°$. (*b*) With the phase set as in part *a*, evaluate the intensities (relative to the maximum) in the broadside and endfire directions.

14.26 In a linear endfire array of *n* elements, a choice of current phasing that improves the directivity is given by the Hansen-Woodyard condition:

$$\xi = \pm\left(\frac{2\pi d}{\lambda} + \frac{\pi}{n}\right)$$

where the plus or minus sign choices give maximum radiation along $\phi =$ 180° and 0°, respectively. Applying this phasing may not necessarily lead to unidirectional endfire operation (zero backward radiation), but it will with the proper choice of element spacing, *d*. (*a*) Determine this required spacing as a function of *n* and λ. (*b*) Show that the spacing as found in part *a* approaches $\lambda/4$ for a large number of elements. (*c*) Show that an even number of elements is required.

14.27 Consider an *n*-element broadside linear array. Increasing the number of elements has the effect of narrowing the main beam. Demonstrate this by evaluating the separation in ϕ between the zeros on either side of the principal maximum at $\phi = 90°$. Show that for large *n* this separation is approximated by $\Delta\phi \doteq 2\lambda/L$, where $L \doteq nd$ is the overall length of the array.

14.28 A large ground-based transmitter radiates 10 kW and communicates with a mobile receiving station that dissipates 1 mW on the matched load of its antenna. The receiver (not having moved) now transmits back to the ground station. If the mobile unit radiates 100 W, what power is received (at a matched load) by the ground station?

14.29 Signals are transmitted at a 1-m carrier wavelength between two identical half-wave dipole antennas spaced by 1 km. The antennas are oriented such that they are exactly parallel to each other. (*a*) If the transmitting antenna radiates 100 W, how much power is dissipated by a matched load at the receiving antenna? (b) Suppose the receiving antenna is rotated by 45° while the two antennas remain in the same plane. What is the received power in this case?

14.30　A half-wave dipole antenna is known to have a maximum effective area, given as A_{max}. (a) Write the maximum directivity of this antenna in terms of A_{max} and wavelength λ. (b) Express the current amplitude, I_0, needed to radiate total power, P_r, in terms of P_r, A_{max}, and λ. (c) At what values of θ and ϕ will the antenna effective area be equal to A_{max}?

Vector Analysis

A.1 GENERAL CURVILINEAR COORDINATES

Let us consider a general orthogonal coordinate system in which a point is located by the intersection of three mutually perpendicular surfaces (of unspecified form or shape),

$$
\begin{aligned}
u &= \text{constant} \\
v &= \text{constant} \\
w &= \text{constant}
\end{aligned}
$$

where u, v, and w are the variables of the coordinate system. If each variable is increased by a differential amount and three more mutually perpendicular surfaces are drawn corresponding to these new values, a differential volume is formed which approximates a rectangular parallelepiped. Because u, v, and w need not be measures of length, such as the angle variables of the cylindrical and spherical coordinate systems, each must be multiplied by a general function of u, v, and w in order to obtain the differential sides of the parallelepiped. Thus we define the scale factors h_1, h_2, and h_3 each as a function of the three variables u, v, and w and write the lengths of the sides of the differential volume as

$$
\begin{aligned}
dL_1 &= h_1\, du \\
dL_2 &= h_2\, dv \\
dL_3 &= h_3\, dw
\end{aligned}
$$

In the three coordinate systems discussed in Chapter 1, it is apparent that the variables and scale factors are

$$
\begin{array}{lllll}
\text{Rectangular:} & u = x & v = y & w = z & \\
& h_1 = 1 & h_2 = 1 & h_3 = 1 & \\
\text{Cylindrical:} & u = \rho & v = \phi & w = z & \\
& h_1 = 1 & h_2 = \rho & h_3 = 1 & \text{(A-1)} \\
\text{Spherical:} & u = r & v = \theta & w = \phi & \\
& h_1 = 1 & h_2 = r & h_3 = r\sin\theta &
\end{array}
$$

The choice of u, v, and w has been made so that $\mathbf{a}_u \times \mathbf{a}_v = \mathbf{a}_w$ in all cases. More involved expressions for h_1, h_2, and h_3 are to be expected in other less familiar coordinate systems.[1]

A.2 DIVERGENCE, GRADIENT, AND CURL IN GENERAL CURVILINEAR COORDINATES

If the method used to develop divergence in Sections 3.4 and 3.5 is applied to the general curvilinear coordinate system, the flux of the vector $\mathbf{D}$ passing through the surface of the parallelepiped whose unit normal is $\mathbf{a}_u$ is

$$D_{u0} dL_2 dL_3 + \frac{1}{2}\frac{\partial}{\partial u}(D_u dL_2 dL_3)\, du$$

or

$$D_{u0} h_2 h_3 dv\, dw + \frac{1}{2}\frac{\partial}{\partial u}(D_u h_2 h_3 dv\, dw)\, du$$

and for the opposite face it is

$$-D_{u0} h_2 h_3 dv\, dw + \frac{1}{2}\frac{\partial}{\partial u}(D_u h_2 h_3 dv\, dw)\, du$$

giving a total for these two faces of

$$\frac{\partial}{\partial u}(D_u h_2 h_3 dv\, dw)\, du$$

Because u, v, and w are independent variables, this last expression may be written as

$$\frac{\partial}{\partial u}(h_2 h_3 D_u)\, du\, dv\, dw$$

and the other two corresponding expressions obtained by a simple permutation of the subscripts and of u, v, and w. Thus the total flux leaving the differential volume is

$$\left[\frac{\partial}{\partial u}(h_2 h_3 D_u) + \frac{\partial}{\partial v}(h_3 h_1 D_v) + \frac{\partial}{\partial w}(h_1 h_2 D_w)\right] du\, dv\, dw$$

and the divergence of $\mathbf{D}$ is found by dividing by the differential volume

$$\nabla \cdot \mathbf{D} = \frac{1}{h_1 h_2 h_3}\left[\frac{\partial}{\partial u}(h_2 h_3 D_u) + \frac{\partial}{\partial v}(h_3 h_1 D_v) + \frac{\partial}{\partial w}(h_1 h_2 D_w)\right] \qquad (A.2)$$

The components of the gradient of a scalar V may be obtained (following the methods of Section 4.6) by expressing the total differential of V,

$$dV = \frac{\partial V}{\partial u} du + \frac{\partial V}{\partial v} dv + \frac{\partial V}{\partial w} dw$$

---

[1] The variables and scale factors are given for nine orthogonal coordinate systems on pp. 50–59 in J. A. Stratton, *Electromagnetic Theory*. New York: McGraw-Hill, 1941. Each system is also described briefly.

in terms of the component differential lengths, $h_1 du$, $h_2 dv$, and $h_3 dw$,

$$dV = \frac{1}{h_1}\frac{\partial V}{\partial u}h_1\,du + \frac{1}{h_2}\frac{\partial V}{\partial v}h_2\,dv + \frac{1}{h_3}\frac{\partial V}{\partial w}h_3\,dw$$

Then, because

$$d\mathbf{L} = h_1\,du\,\mathbf{a}_u + h_2\,dv\,\mathbf{a}_v + h_3\,dw\,\mathbf{a}_w \qquad \text{and} \qquad dV = \nabla V \cdot d\mathbf{L}$$

we see that

$$\nabla V = \frac{1}{h_1}\frac{\partial V}{\partial u}\mathbf{a}_u + \frac{1}{h_2}\frac{\partial V}{\partial v}\mathbf{a}_v + \frac{1}{h_3}\frac{\partial V}{\partial w}\mathbf{a}_w \tag{A.3}$$

The components of the curl of a vector $\mathbf{H}$ are obtained by considering a differential path first in a $u =$ constant surface and finding the circulation of $\mathbf{H}$ about that path, as discussed for rectangular coordinates in Section 7.3. The contribution along the segment in the $\mathbf{a}_v$ direction is

$$H_{v0}h_2\,dv - \frac{1}{2}\frac{\partial}{\partial w}(H_v h_2\,dv)\,dw$$

and that from the oppositely directed segment is

$$-H_{v0}h_2\,dv - \frac{1}{2}\frac{\partial}{\partial w}(H_v h_2\,dv)\,dw$$

The sum of these two parts is

$$-\frac{\partial}{\partial w}(H_v h_2\,dv)\,dw$$

or

$$-\frac{\partial}{\partial w}(h_2 H_v)\,dv\,dw$$

and the sum of the contributions from the other two sides of the path is

$$\frac{\partial}{\partial v}(h_3 H_w)\,dv\,dw$$

Adding these two terms and dividing the sum by the enclosed area, $h_2 h_3 dv\,dw$, we see that the $\mathbf{a}_u$ component of curl $\mathbf{H}$ is

$$(\nabla \times \mathbf{H})_u = \frac{1}{h_2 h_3}\left[\frac{\partial}{\partial v}(h_3 H_w) - \frac{\partial}{\partial w}(h_2 H_v)\right]$$

and the other two components may be obtained by cyclic permutation. The result is expressible as a determinant,

$$\nabla \times \mathbf{H} = \begin{vmatrix} \dfrac{\mathbf{a}_u}{h_2 h_3} & \dfrac{\mathbf{a}_v}{h_3 h_1} & \dfrac{\mathbf{a}_w}{h_1 h_2} \\[2mm] \dfrac{\partial}{\partial u} & \dfrac{\partial}{\partial v} & \dfrac{\partial}{\partial w} \\[2mm] h_1 H_u & h_2 H_v & h_3 H_w \end{vmatrix} \tag{A.4}$$

The Laplacian of a scalar is found by using (A.2) and (A.3):

$$\nabla^2 V = \nabla \cdot \nabla V = \frac{1}{h_1 h_2 h_3}\left[\frac{\partial}{\partial u}\left(\frac{h_2 h_3}{h_1}\frac{\partial v}{\partial u}\right) + \frac{\partial}{\partial v}\left(\frac{h_3 h_1}{h_2}\frac{\partial V}{\partial v}\right)\right.$$

$$\left. + \frac{\partial}{\partial w}\left(\frac{h_1 h_2}{h_3}\frac{\partial V}{\partial w}\right)\right] \tag{A.5}$$

Equations (A.2) to (A.5) may be used to find the divergence, gradient, curl, and Laplacian in any orthogonal coordinate system for which h_1, h_2, and h_3 are known.

Expressions for $\nabla \cdot \mathbf{D}$, ∇V, $\nabla \times \mathbf{H}$, and $\nabla^2 V$ are given in rectangular, circular cylindrical, and spherical coordinate systems at the end of the book.

A.3 VECTOR IDENTITIES

The vector identities that follow may be proved by expansion in rectangular (or general curvilinear) coordinates. The first two identities involve the scalar and vector triple products, the next three are concerned with operations on sums, the following three apply to operations when the argument is multiplied by a scalar function, the next three apply to operations on scalar or vector products, and the last four concern the second-order operations.

$$(\mathbf{A} \times \mathbf{B}) \cdot \mathbf{C} \equiv (\mathbf{B} \times \mathbf{C}) \cdot \mathbf{A} \equiv (\mathbf{C} \times \mathbf{A}) \cdot \mathbf{B} \tag{A.6}$$

$$\mathbf{A} \times (\mathbf{B} \times \mathbf{C}) \equiv (\mathbf{A} \cdot \mathbf{C})\mathbf{B} - (\mathbf{A} \cdot \mathbf{B})\mathbf{C} \tag{A.7}$$

$$\nabla \cdot (\mathbf{A} + \mathbf{B}) \equiv \nabla \cdot \mathbf{A} + \nabla \cdot \mathbf{B} \tag{A.8}$$

$$\nabla (V + W) \equiv \nabla V + \nabla W \tag{A.9}$$

$$\nabla \times (\mathbf{A} + \mathbf{B}) \equiv \nabla \times \mathbf{A} + \nabla \times \mathbf{B} \tag{A.10}$$

$$\nabla \cdot (V\mathbf{A}) \equiv \mathbf{A} \cdot \nabla V + V \nabla \cdot \mathbf{A} \tag{A.11}$$

$$\nabla (VW) \equiv V \nabla W + W \nabla V \tag{A.12}$$

$$\nabla \times (V\mathbf{A}) \equiv \nabla V \times \mathbf{A} + V \nabla \times \mathbf{A} \tag{A.13}$$

$$\nabla \cdot (\mathbf{A} \times \mathbf{B}) \equiv \mathbf{B} \cdot \nabla \times \mathbf{A} - \mathbf{A} \cdot \nabla \times \mathbf{B} \tag{A.14}$$

$$\nabla (\mathbf{A} \cdot \mathbf{B}) \equiv (\mathbf{A} \cdot \nabla)\mathbf{B} + (\mathbf{B} \cdot \nabla)\mathbf{A} + \mathbf{A} \times (\nabla \times \mathbf{B})$$
$$+ \mathbf{B} \times (\nabla \times \mathbf{A}) \tag{A.15}$$

$$\nabla \times (\mathbf{A} \times \mathbf{B}) \equiv \mathbf{A}\nabla \cdot \mathbf{B} - \mathbf{B}\nabla \cdot \mathbf{A} + (\mathbf{B} \cdot \nabla)\mathbf{A} - (\mathbf{A} \cdot \nabla)\mathbf{B} \tag{A.16}$$

$$\nabla \cdot \nabla V \equiv \nabla^2 V \tag{A.17}$$

$$\nabla \cdot \nabla \times \mathbf{A} \equiv 0 \tag{A.18}$$

$$\nabla \times \nabla V \equiv 0 \tag{A.19}$$

$$\nabla \times \nabla \times \mathbf{A} \equiv \nabla (\nabla \cdot \mathbf{A}) - \nabla^2 \mathbf{A} \tag{A.20}$$

Units

We describe first the International System (abbreviated SI, for Système International d'Unités), which is used in this book and is now standard in electrical engineering and much of physics. It has also been officially adopted as the international system of units by many countries, including the United States.[1]

The fundamental unit of length is the meter, which was defined in the latter part of the nineteenth century as the distance between two marks on a certain platinum-iridium bar. The definition was improved in 1960 by relating the meter to the wavelength of the radiation emitted by the rare gas isotope krypton-86 under certain specified conditions. This so-called krypton meter was accurate to four parts per billion, a value leading to negligible uncertainties in constructing skyscrapers or building highways, but capable of causing an error greater than one meter in determining the distance to the moon. The meter was redefined in 1983 in terms of the velocity of light. At that time the velocity of light was specified to be an auxiliary constant with an *exact* value of 299,792,458 meters per second. As a result, the latest definition of the meter is the distance light travels in a vacuum in 1/299,792,458 of a second. If greater accuracy is achieved in measuring c, that value will remain 299,792,458 m/s, but the length of the meter will change.

It is evident that our definition of the meter is expressed in terms of the second, the fundamental unit of time. The second is defined as 9,192,631,770 periods of the transition frequency between the hyperfine levels $F = 4$, $m_F = 0$, and $F = 3$, $m_F = 0$ of the ground state $^2s_{1/2}$ of the atom of cesium 133, unperturbed by external fields.

---

[1] The International System of Units was adopted by the Eleventh General Conference on Weights and Measures in Paris in 1960, and it was officially adopted for scientific usage by the National Bureau of Standards in 1964. It is a metric system, which interestingly enough is the only system which has ever received specific sanction from Congress. This occurred first in 1966 and then again in 1975 with the Metric Conversion Act, which provides for "voluntary conversion" to the metric system. No specific time was specified, however, and we can assume that it will still be a few years before the bathroom scale reads mass in kilograms and Miss America is a 90–60–90.

This definition of the second, complex though it may be, permits time to be measured with an accuracy better than one part in 10^{13}.

The standard mass of one kilogram is defined as the mass of an international standard in the form of a platinum-iridium cylinder at the International Bureau of Weights and Measures at Sèvres, France.

The unit of temperature is the kelvin, defined by placing the triple-point temperature of water at 273.16 kelvins.

A fifth unit is the candela, in which one candela is the *luminous intensity* in a given direction of a source that emits monochromatic radiation at frequency 5.40 x 10^{14} Hz (556 nm wavelength in free space), in which the radiation intensity in that direction is 1/683 W/Sr.

The last of the fundamental units is the ampere. Before explicitly defining the ampere, we must first define the newton. It is defined in terms of the other fundamental units from Newton's third law as the force required to produce an acceleration of one meter per second per second on a one-kilogram mass. We now may define the ampere as that constant current, flowing in opposite directions in two straight parallel conductors of infinite length and negligible cross section, separated one meter in vacuum, that produces a repulsive force of 2×10^{-7} newton per meter length between the two conductors. The force between the two parallel conductors is known to be

$$F = \mu_0 \frac{I^2}{2\pi d}$$

and thus

$$2 \times 10^{-7} = \mu_0 \frac{1}{2\pi}$$

or

$$\mu_0 = 4\pi \times 10^{-7} \quad (\text{kg} \cdot \text{m/A}^2 \cdot \text{s}^2, \text{or H/m})$$

We thus find that our definition of the ampere has been formulated in such a way as to assign an exact, simple, numerical value to the permeability of free space.

Returning to the International System, the units in which the other electric and magnetic quantities are measured are given in the body of the text at the time each quantity is defined, and all of them can be related to the basic units already defined. For example, our work with the plane wave in Chapter 11 shows that the velocity with which an electromagnetic wave propagates in free space is

$$c = \frac{1}{\sqrt{\mu_0 \epsilon_0}}$$

and thus

$$\epsilon_0 = \frac{1}{\mu_0 c^2} = \frac{1}{4\pi 10^{-7} c^2} = 8.854\,187\,817 \times 10^{-12}\,\text{F/m}$$

It is evident that the numerical value of ϵ_0 depends upon the defined value of the velocity of light in vacuum, 299,792,458 m/s.

The units are also given in Table B.1 for easy reference. They are listed in the same order in which they are defined in the text.

Table B.1 Names and units of the electric and magnetic quantities in the International System (In the order in which they appear in the text)

Symbol	Name	Unit	Abbreviation
v	Velocity	meter/second	m/s
F	Force	newton	N
Q	Charge	coulomb	C
r, R	Distance	meter	m
ϵ_0, ϵ	Permittivity	farad/meter	F/m
E	Electric field intensity	volt/meter	V/m
ρ_v	Volume charge density	coulomb/meter^3	C/m^3
v	Volume	meter^3	m^3
ρ_L	Linear charge density	coulomb/meter	C/m
ρ_S	Surface charge density	coulomb/meter^2	C/m^2
Ψ	Electric flux	coulomb	C
D	Electric flux density	coulomb/meter^2	C/m^2
S	Area	meter^2	m^2
W	Work, energy	joule	J
L	Length	meter	m
V	Potential	volt	V
p	Dipole moment	coulomb-meter	C $\cdot$ m
I	Current	ampere	A
J	Current density	ampere/meter^2	A/m^2
μ_e, μ_h	Mobility	meter^2/volt-second	m^2/V $\cdot$ s
e	Electronic charge	coulomb	C
σ	Conductivity	siemens/meter	S/m
R	Resistance	ohm	Ω
P	Polarization	coulomb/meter^2	C/m^2
$\chi_{e,m}$	Susceptibility		
C	Capacitance	farad	F
R_s	Sheet resistance	ohm per square	Ω
H	Magnetic field intensity	ampere/meter	A/m
K	Surface current density	ampere/meter	A/m
B	Magnetic flux density	tesla (or weber/meter^2)	T (or Wb/m^2)
μ_0, μ	Permeability	henry/meter	H/m
Φ	Magnetic flux	weber	Wb
V_m	Magnetic scalar potential	ampere	A
A	Vector magnetic potential	weber/meter	Wb/m
T	Torque	newton-meter	N $\cdot$ m
m	Magnetic moment	ampere-meter^2	A $\cdot$ m^2
M	Magnetization	ampere/meter	A/m
$\mathcal{R}$	Reluctance	ampere-turn/weber	A $\cdot$ t/Wb
L	Inductance	henry	H
M	Mutual inductance	henry	H

(Continued)

Table B.1   (Continued)

Symbol	Name	Unit	Abbreviation
ω	Radian frequency	radian/second	rad/s
c	Velocity of light	meter/second	m/s
λ	Wavelength	meter	m
η	Intrinsic impedance	ohm	Ω
k	Wave number	meter^{-1}	m^{-1}
α	Attenuation constant	neper/meter	Np/m
β	Phase constant	radian/meter	rad/m
f	Frequency	hertz	Hz
S	Poynting vector	watt/meter^2	W/m^2
P	Power	watt	W
δ	Skin depth	meter	m
Γ	Reflection coefficient		
s	Standing wave ratio		
γ	Propagation constant	meter^{-1}	m^{-1}
G	Conductance	siemen	S
Z	Impedance	ohm	Ω
Y	Admittance	siemen	S
Q	Quality factor		

Finally, other systems of units have been used in electricity and magnetism. In the electrostatic system of units (esu), Coulomb's law is written for free space,

$$F = \frac{Q_1 Q_2}{R^2} \quad \text{(esu)}$$

The permittivity of free space is assigned the value of unity. The gram and centimeter are the fundamental units of mass and distance, and the esu system is therefore a cgs system. Units bearing the prefix *stat-* belong to the electrostatic system of units.

In a similar manner, the electromagnetic system of units (emu) is based on Coulomb's law for magnetic poles, and the permeability of free space is unity. The prefix *ab-* identifies emu units. When electric quantities are expressed in esu units, magnetic quantities are expressed in emu units, and both appear in the same equation (such as Maxwell's curl equations), and the velocity of light appears explicitly. This follows from noting that in esu $\epsilon_0 = 1$, but $\mu_0 \epsilon_0 = 1/c^2$, and therefore $\mu_0 = 1/c^2$, and in emu $\mu_0 = 1$, and hence $\epsilon_0 = 1/c^2$. Thus, in this intermixed system known as the Gaussian system of units,

$$\nabla \times \mathbf{H} = 4\pi \mathbf{J} + \frac{1}{c}\frac{\partial \mathbf{D}}{\partial t} \quad \text{(Gaussian)}$$

Other systems include the factor 4π explicitly in Coulomb's law, and it then does not appear in Maxwell's equations. When this is done, the system is said to be rationalized. Hence the Gaussian system is an unrationalized cgs system (when rationalized it is known as the Heaviside–Lorentz system), and the International System we have used throughout this book is a rationalized mks system.

Table B.2   Conversion of International to Gaussian and other units
(use $c = 2.997\ 924\ 58 \times 10^8$)

Quantity	1 mks unit	= Gaussian units	= Other units
d	1 m	10^2 cm	39.37 in.
F	1 N	10^5 dyne	0.2248 lb_f
W	1 J	10^7 erg	0.7376 ft-lb_f
Q	1 C	$10c$ statC	0.1 abC
ρ_v	1 C/m^3	$10^{-5}c$ statC/cm^3	10^{-7} abC/cm^3
D	1 C/m^2	$4\pi\ 10^{-3}c$ (esu)	$4\pi\ 10^{-5}$ (emu)
E	1 V/m	$10^4/c$ statV/cm	10^6 abV/cm
V	1 V	$10^6/c$ statV	10^8 abV
I	1 A	0.1 abA	$10c$ statA
H	1 A/m	$4\pi 10^{-3}$ oersted	$0.4\pi c$ (esu)
V_m	1 A · t	0.4π gilbert	$40\pi c$ (esu)
B	1 T	10^4 gauss	$100/c$ (esu)
Φ	1 Wb	10^8 maxwell	$10^6/c$ (esu)
A	1 Wb/m	10^6 maxwell/cm	
R	1 Ω	10^9 abΩ	$10^5/c^2$ statΩ
L	1 H	10^9 abH	$10^5/c^2$ statH
C	1 F	$10^{-5}c^2$ statF	10^{-9} abF
σ	1 S/m	10^{-11} abS/cm	$10^{-7}c^2$ statS/cm
μ	1 H/m	$10^7/4\pi$ (emu)	$10^3/4\pi c^2$ (esu)
ϵ	1 F/m	$4\pi 10^{-7}c^2$ (esu)	$4\pi 10^{-11}$ (emu)

Table B.2 gives the conversion factors between the more important units of the International System (or rationalized mks system) and the Gaussian system, and several other assorted units.

Table B.3 lists the prefixes used with any of the SI units, their abbreviations, and the power of ten each represents. Those checked are widely used. Both the prefixes and their abbreviations are written without hyphens, and therefore 10^{-6} F = 1 microfarad = 1μF = 1000 nanofarads = 1000 nF, and so forth.

Table B.3   Standard prefixes used with SI units

Prefix	Abbrev.	Meaning	Prefix	Abbrev.	Meaning
atto-	a-	10^{-18}	deka-	da-	10^1
femto-	f-	10^{-15}	hecto-	h-	10^2
pico-	p-	10^{-12}	kilo-	k-	10^3
nano-	n-	10^{-9}	mega-	M-	10^6
micro-	μ-	10^{-6}	giga-	G-	10^9
milli-	m-	10^{-3}	tera-	T-	10^{12}
centi-	c-	10^{-2}	peta-	P-	10^{15}
deci-	d-	10^{-1}	exa-	E-	10^{18}

C APPENDIX

Material Constants

Table C.1 lists typical values of the relative permittivity ϵ_r' or dielectric constant for common insulating and dielectric materials, along with representative values for the loss tangent. The values should only be considered representative for each material, and they apply to normal temperature and humidity conditions and to very low audio frequencies. Most of them have been taken from *Reference Data for Radio Engineers.*[1] *The Standard Handbook for Electrical Engineers,*[2] and von Hippel,[3] and these volumes may be referred to for further information on these and other materials.

Table C.2 gives the conductivity for a number of metallic conductors, for a few insulating materials, and for several other materials of general interest. The values have been taken from the references listed previously, and they apply at zero frequency and at room temperature. The listing is in the order of decreasing conductivity.

Some representative values of the relative permeability for various diamagnetic, paramagnetic, ferrimagnetic, and ferromagnetic materials are listed in Table C.3. They have been extracted from the references listed previously, and the data for the ferromagnetic materials is only valid for very low magnetic flux densities. Maximum permeabilities may be an order of magnitude higher.

Values are given in Table C.4 for the charge and rest mass of an electron, the permittivity and permeability of free space, and the velocity of light.[4]

---

[1] M.E. Van Valkenburg,ed., Reference Data for Engineers, 9th ed., Newnes, Boston, 2002.

[2] See References for Chapter 5.

[3] von Hippel, A. R. *Dielectric Materials and Applications.* Cambridge, Mass. and New York: The Technology Press of the Massachusetts Institute of Technology and John Wiley & Sons, 1954.

[4] Cohen, E. R., and B. N. Taylor. *The 1986 Adjustment of the Fundamental Physical Constants.* Elmsford, N.Y.: Pergamon Press, 1986.

Table C.1 ϵ_r' and ϵ''/ϵ'

Material	ϵ_r'	ϵ''/ϵ'
Air	1.0005	
Alcohol, ethyl	25	0.1
Aluminum oxide	8.8	0.000 6
Amber	2.7	0.002
Bakelite	4.74	0.022
Barium titanate	1200	0.013
Carbon dioxide	1.001	
Ferrite (NiZn)	12.4	0.000 25
Germanium	16	
Glass	4–7	0.002
Ice	4.2	0.05
Mica	5.4	0.000 6
Neoprene	6.6	0.011
Nylon	3.5	0.02
Paper	3	0.008
Plexiglas	3.45	0.03
Polyethylene	2.26	0.000 2
Polypropylene	2.25	0.000 3
Polystyrene	2.56	0.000 05
Porcelain (dry process)	6	0.014
Pyranol	4.4	0.000 5
Pyrex glass	4	0.000 6
Quartz (fused)	3.8	0.000 75
Rubber	2.5–3	0.002
Silica or SiO_2 (fused)	3.8	0.000 75
Silicon	11.8	
Snow	3.3	0.5
Sodium chloride	5.9	0.000 1
Soil (dry)	2.8	0.05
Steatite	5.8	0.003
Styrofoam	1.03	0.000 1
Teflon	2.1	0.000 3
Titanium dioxide	100	0.001 5
Water (distilled)	80	0.04
Water (sea)		4
Water (dehydrated)	1	0
Wood (dry)	1.5–4	0.01

Table C.2 σ

Material	σ, S/m	Material	σ, S/m
Silver	6.17×10^7	Nichrome	0.1×10^7
Copper	5.80×10^7	Graphite	7×10^4
Gold	4.10×10^7	Silicon	2300
Aluminum	3.82×10^7	Ferrite (typical)	100
Tungsten	1.82×10^7	Water (sea)	5
Zinc	1.67×10^7	Limestone	10^{-2}
Brass	1.5×10^7	Clay	5×10^{-3}
Nickel	1.45×10^7	Water (fresh)	10^{-3}
Iron	1.03×10^7	Water (distilled)	10^{-4}
Phosphor bronze	1×10^7	Soil (sandy)	10^{-5}
Solder	0.7×10^7	Granite	10^{-6}
Carbon steel	0.6×10^7	Marble	10^{-8}
German silver	0.3×10^7	Bakelite	10^{-9}
Manganin	0.227×10^7	Porcelain (dry process)	10^{-10}
Constantan	0.226×10^7	Diamond	2×10^{-13}
Germanium	0.22×10^7	Polystyrene	10^{-16}
Stainless steel	0.11×10^7	Quartz	10^{-17}

Table C.3 μ_r

Material	μ_r	Material	μ_r
Bismuth	0.999 998 6	Powdered iron	100
Paraffin	0.999 999 42	Machine steel	300
Wood	0.999 999 5	Ferrite (typical)	1000
Silver	0.999 999 81	Permalloy 45	2500
Aluminum	1.000 000 65	Transformer iron	3000
Beryllium	1.000 000 79	Silicon iron	3500
Nickel chloride	1.000 04	Iron (pure)	4000
Manganese sulfate	1.000 1	Mumetal	20 000
Nickel	50	Sendust	30 000
Cast iron	60	Supermalloy	100 000
Cobalt	60		

Table C.4   Physical constants

Quantity	Value
Electron charge	$e = (1.602\ 177\ 33 \pm 0.000\ 000\ 46) \times 10^{-19}$ C
Electron mass	$m = (9.109\ 389\ 7 \pm 0.000\ 005\ 4) \times 10^{-31}$ kg
Permittivity of free space	$\epsilon_0 = 8.854\ 187\ 817 \times 10^{-12}$ F/m
Permeability of free space	$\mu_0 = 4\pi \times 10^{-7}$ H/m
Velocity of light	$c = 2.997\ 924\ 58 \times 10^8$ m/s

The Uniqueness Theorem

Let us assume that we have two solutions of Laplace's equation, V_1 and V_2, both general functions of the coordinates used. Therefore

$$\nabla^2 V_1 = 0$$

and

$$\nabla^2 V_2 = 0$$

from which

$$\nabla^2(V_1 - V_2) = 0$$

Each solution must also satisfy the boundary conditions, and if we represent the given potential values on the boundaries by V_b, then the value of V_1 on the boundary V_{1b} and the value of V_2 on the boundary V_{2b} must both be identical to V_b,

$$V_{1b} = V_{2b} = V_b$$

or

$$V_{1b} - V_{2b} = 0$$

In Section 4.8, Eq. (43), we made use of a vector identity,

$$\nabla \cdot (V\mathbf{D}) \equiv V(\nabla \cdot \mathbf{D}) + \mathbf{D} \cdot (\nabla V)$$

which holds for any scalar V and any vector $\mathbf{D}$. For the present application we shall select $V_1 - V_2$ as the scalar and $\nabla(V_1 - V_2)$ as the vector, giving

$$\nabla \cdot [(V_1 - V_2)\nabla(V_1 - V_2)] \equiv (V_1 - V_2)[\nabla \cdot \nabla(V_1 - V_2)]$$
$$+ \nabla(V_1 - V_2) \cdot \nabla(V_1 - V_2)$$

which we shall integrate throughout the volume *enclosed* by the boundary surfaces specified:

$$\int_{vol} \nabla \cdot [(V_1 - V_2) \nabla (V_1 - V_2)]dv$$
$$\equiv \int_{vol}(V_1 - V_2)[\nabla \cdot \nabla (V_1 - V_2)]dv + \int_{vol}[\nabla (V_1 - V_2)]^2 dv \tag{D.1}$$

The divergence theorem allows us to replace the volume integral on the left side of the equation with the closed surface integral over the surface surrounding the volume. This surface consists of the boundaries already specified on which $V_{1b} = V_{2b}$, and therefore

$$\int_{vol}\nabla \cdot [(V_1 - V_2)\nabla (V_1 - V_2)]dv = \oint_S [(V_{1b} - V_{2b}) \nabla (V_{1b} - V_{2b})]\cdot d\mathbf{S} = 0$$

One of the factors of the first integral on the right side of (D.1) is $\nabla \cdot \nabla (V_1 - V_2)$, or $\nabla^2(V_1 - V_2)$, which is zero by hypothesis, and therefore that integral is zero. Hence the remaining volume integral must be zero:

$$\int_{vol}[\nabla (V_1 - V_2)]^2 dv = 0$$

There are two reasons why an integral may be zero: either the integrand (the quantity under the integral sign) is everywhere zero, or the integrand is positive in some regions and negative in others, and the contributions cancel algebraically. In this case the first reason must hold because $[\nabla (V_1 - V_2)]^2$ cannot be negative. Therefore

$$[\nabla (V_1 - V_2)]^2 = 0$$

and

$$\nabla (V_1 - V_2) = 0$$

Finally, if the gradient of $V_1 - V_2$ is everywhere zero, then $V_1 - V_2$ cannot change with any coordinates, and

$$V_1 - V_2 = \text{constant}$$

If we can show that this constant is zero, we shall have accomplished our proof. The constant is easily evaluated by considering a point on the boundary. Here $V_1 - V_2 = V_{1b} - V_{2b} = 0$, and we see that the constant is indeed zero, and therefore

$$V_1 = V_2$$

giving two identical solutions.

The uniqueness theorem also applies to Poisson's equation, for if $\nabla^2 V_1 = -\rho_v/\epsilon$ and $\nabla^2 V_2 = -\rho_v/\epsilon$, then $\nabla^2(V_1 - V_2) = 0$ as before. Boundary conditions still require that $V_{1b} - V_{2b} = 0$, and the proof is identical from this point.

This constitutes the proof of the uniqueness theorem. Viewed as the answer to a question, "How do two solutions of Laplace's or Poisson's equation compare if they both satisfy the same boundary conditions?" the uniqueness theorem should please us because it ensures that the answers are identical. Once we can find any method of solving Laplace's or Poisson's equation subject to given boundary conditions, we have solved our problem once and for all. No other method can ever give a different answer.

Origins of the Complex Permittivity

As we learned in Chapter 5, a dielectric can be modeled as an arrangement of atoms and molecules in free space, which can be polarized by an electric field. The field forces positive and negative bound charges to separate against their Coulomb attractive forces, thus producing an array of microscopic dipoles. The molecules can be arranged in an ordered and predictable manner (such as in a crystal) or may exhibit random positioning and orientation, as would occur in an amorphous material or a liquid. The molecules may or may not exhibit permanent dipole moments (existing before the field is applied), and if they do, they will usually have random orientations throughout the material volume. As discussed in Section 5.7, the displacement of charges in a regular manner, as induced by an electric field, gives rise to a macroscopic polarization, $\mathbf{P}$, defined as the dipole moment per unit volume:

$$\mathbf{P} = \lim_{\Delta v \to 0} \frac{1}{\Delta v} \sum_{i=1}^{N\Delta v} \mathbf{p}_i \qquad (E.1)$$

where N is the number of dipoles per unit volume and $\mathbf{p}_i$ is the dipole moment of the ith atom or molecule, found through

$$\mathbf{p}_i = Q_i \mathbf{d}_i \qquad (E.2)$$

Q_i is the positive one of the two bound charges composing dipole i, and $\mathbf{d}_i$ is the distance between charges, expressed as a vector from the negative to the positive charge. Again, borrowing from Section 5.7, the electric field and the polarization are related through

$$\mathbf{P} = \epsilon_0 \chi_e \mathbf{E} \qquad (E.3)$$

where the electric susceptibility, χ_e, forms the more interesting part of the dielectric constant:

$$\epsilon_r = 1 + \chi_e \qquad (E.4)$$

Therefore, to understand the nature of ϵ_r, we need to understand χ_e, which in turn means that we need to explore the behavior of the polarization, **P**.

Here, we consider the added complications of how the dipoles respond to a time-harmonic field that propagates as a wave through the material. The result of applying such a forcing function is that *oscillating* dipole moments are set up, and *these in turn establish a polarization wave that propagates through the material*. The effect is to produce a polarization function, $\mathbf{P}(z,t)$, having the same functional form as the driving field, $\mathbf{E}(z,t)$. The molecules themselves do not move through the material, but their oscillating dipole moments collectively exhibit wave motion, just as waves in pools of water are formed by the up-and-down motion of the water. From here, the description of the process gets complicated and in many ways beyond the scope of our present discussion. We can form a basic qualitative understanding, however, by considering the classical description of the process, which is that the dipoles, once oscillating, behave as microscopic antennas, re-radiating fields that in turn co-propagate with the applied field. Depending on the frequency, there will be some phase difference between the incident field and the radiated field at a given dipole location. This results in a net field (formed through the superposition of the two) that now interacts with the next dipole. Radiation from this dipole adds to the previous field as before, and the process repeats from dipole to dipole. The accumulated phase shifts at each location are manifested as a net slowing down of the phase velocity of the resultant wave. Attenuation of the field may also occur which, in this classical model, can be accounted for by partial phase cancellation between incident and radiated fields.

For our classical description, we use the Lorentz model, in which the medium is considered as an ensemble of identical fixed electron oscillators in free space. The Coulomb binding forces on the electrons are modeled by springs that attach the electrons to the positive nuclei. We consider electrons for simplicity, but similar models can be used for any bound charged particle. Figure E.1 shows a single oscillator, located at position z in the material, and oriented along x. A uniform plane wave, assumed to be linearly polarized along x, propagates through the material in the z direction. The electric field in the wave displaces the electron of the oscillator in the x direction through a distance represented by the vector **d**; a dipole moment is thus established,

$$\mathbf{p}(z,t) = -e\mathbf{d}(z,t) \tag{E.5}$$

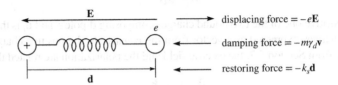

Figure E.1   Atomic dipole model, with the Coulomb force between positive and negative charge modeled by that of a spring having spring constant k_s. An applied electric field displaces the electron through distance d, resulting in dipole moment $\mathbf{p} = -e\mathbf{d}$.

where the electron charge, e, is treated as a positive quantity. The applied force is

$$\mathbf{F}_a(z,t) = -e\mathbf{E}(z,t) \tag{E.6}$$

We need to remember that $\mathbf{E}(z,t)$ at a given oscillator location is the *net* field, composed of the original applied field plus the radiated fields from all other oscillators. The relative phasing between oscillators is precisely determined by the spatial and temporal behavior of $\mathbf{E}(z,t)$.

The restoring force on the electron, $\mathbf{F}_r$, is that produced by the spring, which is assumed to obey Hooke's law:

$$\mathbf{F}_r(z,t) = -k_s\mathbf{d}(z,t) \tag{E.7}$$

where k_s is the spring constant (not to be confused with the propagation constant). If the field is turned off, the electron is released and will oscillate about the nucleus at the *resonant frequency,* given by

$$\omega_0 = \sqrt{k_s/m} \tag{E.8}$$

where m is the mass of the electron. The oscillation, however, will be damped since the electron will experience forces and collisions from neighboring oscillators. We model these as a velocity-dependent damping force:

$$\mathbf{F}_d(z,t) = -m\gamma_d\mathbf{v}(z,t) \tag{E.9}$$

where $\mathbf{v}(z,t)$ is the electron velocity. Associated with this damping is the *dephasing* process among the electron oscillators in the system. Their relative phasing, once fixed by the applied sinusoidal field, is destroyed through collisions and dies away exponentially until a state of totally random phase exists between oscillators. The $1/e$ point in this process occurs at the *dephasing time* of the system, which is inversely proportional to the damping coefficient, γ_d (in fact it is $2/\gamma_d$). We are, of course, driving this damped resonant system with an electric field at frequency ω. We can therefore expect the response of the oscillators, measured through the magnitude of $\mathbf{d}$, to be frequency-dependent in much the same way as an RLC circuit is when driven by a sinusoidal voltage.

We can now use Newton's second law and write down the forces acting on the single oscillator of Figure E.1. To simplify the process a little we can use the complex form of the electric field:

$$\mathbf{E}_c = \mathbf{E}_0 e^{-jkz} e^{j\omega t} \tag{E.10}$$

Defining $\mathbf{a}$ as the acceleration vector of the electron, we have

$$m\mathbf{a} = \mathbf{F}_a + \mathbf{F}_r + \mathbf{F}_d$$

or

$$m\frac{\partial^2 \mathbf{d}_c}{\partial t^2} + m\gamma_d\frac{\partial \mathbf{d}_c}{\partial t} + k_s\mathbf{d}_c = -e\,\mathbf{E}_c \tag{E.11}$$

Note that since we are driving the system with the complex field, $\mathbf{E}_c$, we anticipate a displacement wave, $\mathbf{d}_c$, of the form

$$\mathbf{d}_c = \mathbf{d}_0 e^{-jkz} e^{-j\omega t} \tag{E.12}$$

With the waves in this form, time differentiation produces a factor of $j\omega$. Consequently (E.11) can be simplified and rewritten in phasor form:

$$-\omega^2 \mathbf{d}_s + j\omega \gamma_d \mathbf{d}_s + \omega_0^2 \mathbf{d}_s = -\frac{e}{m} \mathbf{E}_s \qquad \text{(E.13)}$$

where (E.4) has been used. We now solve (E.13) for $\mathbf{d}_s$, obtaining

$$\mathbf{d}_s = \frac{-(e/m)\mathbf{E}_s}{(\omega_0^2 - \omega^2) + j\omega \gamma_d} \qquad \text{(E.14)}$$

The dipole moment associated with displacement $\mathbf{d}_s$ is

$$\mathbf{p}_s = -e \mathbf{d}_s \qquad \text{(E.15)}$$

The polarization of the medium is then found, assuming that all dipoles are identical. Eq. (E.1) thus becomes

$$\mathbf{P}_s = N\mathbf{p}_s$$

which, when using (E.14) and (E.15), becomes

$$\mathbf{P}_s = \frac{Ne^2/m}{(\omega_0^2 - \omega^2) + j\omega \gamma_d} \mathbf{E}_s \qquad \text{(E.16)}$$

Now, using (E.3) we identify the susceptibility associated with the resonance as

$$\chi_{\text{res}} = \frac{Ne^2}{\epsilon_0 m} \frac{1}{(\omega_0^2 - \omega^2) + j\omega \gamma_d} = \chi'_{\text{res}} - j\chi'_{\text{res}} \qquad \text{(E.17)}$$

The real and imaginary parts of the permittivity are now found through the real and imaginary parts of χ_{res}: Knowing that

$$\epsilon = \epsilon_0(1 + \chi_{\text{res}}) = \epsilon' - j\epsilon''$$

we find

$$\epsilon' = \epsilon_0(1 + \chi'_{\text{res}}) \qquad \text{(E.18)}$$

and

$$\epsilon'' = \epsilon_0 \chi''_{\text{res}} \qquad \text{(E.19)}$$

The preceding expressions can now be used in Eqs. (44) and (45) in Chapter 11 to evaluate the attenuation coefficient, α, and phase constant, β, for the plane wave as it propagates through our resonant medium.

The real and imaginary parts of χ_{res} as functions of frequency are shown in Figure E.2 for the special case in which $\omega \doteq \omega_0$. Eq. (E.17) in this instance becomes

$$\chi_{\text{res}} \doteq -\frac{Ne^2}{\epsilon_0 m \omega_0 \gamma_d}\left(\frac{j + \delta_n}{1 + \delta_n^2}\right) \qquad \text{(E.20)}$$

where the *normalized detuning* parameter, δ_n, is

$$\frac{2}{\gamma_d}(\omega - \omega_0) \qquad \text{(E.21)}$$

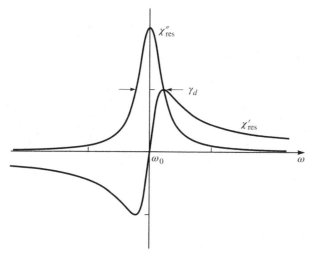

Figure E.2 Plots of the real and imaginary parts of the resonant susceptibility, χ_{res}, as given by Eq. (E.20). The full-width at half-maximum of the imaginary part, χ''_{res}, is equal to the damping coefficient, γ_d.

Key features to note in Figure E.2 include the symmetric χ''_e function, whose full width at its half-maximum amplitude is γ_d. Near the resonant frequency, where χ''_{res} maximizes, wave attenuation maximizes as seen from Eq. (44), Chapter 11. Additionally, we see that away from resonance, attenuation is relatively weak, and the material becomes transparent. As Figure E.2 shows, there is still significant variation of χ'_{res} with frequency away from resonance, which leads to a frequency-dependent refractive index; this is expressed approximately as

$$n \doteq \sqrt{1 + \chi'_{res}} \qquad \text{(away from resonance)} \qquad \text{(E.22)}$$

This frequency-dependent n, arising from the material resonance, leads to phase and group velocities that also depend on frequency. Thus, group dispersion, leading to pulse-broadening effects as discussed in Chapter 12, can be directly attributable to material resonances.

Somewhat surprisingly, the classical "spring model" described here can provide very accurate predictions on dielectric constant behavior with frequency (particularly off-resonance) and can be used to a certain extent to model absorption properties. The model is insufficient, however, when attempting to describe the more salient features of materials; specifically, it assumes that the oscillating electron can assume any one of a continuum of energy states, when, in fact, energy states in any atomic system are quantized. As a result, the important effects arising from transitions between discrete energy levels, such as spontaneous and stimulated absorption and emission, are not included in our classical spring system. Quantum mechanical models must be used

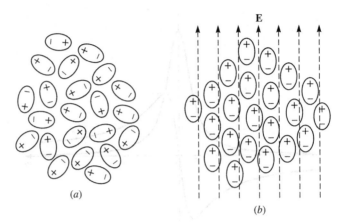

Figure E.3 Idealized sketches of ensembles of polar molecules under conditions of (*a*) random orientation of the dipole moments, and (*b*) dipole moments aligned under the influence of an applied electric field. Conditions in (*b*) are greatly exaggerated, since typically only a very small percentage of the dipoles align themselves with the field. But still enough alignment occurs to produce measurable changes in the material properties.

to fully describe the medium polarization properties, but the results of such studies often reduce to those of the spring model when field amplitudes are very low.

Another way that a dielectric can respond to an electric field is through the orientation of molecules that possess permanent dipole moments. In such cases, the molecules must be free to move or rotate, and so the material is typically a liquid or a gas. Figure E.3 shows an arrangement of polar molecules in a liquid (such as water) in which there is no applied field (Figure E.3*a*) and where an electric field is present (Figure E.3*b*). Applying the field causes the dipole moments, previously having random orientations, to line up, and so a net material polarization, **P**, results. Associated with this, of course, is a susceptibility function, χ_e, through which **P** relates to **E**.

Some interesting developments occur when the applied field is time-harmonic. With field periodically reversing direction, the dipoles are forced to follow, but they do so against their natural propensity to randomize, owing to thermal motion. Thermal motion thus acts as a "restoring" force, effectively opposing the applied field. We can also think of the thermal effects as viscous forces that introduce some difficulty in "pushing" the dipoles back and forth. One might expect (correctly) that polarizations of greater amplitude in each direction can be attained at lower frequencies, because enough time is given during each cycle for the dipoles to achieve complete alignment. The polarization amplitude will weaken as the frequency increases because there is no longer enough time for complete alignment during each cycle. This is the basic description of the *dipole relaxation* mechanism for the complex permittivity. There is no resonant frequency associated with the process.

The complex susceptibility associated with dipole relaxation is essentially that of an "overdamped" oscillator, and is given by

$$\chi_{\text{rel}} = \frac{Np^2/\epsilon_0}{3k_B T(1 + j\omega\tau)} \tag{E.23}$$

where p is the permanent dipole moment magnitude of each molecule, k_B is Boltzmann's constant, and T is the temperature in degees Kelvin. τ is the thermal randomization time, defined as the time for the polarization, **P**, to relax to $1/e$ of its original value when the field is turned off. χ_{rel} is complex, and so it will possess absorptive and dispersive components (imaginary and real parts) as we found in the resonant case. The form of Eq. (E.23) is identical to that of the response of a series RC circuit driven by a sinusoidal voltage (where τ becomes RC).

Microwave absorption in water occurs through the relaxation mechanism in polar water molecules, and is the primary means by which microwave cooking is done, as discussed in Chapter 11. Frequencies near 2.5 GHz are typically used, since these provide the optimum penetration depth. The peak water absorption arising from dipole relaxation occurs at much higher frequencies, however.

A given material may possess more than one resonance and may have a dipole relaxation response as well. In such cases, the net susceptibility is found in frequency domain by the direct sum of all component susceptibilities. In general, we may write:

$$\chi_e = \chi_{\text{rel}} + \sum_{i=1}^{n} \chi_{\text{res}}^i \tag{E.24}$$

where χ_{res}^i is the susceptibility associated with the ith resonant frequency, and n is the number of resonances in the material. The reader is referred to the references for Chapter 11 for further reading on resonance and relaxation effects in dielectrics.

F

APPENDIX

Answers to Odd-Numbered Problems

Chapter 1

1.1 $C = 4.5\,\mathbf{a}_y$ (N) $\mathbf{D} = 2\mathbf{a}_x + 1.5\mathbf{a}_y\,D = 2.5$
(generally ENE)

1.3 $(7.8, -7.8, 3.9)$

1.5 (a) $\mathbf{a}_y$, $-0.87(\mathbf{a}_x + 0.58\mathbf{a}_y)$, $+ 0.87(\mathbf{a}_x - 0.58\mathbf{a}_y)$
(b) $-\mathbf{a}_y$, $-0.87(\mathbf{a}_x - 0.58\mathbf{a}_y)$, $+ 0.87(\mathbf{a}_x + 0.58\mathbf{a}_y)$

1.7 (a) (1) the plane $z = 0$, with $|x| < 2$, $|y| < 2$; (2) the plane $y = 0$ with $|x| < 2$, $|z| < 2$; (3) the plane $x = 0$, with $|y| < 2$, $|z| < 2$; (4) the plane $x = \pi/2$, with $|y| < 2$, $|z| < 2$ (b) the plane $2z = y$, with $|x| < 2$, $|y| < 1$ (c) the plane $y = 0$, with $|x| < 2$, $|z| < 2$

1.9 (a) $0.6\mathbf{a}_x + 0.8\mathbf{a}_y$ (b) $53°$ (c) 26

1.11 (a) $(-0.3, 0.3, 0.4)$ (b) 0.05
(c) 0.12 (d) $78°$

1.13 (a) $(0.93, 1.86, 2.79)$ (b) $(9.07, -7.86, 2.21)$
(c) $(0.02, 0.25, 0.26)$

1.15 (a) $(0.08, 0.41, 0.91)$ (b) $(0.30, 0.81, 0.50)$
(c) 30.3 (d) 32.0

1.17 (a) $(0.664, -0.379, 0.645)$ (b) $(-0.550, 0.832, 0.077)$ (c) $(0.168, 0.915, 0.367)$

1.19 (a) $\mathbf{F} = A(\rho^2 + z^2)^{-3/2}\,(\rho\mathbf{a}_\rho + z\mathbf{a}_z)$
(b) $\mathbf{F} = A(x^2 + y^2 + z^2)^{-3/2}(x\mathbf{a}_x + y\mathbf{a}_y + z\mathbf{a}_z)$

1.21 (a) $-6.66\mathbf{a}_\rho - 2.77\mathbf{a}_\phi + 9\mathbf{a}_z$ (b) $-0.59\mathbf{a}_\rho + 0.21\mathbf{a}_\phi - 0.78\mathbf{a}_z$ (c) $-0.90\mathbf{a}_\rho - 0.44\mathbf{a}_z$

1.23 (a) 6.28 (b) 20.7 (c) 22.4 (d) 3.21

1.25 $\mathbf{H} = (A/\rho)\mathbf{a}_\phi$

1.27 (a) xy plane, on which $\mathbf{E} = -A/\rho^3\,\mathbf{a}_z$
(b) z axis, on which $\mathbf{E} = 2A/z^3\,\mathbf{a}_z\,$V/m

1.29 (a) $F_r = 10a\cos^2\theta$ (b) $F_\theta = -10a\cos\theta\sin\theta$

Chapter 2

2.1 $(1.19, 1.19)$

2.3 $21.5\mathbf{a}_x\,\mu$N

2.5 -10.1 nC

2.7 (a) $\mathbf{E}(z) = \dfrac{q\mathbf{a}_z}{4\pi\epsilon_0 z^2}\left[\left(1 - \dfrac{d}{2z}\right)^{-2} - \left(1 + \dfrac{d}{2z}\right)^{-2}\right]$V/m

(b) $\mathbf{E}(z = 0) = -\dfrac{2q}{\pi\epsilon_0 d^2}\,\mathbf{a}_z\,$V/m

(c) $\mathbf{E}(\rho) = \dfrac{-qd}{4\pi\epsilon_0(\rho^2 + d^2/4)^{3/2}}\,\mathbf{a}_z\,$V/m

(d) Same as part b

(e) $\mathbf{E}(\rho \gg d) \doteq -\dfrac{qd}{4\pi\epsilon_0\rho^3}\,\mathbf{a}_z\,$V/m

2.9 (a) $(x + 1) = 0.56\,[(x + 1)^2 + (y - 1)^2 + (z - 3)^2]^{1.5}$ (b) 1.69 or 0.31

2.11 (a) $-1.63\,\mu$C (b) $-30.11\mathbf{a}_x - 180.63\mathbf{a}_y - 150.53\mathbf{a}_z$
V/m (c) $-183.12\mathbf{a}_\rho - 150.53\mathbf{a}_z\,$V/m
(d) $-237.1\,\mathbf{a}_r\,$V/m

2.13 (a) 82.1 pC (b) 4.24 cm

2.15 (a) 3.35 pC (b) 124 μC/m^3

2.17 (a) $\mathbf{E}(\rho) = -\dfrac{\rho_0}{2\pi\epsilon_0\rho}\left[1 - \left(1 + \dfrac{d^2}{4\rho^2}\right)^{-1/2}\right]\mathbf{a}_z\,$V/m

(b) $\mathbf{E}(\rho \gg d) \doteq -\dfrac{qd}{4\pi\epsilon_0\rho^3}\,\mathbf{a}_z\,$V/m

2.19 (a) $\mathbf{E} = \frac{\rho_0}{2\pi\epsilon_0}\left[1-\left(1+\frac{\ell^2}{4\rho^2}\right)^{-1/2}\right]\mathbf{a}_\rho$ V/m

(b) $\mathbf{E}\,(\ell\to\infty) \doteq \frac{\rho_0}{2\pi\epsilon_0}\,\mathbf{a}_\rho$ V/m (independent of radius)

2.21 $\mathbf{E} = -\frac{\rho_0}{\pi\epsilon_0 a}\,\mathbf{a}_x$ V/m

2.23 (a) 8.1 kV/m (b) −8.1 kV/m

2.25 (a) $\mathbf{E}(z) = -\frac{\rho_{s0}}{\pi\epsilon_0 z}\left[1-(1+a^2/z^2)^{-1/2}\right]\mathbf{a}_y$ V/m

(b) $\mathbf{E}(z\gg a) \doteq -\frac{\rho_{s0}a^2}{2\pi\epsilon_0 z^3}\,\mathbf{a}_y$ V/m

2.27 (a) $y^2 - x^2 = 4xy - 19$ (b) $0.99\mathbf{a}_x + 0.12\mathbf{a}_y$

2.29 (a) 12.2 (b) $-0.87\mathbf{a}_x - 0.50\mathbf{a}_y$
(c) $y = (1/5)\ln\cos 5x + 0.13$

Chapter 3

3.1 (a) $\bar{F} = [Q_1 Q_2/4\pi\epsilon_0 R^2]\,\mathbf{a}_r$ (b) Same as part (a)
(c) 0 (d) Force will become attractive!

3.3 (a) q

3.5 (a) $\mathbf{E}(z) = \pm\frac{q}{2\epsilon_0 A}[1 - \exp(-|z|/d)]\mathbf{a}_z$ V/m +
for $z>0$, $-$ for $z<0$

(b) q would be the charge contained within a volume of infinite length in z, and of cross-sectional area A.

3.7 (a) $\mathbf{E}(r) = \frac{q}{\epsilon_0}\left[\frac{1-e^{-br}}{r^2}\right]\mathbf{a}_r$ V/m (b) $4\pi q$

3.9 (a) $\mathbf{E}_I = \rho_0 r^2/(4\epsilon_0 a)\,\mathbf{a}_r$ V/m (b) $\mathbf{E}_{II} = \rho_0 a^3/(4\epsilon_0 r^2)$
$\mathbf{a}_r$ V/m (c) $\rho_s = -\rho_0 a^3/(4b^2)$ C/m^2
(d) $-\rho_0^2 a^6/(16\epsilon_0 b^4)\,\mathbf{a}_r$ N/m^2

3.11 (a) $\mathbf{E} \doteq \rho_0 b^2/(2\epsilon_0\rho)\,\mathbf{a}_\rho$ V/m (b) Same as part a

3.13 (a) $D_r\,(r<2) = 0$; $D_r(r=3) = 8.9\times 10^{-9}$C/m^2;
$D_r(r=5) = 6.4\times 10^{-10}$C/m^2 (b) $\rho_{s0} = -(4/9)$
$\times 10^{-9}$C/m^2

3.15 (a) $[(8\pi L)/3]\,[\rho_1^3 - 10^{-9}]\,\mu$C where ρ_1 is in meters
(b) $4(\rho_1^3 - 10^{-9})/(3\rho_1)\mu$C/m^2 where ρ_1 is in meters
(c) $D_\rho\,(0.8$ mm$) = 0$; $D_\rho\,(1.6$ mm$) = 3.6\times 10^{-6}$
μCm^2; $D_\rho(2.4$ mm$) = 3.9\times 10^{-6}\mu$C/m^2

3.17 Surfaces are spherical shells of radii $r_0 = 2ma$,
where m is an integer.

3.19 113 nC

3.21 (a) Spherical volume charge of radius a and of
density ρ_0, surrounded by spherical surface charge
layers $\rho_{s1} = -a\rho_0/3$ at $r = a$ and $\rho_{s2} = a^3\rho_0/(3b^2)$
at $r = b$. (b) $Q_{tot} = 4\pi a^3\rho_0/3$

3.23 (a) (proof) (b) $\rho_{v0} = 3Q/(4\pi\,a^3)$ $(0 < r < a)$;
$D_r = Qr/4\pi\,a^3)$ and $\nabla\cdot\mathbf{D} = 3Q/(4\pi\,a^3)$ $(0 <r <a)$;
$D_r = Q/(4\pi\,r^2)$ and $\nabla\cdot\mathbf{D} = 0$ $(r>a)$

3.25 (a) 17.50 C/m^3 (b) $5\mathbf{a}_r$ C/m^2 (c) 320π C (d) 320π C

3.27 (a) $\rho_v = -\rho_0\exp(-\alpha z)$ C/m^3
(b) $Q = -\frac{A\rho_0}{\alpha}[1-\exp(-\alpha d)]$C

3.29 (a) 3.47 C (b) 3.47 C

3.31 −3.91 C

Chapter 4

4.1 (a) $-q\delta E_x$ (b) $(-q\delta/\sqrt{2})(E_x + E_y)$
(c) $(-q\delta/\sqrt{3})(E_x + E_y+ E_z)$

4.3 (a) $(-q\delta/\sqrt{2})(E_\rho + E_\phi)$ (b) $\Delta\phi = \delta/(b\sqrt{2})$

4.5 (a) $A\alpha$ (b) $-A\alpha$ (c) 0

4.7 (a) $E_0 b/a$

4.9 (a) $V_0 e^{-r/a}$ (b) V_0 is the potential at the origin.

4.11 (a) $V = 2A\sin\theta_0\sin(\Delta\theta/2)$ where $\Delta\theta = \tan^{-1}(\ell/r)$
(b) $V \doteq (A\ell/r)\sin\theta_0$

4.13 576 pJ

4.15 −68.4 V

4.17 (a) –3.026 V (b) –9.678 V

4.19 (a) ρ_0/ϵ_0 (b) Same

4.21 (a) $\mathbf{E} = \frac{V_0}{r^2\left(\frac{1}{a}-\frac{1}{b}\right)}\,\mathbf{a}_r$ V/m

(b) zero in both regions

(c) $W_E = \frac{2\pi\epsilon_0 V_0^2}{\left(\frac{1}{a}-\frac{1}{b}\right)}$ J

(d) $\rho_{sa} = \frac{\epsilon_0 V_0}{a^2\left(\frac{1}{a}-\frac{1}{b}\right)}$ C/m^2

(e) V_0

4.23 (a) $\mathbf{E} = \frac{a\rho_0}{\epsilon_0}\,e^{-\rho/a}\,\mathbf{a}_\rho$ V/m

(b) $\rho_v = \frac{a\rho_0}{\epsilon_0}\left(1-\frac{\rho}{a}\right)e^{-\rho/a}$ C/m^3

(c) and (d) $W_E = \frac{\pi a^4\rho_0^2}{4\epsilon_0}$ J

4.25 (a) $\rho_v = \frac{2\rho_0 a}{r}\left(1-\frac{r^2}{a^2}\right)e^{-r^2/a^2}$ C/m^3

(b) 0

(c) $V_0 = \frac{\rho_0 a^3\sqrt{\pi}}{2\epsilon_0}$ V

(d) $W_E = \frac{\pi a^5\rho_0^2}{4\epsilon_0}\sqrt{\frac{\pi}{2}}$ J

4.27 (proof)

4.29 1.31 V

4.31 (a) 387 pJ (b) 207 pJ

4.33 (a) $(5\times 10^{-6})/(4\pi r^2)\,\mathbf{a}_r$ C/m^2
(b) 2.81 J (c) 4.45 pF

4.35 (a) 0.779 μJ (b) 1.59 μJ

Chapter 5

5.1 (a) −1.23 MA (b) 0 (c) 0, as expected

5.3 $I = 3\Omega Q/(4\pi b)$ A

5.5 $\beta = \omega/v$

5.7 (a) mass flux density in (kg/m^2 − s) and mass density in (kg/m^3) (b) −550 g/m^3 − s

5.9 (a) 0.28 mm (b) 6.0 × 10^{7}A/m^2

5.11 (a) $\mathbf{E} = [(9.55)/\rho l]\mathbf{a}_\rho$ V/m, $V = (4.88)/l$ V and $R = (1.63)/l \ \Omega$, (b) 14.64/l W

5.13 (a) 0.147 V (b) 0.144 V

5.15 (a) $\mathbf{J} = I_0/(2\pi r) \ \mathbf{a}_r$ A/m^2 (b) $\mathbf{E} = I_0/(2\pi a \sigma_0 r) \ \mathbf{a}_r$ V/m (c) $V_0 = I_0 \ln(b/a)/(2\pi a \sigma_0)$ V (d)$R = \ln(b/a)/(2\pi a \sigma_0)$ ohms

5.17 $R = \ln(b/a)/(2\pi a \sigma_0)$ ohms

5.19 $R = 3d/(2\pi \sigma_0 a^3)$ ohms

5.21 (a) 1.20 kV (b) $\mathbf{E}_p = 723\mathbf{a}_x − 18.9\mathbf{a}_y$ V/m

5.23 (a) 289.5 V (b) $z/[(x − 1)^2 + y^2 + z^2]^{1.5} − z/[(x + 1)^2 + y^2 + z^2]^{1.5} = 0.222$

5.25 (a) 4.7 × 10^{-5}S/m (b) 1.1 × 10^{-3}S/m (c) 1.2 × 10^{-2}S/m

5.27 (a) 6.26 pC/m^2 (b) 1.000176

5.29 (a) $\mathbf{E} = [(144.9)/\rho]\mathbf{a}_\rho$ V/m, $\mathbf{D} = (3.28\mathbf{a}_\rho)/\rho$ nC/m^2 (b) $V_{ab} = 192$ V, $\chi_e = 1.56$ (c) $[(5.0 \times 10^{-29})/\rho]\mathbf{a}_\rho$ C · m

5.31 (a) $\mathbf{E} = \dfrac{E_0}{\epsilon_r} \left[(\cos^2 \phi + \epsilon_r \sin^2 \phi)\mathbf{a}_x + (1 − \epsilon_r) \sin\phi \cos\phi \ \mathbf{a}_y\right]$ V/m (b) $\epsilon_r = 2.41$

5.33 $125\mathbf{a}_x + 175\mathbf{a}_y$ V/m

5.35 (a) $\mathbf{E}_2 = \mathbf{E}_1$ (b) $W_{E1} = 45.1 \ \mu$J, $W_{E2} = 338 \ \mu$J

Chapter 6

6.1 $b/a = \exp(2\pi d/W)$

6.3 barium titanate

6.5 $f = d[(d − b) + b/\epsilon_r]^{-1}$

6.7 $C = 4a^2 \epsilon_0/d$

6.9 $C = 3\pi \epsilon_0 a^2/(2d)$ F

6.11 (a) $\dfrac{\epsilon_{r1}}{\epsilon_{r2}} = \dfrac{a + b}{2a}$

(b) $C = \dfrac{2\pi \epsilon_{r1} \epsilon_0}{\ln(b/a) − (1 − \epsilon_{r2}/\epsilon_{r1}) \ln(\epsilon_{r1}/\epsilon_{r2})}$ F

6.13 $K_1 = 23.0$, $\rho_L = 8.87$ nC/m, $a = 13.8$ m, $C = 35.5$ pF

6.15 (a) 47.3 nC/m^2 (b) −15.8 nC/m^2 (c) 24.3 pF/m

6.17 Exact value: 57 pF/m

6.19 Exact value: 11ϵ_0 F/m

6.21 (b) $C \approx 110$ pF/m (c) Result would not change.

6.23 (a) 3.64 nC/m (b) 206 mA

6.25 (proof)

6.27 $f(x, y) = −4e^{2x} + 3x^2$, $V(x, y) = 3(x^2 − y^2)$

6.29 $Q = 120\pi$ C

6.31 $V_1(z) = \dfrac{V_0 z}{b + (\epsilon_{r1}/\epsilon_{r2})(d − b)}$ (0 < z < b)

$V_2(z) = V_0 \left[1 − \dfrac{(d − z)}{(\epsilon_{r2}/\epsilon_{r1}) b + (d − b)}\right]$ (b < z < d)

6.33 (a) yes, yes, yes, no (b) At the 100 V surface, no for all. At the 0 V surfaces, yes, except for $V_1 + 3$. (c) Only V_2 is

6.35 (a) $V(\theta) = 100 \ln \tan(\theta/2) + 10$ V (b) $\mathbf{E} = 100\mathbf{a}_y$ V/m

6.37 (a) 1.01 cm (b) 22.8 kV/m (c) 3.15

6.39 (a) $(−2.00 × 10^4)\phi + 3.78 × 10^3$ V (b) $[(2.00 × 10^4)/\rho]\mathbf{a}_\phi$ V/m (c) $(2.00 × 10^4 \epsilon_0/\rho)\mathbf{a}_\phi$ C/m^2 (d) $[(2.00 × 10^4)/\rho]$ C/m^2 (e) 84.7 nC (f) $V(\phi) = 28.7\phi + 194.9$ V, $\mathbf{E} = −(28.7)/\rho\mathbf{a}_\phi$ V/m, $\mathbf{D} = −(28.7\epsilon_0)/\rho\mathbf{a}_\phi$ C/m^2, $\rho_s = (28.7\epsilon_0)/\rho$ C/m^2, $Q_b = 122$ pC (g) 471 pF

6.41 (a) 12.5 mm (b) 26.7 kV/m (c) 4.23 (with given $\rho_s = 1.0 \ \mu$C/m^2)

6.43 $V_1(r) = V_0 \left[1 − \dfrac{\left(\frac{1}{a} − \frac{1}{r}\right)}{\frac{\epsilon_1}{\epsilon_2}\left(\frac{1}{c} − \frac{1}{b}\right) + \left(\frac{1}{a} − \frac{1}{c}\right)}\right]$ (a ≤ r ≤ c)

$V_2(r) = V_0 \left[\dfrac{\left(\frac{1}{r} − \frac{1}{b}\right)}{\left(\frac{1}{c} − \frac{1}{b}\right) + \frac{\epsilon_1}{\epsilon_2}\left(\frac{1}{a} − \frac{1}{c}\right)}\right]$ (c ≤ r ≤ b)

6.45 (a) $833.3r^{-.4}$ V (b) $833.3r^{-.4}$ V

Chapter 7

7.1 (a) $−294\mathbf{a}_x + 196\mathbf{a}_y \ \mu$A/m (b) $−127\mathbf{a}_x + 382\mathbf{a}_y \ \mu$A/m (c) $−421\mathbf{a}_x + 578\mathbf{a}_y \ \mu$A/m

7.3 $\mathbf{H} = \dfrac{\pi a^2 I}{2\pi(a^2 + z^2)^{3/2}} \ \mathbf{a}_z$ A/m

7.5 $|\mathbf{H}| = \dfrac{I}{2\pi}\left[\left(\dfrac{2}{y^2 + 2y + 5} − \dfrac{2}{y^2 − 2y + 5}\right)^2 + \left(\dfrac{(y − 1)}{y^2 − 2y + 5} − \dfrac{(y + 1)}{y^2 + 2y + 5}\right)^2\right]^{1/2}$

7.7 (a) $\mathbf{H} = I/(2\pi^2 z) (\mathbf{a}_x − \mathbf{a}_y)$ A/m (b) 0

7.9 (a) $I = 2a^2 \rho_0 \Omega$ A (b) $\mathbf{H} = 2a\rho_0 \Omega/3 \ \mathbf{a}_z$ A/m

7.11 2.0 A/m, 933 mA/m, 360 mA/m, 0

7.13 (e) $H_z(a < \rho < b) = k_b$; $H_z(\rho > b) = 0$

7.15 $\mathbf{H} = 0$ (r < a) $\mathbf{H} = −I/(2\pi r \sin \theta) \ \mathbf{a}_\phi$ A/m (r > a)

7.17 (a) $\mathbf{J} = \sigma_0 E_0 \rho \ \mathbf{a}_z$ A/m^2 (b) $E_0 = 3I/(2\pi \sigma_0 a^3)$ V/m (c) $\mathbf{J} = 3I\rho/(2\pi a^3) \ \mathbf{a}_z$ A/m^2 (d) $\mathbf{H} = I\rho^2/(2\pi a^3) \ \mathbf{a}_\phi$ A/m (ρ ≤ a) $\mathbf{H} = I/(2\pi\rho) \ \mathbf{a}_\phi$ A/m (ρ ≥ a)

7.19 (a) $\mathbf{K} = -I\mathbf{a}_r/2\pi r$ A/m ($\theta = \pi/2$)
(b) $\mathbf{J} = I\,\mathbf{a}_r\,/\,[2\pi r^2(1 - 1/\sqrt{2})]$ A/m^2 ($\theta < \pi/4$)
(c) $\mathbf{H} = I\mathbf{a}_\phi/[2\pi\,r\sin\theta]$ A/m ($\pi/4 < \theta < \pi/2$)
(d) $\mathbf{H} = I(1 - \cos\theta)\mathbf{a}_\phi\,/\,[2\pi r\sin\theta\,(1 - 1/\sqrt{2})]$
A/m ($\theta < \pi/4$)

7.21 (a) $\mathbf{I} = 2\pi b a^3/3$ A (b) $\mathbf{H}_{\text{in}} = b\rho^2/3\mathbf{a}_\phi$ A/m
(c) $\mathbf{H}_{\text{out}} = ba^3/3\rho\,\mathbf{a}_\phi$ A/m

7.23 (a) σ would have a $1/\rho$ dependence
(b) $\nabla \times \mathbf{H} = \mathbf{J} = H_0/\rho\,\mathbf{a}_z$ A/m^2

7.25 (a) -259 A (b) -259 A

7.27 (a) $2(x + 2y)/z^3\mathbf{a}_x + 1/z^2\mathbf{a}_x$ A/m (b) same as part (a)
(c) 1/8 A

7.29 (a) $1.59 \times 10^7\mathbf{a}_z$ A/m^2
(b) $7.96 \times 10^6\rho\mathbf{a}_\phi$ A/m, $10\rho\mathbf{a}_\phi$ Wb/m^2
(c) as expected
(d) $1/(\pi\rho)\mathbf{a}_\phi$ A/m, $\mu_0/(\pi\rho)\mathbf{a}_\phi$ Wb/m^2
(e) as expected

7.31 (a) 0.392 μWb (b) 1.49 μWb (c) 27 μWb

7.33 proof

7.35 (a) -40ϕ A (2 $<\rho<$ 4), 0 ($\rho > 4$)
(b) $40\,\mu_0\ln(3/\rho)\mathbf{a}_z$ Wb/m

7.37 $[120 - (400/\pi)\phi]$ A ($0 < \phi < 2\pi$)

7.39 (a) $-30\mathbf{a}_y$ A/m (b) $30y - 6$ A (c) $-30\mu_0\mathbf{a}_y$ Wb/m^2
(d) $\mu_0(30x - 3)\mathbf{a}_z$ Wb/m

7.41 (a) Use $\mathbf{A}(\rho) = \mu_0 I/(2\pi)\ln(b/\rho)\,\mathbf{a}_z$
(b) $\mathbf{A}(\rho) = \mu_0\pi\rho\,nI/(2\pi)\,\mathbf{a}_\phi$

7.43 $A_z = \dfrac{\mu_0 I}{96\pi}\left[\left(\dfrac{\rho^2}{a^2} - 25\right) + 98\ln\left(\dfrac{5a}{\rho}\right)\right]$ Wb/m

Chapter 8

8.1 (a) (.90, 0, $-$.135) (b) $3 \times 10^5\mathbf{a}_x - 9 \times 10^4\mathbf{a}_z$ m/s
(c) 1.5×10^{-5}J

8.3 (a) $.70\mathbf{a}_x + .70\mathbf{a}_y - .12\mathbf{a}_z$ (b) 7.25 fJ

8.5 $\mathbf{T} = -2\pi a^2\ell K_0 B_0\,\mathbf{a}_z$ N-m

8.7 $\mathbf{F} = -\dfrac{\mu_0 I_1 I_2}{2b}\mathbf{a}_z$ N/m

8.9 $4\pi \times 10^{-5}$ N/m

8.11 proof

8.13 (a) $d\mathbf{F} = \dfrac{\mu_0 I_1 I_2}{2\pi(x^2 + z_0^2)}\left[x(\mathbf{a}_x - \mathbf{a}_y) - z_0\mathbf{a}_z\right]dx$ N

(b) $F_z = -\dfrac{\mu_0 I_1 I_2}{2}$ N

8.15 $(6 \times 10^{-6})\,[b - 2\tan^{-1}(b/2)]\mathbf{a}_y$ N $\cdot$ m

8.17 $\Delta w/w = \Delta m/m = 1.3 \times 10^{-6}$

8.19 (a) $77.6y\mathbf{a}_z$ kA/m (b) 5.15×10^{-6}H/m (c) 4.1
(d) $241y\mathbf{a}_z$ kA/m (e) $77.6\mathbf{a}_x$ kA/m^2
(f) $241\mathbf{a}_x$ kA/m^2 (g) $318\mathbf{a}_x$ kA/m^2

8.21 (Use $\chi_m = .003$) (a) 47.7 A/m
(b) 6.0 A/m (c) 0.288 A/m

8.23 (a) 637 A/m, 1.91×10^{-3} Wb/m^2, 884 A/m
(b) 478 A/m, 2.39×10^{-3} Wb/m^2, 1.42×10^3A/m
(c) 382 A/m, 3.82×10^{-3} Wb/m^2, 2.66×10^3A/m

8.25 (a) $1.91/\rho$A/m   (0 $<\rho< \infty$)
(b) $(2.4 \times 10^{-6}/\rho)\mathbf{a}_\phi$ T ($\rho < .01$), $(1.4 \times 10^{-5}/\rho)\mathbf{a}_\phi$
T (.01 $<\rho< .02$), $(2.4 \times 10^{-6}/\rho)\mathbf{a}_\phi$ T ($\rho > .02$)
(ρ in meters)

8.27 (a) $-4.83\mathbf{a}_x - 7.24\mathbf{a}_y + 9.66\mathbf{a}_z$ A/m
(b) $54.83\mathbf{a}_x - 22.76\mathbf{a}_y + 10.34\mathbf{a}_z$ A/m
(c) $54.83\mathbf{a}_x - 22.76\mathbf{a}_y + 10.34\mathbf{a}_z$ A/m
(d) $-1.93\mathbf{a}_x - 2.90\mathbf{a}_y + 3.86\mathbf{a}_z$ A/m
(e) 102° (f) 95°

8.29 10.5 mA

8.31 (a) 2.8×10^{-4} Wb (b) 2.1×10^{-4} Wb
(c) $\approx 2.5 \times 10^{-4}$ Wb

8.33 (a) $23.9/\rho$ A/m (b) $3.0 \times 10^{-4}/\rho$ Wb/m^2
(c) 5.0×10^{-7} Wb
(d) $23.9/\rho$ A/m, $6.0 \times 10^{-4}/\rho$ Wb/m^2,
1.0×10^{-6} Wb (e) 1.5×10^{-6} Wb

8.35 (a) $20/(\pi r\sin\theta)\mathbf{a}_\phi$ A/m (b) 1.35×10^{-4} J

8.37 (a) $L_1 = N_1^2/\mathcal{R}$, $L_2 = N_2^2/\mathcal{R}$
(b) $M = N_1 N_2/\mathcal{R}$

8.39 (a) $(1/2)wd\,\mu_0\,K_0^2$ J/m
(b) $\mu_0 d/w$ H/m (c) $\Phi = \mu_0 d\,K_0$ Wb

8.41 (a) 33 μH (b) 24 μH

8.43 (b) $L_{\text{int}} = \dfrac{2W_H}{I^2}$

$= \dfrac{\mu_0}{8\pi}\left[\dfrac{d^4 - 4a^2c^2 + 3c^4 + 4c^4\ln(a/c)}{(a^2 - c^2)^2}\right]$ H/m

8.45 proof

Chapter 9

9.1 (a) $-5.33\sin 120\pi\,t$ V (b) $21.3\sin(120\pi\,t)$ mA

9.3 (a) $-1.13 \times 10^5[\cos(3 \times 10^8 t - 1) - \cos(3 \times 10^8 t)]$ V
(b) 0

9.5 (a) -4.32 V (b) -0.293 V

9.7 (a) $(-1.44)/(9.1 + 39.6t)$ A

(b) $-1.44\left[\dfrac{1}{61.9 - 39.6t} + \dfrac{1}{9.1 + 39.6t}\right]$ A

9.9 $2.9 \times 10^3[\cos(1.5 \times 10^8 t - 0.13x) - \cos(1.5 \times 10^8 t)]$ W

9.11 (a) $\left(\dfrac{10}{\rho}\right)\cos(10^5 t)\mathbf{a}_\rho$ A/m^2 (b) $8\pi\cos(10^5 t)$ A
(c) $-0.8\pi\sin(10^5 t)$ A (d) 0.1

9.13 (proof)

9.15 (a) Magnetostatic case (b) Electrostatic case

9.17 $a = 66 \text{ m}^{-1}$

9.19 proof

9.21 (a) $\pi \times 10^9 \text{sec}^{-1}$

(b) $\frac{500}{\rho} \sin(10\pi z) \sin(\omega t)\mathbf{a}_\rho$ V/m

9.23 $\mathbf{A} = (\beta E_0/\omega)y \cos(\omega t - \beta z)\mathbf{a}_z$ Wb/m

9.25 (b) $\mathbf{B} = \left(t - \frac{z}{c}\right)\mathbf{a}_y$ T $\mathbf{H} = \frac{1}{\mu_0}\left(t - \frac{z}{c}\right)\mathbf{a}_y$ A/m

$\mathbf{E} = (ct - z)\mathbf{a}_x$ V/m $\mathbf{D} = \epsilon_0(ct - z)\mathbf{a}_x$ C/m^2

Chapter 10

10.1 $\alpha = 0.080 \text{ m}^{-1} \beta = 2.25$ rad/m $\lambda = 2.80$ m
$Z_0 = 93.6 - j3.34$ ohms

10.3 63 nJ

10.5 (a) $V_{sT}(z) = 2V_0 \cos(\phi/2)e^{-\alpha z}e^{j\phi/2}e^{-j\beta z}$
$V(z, t) = 2V_0 \cos(\phi/2)e^{-\alpha z}\cos(\omega t - \beta z + \phi/2)$ V

(b) $I_{sT}(z) = \frac{2V_0}{Z_0} \cos(\phi/2)e^{-\alpha z}e^{j\phi/2 - \delta}e^{-j\beta z}$

$I(z,t) = \frac{2V_0}{Z_0} \cos(\phi/2)e^{-\alpha z}\cos(\omega t - \beta z + \phi/2 - \delta)$ A

(c) $P = \frac{2V_0^2}{Z_0} \cos^2(\phi/2) \cos \delta \, e^{-2\alpha z}$ W

10.7 7.9 mW

10.9 $P_{rec} = 6.8$ dBm $= 4.8$ mW

10.11 (a) V_0^2/R_L
(b) $R_L V_0^2/(R\ell + R_L)^2$
(c) V_0^2/R_L
(d) $(V_0^2/R_L) \exp(-2\ell\sqrt{RG})$

10.13 (a) $A = 1.15 \times 10^{-4}$ ohms / m-$\sqrt{\text{Hz}}$,
$R = 1.15$ ohms / m
(b) $P_L = 89$ mW

10.15 (a) 104 V (b) $52.6 - j123$ V

10.17 $P_{25} = 2.28$ W, $P_{100} = 1.16$ W

10.19 16.5 W

10.21 (a) $s = 2.62$ (b) $Z_L = 1.04 \times 10^3 + j \, 69.8 \, \Omega$
(c) $z_{max} = -7.2$ mm

10.23 (a) 0.037λ or 0.74 m (b) 2.61
(c) 2.61 (d) $0.463 \, \lambda$ or 9.26 m

10.25 (a) $495 + j \, 290 \, \Omega$ (b) $j \, 98 \, \Omega$

10.27 (a) 2.6 (b) $11 - j \, 7.0$ mS (c) $0.213 \, \lambda$

10.29 $47.8 + j \, 49.3 \, \Omega$

10.31 (a) 3.8 cm (b) 14.2 cm

10.33 (a) $d = 0.25\lambda \, d_1 = 0.094\lambda$
(b) $d = 0.25\lambda \, d_1 = 0.344\lambda$

10.35 (a) 39.6 cm (b) 24 pF

10.37 $V_L = (1/3) \, V_0 \, (l/v < t < \infty)$ and is zero for
$t < l/v$. $I_B = (V_0/100)$ A for $0 < t < 2l/v$ and is
$(V_0/75)$ for $t > 2l/v$

10.39 $\frac{l}{v} < t < \frac{5l}{4v}$: $V_1 = 0.44 \, V_0$

$\frac{3l}{v} < t < \frac{13l}{4v}$: $V_2 = -0.15 \, V_0$

$\frac{5l}{v} < t < \frac{21l}{4v}$: $V_3 = 0.049 \, V_0$

$\frac{7l}{v} < t < \frac{29l}{4v}$: $V_4 = -0.017 \, V_0$

Voltages in between these times are zero.

10.41 $0 < t < \frac{1}{2v}$: $V_L = 0$

$\frac{l}{2v} < t < \frac{3l}{2v}$: $V_L = \frac{V_0}{2}$

$t > \frac{3l}{2v}$: $V_L = V_0$

10.43 $0 < t < 2l/v$: $V_{R_L} = V_0/2$

$t > 2l/v$: $V_{R_L} = 3V_0/4$

$0 < t < l/v$: $V_{R_g} = 0, I_B = 0$

$t > l/v$: $V_{R_g} = V_0/4, I_B = 3V_0/4Z_0$

Chapter 11

11.1 proof

11.3 (a) 0.33 rad/m (b) 18.9 m (c) $-3.76 \times 10^3 \mathbf{a}_z$ V/m

11.5 $I \propto EE^* = 4E_0^2 \cos^2\left(\frac{\Delta\omega}{2}t - k_m z\right)$ where $\Delta\omega$
$= \omega_2 - \omega_1, k_m = \frac{k_1 + k_2}{2}$ velocity $= c\Delta\omega/(2\omega_0)$

11.7 (a) 0.12δ (b) 0.35δ (c) 3.45δ

11.9 (a) $\mathbf{H}_s = \frac{-E_0}{\omega\mu_0}(\beta\mathbf{a}_x + j\alpha\mathbf{a}_z)e^{-\alpha x} \, e^{-\beta iz}$
(b) $\alpha = \sqrt{\beta^2 - \omega^2 \mu_0\epsilon_0}$

11.11 (a) $\ell.t. = 67.5$ (good conductor) (b) $\alpha \doteq \beta$
$\doteq 34.4$ m, $\eta \doteq 11.5(1 + j)$ ohms
(c) $\mathbf{E}_s(z) = 100e^{-\alpha z} \, e^{-j\beta z} \, \mathbf{a}_x$ $\mathscr{E}(z, t) = 100e^{-\alpha z}$
$\cos(\omega t - \beta z) \, \mathbf{a}_x \, \alpha$ and β as in part a
(d) $\mathbf{H}_s(z) = 6.15e^{-j0.79} \, e^{-\alpha z} \, e^{-j\beta z} \, \mathbf{a}_y \, \mathscr{H}(z, t)$
$= 6.15e^{-\alpha z} \cos(\omega t - \beta z - 0.79)\mathbf{a}_y$
(e) $<\mathbf{S}>(z) = 216e^{-2\alpha z} \, \mathbf{a}_z$ W/m^2 (f) $z(25\%)$
$= 2.0$ cm (g) 8.69 dB

11.13 $\mu = 2.28 \times 10^{-6}$ H/m, $\epsilon' = 1.07 \times 10^{-11}$ F/m, and
$\epsilon'' = 2.90 \times 10^{-12}$ F/m

11.15 (a) $\lambda = 3$ cm, $\alpha = 0$ (b) $\lambda = 2.95$ cm, $\alpha = 9.24$
$\times 10^{-2}$ Np/m (c) $\lambda = 1.33$ cm, $\alpha = 335$ Np/m

11.17 (a) $\mathbf{E} = V_0/\ell \, \mathbf{a}_z$

(b) $\mathbf{H} = \frac{\sigma\rho V_0}{2\ell} \, \mathbf{a}_\phi$ A/m $(0 < \rho < a)$

(c) $\mathbf{S} = -\frac{\sigma\rho V_0^2}{2\ell^2} \, \mathbf{a}_\rho$ W/m^2

(d) and (e) $P = \frac{V_0^2}{R}$ W

11.19 (a) $\omega = 4 \times 10^8$ rad/s (b) $\mathbf{H} \, (\rho, z, t) = (4.0/\rho)$ cos
$(4 \times 10^8 t - 4z)\mathbf{a}_\phi$ A/m (c) $\langle S \rangle = (2.0 \times 10^{-3}/\rho^2)$
$\cos^2(4 \times 10^8 t - 4z)\mathbf{a}_z$ W/m^2 (d) P = 5.7 kW

11.21 (a) $H_{\phi 1}(\rho) = (54.5/\rho)(10^4\rho^2 - 1)$ A/m ($.01 < \rho$ $< .012$), $H_{\phi 2}(\rho) = (24/\rho)$ A/m ($\rho > .012$), $H_\phi = 0$ ($\rho < .01$m) (b) $\mathbf{E} = 1.09\mathbf{a}_z$ V/m (c) $\langle S \rangle$ $= -(59.4/\rho)(10^4\rho^2 - 1)\mathbf{a}_\rho$ W/m^2 ($.01 < \rho < .012$ m), $- (26/\rho)\mathbf{a}_\rho$ W/m^2 ($\rho > 0.12$ m)

11.23 (a) 1.4×10^{-3} Ω/m (b) 4.1×10^{-2} Ω/m (c) 4.1×10^{-1} Ω/m

11.25 $f = 1$ GHz, $\sigma = 1.1 \times 10^5$ S/m

11.27 (a) 4.7×10^{-8} (b) 3.2×10^3 (c) 3.2×10^3

11.29 (a) $\mathbf{H}_s = (E_0/\eta_0)(\mathbf{a}_y - j\mathbf{a}_x)e^{-j\beta z}$ (b) $\langle S \rangle = (E_0^2/\eta_0)\mathbf{a}_z$ W/m^2 (assuming E_0 is real)

11.31 (a) $L = 14.6$ λ (b) Left

11.33 (a) $\mathbf{H}_s = (1/\eta)[-18e^{j\phi}\mathbf{a}_x + 15\mathbf{a}_y]e^{-j\beta z}$ A/m (b) $\langle S \rangle = 275$ Re $\{(1/\eta^*)\}$ W/m^2

Chapter 12

12.1 0.01%

12.3 0.056 and 17.9

12.5 (a) 4.7×10^8 Hz (b) $691 + j177$ Ω (c) -1.7 cm

12.7 (a) $s_1 = 1.96$, $s_2 = 2$, $s_3 = 1$ (b) -0.81 m

12.9 (a) 6.25×10^{-2} (b) 0.938 (c) 1.67

12.11 $641 + j501$ Ω

12.13 Reflected wave: left circular polarization; power fraction = 0.09. Transmitted wave: right circular polarization; power fraction = 0.91

12.15 (a) 2.55 (b) 2.14 (c) 0.845

12.17 2.41

12.19 (a) $d_1 = d_2 = d_3 = 0$ or $d_1 = d_3 = 0$, $d_2 = \lambda/2$ (b) $d_1 = d_2 = d_3 = \lambda/4$

12.21 (a) Reflected power: 15%. Transmitted power: 85% (b) Reflected wave: s-polarized. Transmitted wave: Right elliptically polarized.

12.23 $n_0 = (n_1/n_2)\sqrt{n_1^2 - n_2^2}$

12.25 0.76 (-1.19 dB)

12.27 2

12.29 4.3 km

Chapter 13

13.1 (a) 1.14 mm (b) 1.14 mm (c) 1.47 mm

13.3 14.2 pF/m, 0.786 μH/m, 0, 0.023 Ω/m

13.5 (a) 1.23 (b) 1.99 (c) 1.33

13.7 (a) 2.8 (b) 5.85×10^{-2}

13.9 (a) 4.9 (b) 1.33

13.11 9

13.13 9

13.15 1.5 ns

13.17 (a) 12.8 GHz (b) Yes

13.19 (a) 2.5 GHz $< f <$ 3.75 GHz (air-filled) (b) 3.75 GHs $< f <$ 4.5 GHz (air-filled)

13.21 $a = 1.1$ cm, $b = 0.90$ cm

13.23 proof

13.25 72 cm

13.27 3.32

13.29 (a) $\theta_{\min} = \sin^{-1}(n_3/n_1)$ (b) v_p, max $= c/n_3$

13.31 greater than

Chapter 14

14.1 (a) $-0.284\mathbf{a}_x - 0.959\mathbf{a}_z$ (b) 0.258

14.3 (a) $-j(1.5 \times 10^{-2})e^{-j1000}\,\mathbf{a}_z$ V/m (b) $-j(1.5 \times 10^{-2})e^{-j1000}\,\mathbf{a}_y$ V/m (c) $-j(1.5 \times 10^{-2})(\mathbf{a}_y + \mathbf{a}_z)$ V/m (d) $-(1.24 \times 10^{-2})(\mathbf{a}_y + \mathbf{a}_z)$ V/m (e) 1.75×10^{-2} V/m

14.5 proof

14.7 (a) 0.711 Ω (b) 0.178 Ω (c) 0.400 Ω

14.9 (a) 85.4 A (b) 5.03 A

14.11 (a) 85.4 A (b) 7.1 A

14.13 (a) $0.2e^{-j1000\pi}$ V/m (b) $0.2e^{-j1000\pi}e^{j0.5\pi}$ V/m (c) 0

14.15 Primary maxima: $\theta = \pm 90°$, relative magnitude 1.00. Secondary maxima: $\theta = \pm 33.8°$ and $\theta = \pm 146.2°$, relative magnitude 0.186. $S_s = 7.3$ dB

14.17 (a) 36.5 W (b) 4.8 μW

14.19 $\xi = 0$, $d = \lambda$

14.21 (a) $\pm 48.2°$ (b) $\pm 60°$

14.23 (a) $+\pi/2$ (b) 0

14.25 (a) $\xi = -\pi/2$ (b) 5.6% of maximum (12.6 dB down)

14.27 proof

14.29 (a) 1.7 μW (b) 672 nW

图 书 资 源 支 持

感谢您一直以来对清华版图书的支持和爱护。为了配合本书的使用，本书提供配套的资源，有需求的读者请扫描下方的"书圈"微信公众号二维码，在图书专区下载，也可以拨打电话或发送电子邮件咨询。

如果您在使用本书的过程中遇到了什么问题，或者有相关图书出版计划，也请您发邮件告诉我们，以便我们更好地为您服务。

我们的联系方式：

地　　址：北京市海淀区双清路学研大厦 A 座 701

邮　　编：100084

电　　话：010－62770175－4608

资源下载：http://www.tup.com.cn

客服邮箱：tupjsj@vip.163.com

QQ：2301891038（请写明您的单位和姓名）

资源下载、样书申请

书圈

扫一扫，获取最新目录

用微信扫一扫右边的二维码，即可关注清华大学出版社公众号"书圈"。